An Introduction to Elementary Particle Phenomenology
(Second Edition)

Online at: https://doi.org/10.1088/978-0-7503-5759-3

An Introduction to Elementary Particle Phenomenology (Second Edition)

Philip G Ratcliffe

Department of Science and High Technology, University of Insubria in Como, Como, Italy

and

Istituto Nazionale di Fisica Nucleare, Sezione di Milano -Bicocca, Milano, Italy

IOP Publishing, Bristol, UK

ISBN   978-0-7503-5759-3 (ebook)
ISBN   978-0-7503-5757-9 (print)
ISBN   978-0-7503-5760-9 (myPrint)
ISBN   978-0-7503-5758-6 (mobi)

DOI   10.1088/978-0-7503-5759-3

Version: 20250101

IOP ebooks

British Library Cataloguing-in-Publication Data: A catalogue record for this book is available from the British Library.

Published by IOP Publishing, wholly owned by The Institute of Physics, London

IOP Publishing, No.2 The Distillery, Glassfields, Avon Street, Bristol, BS2 0GR, UK

US Office: IOP Publishing, Inc., 190 North Independence Mall West, Suite 601, Philadelphia, PA 19106, USA

This second edition is dedicated, as ever,

with love to my wife Sandra and children Alberto and Alice.

Contents

Preface

Preface to the first edition

The present volume is based on lecture notes prepared for an introductory course on the *Phenomenology of Elementary Particles*, held within the framework of the Master's equivalent degree course in Physics at Insubria University in Como. The lectures were first delivered during the academic year 2005/06 and have since undergone much evolution. The notes have been augmented and edited with the aim of being as up-to-date and self-contained as is reasonably possible and therefore of more general utility.

The volume is thus primarily intended for use by students with a basic knowledge of classical electromagnetism, quantum mechanics and special relativity but not necessarily, for example, of quantum field theory. However, it should also hopefully represent a useful reference text and study aid for other similar courses. With this in mind a number of appendices have been included to cover material normally taught in earlier courses. In particular, they include certain more advanced topics in quantum mechanics such as relativistic wave equations, isospin and quantum oscillation; the appendix on scattering takes the reader through the development of the formalism from elastic electron–proton scattering up to deeply inelastic scattering covering also the technique of partial-wave expansion and the Breit–Wigner resonance formalism.

Preface to the second edition

Some ten years after the first edition was published, there have now been numerous advances in our knowledge and understanding of particle physics. While at that time, the Higgs boson had been observed experimentally only relatively recently and thus with only limited data and statistics, the two major experiments at LHC have now gathered a great deal of important information as to its nature and couplings to the particles of the Standard Model. The large variety of dedicated neutrino experiments have also produced interesting and detailed results on oscillation phenomena, the Pontecorvo–Maki–Nakagawa–Sakata (PMNS) mixing matrix and the associated non-zero masses.

In the quark sector, access to many beauty-meson states and decay channels and also top-quark physics, not to mention CP-violating effects, provided by LHC has enriched the status of the Cabibbo–Kobayashi–Maskawa (CKM) mixing matrix. The experimental study of exotic mesonic and baryonic states, so-called tetraquarks, pentaquarks and hexaquarks, has also produced definite though not unambiguous signals. There is much debate as to the true nature of the objects; they may simply be the multi-quark states just mentioned or more complex molecule-like bound states of mesons and/or baryons. Not only LHC, but also Relativistic Heavy-Ion Collider (RHIC) have very important heavy-ion programmes aimed at studying the predicted exotic phase of very dense and hot matter known as quark–gluon plasma.

The increasing precision of many measurements of the Standard-Model parameters has also led to the appearance of a small number of discrepancies. While these are generally not yet particularly significant, they could nevertheless be first indications of physics beyond the Standard Model, of which there are various differing aspects. While still awaiting a positive experimental signal, the search for particle-physics solutions to the dark-matter problem has become a major area of interest both from the theoretical and experimental points of view. LHC also provides the opportunity to search for the new particle states envisioned in the various grand unified or supersymmetric theories that have been proposed. Indeed, the very important question of the future of particle phenomenology (i.e. both experimental and theoretical) is a more-than-ever hotly debated theme. This then forms the basis of an additional final chapter, in which various aspects of possible physics beyond the Standard Model and the phenomena that it might engender are examined.

Taking advantage of this new edition, the appendices have also been extended to cover more of the exquisitely theoretical questions necessary for the derivation of the physical results used here. The flow of physical ideas will thus not be unduly interrupted, while the curious or skeptical reader may choose to follow a more technical path at his or her leisure.

This second edition thus aims to bring the volume up-to-date on the many new developments of the last decade and hopefully provide the reader not only with more of the necessary phenomenological tools and techniques, but also bring some insight into the future directions that this still expanding field might take.

Author biography

Philip G Ratcliffe

Philip G Ratcliffe gained his first degree from Trinity College, Cambridge in 1976, where he had also been awarded a senior scholarship. After attending the Diploma course of Imperial College in Mathematical Physics, he went on to obtain his doctoral degree in 1983 at the International School for Advanced Studies (*Scuola Internazionale Superiore di Studi Avanzati*, SISSA), Trieste, Italy, with a thesis entitled 'The Role of Spin Physics in Hadronic Interactions at Short Distances'.

He has been a Research Associate at the Cavendish Laboratory, Cambridge; Queen Mary & Westfield College, London and with the Italian National Institute for Nuclear Physics (INFN) in Milan and Turin. He is currently Associate Professor of Nuclear and Subnuclear Physics at the University of Insubria in Como, Italy, with continuing association to the INFN.

IOP Publishing

An Introduction to Elementary Particle Phenomenology (Second Edition)

Philip G Ratcliffe

Chapter 1

Introduction

'What I am going to tell you about is what we teach our physics students in the third or fourth year of graduate school…. It is my task to convince you not to turn away because you don't understand it. You see my physics students don't understand it…. That is because I don't understand it. Nobody does.'

Richard P Feynman

QED: The Strange Theory of Light and Matter

1.1 Supplementary reading

A short list of suggested supplementary reading material follows. The books by (Griffiths 2008, Perkins 2000 and Martin and Shaw 1997) are particularly recommended for style, clarity and completeness and their relevance to the subject matter treated here. The volume by Griffiths (2008), being rather recent, also covers many of the newer topics. For a modern, more complete but still accessible presentation of the parton model, the book by Roberts is an excellent review. While no attempt will be made to deal seriously with the (rather complex) technicalities of quantum field theory, the important results and implications will, however, be frequently exploited; the reader seeking further explanation is directed to the book by Ryder (1996) as it does make contact with particle phenomenology.

Two more recent and very complete publications should also be mentioned. The first is an open-access resource edited by Schopper *et al* (2020), which provides a very up-to-date and complete reference source. It is available online in three separate parts: 1—Theory and Experiments, 2—Detectors for Particles and Radiation and

3—Accelerators and Colliders. The second, a recent textbook by Rubbia (2022), is very complete and highly recommendable as an excellent and comprehensive reference on the subject.

We should finally not forget the invaluable Particle Data Group (PDG) *Review of Particle Physics* (Navas *et al* 2024); here we shall cite the most recent issue as of going to press. It is revised and updated every two years and may also be found on-line (the online version is updated annually). Indeed, whenever reference is made to the PDG Review (which will be very often here), the most up-to-date information will always be found online. Besides providing an exhaustive compendium of high-energy particle-physics data, it contains many clear and concise review articles on various aspects of both theory and experiment, including such topics as: detector and accelerator specifications, quantum chromodynamics (QCD), the quark–parton model, the theory underlying the Standard Model.

In addition, at the end of each chapter of the present text the reader will find a more-or-less comprehensive list of cited works (both original papers and more general pedagogical review articles), which may be consulted for further study.

1.2 Aims and philosophy

The specific topic of these notes is the phenomenology of particle physics from low to the very highest accessible energies today. The need for such a course is continually growing as the gulf between experiment and theory continues inexorably to widen.

1.2.1 On experiment and theory

It is an unfortunate fact of modern physics (not to say life in general) that extreme specialisation or compartmentalisation forces researchers very early in their careers to take on the exclusive epithet of either 'experimenter' or 'theoretician'. The problem then is that, having donned the cloak and mask of one or other character, identification with the chosen rôle often becomes so pre-eminent as to exclude contact with the other. This is a great loss for both fields. Indeed, the present-day situation might best be summed up in the title of a biographical account of the life and times of Enrico Fermi: '*The Last Man Who Knew Everything*' (Schwartz, 2017).

The following is an excerpt from the work of Roger Bacon (Franciscan monk and controversial theologian, considered to be one of the most influential philosophers of the thirteenth century), see Halsall's Internet Medieval Sourcebook (1996):

> '*On Experimental Science, 1268*
>
> *Having laid down the main points of the wisdom of the Latins as regards language, mathematics and optics, I wish now to review the principles of wisdom from the point of view of experimental science, because without experiment it is impossible to know anything thoroughly.*
>
> *There are two ways of acquiring knowledge, one through reason, the other by experiment. Argument reaches a conclusion and compels us to admit it, but it neither makes us certain nor so annihilates doubt that the*

mind rests calm in the intuition of truth, unless it finds this certitude by way of experience. Thus, many have arguments toward attainable facts, but because they have not experienced them, they overlook them and neither avoid a harmful nor follow a beneficial course. Even if a man that has never seen fire, proves by good reasoning that fire burns, and devours and destroys things, nevertheless the mind of one hearing his arguments would never be convinced, nor would he avoid fire until he puts his hand or some combustible thing into it in order to prove by experiment what the argument taught. But after the fact of combustion is experienced, the mind is satisfied and lies calm in the certainty of truth. Hence argument is not enough, but experience is.

This is evident even in mathematics, where demonstration is the surest. The mind of a man that receives that clearest of demonstrations concerning the equilateral triangle without experiment will never stick to the conclusion nor act upon it till confirmed by experiment by means of the intersection of two circles from either section of which two lines are drawn to the ends of a given line. Then one receives the conclusion without doubt. What Aristotle says of the demonstration by the syllogism being able to give knowledge, can be understood if it is accompanied by experience, but not of the bare demonstration. What he says in the first book of the Metaphysics, that those knowing the reason and cause are wiser than the experienced, he speaks concerning the experienced who know the bare fact only without the cause. But I speak here of the experienced that know the reason and cause through their experience. And such are perfect in their knowledge, as Aristotle wishes to be in the sixth book of the Ethics, whose simple statements are to be believed as if they carried demonstration, as he says in that very place.'

Roger Bacon

To be fair, it is not so much the distance separating the two spheres of activity, rather the depth of specialist knowledge required to be at the forefront in either today. The aim here, through the approach adopted, will be to illustrate the vital interplay between experiment and theory: at various points in history, one pushes the other to make new discoveries, which then return in a never-ending virtuous spiral. The present volume is then also an attempt to strengthen the links and demonstrate, by example, how the continuous and very dynamical interaction between experimental discovery (new particles, interactions, phenomena, detector techniques etc) and the theoretical developments (new theories, approaches, calculational techniques etc) has led and still leads to overall progress in the field of elementary particle physics.

On the one hand, improved experimental techniques (detectors and data analysis) provide new information, often in the form of true discovery of, e.g., a new particle or interaction or simply a hitherto unobserved phenomenon, which requires modifications or additions to the extant theoretical framework. On the other, as theories grow and become more complete, they typically suggest the existence of new particles, states or interaction phenomena, which can then be experimentally investigated in order to test the new ideas. Indeed, this is the only known and accepted method to verify the correctness of a theory.

> *'It doesn't matter how beautiful your theory is, it doesn't matter how smart you are. If it doesn't agree with experiment, it's wrong.'*
>
> Richard P Feynman

1.2.2 Ockham's razor

A guiding principle in scientific endeavour is simplification or, as is often considered a sort of *holy grail* in elementary particle theory, *unification*. The concept is embodied in the term *Ockham's razor* ('Occam' is the latinisation), which describes the principle, espoused by William of Ockham (yet another Franciscan monk and controversial theologian, considered to be one of the most influential philosophers of the fourteenth century), that: *'Numquam ponenda est pluralitas sine necessitate (Plurality must never be postulated unless necessary')*. In other words, the simplest theories with least parameters, particles or states should preferred over those in which there is unnecessary proliferation. Put more simply, but still in Latin, *'lex parsimoniae'* ('law of parsimony').

The notion of the razor lies in the observation that for any given set of facts there is an infinite number of different theories that could explain them. A data set of ten points can (almost) always be described by a ten-parameter fit, but such a description almost never conveys any real physical understanding, whereas a successful one- or two-parameter description almost certainly contains some meaningful physics.

> *'...with four parameters I can fit an elephant, and with five I can make him wiggle his trunk.'*
>
> John von Neumann—see Freeman Dyson (Dyson 2004)

Unnecessary complications should therefore always be *discarded* in favour of simplicity, a concept typically related to a perception of the *beauty* of a theory.

Progress in physics is often made by reforming or even completely overthrowing long-standing principles or theories; however, this should not in any way be taken as in contradiction with Ockham's razor. The razor merely advocates, at any given moment, adoption of a minimal theory; it does not indicate future directions for research. However, it does suggest one further test to perform: any new description or theory of known phenomena should also provide a clear *prediction* and *precise* for some, as yet, unknown or unstudied physical process.

> *'You can recognize truth by its beauty and simplicity. When you get it right, it is obvious that it is right— at least if you have any experience— because usually what happens is that more comes out than goes in.'*
>
> Richard P Feynman

1.3 Prerequisites

A certain basic knowledge of the theoretical foundations (in particular, notions of quantum mechanics and special relativity) is assumed and will be necessary for full appreciation of the subjects treated. As already noted, the main emphasis will be placed on the phenomenological aspects and on the experimental manifestation of the underlying dynamics and symmetries. In particular, the rôle of symmetries (exact and approximate) and their violation will be central to many of the discussions. In this context, as general supplementary reading, I can also strongly recommend the Dirac Memorial Lecture delivered by Steven Weinberg (1987).

While the formulation and physical significance of the Dirac equation is very briefly outlined in the appendices, no attempt will be made to enter into the realms of quantum field theory. The formalism of Dirac spinors and their relation to the symmetries of spin, parity and charge-conjugation will, however, be necessary. In connection with relativistic kinematics (almost always necessary here), the standard four-vector notation will be adopted throughout. Moreover, in order to describe some of the phenomena that have played important rôles in the growth of our knowledge of particle interactions, the notion of Feynman diagrams is introduced in an fairly intuitive manner.

As a final question of formalism, the *bra–ket* representation (introduced by Dirac) is now universally adopted as a convenient, versatile and powerful short-hand for describing quantum-mechanical systems. It will thus be the main language employed here from the very start of the volume. Readers unfamiliar with the usage would be well advised to consult a suitable text.

1.4 Conventions and notation

Equality and equivalence: Borrowing from common programming-language usage, defining equations will be indicated with the notation '$:=$', whereby the right-hand side defines the left-hand side (and *vice versa* for '$=:$'), whereas the notation '$\triangleq$' will be used to indicate equivalence, though not necessarily strict equality (such as between operators, or physical quantities expressed in different units).

Extended 'bar' notation: The momentum-space measure naturally conjugate to dx is $dp/(2\pi)$ and in many cases the Dirac δ-function for momenta is also accompanied by a multiplicative 2π factor; thus, $2\pi\,\delta(p)$. Since most of such 2π factors generally simplify in the final expressions, for clarity of the intermediate steps by avoiding such inessential factors, we shall adopt the following extended 'bar' notation (in analogy with the standard $\hbar$ and common $\lambdabar$):

$$\dbar := \frac{d}{2\pi} \quad \text{and} \quad \bar{\delta} := 2\pi\,\delta(p). \tag{1.4.1}$$

In particular, we then have

$$\int \dbar p'\,\bar{\delta}(p' - p) = 1. \tag{1.4.2}$$

Natural units: For clarity of notation, we shall generally adopt the universal 'natural' units of the high-energy physicist, in which the reduced Planck constant $\hbar$ and the velocity of light c are set to unity and therefore disappear from all expressions. For example, a velocity then becomes a pure number, expressed as a fraction of the velocity of light. The quantity $\hbar c$ becomes a very useful reference value: $\hbar c \simeq 197.3\,\text{MeV·c}$ (corresponding to one unit of angular momentum). Energy, momentum and mass are then expressed in the same units (typically an energy), while time and distance are given as inverse energies.

However, in order to fully appreciate the quantum or relativistic nature of a given situation, it will occasionally be useful to render the dependence on these two parameters explicit; in such cases $\hbar$ and c will be temporarily reinstated. Indeed, this restoration may always be achieved by noting the normal dimensions of the object under study: with the natural system all quantities have dimensions of powers of energy (say) and via multiplication by suitable powers of $\hbar$ and c will regain their true physical dimensions.

Indices: To aid the correct identification of the many indexed objects, we shall attempt to adhere as strictly as possible to the following convention: indices from the middle of the Greek alphabet (μ, ν, ρ, σ,...) will be used to indicate Lorentz four-vector indices (thus running from 0 to 3); mid-alphabet Latin letters ($i, j, k, l,...$) will usually stand for purely spatial indices (running from 1 to 3); letters from the beginning of the Greek alphabet (α, β, γ, δ,...) will be used as Dirac-space indices (i.e. for Dirac spinors or γ-matrix elements); and from the beginning of the Latin alphabet (a, b, c,...) will typically be used for internal spaces of various types. Unindexed four vectors will be printed in a normal math-italic font while unindexed three vectors will appear in bold face.

Symmetries and typeface: In a similar fashion, we shall also exploit various typefaces in the designation of symmetries, symmetry transforms and their related quantum numbers. As an example, consider the case of *spatial inversion* or *parity* (which we shall see plays a very important rôle in particle physics). The symbol used to generically denote the symmetry here will be 'P', while the symmetry transformation operator will be denoted '$\mathcal{P}$' and the related quantum number (or eigenvalue) 'P'. Similarly, for *temporal inversion* we shall adopt 'T', '$\mathcal{T}$' and 'T', while for *charge conjugation* 'C', '$\mathcal{C}$' and 'C'.

References

Cahn R N and Goldhaber G 2009 *The Experimental Foundations of Particle Physics* 2nd edition (Cambridge: Cambridge University Press)

Close F E 1979 *An Introduction to Quarks and Partons* (New York: Academic Press)

Dyson F 2004 *Nature* **427** 297

Griffiths D 2008 *Introduction to Elementary Particles* 2nd edition (Hoboken, NJ: Wiley VCH)

Halzen F and Martin A D 1984 *Quarks and Leptons* (New York: John Wiley & Sons)

Halsall P 1996 *The Internet Medieval Sourcebook* http://www.fordham.edu/halsall/sbook.html

Martin B R and Shaw G 1997 *Particle Physics* 2nd edition (New York: John Wiley & Sons)

Navas S *et al* (Particle Data Group) 2024 *Phys. Rev.* **D110** 030001

Perkins D H 2000 *Introduction to High Energy Physics* 4th edition (Cambridge: Cambridge University Press)

Povh B, Rith K, Scholz C and Zetsche F 1995 *Particles and Nuclei* (Berlin: Springer-Verlag)

Roberts R G 1990 *The Structure of the Proton: Deep Inelastic Scattering* (Cambridge: Cambridge University Press)

Rubbia A 2022 *Phenomenology of Particle Physics* (Cambridge: Cambridge University Press)

Ryder L H 1996 *Quantum Field Theory* 2nd edition (Cambridge: Cambridge University Press)

Schopper H *et al* (eds.) 2020 *Particle Physics Reference Library* vols 1–3 (Berlin: Springer Nature)

Schwartz D N 2017 *The Last Man Who Knew Everything* (New York: Basic Books)

Weinberg S 1987 *Elementary Particles and the Laws of Physics* ed R MacKenzie and P Doust (Cambridge: Cambridge University Press) p 61

IOP Publishing

An Introduction to Elementary Particle Phenomenology
(Second Edition)

Philip G Ratcliffe

Chapter 2

Symmetries (discrete and continuous)

*'So our problem is to explain
where symmetry comes from.
Why is nature so nearly symmetrical?
No one has any idea why.'*

Richard P Feynman

In this chapter we shall be concerned mainly with discrete symmetries, more precisely: charge-conjugation (C), parity (P) and time-reversal (T) invariances and, moreover, their *violation*. However, since it turns out to have a central rôle in their violation, we shall be forced to examine the weak interaction in some depth.

Experimental study of the weak nuclear interaction opened up a veritable Pandora's box: it brought to light surprising and unexpected phenomena never before experienced and forced reconsideration of many aspects of physics that were previously taken for granted. It also perhaps marked the true birth of elementary particle physics, as distinct from nuclear physics: the theoretical description of β-decay required a new fundamental interaction and a new fundamental particle, the neutrino.

It will then also be natural that some discussion be presented here of the so-called *flavour* symmetries (continuous) present in the quark sector. In particular, in order to explain CP violation, it will be necessary to examine the question of *quark mixing* and Cabibbo theory (1963), with its extension to the full three-generation picture due to Kobayashi and Maskawa (1973). Again, we shall find that it is the weak interaction that plays the major role in the relevant phenomenology. However, the principal topics here will remain the symmetries of C, P and T.

doi:10.1088/978-0-7503-5759-3ch2

2.1 Beta decay

Here we shall present the original Fermi formulation of the weak interaction in the context of nuclear β-decay. We begin by describing the various forms of β-decay and the known phenomenology.

2.1.1 Early developments in β-decay

The early studies of β-decay by Chadwick (1914) witnessed apparent two-body decays, in which the electric charge of the parent nucleus changed by one unit (a neutron was transformed into a proton), accompanied by the emission of an electron (the β-particle) with a continuous spectrum of energies ranging from the lowest detectable up to 20 or 30 keV[1] (positron emission was later observed too, as in a sort of inverse process). This was in sharp contrast to e.g. the manifestly two-body decays of α- and γ-emission, with their consequent *mono*-energetic spectra. Both energy and momentum conservation thus *appeared* to be violated in β-decay. Moreover, it was evident from a knowledge of the spin quantum numbers involved that angular momentum could not be conserved either, if this were a strictly two-body decay, truly involving the emission of just a single spin-half object. All known β-decays involved an integer change of nuclear spin.

The resolution[2], first proposed by Wolfgang Pauli (1930), was that this might, in fact, be a *three*-body decay, the third particle being a new particle, the neutrino[3]. Pauli announced his idea in a letter to the attendees of the Gauverein meeting (Tübingen, Dec. 1930).

> '*Dear Radioactive Ladies and Gentlemen!*
>
> *… I have hit upon a desperate remedy to save …the law of conservation of energy. …there could exist in the nuclei electrically neutral particles that I wish to call neutrons, which have spin 1/2…*
>
> *The continuous beta spectrum would then make sense with the assumption that in beta decay, in addition to the electron, a neutron is emitted such that the sum of the energies of neutron and electron is constant…*
>
> *But so far I do not dare to publish anything about this idea, and trustfully turn first to you, dear radioactive ones, with the question of how likely it is to find experimental evidence for such a neutron…*
>
> *I admit that my remedy may seem almost improbable because one probably would have seen those neutrons, if they exist, for a long time. But nothing ventured, nothing gained…*

---

[1] For example, the Q-value for the β^--decay of ^{19}B is $\sim$27 MeV.

[2] For quite some time the idea was not taken seriously by many physicists. Believing it to be too speculative, Pauli did not publish the idea until 1934. Bohr, for example, openly opposed such an interpretation of β-decay, preferring to entertain the possibility that energy, linear and angular momentum might all not be conserved.

[3] Pauli himself initially called the new particle a '*neutron*'. Later on, when what is now known as the neutron was discovered (see section 4.1.2), Edoardo Amaldi jokingly coined the name '*neutrino*' (as a 'small' or light neutral particle) in a humorous conversation with Fermi, who then adopted the term for Pauli's particle (see reference [277] in Amaldi 1984).

Thus, dear radioactive ones, scrutinise and judge...'
Wolfgang Pauli (Zürich, 4th. Dec. 1930)

A third emitted particle immediately solved the problem of energy conservation, but in order to also satisfy angular-momentum conservation, the neutrino also had to be a spin-half fermion. The observed end-point of the electron spectrum corresponded (within experimental precision) to the electron carrying away *all* of the energy released (i.e. the *visible* mass difference), which indicated that the neutrino should have negligible mass. In first approximation therefore, it was assumed to have precisely zero mass (see e.g. Fermi 1934a).

In 1933, Enrico Fermi[4] devised a theory of the weak nuclear interaction. Such decays were described via a local four-point interaction transmuting a neutron inside the nucleus into a proton, together with the simultaneous emission of an electron and (what is now known as) an electron *anti*neutrino $(\bar{\nu}_e)$[5].

There are three basic types of β-decays: in its simplest form, known more precisely as β^--decay. It may occur for both free and bound neutrons, as in

$$n \rightarrow p + e^- + \bar{\nu}_e \qquad (m_n > m_p + m_e). \qquad (2.1.1)$$

For bound neutrons in many nuclei (e.g. those at the bottom of the valley of stability), this is not energetically allowed, as the daughter nucleus would be heavier than the parent. The inverse nucleon transition occurs too, known as β^+-decay, in which a proton transmutes into a neutron, emitting a positron and an electron neutrino (ν_e). Since the proton is lighter than the neutron, such a process is possible *only* inside a nucleus (but again not near to the bottom of the valley of stability), the final-state neutron may occupy a lower energy level than the original proton; in other words, the binding energy of the daughter nucleus may be greater than that of the parent:

$$p + \text{binding energy} \rightarrow n + e^+ + \nu_e. \qquad (2.1.2)$$

Finally, a process known as *electron capture* (EC) or K-capture is also possible. An atomic electron from the lowest (K) shell may be captured by a proton (bound inside a nucleus), which thereby transmutes into a neutron, simultaneously emitting a neutrino:

$$p + e^- + \text{binding energy} \rightarrow n + \nu_e, \qquad (2.1.3)$$

---

[4] The 1938 Nobel Prize in Physics was awarded to Enrico Fermi 'for his demonstrations of the existence of new radioactive elements produced by neutron irradiation, and for his related discovery of nuclear reactions brought about by slow neutrons'.

[5] Naturally, Fermi called it simply a *neutrino*. However, to respect angular-momentum conservation (orbital angular momentum and boson spin can only compensate integer spin), fermions can be produced or annihilated only as particle–antiparticle pairs (e.g. e^+e^-, $\mu^+\mu^-$, $q\bar{q}$ etc). This constraint is also implicit in the Dirac equation. It is therefore considered more natural today to associate an *antineutrino* with the electron emitted in β-decay and a neutrino with positron emission.

where again the binding energy invoked is that of the nucleons involved[6]. All three forms are intimately related, as the relevant Feynman diagrams[7], shown in figure 2.1, make manifest: they are just space–time rotations and/or reflections of the same basic process.

Examples of the three basic types of β-decays are

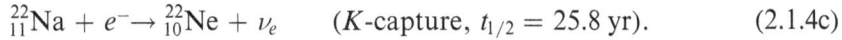

$$^{137}_{55}\text{Cs} \rightarrow {}^{137}_{56}\text{Ba} + e^- + \bar{\nu}_e \qquad (\beta^-\text{-decay, } t_{1/2} = 30.2 \text{ yr}), \qquad (2.1.4a)$$

$$^{22}_{11}\text{Na} \rightarrow {}^{22}_{10}\text{Ne} + e^+ + \nu_e \qquad (\beta^+\text{-decay, } t_{1/2} = 2.89 \text{ yr}), \qquad (2.1.4b)$$

$$^{22}_{11}\text{Na} + e^- \rightarrow {}^{22}_{10}\text{Ne} + \nu_e \qquad (K\text{-capture, } t_{1/2} = 25.8 \text{ yr}). \qquad (2.1.4c)$$

In all cases the nuclear mass number (A) does not change, but the charge or atomic number (Z) does, by ± 1.

As described by the proton–neutron asymmetry term in the Bethe–Weizsäcker semi-empirical mass formula (von Weizsäcker 1935, Bethe and Bacher 1936), for fixed A, the nuclear mass is a quadratic function of ($Z - N$). We may thus expect cascades or chains of $\beta^\pm$-decays down either side of the valley so-formed until the lightest isobar is reached, which will then be stable against further such decay, see figure 2.2(a).

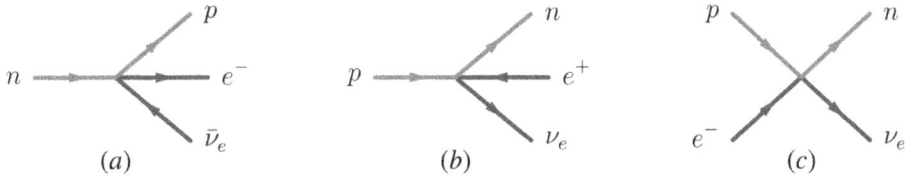

Figure 2.1. The Feynman diagrams depicting: (a) β^--decay, (b) β^+-decay and (c) K-capture.

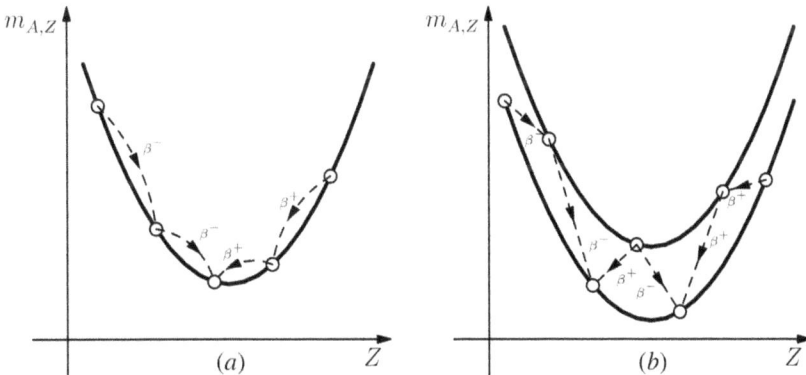

Figure 2.2. The two possible types of β-decay chains for (a) A odd and (b) A even.

---

[6] Recall that the atomic binding energy is negligible.

[7] Feynman diagrams will be explained better later; for the time being they are to be considered just useful pictorial but faithful representations of the processes under consideration.

The empirical mass formula also contains a nucleon-pairing term: nuclei with even numbers of protons and/or neutrons are more strongly bound. Thus, even–even nuclei are lightest, odd–odd heaviest, with odd–even falling in between. The β-decay chains for odd and even A must thus be distinguished. In the case A odd, either one of N or Z is odd and the other even and so the pairing contribution is constant (always zero). For A even, either both are odd or both even and thus there are two separate mass parabolæ (the series for even–even lying below the odd–odd), see figure 2.2(b). If the two lowest-lying isobars in a given A-odd chain are too close in mass for the heavier to decay, then both are β-stable (e.g. $^{123}_{51}\text{Sb}$ and $^{123}_{52}\text{Te}$. In the A-even case, there may be up to three β-stable isobars (e.g. $^{96}_{40}\text{Zr}$, $^{96}_{42}\text{Mo}$ and $^{96}_{44}\text{Ru}$).

Double β-decay

Moreover, so-called *double* β-decay is possible, in which Z changes by two units, with emission of two electron–neutrino pairs. Such a process was predicted as long ago as by Maria Goeppert-Mayer (1935). Being doubly weak and therefore highly suppressed (all measured half-lives are in excess of 10^{19} yr), any given double β-decay mode will be observable if and only if the single-decay mode for that nucleus is prohibited. For example, in figure 2.2(b) the two lowest-lying states are both stable against single β-decay, but the higher-mass nucleus of the two (that on the left) may decay via the double β-decay mode into the lower. It turns out that 35 naturally occurring isotopes are capable of double β-decay, as in the example above. In any case, just 11 such decays have been observed. The double β-decay of $^{96}_{40}\text{Zr}$ into $^{96}_{42}\text{Mo}$ is just such a case, with half-life $\sim 2 \times 10^{19}$ yr. See section 6.1.2 for a more detailed discussion, including the possibility of neutrinoless double β-decay.

2.1.2 Fermi's theory of β-decay

As already mentioned, the theory describing β-decay (and, more in general, the weak nuclear interaction) was devised by Fermi (1933).[8] Let us first examine the kinematics: since it is relatively very massive, the final-state nucleus may be taken to have negligible recoil energy (the recoil momentum may, as usual though, be appreciable) and therefore the electron–neutrino pair have *kinetic* energy

$$E_e^{\text{kin}} + E_\nu^{\text{kin}} \simeq Q, \qquad (2.1.5)$$

where Q is the energy released (or so-called *Q-value*). As usual, Q is defined as the mass difference between the parent and daughter nuclei minus the final electron mass (assuming a zero-mass neutrino):

$$Q := (M_i - M_f - m_e)c^2; \qquad (2.1.6)$$

i.e. it indeed represents the kinetic energy available to the electron–neutrino pair. Note that since the electron emitted also corresponds to the extra electron of the final-state *atom*, Q is also given by the neutral *atomic*-mass differences (neglecting, as

---

[8] An initial paper by Fermi was rejected for publication by *Nature*, as containing 'speculations' that were deemed 'too remote from reality.'

always, the electron binding energy). However, it will be convenient in what follows to define the *total* energy available to the electron–neutrino pair (usually indicated E_0), which is just the mass difference between the parent and daughter nuclei:

$$E_0 := (M_i - M_f)\, c^2 = Q + m_e c^2. \tag{2.1.7}$$

Exercise 2.1.1. *The decay* $^{210}\mathrm{Bi} \to {}^{210}\mathrm{Po} + e^- + \bar{\nu}_e$ *has a Q-value of 1.16* MeV; *calculate the maximum recoil kinetic energy of the daughter nucleus.*

Do the same for $^{22}_{11}\mathrm{Na} \to {}^{22}_{10}\mathrm{Ne} + e^+ + \nu_e$, $Q = 4.38$ MeV *(take* $m_\nu = 0$*).*
What is the minimum recoil energy?

The rate of such a decay may be calculated in quantum mechanics using Fermi's golden rule, equation (C.6.47), which here reads

$$\Gamma = \frac{1}{\tau} = \frac{2\pi}{\hbar} \int \mathrm{d}n_f\, |\mathcal{M}_{fi}|^2, \tag{2.1.8}$$

where τ is the mean lifetime for this decay. The transition matrix element is

$$\mathcal{M}_{fi} = \int \mathrm{d}^3x\; \psi_f^*(x)\, \mathcal{H}_{int}(x)\, \psi_i(x), \tag{2.1.9}$$

where $\psi_{i,f}$ are the initial- and final-state wave-functions and $\mathcal{H}_{int}$ is the relevant interaction Hamiltonian. The initial-state wave-function is just that of the parent nucleus (Ψ_i), while that of the final state is the product of the wave-functions describing the daughter nucleus (Ψ_f) and those of the emitted electron (ψ_e) and antineutrino (ψ_ν).

The remaining problem is the choice of operator to represent the (new) interaction Hamiltonian. The solution adopted by Fermi was simply the identity operator, thus:

$$\mathcal{M}_{fi} = \int \mathrm{d}^3x \Psi_f^*(x)\, \psi_e^*(x)\, \psi_\nu^*(x) \cdot G_F \cdot \Psi_i(x), \tag{2.1.10}$$

where G_F is a constant defining the strength of the interaction (now known as the Fermi constant). Such a form is significantly different to that used for the electromagnetic interaction, which is viewed as mediated by the exchange of a (virtual) photon propagating between two distinct space–time points. However, at those time and distance scales to which β-decay is sensitive, there was no experimental evidence for similar propagation and the interaction appeared to take place at a single space–time point.

Exercise 2.1.2. *We now know that the weak interaction is in fact mediated by the* $W^\pm$ *and* Z^0 *bosons; the former has a mass around* ∼80 *GeV and the latter* ∼91 *GeV. By appealing to the uncertainty principle, estimate the maximum distance that such particles may propagate as virtual intermediate states. Compare this with the wavelength associated with the typical momenta involved in* β-decay.

To describe particles of well-defined linear momenta, the wave-functions for the electron–neutrino pair should naturally be taken as plane waves:

$$\psi_{e,\,\nu}(x) = \frac{e^{ik_{e,\,\nu}\cdot x}}{\sqrt{V}}, \tag{2.1.11}$$

where we have adopted the standard normalisation of one particle in a volume V, which will, of course, cancel in the final answer. We thus have

$$\mathcal{M}_{fi} = \frac{G_F}{V}\int d^3x\; e^{-i(k_e + k_\nu)\cdot x}\,\Psi_f^*(x)\,\Psi_i(x), \tag{2.1.12}$$

where $k_{e,\nu}$ are the electron and neutrino wave-numbers ($\hbar k = p$). A further simplifying approximation may be made: the momenta of the electron and neutrino are limited to only a few MeV and the corresponding wavelengths are therefore much larger than the size of the nucleus, which is the integration region determined by the presence of $\Psi_i(x)$ and $\Psi_f(x)$. The exponential will thus not vary appreciably under the integral and may be taken as constant (i.e. unity). We then have

$$\mathcal{M}_{fi} \simeq \frac{G_F}{V}\int d^3x\; \Psi_f^*(x)\,\Psi_i(x), \tag{2.1.13}$$

describing the overlap between the initial- and final-state nuclear wave-functions.

In certain cases, where the final-state proton (neutron) occupies the same nuclear level as the initial-state neutron (proton), the two wave-functions are (almost) identical and thus, to a good approximation, the integral is unity. Otherwise, with some knowledge of the approximate wave-functions, it may be estimated rather well and, in any case, to a very good approximation is constant, independent of the electron and neutrino energies.

For a particle of mass m and momentum p, with the chosen normalisation to a box of volume V, the density of states dn/dE is derived by considering

$$d^3n = \frac{V}{(2\pi\hbar)^3}\,d^3p = \frac{V}{(2\pi\hbar)^3}\,p^2 d\Omega\,dp. \tag{2.1.14}$$

The angular integrals may be performed by assuming spherical symmetry and give $\int d\Omega = 4\pi$. The single-particle density of states is therefore

$$\frac{dn}{dE} = \frac{V}{(2\pi\hbar)^3}\,4\pi p^2\,\frac{dp}{dE}. \tag{2.1.15}$$

Consider now a final-state electron of given energy E_e; this corresponds to a neutrino energy $E_\nu = E_0 - E_e$ and therefore $dn_\nu/dE_0 = dn_\nu/dE_\nu$ (for fixed E_e). There are, however, $\int dE_e\,dn_e/dE_e$ such possible electron states and thus, for the two-particle density of states in consideration here, we have

$$\frac{d^2n(E_0, E_e)}{dE_e\,dE_0} = \frac{dn_e(E_e)}{dE_e}\frac{dn_\nu(E_\nu)}{dE_\nu}\bigg|_{E_e + E_\nu = E_0}$$

$$= \left[\frac{4\pi V}{(2\pi\hbar)^3}\right]^2 p_e^2\,\frac{dp_e}{dE_e}\,p_\nu^2\,\frac{dp_\nu}{dE_\nu}\bigg|_{E_e + E_\nu = E_0}, \tag{2.1.16}$$

where the integral over E_e still remains to be performed.

Using relativistic kinematics, the mass-shell relation is $E^2 = p^2c^2 + m^2c^4$ and therefore $E \, dE = pc^2 dp$. Thus, for a given final-particle energy–momentum, we have

$$p^2 \frac{dp}{dE} = \frac{E \, p}{c^2} = \frac{E \, \sqrt{E^2 - m^2c^4}}{c^3}. \tag{2.1.17}$$

Finally, assuming a zero-mass neutrino, so that $|\boldsymbol{p}_\nu| = E_\nu$, and using $E_\nu = E_0 - E_e$,

$$\frac{d^2n(E_0, E_e)}{dE_e \, dE_0} = \left[\frac{4\pi V}{(2\pi\hbar)^3} \right]^2 \frac{E_e \, \sqrt{E_e^2 - m_e^2 c^4}}{c^3} \frac{(E_0 - E_e)^2}{c^3}. \tag{2.1.18}$$

The complete expression must now be integrated over E_e from the rest-mass energy $m_e c^2$ up to E_0. It is customary here to normalise the energies to the electron rest-mass energy and thus the integral to be performed is

$$f(E_0) = \int_1^{x_{max}} dx \, x \, \sqrt{x^2 - 1} \, (x_{max} - x)^2, \tag{2.1.19}$$

where $x_{max} = E_0/m_e c^2$. For $x_{max} \gg 1$, $f(E_0) \simeq x_{max}^5/30$. And so the final result for the decay rate is

$$\Gamma = \frac{1}{\tau} \simeq \frac{G_F^2 \, E_0^5 \, K}{60 \, \pi^3 \, \hbar^7 c^6}, \tag{2.1.20}$$

where K is the nuclear wave-function overlap integral in (2.1.13). The E_0^5 dependence (arising purely from the phase-space integral) is known as Sargent's rule (1933), originally only a simple empirical observation, which is thus explained theoretically.[9] Note that such a power dependence, while important, is not as strong as the exponential variation observed in α-decay.

There is an important correction to the above expression, due to the Coulomb interaction between the charges of the emitted electron or positron and the daughter nucleus (Fermi 1933, see the 1934 papers). Put simply, we note that, since the nuclear Coulomb force is repulsive for a positron, it will be ejected with greater final energy than an electron, where it is attractive. More precisely, the plane-wave solutions used will be distorted near to the nucleus. In a non-relativistic approximation ($E_e \ll m_e c^2$), we then find the following corrective factor for Γ (known as the Fermi function):

$$F(Z_f, E_e) \simeq \frac{\eta}{1 - e^{-\eta}}, \quad \text{with } \eta = \pm 2\pi Z_f \, \alpha/\beta_e, \tag{2.1.21}$$

where the plus (minus) sign applies to electrons (positrons) and $\beta_e = v_e/c$ is the asymptotic velocity of the electron or positron emitted.

---

[9] Note that Sargent's rule ($\Gamma \propto Q^5$) is indeed a generally valid law for any *three-body* decay, depending only on the final-state phase-space integral (see e.g. equation (2.1.19)).

2.1.3 The 'forbidden' β-decays

In evaluating the matrix-element integral in (2.1.12), we have so far approximated the electron and neutrino wave-functions with unity; i.e. we have effectively performed the following Taylor expansion

$$e^{-i(k_e + k_\nu)\cdot x} = 1 + \left[-i(k_e + k_\nu) \cdot x \right] + \frac{1}{2!}\left[-i(k_e + k_\nu) \cdot x \right]^2 + \dots \quad (2.1.22)$$

and, noting that inside the nuclear volume $|(k_e + k_\nu) \cdot x| \ll 1$, we have neglected all terms beyond the very first. Therefore, should the resulting overlap integral equation (2.1.13) vanish, the decay process may still be possible, but we should consider the contribution of some higher-order term in the above expansion. The leading-order vanishing will typically be due to violation of a selection rule (i.e. a symmetry or conservation law); e.g. when there should be a change of either the parity or orbital angular momentum between parent and daughter nuclei.

In this case, we speak of *first-forbidden*, *second-forbidden* etc decays, according to the order of the first term necessary to obtain a non-vanishing matrix element. They are not, of course, truly forbidden, but are relatively more and more suppressed. Note moreover, that extra energy dependence is then introduced into the spectrum function, which can therefore change shape significantly. Two examples of so-called *forbidden* β-decays are shown in table 2.1. Note that $k \cdot x$ is equivalent to angular momentum; so that each power of $k \cdot x$ adds one unit to the possible nuclear-spin variation, but each power of x also changes the parity.

Table 2.1. Two examples of 'forbidden' β-decays and their relevant parameters.

Decay	Forbidden	$t_{1/2}$	Q
^{198}Au (β^-) ^{198}Hg	1st	2.7 d	1.37 MeV
^{36}Cl (β^-) ^{36}Ar	2nd	3×10^5 yr	0.71 MeV

A final consideration to be made on possible β-decay suppression effects regards the rôle of the radial quantum number n: orthogonality of the radial wave-functions leads to the vanishing of matrix elements also involving initial and final single-particle nuclear wave-functions of differing n. For many β-decays this is not a strong restriction, since wave-functions of the initial neutron (proton) and the final-state proton (neutron) usually have precisely the same radial dependence; the question is thus relatively unimportant for isotopes that are proton-rich or close to the stability line. On the other hand, the radial selection rule can play a significant rôle for heavier neutron-rich nuclei, in particular, for the region $N > 126$ and $Z < 82$ (i.e. with Z smaller than $^{208}_{82}$Pb, but A the same or larger. Note that this all renders experimental investigation of the validity of the radial selection rule difficult.

2.1.4 The Kurie plot

The simple shape of the (uncorrected) spectrum,

$$P(E_e) \propto F(Z_f, E_e) \, E_e \sqrt{E_e^2 - m_e^2 c^4} \, (E_0 - E_e)^2, \quad (2.1.23)$$

suggests a rather useful method of plotting the data (Kurie *et al* 1936). Taking $P(E_e)$ now as the *measured* electron spectrum, let us define the following function:

$$K(E_e) := \sqrt{\frac{P(E_e)}{F(Z_f, E_e) \cdot E_e \cdot \sqrt{E_e^2 - m_e^2 c^4}}}, \qquad (2.1.24)$$

From comparison with equation (2.1.23), this should clearly give

$$K(E_e) \propto E_0 - E_e \qquad (2.1.25)$$

We thus expect a straight line intersecting the abscissa at $E_e = E_0$; see figure 2.3.

Such a plot is useful in the search for a non-zero neutrino mass, a possibility already contemplated by Fermi (1933). So far we have assumed a zero-mass electron neutrino; if instead we insert a mass term for the neutrino, then the expected form of the Kurie plot becomes

$$K(E_e) \propto \sqrt{(E_0 - E_e)\sqrt{(E_0 - E_e)^2 - m_\nu^2 c^4}}, \qquad (2.1.26)$$

where by m_ν we mean the mass of the electron antineutrino. This intersects the abscissa at $E_e = E_0 - m_\nu c^2$, leading to a curve of the form shown by the heavier dotted line in figure 2.3. However, the effects of finite experimental resolution and Fermi motion of the decaying nucleus etc combine to smear the measured distribution into the form of the lighter dotted line in figure 2.3.

Exercise 2.1.3. *Derive expression* (2.1.26).

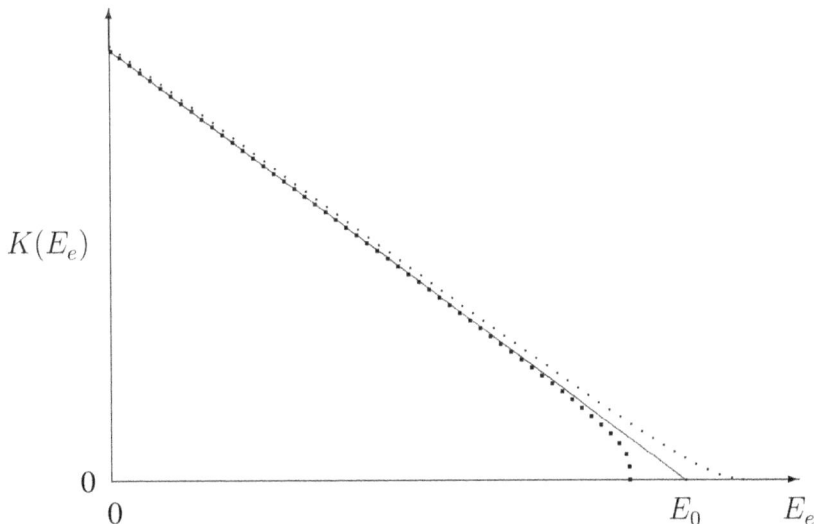

Figure 2.3. Example Kurie plots of the β-decay spectrum: the ideal zero-mass neutrino case (continuous line), ideal non-zero mass (heavy dotted line) and realistic non-zero mass (light dotted line).

2.1.5 Neutrino mass limits

We now list some of the methods used to obtain limits on the (principally electron) neutrino mass. We start with the method exploiting the preceding discussion.

Tritium decay limits
The best candidate for a possible measurement of the electron–antineutrino mass is the β^--decay of tritium into helium-3:

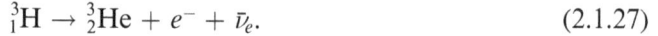

$$^3_1\text{H} \rightarrow ^3_2\text{He} + e^- + \bar{\nu}_e. \tag{2.1.27}$$

The preference for this decay lies in both the favourable half-life, $t_{1/2} = 12.32$ years, and the very low Q-value,

$$Q = 18.5737 \pm 0.000\,25 \text{ keV}, \tag{2.1.28}$$

where $Q = E_0 - m_e$ and E_0 is the mass difference between ^3_1H and ^3_2He nuclei. Since the recoil energy of the ^3_2He nucleus is negligible, this is the maximum kinetic energy of the electron in the limit of zero neutrino mass. Using this decay, the Karlsruhe Tritium Neutrino (KATRIN) collaboration is the presently most sensitive experiment to place an upper limit on the neutrino mass (Aker *et al* 2022):

$$m_\nu < 0.8 \text{ eV} \quad (90\% \text{ CL}). \tag{2.1.29}$$

Supernova limits
It is interesting to note that a comparable, though less stringent, limit is provided by the finite spread in arrival times of the few neutrinos recorded after the supernova SN 1987A in the Large Magellanic Cloud, approximately 51.4 kpc ($\sim$170 thousand light years) from Earth. Bursts of neutrinos were independently detected at three separate neutrino observatories: Kamiokande II, IMB and Baksan. In total, 24 neutrinos were detected over a time interval of less than 13 s, a significant deviation from the observed background level; 11 were detected by Kamiokande II, 8 by IMB and 5 by Baksan. Analysis of the energy and arrival-time spread provided an upper limit of about 11 eV (Bahcall and Glashow, 1987); although a more recent analysis sets a 95% CL limit of 5.7 eV (Loredo and Lamb, 2002).

Cosmological limits
Finally, we should mention the cosmological limits coming from the Planck Collaboration study of the cosmic microwave background (CMB) spectrum (see, for example, Aghanim *et al* 2021). The latest estimates, combining a variety of data, give

$$\sum m_\nu < 0.12 \text{ eV} \quad \text{at the 95\% CL}. \tag{2.1.30}$$

We note in passing that the same collaboration also obtains the following estimate of the number of neutrino species:

$$N_{\text{eff}} = 2.96^{+0.34}_{-0.33}, \quad \text{again at the 95\% CL.} \tag{2.1.31}$$

2.2 Parity violation in weak interactions

Within the realm of particle physics the first and most notable phenomenological manifestation of symmetry (or lack thereof) is perhaps that of the rôle of parity in the weak interaction. Of course, symmetry and its importance in general was recognised much earlier: Einstein's[10] development of the theory of relativity (both special and general) rests on notions of symmetry (with respect to the choice of reference frames), while the seminal work of Emmy Noether (1918) on the relationship between continuous symmetries and conserved quantities lies at the very foundation of all modern theory, both classical and quantum. In relation to the specific subject of the following section, the pioneering work of Eugene Wigner (1927)[11] is central to our understanding of the deep relations between rotational symmetry and angular momentum,

Parity, however, holds a special place as the first, simplest and previously unquestioned[12] symmetry to be found violated in Nature (Wu *et al* 1957). This opened the window onto a completely new perspective: the *breaking* of symmetries. Just over half a century later it is now quite normal to seek violation of symmetries and indeed to exploit the natural violations that can occur at the quantum level to explain, at least in part, the phenomenology of the particles populating the world we see and experience.

2.2.1 Noether's theorem

Let us first briefly recall Noether's theorem (1918): put simply, this states that to each and every continuous symmetry of the Lagrangian (or Hamiltonian) there corresponds a conserved quantity and *vice versa*. Some classic well-known examples are:

- temporal-translation invariance and energy conservation,
- spatial-translation invariance and momentum conservation,
- rotational invariance and angular-momentum conservation, etc.

A few observations on the theorem will be useful for much of what follows. The concepts of both symmetry and a conserved quantum number will be central in many of the topics we shall address here.

First of all, while the original context was *continuous* symmetries in *classical* field theories, the theorem applies equally to quantum field theory and thus to particle physics. Indeed, it often provides a guiding light in the construction of theories describing the fundamental interactions. A simple example is the conservation of

---

[10] The 1921 Nobel Prize in Physics was awarded to Albert Einstein 'for his services to Theoretical Physics, and especially for his discovery of the law of the photoelectric effect'.

[11] The 1963 Nobel Prize in Physics was awarded one half to Eugene Paul Wigner 'for his contributions to the theory of the atomic nucleus and the elementary particles, particularly through the discovery and application of fundamental symmetry principles' and one quarter each to Maria Goeppert-Mayer and J Hans D. Jensen 'for their discoveries concerning nuclear shell structure'.

[12] It is recounted that when Abdus Salam, as a young researcher, proposed a theory involving parity violation to Wolfgang Pauli, he was unceremoniously dismissed with the remark: 'This young man does not realise the sanctity of parity!'

charge or of particle type (e.g. lepton or baryon, leading to the concepts of lepton and baryon numbers) arising from the global phase (or gauge) invariance of the wave-functions in quantum theories.

Secondly, the theorem may also be extended to *discrete* symmetries, such as spatial or temporal inversion and charge conjugation. There is though a slight difference: whereas continuous symmetries are associated with *additively* conserved quantities, in the discrete case, conservation is *multiplicative*. That is, for example, a particle or system of particles may be assigned a temporal-inversion signature or *T*, which takes on the value '*plus*' or '*minus*' according to the behaviour (i.e. a sign change or not) of the given wave-function under the symmetry operation $\mathcal{T}$. The overall signature under time reversal of a composite system is then the product of the subsystem signatures[13]. We shall, in fact, meet all three of the discrete symmetries already mentioned here: C, P and T. Moreover, we shall find that they are all *violated* in particle physics.

Thirdly, the concepts outlined will also prove to be of great utility even in the case of an *imperfect* symmetry. As the reader might easily imagine, in such a case, we find a *partially* conserved quantity. The best examples of such a situation are probably those of Heisenberg's isospin and its extension by Gell-Mann to the SU(3) of quarks. Already for isospin in nuclear physics, we immediately realise that e.g. the proton and neutron do not have exactly identical masses[14]. Thus, the isospin quantum numbers are not always perfectly conserved. Inverting the logic, we might have arrived at the construction of the quark model (including the third quark flavour, *strangeness*) by observing, not the broken symmetry between, say, the proton, neutron and Λ^0 hyperon, but by appealing to the *partially* conservation of the different quark flavours (up, down and strange). Indeed, it was already well known that the weak interaction violated the conservation of both isospin and strangeness, whereas the strong interaction respected both. In particular, strangeness conservation is violated more than isospin and we find that the approximate p–n–Λ^0 symmetry is broken most by the larger Λ^0–nucleon mass difference.

2.2.2 The τ–θ puzzle

In the early fifties a puzzle arose (Dalitz 1953) involving two new subatomic particles, then called τ and θ. Both were members of the newly found family of so-called *strange* particles, relatively long-lived objects that were being produced in the new accelerator experiments. The long lifetimes of these particles suggested that, although the energetically very favourable final states often only contained strongly interacting particles, they did not decay via the strong interaction; they were thus

---

[13] An important tangible effect of time-reversal invariance is Kramers' degeneracy theorem (1930). That is, the observation that the energy levels of a system containing an *odd* number of fermions must be at least doubly degenerate, provided that no external magnetic fields act. While this has significant repercussions in solid-state physics, we might also mention that, as a consequence, e.g. the deuterium atom (a proton, a neutron and an electron) displays a typical hyperfine splitting.

[14] There are two competing contributions to the difference: the up and down quarks have slightly different masses and the electromagnetic interaction provides a small isospin-violating perturbation.

deemed strange. The τ^+ and the θ^+ are now known to be one and the same particle, the charged kaon or K^+.

In fact, the τ^+ and θ^+ were found to be identical in terms of their mass, charge and other properties (within experimental precision) and were only distinguished by their decay modes. Naturally, their *charge-conjugate* versions or *antiparticles*, with opposite charge, also exist and exhibit identical behaviour. The two particles decayed quite differently and hence their being considered distinct. The τ^+ decayed into three pions ($\pi^+\pi^+\pi^-$ or $\pi^+\pi^0\pi^0$), whereas the θ^+ produced only two ($\pi^+\pi^0$). Indeed, it was *only* the different decay modes that distinguished them and suggested the need for two separate particles. The fact that their lifetimes were also very similar (identical within errors) rendered the idea of two distinct yet almost identical states (we might say 'twins') very puzzling; no other known pair of particles displayed such striking similarity.

The intriguing necessity for two distinct particles arose from the realisation that the parity of the two final states must be different (Dalitz 1953). However, in 1956, based on such observations of the charged-kaon decays, Tsung-Dao Lee and Chen Ning Yang were led to make the (then extravagant) proposition that parity conservation might be violated[15]. They further suggested that if the answer to the τ–θ puzzle were indeed parity violation, then such an effect might also be observed in the spatial distribution of the β-decay of polarised nuclei. In essence, they proposed measuring the dependence of the decay rate on a pseudoscalar quantity such as $\boldsymbol{p}\cdot\boldsymbol{s}$, where $\boldsymbol{p}$ and $\boldsymbol{s}$ are, say, the final electron momentum and the spin of the decaying nucleus.

Let us first examine why the final-state parity assignments must differ (Dalitz 1953). Experiments indicated that the spin of both objects was zero; thus, the question of angular-momentum conservation is rather simple since pions too have spin zero: the total orbital angular momentum of the final state must then also be zero. With zero orbital contribution to the total angular momentum of the system, the overall parity is just the product of the intrinsic parities of the final-state particles; recall that the parity of the spatial part of the wave-function is simply $(-1)^L$. This is trivial in the two-pion case, whereas for three pions, although they may individually have non-zero orbital angular momentum, the net cancellation required implies overall even parity of the spatial part. The parities of the final states are thus determined by the *intrinsic parities* of the pions they contain. Note finally that it was in this context that Richard Dalitz (1953) introduced a special two-dimensional representation of the three-body final state (see appendix C.7).

2.2.3 Intrinsic parity and its measurement

All particles either naturally possess or may be assigned an intrinsic parity. In the case of fermions, a consequence of the Dirac equation (see appendix B.1) is that fermion and antifermion have *opposite* parities. Since fermion number is conserved

---

[15] The 1957 Nobel Prize in Physics was awarded equally to Chen Ning Yang and Tsung-Dao Lee 'for their penetrating investigation of the so-called parity laws which has led to important discoveries regarding the elementary particles'.

(only fermion–antifermion pairs may be created or annihilated), the absolute value of, say, the electron parity is undetectable and irrelevant. By convention, the parity of fermions (antifermions) is taken as positive (negative) although no physical significance may be attached to either separately. However, the intrinsic parity of a fermion–antifermion pair is meaningful and is thus predicted to be *negative*.

No equivalent statement exists for the parity of bosons, which must be determined experimentally or deduced via theoretical arguments; given that bosons may be created and annihilated, their individual parities are meaningful. Recall that the parity of a composite state is just the product of the parities of the parts. The parity of a q$\bar{\text{q}}$ meson (a boson) is thus the product of the intrinsic parities of the quarks of which it is composed and the parity of the spatial wave-function describing their relative orbital motion. In other words,

$$P_\pi = P_q P_{\bar{q}} (-1)^L = (-1)^{L+1}, \qquad (2.2.1)$$

where L is the orbital quantum number. This has the immediate consequence that a pion, being the lowest-mass, zero-spin q$\bar{\text{q}}$ state (and therefore presumably having zero internal orbital angular momentum[16] and consequently the quark and antiquark spins antiparallel), should have *negative* intrinsic parity. This is indeed experimentally verified; i.e. it is pseudoscalar.

The measurement of $P_\pi = -1$ is conceptually rather simple: consider the associated production (via the parity-conserving strong interaction) of a neutron pair from a low-energy, negatively charged pion incident on a deuteron (Chinowsky and Steinberger, 1954):

$$\pi^- d \rightarrow nn. \qquad (2.2.2)$$

The deuteron is a spin-one nucleus ($L_{pn} = 0$ and $S_{pn} = 1$) of *positive parity* and the pion has zero spin, whereas the proton and neutron both have spin one-half. In any case, the two nucleons are, with respect to parity, mere spectators in this process. What count therefore are the relative initial- and final-state orbital angular momenta. Now, the process is actually that of *K-capture*: the pion is initially trapped, forming an excited pionic atom, and then rapidly cascades down to the lowest Bohr orbit, i.e. an s-wave, from which it finally interacts with the deuteron. The total angular momentum of the initial state is thus one, but with $L_{\pi d} = 0$ and therefore positive spatial parity. The parity of the initial and (so too) the final states is therefore precisely the pion parity: $P_\pi = (-1)^{L_{nn}}$.

The orbital angular momentum of the final state could, in principle, be measured by studying (statistically) the angular distribution of the neutrons produced[17]. However, it is easier (and more instructive) to appeal to the Fermi–Dirac statistics obeyed by a system of two *identical* fermions. The final nn system must have unit total angular momentum and this can be constructed from either a sum of the two neutron spins giving zero (singlet) and orbital motion $L = 1$ or a sum of spins giving

---

[16] Non-zero internal motion in a compound system would imply non-negligible internal kinetic energy, which in turn would contribute to an increased mass of the composite particle.

[17] Recall that the angular distribution of the final state is governed by the Legendre polynomial $P_L(\cos\theta)$.

one (triplet) and orbital motion $L = 0$, 1 or 2 (recall that angular momenta are to be added vectorially). The spin-singlet state has the following spin wave-function:

$$\frac{1}{\sqrt{2}}\big(|\uparrow\downarrow\rangle - |\downarrow\uparrow\rangle\big). \tag{2.2.3a}$$

Note, in particular, that it is antisymmetric under interchange of the neutrons. Since they are identical fermions, the spatial part must then be symmetric and thus L even. The total spin can thus only be even and therefore one is excluded. For the triplet state, on the other hand, we have the following three possible spin wave-functions:

$$|\uparrow\uparrow\rangle, \qquad \frac{1}{\sqrt{2}}\big(|\uparrow\downarrow\rangle + |\downarrow\uparrow\rangle\big), \qquad |\downarrow\downarrow\rangle. \tag{2.2.3b}$$

These are all symmetric under interchange and thus the spatial part must be antisymmetric, giving L odd, which will only accommodate $L = 1$, combining with $S = 1$ to provide total spin one. We must therefore have $S = 1$, $L = 1$, negative spatial parity and $P_\pi = -1$.

Exercise 2.2.1. *By considering the intrinsic parities of the proton and neutron (conventionally taken to be positive), together with their known orbital and spin alignments inside the deuteron, show that we do indeed expect $P_d = +1$.*

2.2.4 The physical consequences of parity violation

As mentioned earlier, the idea was to measure the dependence of, say, a decay rate on a *pseudoscalar* quantity such as $\boldsymbol{p} \cdot \boldsymbol{s}$. The reason for this is quite simple. Let us suppose that the transition matrix element or quantum amplitude for some such rate (or, of course, interaction cross-section) takes the general form

$$\mathcal{M} \propto A + B\,\boldsymbol{p} \cdot \boldsymbol{s}, \tag{2.2.4}$$

with A and B scalar quantities that do *not* depend (linearly) on either $\boldsymbol{p}$ or $\boldsymbol{s}$. Then, since $\boldsymbol{p}$ is a polar vector, it changes sign under the action of parity inversion, whereas $\boldsymbol{s}$, a pseudovector or axial-vector, does not. The B term will thus change sign with respect to the A term and so

$$|\mathcal{M}^P| \neq |\mathcal{M}|, \tag{2.2.5}$$

where $\mathcal{M}^P$ stands for the corresponding matrix element under parity inversion. While there was no guarantee that this should be the correct description, such a difference would be precisely the manifestation of parity violation that Lee and Yang (1956) sought.

Now, although it is not possible to apply $\mathcal{P}$ as such experimentally, the presence of the parity-violating form described may be detected from the spatial dependence it implies. For example, for a fixed spin $\boldsymbol{s}$, we also have that

$$|\mathcal{M}(-\boldsymbol{p})| \neq |\mathcal{M}(+\boldsymbol{p})|. \tag{2.2.6a}$$

And so it is sufficient to simply examine, e.g., the decay rate for final-state electron momenta parallel and antiparallel to the nucleon-spin axis. Note that the two statements are entirely equivalent. A final-state electron observed moving, say, parallel to the nucleon-spin axis, would move in the opposite direction in the parity-inverted experiment. Note also that analogously, for a fixed electron momentum p, we have

$$|\mathcal{M}(-s)| \neq |\mathcal{M}(+s)|. \tag{2.2.6b}$$

2.3 Parity violation in β-decay

At that time based at Columbia University, Chien-Shiung Wu was considered a leading world expert on β-decay and Lee also worked at Columbia. Shortly after detailed discussions with Lee and Yang in 1956, the invitation to perform an experiment to detect such parity violation was taken up by Wu and collaborators (1957) at the National Bureau of Standards low-temperature laboratories, where the necessary advanced cooling systems were available[18].

The basic requirement of polarising the initial-state decaying nucleus then led Wu *et al* (1957) to select ^{60}Co as it has a nuclear spin of $J = 5$ in natural units (with a large magnetic moment too). The experiment was by no means simple; in order to substantially polarise the cobalt specimen and avoid depolarisation by thermal motion, exceedingly low temperatures and thus new refrigeration techniques were necessary. To this end Wu enlisted the help of Ernest Ambler, a cryogenics expert. Indeed, so convinced was Wu of the importance of such an experiment that she forwent a long-awaited trip to her native China and immediately set about preparing the necessary equipment at the National Bureau of Standards' headquarters in Maryland, where important cryogenic facilities were available. And so, with group members sleeping in the laboratory in order to be ready when the required temperatures were reached, just a few months after the discussions with Lee and Yang (1956), the experiments were performed.

The measurement of any possible variations with respect to p would require either two identical detectors or two independent runs with a single detector being placed first above and then below the decaying specimen. However, the inherent difficulties in control over systematics would render any difference found highly suspect. The observation exhibited in equation (2.2.6b) was thus exploited and the polarisation axis flipped by inverting the applied (vertical) polarising magnetic field. Still, some control over systematics was desirable and so advantage was taken of the full decay chain. The basic β-decay process is

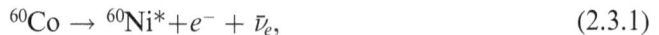

$$^{60}\text{Co} \rightarrow {^{60}}\text{Ni}^* + e^- + \bar{\nu}_e, \tag{2.3.1}$$

where the nickel daughter nucleus ($J = 4$) is produced in an excited state. The subsequent de-excitation transition to the ^{60}Ni ground state is a two-step process involving double γ-emission. Both transitions are of the electric-quadrupole (E4) type

---

[18] An interesting review of the history of β-decay, including parity violation, can be found in Focardi and Ricci (1983).

with the characteristic double-lobe angular distribution. The presence of nuclear polarisation could thus be checked by observation of the relative γ-ray intensities in the polar (i.e. vertical) and equatorial (i.e. horizontal) directions; the emitted photons prefer to lie in the equatorial plane rather than along the polarisation axis.

The procedure thus consisted essentially of cooling the specimen to ~0.01 K and applying a magnetic field along, say, the vertical direction. The refrigeration system was then switched off and continuous read-outs were taken of both the polar β flux and the polar and equatorial γ intensities. This was then repeated with the polarising magnetic field inverted. The results are displayed in figure 2.4 and may be

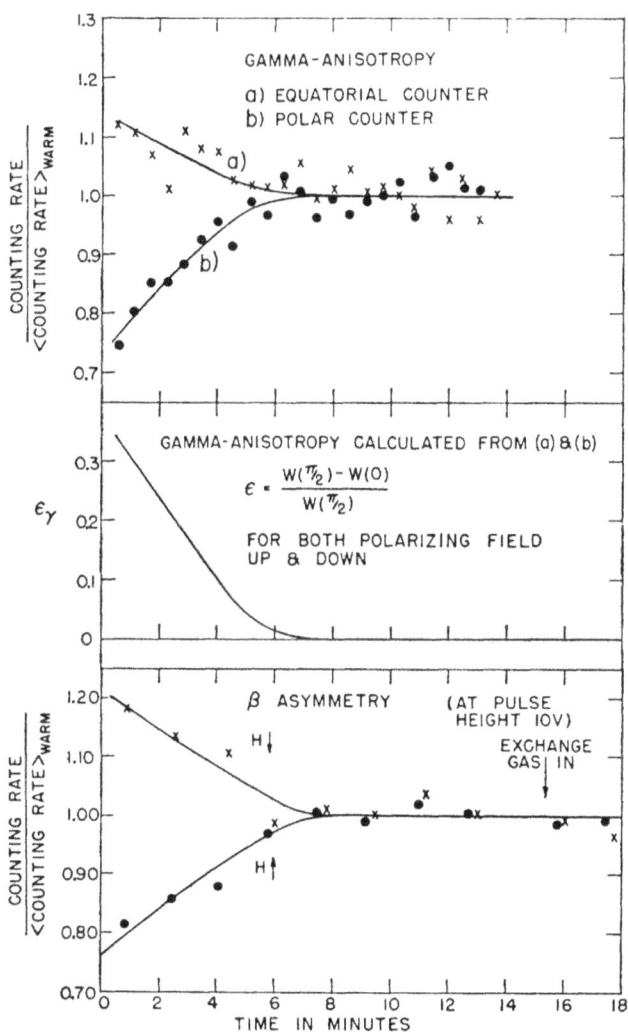

Figure 2.4. Results of the parity violation experiment: the γ anisotropy and the β asymmetry with the polarising magnetic field pointing up and down as functions of the time after switching off the cryostat; figure reproduced with permission from Wu *et al* (1957), copyright (1957) by the American Physical Society.

summarised via the following angular decay distribution for the electrons with respect to the polarisation axis of the ^{60}Co nucleus (as suggested by Lee and Yang 1956):

$$I(\theta) \propto 1 + \alpha \, \frac{v}{c} \cos \theta, \qquad (2.3.2)$$

where, given that $\cos \theta = \pm 1$ here, the asymmetry parameter α is effectively measured directly. Wu *et al* (1957) found α to be negative and gave a lower-limit estimate of $|\alpha| \gtrsim 0.7$. In fact, α is negative for electrons and positive for positrons. More precise measurements subsequently revealed that, in general, $|\alpha| = 1$ (see e.g. Frauenfelder *et al* 1957); i.e. parity is *maximally* violated in β-decay[19].

> '*The sudden liberation of our thinking on the very structure of the physical world was overwhelming.*'
>
> Chien-Shiung Wu

The energies were so low that orbital angular momentum can play no rôle; we may therefore make an argument to infer the electron spin by noting that $\Delta J = -1$ for the nuclei in this decay and therefore the electron–neutrino pair must carry off one unit of spin aligned in the positive z direction, i.e. both spins must be aligned along the same positive z direction. Since the electron tends to move along the negative z direction, it must have helicity[20] -1; in other words, it is *left-handed*. Likewise, in β-decay the positron is always found to have helicity $+1$; i.e. it is *right-handed*.

2.3.1 Muon-decay experiments

The findings of Wu *et al* (1957) were confirmed by parallel experiments on muon decay carried out at almost the same time by Richard Garwin, Leon Lederman and Marcel Weinrich (1957), using a π^+ beam at the University of Columbia cyclotron, and very shortly after by Friedman and Telegdi (1957), also using a π^+ beam at the University of Chicago synchrocyclotron. For completeness, we shall just briefly review the basic concepts. Once again, the idea had been provided by Lee and Yang: they had suggested studying pion decay into muons and the subsequent in-flight muon decay:

$$\begin{aligned} \pi^+ &\to \mu^+ + \nu_\mu \\ &\hookrightarrow e^+ + \nu_e + \bar{\nu}_\mu. \end{aligned} \qquad (2.3.3)$$

---

[19] Although nominated at least seven times, surprisingly, Wu did not receive a Nobel prize for this discovery. Her contribution was, however, recognised via many other honours and awards: she was the first female president of the American Physical Society and the first ever winner of the Wolf Prize in Physics, to mention just two.

[20] Recall that the helicity of a particle is defined as $h := \hat{p} \cdot \hat{s}$ and therefore $-1 < h < +1$. Choosing $\hat{p}$ as the spin-quantisation axis, the two possible electron spin projections give $h = \pm 1$.

In the first decay, owing to parity violation, the muon spin is predominantly aligned along its direction of motion. The muons are then stopped in a carbon target and are assumed to lose little of their polarisation in the stopping process. If this is the case, then again, by virtue of parity violation, the decay electrons will emerge with an angular distribution (with respect to the muon direction) of the form

$$I(\theta) \propto 1 + \alpha \cos \theta. \tag{2.3.4}$$

In fact, rather than moving the detectors, the experimenters exploited the magnetic moment of the muon to precess its spin. The measured distribution suggested a value $\alpha = 0.33 \pm 0.03$. The fact here is that $|\alpha| \neq 1$ can be explained by the non-trivial ($q\bar{q}$) composite nature of the pion.

2.3.2 The helicity of the neutrino

The measurements performed by Wu *et al* (1957), while unequivocally indicating that parity is indeed violated and that the electrons (positrons) produced in β-decay are left (right) handed, did not provide any indication as to the relative spin alignment of the (undetected) neutrinos. Just a few months later, Goldhaber *et al* (1958) thus set out to measure the helicity of the neutrinos produced in β-decay. The method devised is ingenious (see figure 2.5), combining as it does a number of non-trivial physical effects and phenomena.

The process used was K-capture in europium:

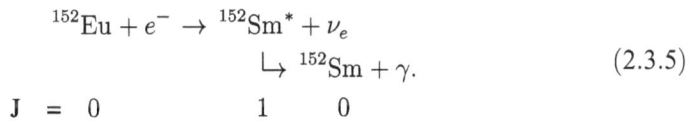

$$\begin{array}{ccc}
^{152}\text{Eu} + e^- & \rightarrow & ^{152}\text{Sm}^* + \nu_e \\
& & \quad\ \hookrightarrow\ ^{152}\text{Sm} + \gamma. \\
\text{J} \quad = \quad 0 & & \quad\ 1 \qquad\ 0
\end{array} \tag{2.3.5}$$

The samarium daughter nucleus subsequently de-excites via γ-emission. The spin actually measured was that of the emitted γ. The analysing magnet and block of magnetised iron surrounding the source served to filter out one or other of the two possible γ helicities: a photon may be absorbed by an atomic electron if and only if their spins are opposite (the electron then flips its spin to conserve angular momentum). Conversely, a photon with spin parallel to the electrons in the block of iron will pass relatively unhindered.

$$\begin{array}{cccc}
J_z: & 1 - \tfrac{1}{2} = \tfrac{1}{2} & 1 + \tfrac{1}{2} = \tfrac{3}{2} & \text{(not allowed)} \\[4pt]
& \gamma \quad e^- & \gamma \quad e^- & \\
\longrightarrow & \Rightarrow + \leftarrow & \Rightarrow + \rightarrow & \\[4pt]
\text{photon direction} & \text{absorbed} & \text{unabsorbed} &
\end{array} \tag{2.3.6}$$

Since the γ-emission process involves a $J = 1$ nucleus decaying into $J = 0$, the photon evidently carries the same spin as the original nucleus, which in turn must be opposite to that of the neutrino.

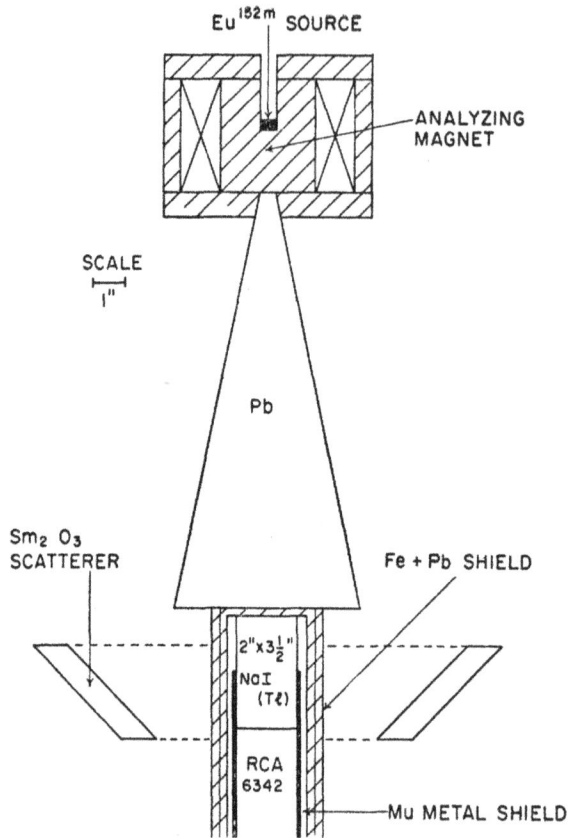

Figure 2.5. Apparatus for analysing the circular polarisation of γ-rays in the experiment to measure the neutrino helicity; figure reproduced with permission from Goldhaber *et al* (1958), copyright (1958) but the American Physical Society.

In order to extract the neutrino helicity, information is also needed on its direction of motion (without directly detecting it). The lower samarium *rescatterer* is only effective for those photons with energy *exactly* corresponding to the first excited state (from which the photons were originally produced). Now, the effect of nuclear recoil in such processes results in the initially emitted photons having slightly *less* than the excitation energy, whereas they need slightly *more* to be absorbed. Therefore, were both the emitting and absorbing samarium nuclei both stationary, there could be *no* absorbtion. However, in the K-capture process, for which the captured electron is essentially at rest, whereas the neutrino is emitted with a non-negligible momentum, the daughter nucleus recoils. Such a recoil may be sufficient, if it is in the right direction, to provide the necessary extra energy to the subsequently emitted photon. Thus, only a photon produced from a nucleus moving in the same direction (and therefore opposite to the neutrino) may be rescattered and finally detected.

To recap, the photon and neutrino must move in opposite directions and also have opposite spin projections; they therefore have the same helicities. By comparing the counting rate with the magnetic field applied in the positive and negative vertical directions, Goldhaber *et al* (1958) were thus able to infer the neutrino helicity or handedness. The results demonstrate that the neutrino too is always left-handed. Similar experiments on antineutrino emission (using, e.g., ^{203}Hg) show that the antineutrino is instead right-handed (as too is the positron emitted in β-decay).

2.3.3 Interpretation

While we might be led to attribute the existence of parity violation just to the neutrino, closer inspection reveals this to be incorrect. The reason that suspicion falls on the neutrino has to do with its mass, or rather lack thereof. A left-handed electron appears so only in certain reference frames. However, if the observer is boosted to a velocity exceeding that of the electron (which is always possible for a massive particle), then it will now appear to have the opposite velocity while maintaining the same spin projection and will thus have effectively flipped its helicity to become right-handed. There can therefore be nothing intrinsically special about a right- or left-handed electron, or indeed any massive fermion. In contrast, a massless neutrino always travels at the speed of light and so a neutrino that is left-handed in some given reference frame appears left-handed in *any* frame. It is thus tempting to attribute parity violation to the non-existence of right-handed neutrinos and left-handed antineutrinos.

Alternatively, rather than postulate the non-existence of right-handed neutrinos, we could simply invoke a very large mass for this particle; if sufficiently large (as is the case e.g. in some grand-unified models), it would then not be able to participate in processes initiated at typical presently accessible energies. However, as will now be shown, this would not be sufficient to explain all of the multitude of parity-violating effects experimentally observed.

What has not yet been mentioned here is that there are also a large number of weak decays that do not involve neutrinos (so-called *non-leptonic* decays), but in which parity is nevertheless strongly violated[21]. A simple example is furnished by the purely hadronic decay channels of the neutral lambda hyperon

$$\Lambda^0 \rightarrow p\pi^-, \, n\pi^0. \tag{2.3.7}$$

By virtue of its non-zero magnetic moment, the Λ^0 may easily be polarised (this is indeed how its magnetic moment is measured) and thus we may measure the correlation between the Λ^0 spin direction and, say, the momentum direction of the emitted nucleon (the nucleon and pion naturally emerge back-to-back in the Λ^0 rest frame). Again, such a quantity, depending on a scalar product $\boldsymbol{p} \cdot \boldsymbol{s}$, violates parity. In this case the asymmetry parameter is $\alpha \approx 0.64$. And again, the reason that $|\alpha| \neq 1$ has to do with the non-trivial internal structure of baryons.

Consequently, the violation of parity must be ascribed to the nature of the weak interaction itself. This will be examined in more depth in the next section; suffice it to

---

[21] Of course, the τ–θ or charged-kaon decays that triggered the idea of parity non-conservation are first examples of non-leptonic parity violation.

say here that the nature of the weak force is such that it only couples left-handed fermions and right-handed antifermions. In other words, at this level it is entirely irrelevant whether or not right-handed neutrinos and/or left-handed antineutrinos exist: they are simply non-interacting or *sterile*.

It should be pointed out that the phenomena examined above are all examples of what is called *direct* parity violation; i.e. they are caused directly by parity violation in the interaction involved in the processes observed. However, the presence of a parity-violating interaction (seen as a perturbation) can also induce parity mixing in e.g. bound states (nucleons, nuclei and even atoms), which may then manifest itself via an apparent parity violation in some process (usually a decay) that is not necessarily weak and that should not therefore violate parity. Such an effect is then said to be *indirect* parity violation and is usually very small.

2.3.4 Closing remarks

> *Discoveries in physics often depend on looking toward a new direction, quite often with the very latest detector technology. Parity non-conservation is an exception. The reason it was not discovered [earlier] was not because it was at the margin of detector technology, but simply because people did not look for it.*

Tsung-Dao Lee

Indeed, such an experiment could have been performed 30 years earlier and it is probable that the effect had actually been seen as early as the twenties, but not recognised. It was noticed later that an experiment performed by Cox *et al* (1928), had, in effect, already observed parity violation in β-decay. Despite the title of the paper *'Apparent Evidence of Polarization in a Beam of β-Rays'*, the authors made little insistence on an explanation based on what they termed 'asymmetrical electrons', although they did note that it seemed the most likely explanation. Follow-up experiments (Chase 1929) even fully confirmed their results. However, it all went essentially unnoticed owing to the general belief that parity could not be violated and that there must therefore have been some not well understood effect or instrumentation error in the experiment.

After all, the electron spin had only been discovered in 1925 by Samuel Goudsmit and George Uhlenbeck and it was not until the 1930s, following the work of Wigner (1927), that the rôle of parity was at all appreciated. In other words, a bold step and no small amount of intuition is sometimes required to venture into the unknown in search of something new. Indeed, it almost goes without saying that already well-combed areas, although easier to search, are unlikely to yield new or important information, which can generally only come from as yet unexplored territories.

2.4 *V–A* formulation of the weak currents

As we have seen in the previous section, parity violation, at least in the case of leptonic interactions, is maximal. The questions that now arise are how this may be explained and what the implications are. We shall therefore need a theoretical description of such processes, in which parity violation can also be implemented in as natural as possible a manner. It should then be possible to obtain predictions for further similar effects.

2.4.1 Fermi and Gamow–Teller transitions

As we have seen, the first attempt at a theoretical construction of the weak nuclear interaction was due to Enrico Fermi (1934b). This was an intrinsically non-relativistic approach based on Schrödinger-type wave-functions and a simple *ansatz* for the interaction involved in β-decay. Moreover, it made no account for spin effects and so could not alone explain the parity violation just described. Later developments were made by Gamow and Teller (1936), who introduced spin effects, which still could not directly explain parity violation. We shall thus now describe the improvements of the early theory of the weak interaction.

The basis for the theory of nuclear β-decay Fermi (1934b) was the *ansatz* of a four-body point-like interaction, presented in equation (2.1.10). The (trivial) spin structure implicit in this form for the matrix element naturally leads to a description of the so-called Fermi transitions, in which the e–ν pair carries zero total angular momentum. Including the two-component spinorial forms due to Pauli allows for a description of the so-called Gamow–Teller transitions, in which the e–ν pair have their spins aligned and so carry one unit of angular momentum. However, there is no apparent relation between the two types of transitions. Phenomenologically, even taking into account a factor three for the triplet final state, a further (phenomenological) factor $\sim 5/4$ is needed if the same constant G_F is to describe both. To be precise, a complete description of the substructure in terms of quarks is required, but we shall see how this works later.

2.4.2 A relativistic formulation

One of the great assets of Paul Dirac's (1928) relativistic formulation of quantum mechanics is that it places severe constraints on the different components, linking spin-dependent and -independent matrix elements. Readers who are not familiar with the Dirac theory are referred to appendix B.1 for a simple introduction. In order to replace the *naïve* non-relativistic Fermi interaction above with a relativistic version, we must introduce the concept of current–current interactions, borrowed from the only complete and elementary theory then in existence: namely, quantum electrodynamics (QED). Thus, it is natural to write the possible interaction terms schematically a ('current–current') form:

$$\propto G_F \int d^3x\, j_1^\dagger(x) \cdot j_2(x), \tag{2.4.1}$$

where $j_{1,2}$ may be any combination of the five possible transition currents formed from spinor bilinears (see appendix B.3.5), provided that the indices may be suitably saturated. This is the most general form compatible with the Dirac construction.

We should point out that, as the initial and final states here are different, the ψ and $\bar{\psi}$ appearing in the currents j_i refer to different particles (e.g. e and ν or n and p, respectively). That is, they are what might be called *transition currents*, describing new and important effects: namely, particle transformation or creation and annihilation. Such processes are not strictly present in non-relativistic quantum mechanics and in QED only as particle–antiparticle pairs of the same type (e.g. e^+e^-, $\mu^+\mu^-$ etc)[22].

It turns out that each current leads to different parity and angular-momentum selection rules and also to different angular distributions. The correct form may thus be experimentally identified. It is found that only the vector, $\bar{\psi}\gamma^\mu\psi$, and axial-vector, $\bar{\psi}\gamma_5\gamma^\mu\psi$, forms (describing Fermi and Gamow–Teller transitions, respectively) contribute to the weak interaction (Sudarshan and Marshak, 1958; Feynman and Gell-Mann, 1958).

Now, the parity assignments of these two currents are precisely those of a standard vector (negative) and axial-vector (positive). However, since the decay distribution is determined by the modulus squared of the transition matrix element, neither taken separately can provoke parity violation. Therefore, a linear combination must be taken, with the consequent interference term being parity odd. The two possible extremes are $V \pm A$ or

$$ j_W^\mu = \bar{\psi}(\gamma^\mu \pm \gamma^\mu\gamma_5)\psi. \tag{2.4.2} $$

The ordering $\gamma^\mu\gamma_5$ is chosen so that the factorised form $\gamma^\mu(1 \pm \gamma_5)$ makes explicit the natural projective property.

Exercise 2.4.1. *Using the Dirac matrix algebra and the spinor structure given, show that the operators $P_{R/L} := \frac{1}{2}(1 \pm \gamma_5)$ project onto right- and left-handed helicity states, respectively; that is, $\frac{1}{2}(1 \pm \gamma_5)\,\psi = \psi_{R/L}$.*

It turns out that the signs of the asymmetries found in the experiments by Wu *et al* (1957), Goldhaber *et al* (1958), Garwin *et al* (1957) and many others are consistent with just one specific choice: *V–A* (Sudarshan and Marshak 1958, Feynman and Gell-Mann 1958, Okubo *et al* 1959). That is, the weak interaction, involving only *left-handed* currents, is *maximally* parity violating. A simple and experimentally verifiable consequence of this form is that parity will only be violated in those nuclear β-decays in which *both* Fermi and Gamow–Teller transitions are possible and play a rôle. Indeed, parity violation is seen to be the result of an *interference* effect and, as such (not to mention the rôle of spin), is evidently a *quantum-mechanical* phenomenon; it cannot be accommodated by classical physics.

---

[22] In a sense, Fermi had already allowed himself the freedom of venturing beyond the constant-number nature of standard non-relativistic quantum mechanics by introducing (in an *ad hoc* and somewhat surreptitious manner) the possibilities of varying the number of particles in a closed system and also of transforming them. Strictly speaking, the description of such processes requires field or second quantisation.

As mentioned above, in other than purely leptonic decays (i.e. when hadrons take part), this maximal violation is not always apparent, owing to the complex internal structure of hadronic matter (in particular, that of the baryons). However, if hadrons are described as bound states of quarks and proper account is made for the quark-spin symmetry (see the quark model later), then the violation is again found to be maximal at the purely quark level and the weak current is still precisely V–A. Let us again emphasise that the V–A form factorises as follows:

$$\bar{\psi}(\gamma^\mu - \gamma^\mu\gamma_5)\psi = 2\bar{\psi}\gamma^\mu\tfrac{1}{2}(1 - \gamma_5)\psi = 2\bar{\psi}_L\gamma^\mu\psi_L, \qquad (2.4.3)$$

where $\psi_L := \tfrac{1}{2}(1 - \gamma_5)\psi$. This demonstrates that only left-handed (elementary) fermions interact weakly; although, of course, the existence of right-handed fermions is not explicitly precluded.

It should also be appreciated that implicit in this description of parity (P) violation is the violation of charge conjugation (C) symmetry (see e.g. Ioffe *et al* 1957, Lee *et al* 1957): the only fermions that may interact weakly are left-handed, whereas antifermions must be right-handed. However, since the two violations go perfectly hand-in-hand, the combined CP symmetry is still respected. In other words, while comparisons of either 'mirror' or 'antiparticle' processes reveal differences, the translation to 'mirror–antiparticle' experiments returns the original observations, e.g., β^+-decay viewed in a mirror is indistinguishable from normal β^--decay. Later on, we shall examine more particular systems in which even CP is violated.

Exercise 2.4.2. *Show how the transformation properties derived earlier lead to a form* $\gamma^\mu(1 + \gamma_5)$ *for antifermionic currents.*

2.5 Cabibbo theory

Before going on to discuss CP violation, we must address the question of universality. That is, can all weak interactions or processes be described with just the one coupling constant G_F? In other words, can we keep Ockham happy? We shall find that the answer to this question reveals yet another surprising aspect of the weak interaction.

2.5.1 Universality

The Fermi description of the weak interaction requires a new coupling constant: G_F. The β-decay matrix element is proportional to G_F and thus the decay rates are proportional to G_F^2. In muon decay all participants are elementary and so there are no other unknown ingredients. Measurement of the muon lifetime thus translates directly into a determination of G_F. We have (Navas *et al* 2024)

$$\tau_\mu = 2.196\,9811(22) \times 10^{-6}\,\text{s}, \qquad (2.5.1)$$

which (in natural units) leads to

$$G_F = 1.166\ 3788(6) \times 10^{-5}\ \text{GeV}^{-2}. \qquad (2.5.2)$$

Note that to arrive at this result, the full electroweak theory (see later) has been used.

Now, already the formulation in terms of Dirac spinors has unified the description of Fermi and Gamow–Teller transitions. However, there still remains the question of the relationship between hadronic and leptonic weak couplings, and even between different hadrons. That is, how does the above value for G_F compare with that deduced, for example, from neutron β-decay or from electron–neutrino scattering? In other words, is it *universal?*

One of the first to consider the likelihood of a universal description for the weak interaction (i.e. in terms of a unique coupling constant) was Giampietro Puppi (1948); the similarity of standard β-decay, nuclear muon-capture and muon-decay processes led to the notion of the Puppi triangle[23], see figure 2.6.

In the case of other purely leptonic processes the answer is simple, the agreement is perfect within errors. Here, though, the theory is rather simple: for purely leptonic processes, the particles participating are all elementary. Indeed, in their paper proposing the *V–A* form, Feynman and Gell-Mann (1958) had also considered such universality, noting that '*the lifetime of the μ agrees [with β-decay determinations] to within the experimental errors of 2%*'.

When dealing with semi-leptonic and even more so for non-leptonic processes, the details of the hadronic bound state (baryon or meson) cloud the issue somewhat. Let us examine in more detail the case of neutron and also nuclear β-decay.

When Nicola Cabibbo was developing his theory of hadronic weak couplings in 1963 there was, as yet, little or no notion of a real physical quark substructure and so the description was constructed purely in terms of the physical baryon or meson

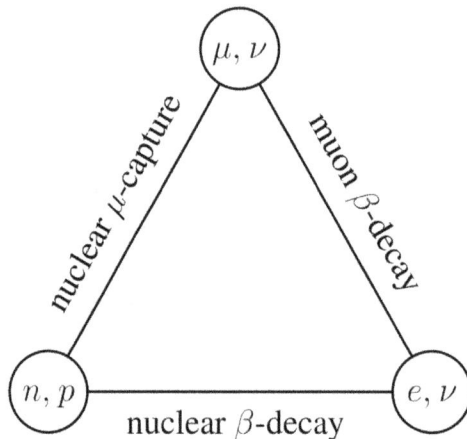

Figure 2.6. The Puppi triangle (1948) representing the universality of standard nuclear β-decay, nuclear muon-capture and muon-decay weak processes.

---

[23] Note that anther open question at that time was whether or not the neutrinos in the two lepton vertices of the Puppi triangle (1948) were just one and the same or necessarily different.

states. And this is how we shall proceed for the time being. As already noted, there are two general types of nuclear transitions (Fermi and Gamow–Teller), which we have found correspond to vector and axial-vector couplings, respectively. In analogy with electromagnetism, it is natural to assign charges g_V and g_A, respectively. The ratio g_A/g_V for the neutron may be determined by measuring various angular and/or spin correlations. The present PDG value is rather precise (Navas *et al* 2024):

$$g_A/g_V = -1.2754 \pm 0.0013. \tag{2.5.3}$$

Note that the overall sign is purely a matter of convention.

As already hinted, the ratio is not unity, as might be hoped by appealing to universality, but is nearer to 5/4. However, given that hadrons are not point-like objects (having evidently non-trivial substructures) and that the stringent relations on particle spins provided by the Dirac formulation only apply to point-like elementary particles (e.g. the electron), it is not surprising that g_V and g_A do not appear to have a simple relationship. We shall see later, however, that the quark model of hadronic structure, with its symmetries, does salvage even this aspect to some extent.

The vector charge g_V, by analogy with electromagnetism, is thus given quite simply as a sort of *weak charge*, which may be set to unity via the normalisation and definition of G_F. On the other hand, g_A may be thought of as a sort of *weak magnetic moment* and, as such, for composite objects cannot be fixed by any normalisation. This means that in order to measure G_F, processes are needed in which the transition is purely of the Fermi type. By noting that the Gamow–Teller transitions involve a spin-one electron–neutrino pair, we see that it is sufficient to consider transitions in which this is prohibited. In general, even if the variation of total nuclear spin ΔJ is zero, since angular momenta add vectorially, it is still possible to have $L_{e\nu} \neq 0$. The exception is the case in which $J = 0$ for both the initial and final states. There are now 23 such decays that have been measured; these are the so-called *superallowed* $J^P = 0^+$ transitions. A recent review may be found in Hardy and Towner (2020). For 15 of these, the experimental precision is of the order of or better than 2‰. Two examples are:

$$^{10}\text{C}(\beta^+)^{10}\text{B*} \quad \text{and} \quad ^{14}\text{O}(\beta^+)^{14}\text{N*}. \tag{2.5.4}$$

Some of the other more precisely known are the β^+-decays of ^{26}Al, ^{34}AC, ^{38}K, ^{42}Sc, ^{46}V, ^{50}Mn and ^{54}Co. In such decays, since the initial and final nuclear wave-functions are almost identical, the wave-function overlap integrals are also essentially free of nuclear uncertainties. They thus provide a second, self-consistent and accurate determination of G_F.

However, the result of comparing the two determinations of G_F (i.e. from μ-decay and superallowed nuclear decays) leads to a small discrepancy. Interestingly, attempts to reconcile the two numbers first looked to improved calculations and the quantum corrections: in particular, that due to the Coulomb repulsion (these are all β^+-decays). However, such more accurate studies actually worsened the situation. The final discrepancy, although small in magnitude, was many standard deviations. Numerically, at the level of decay rates Γ, it was found that the nuclear decays were approximately 4–5% smaller. In terms of the extracted nuclear G_F (since $\Gamma \propto G_F^2$), this implies that it is approximately 2% weaker.

2.5.2 The mixing of weak and mass eigenstate

In attempting to resolve the issue, Cabibbo (1963) examined other known β-decays: namely, that of Λ^0, a neutral spin-half so-called *strange* baryon (or *hyperon*), similar to but a little heavier than the neutron[24]. In this case, it turns out that the discrepancy is enormous. Using these decay rates to extract the Fermi coupling, the value obtained is approximately only 22% of the muon value. Now, $(0.22)^2 \sim 0.05$, which was interpreted by Cabibbo as the missing 5%. The idea was thus to invoke the well-known quantum phenomenon of *mixing*. The actual physical particles detected experimentally correspond to mass eigenstates (technically termed *asymptotic* states), i.e. those that propagate as stationary states in space–time and which do not necessarily coincide with the eigenstates of the weak interaction[25].

To understand this, recall, for example, the effect of an electromagnetic field applied to the hydrogen atom: the perturbing interaction mixes the usual hydrogen levels. In other words, the corresponding new eigenstates are different to those of the free hydrogen atom, being superpositions of the latter.

The strange β-decay considered is $\Lambda^0 \rightarrow pe^- \bar{\nu}$, where the final state is perfectly identical to that of neutron β-decay. Therefore, if the states are properly normalised, it is natural to describe such mixing, which here then is between just two states (n and Λ^0), via an angle:

$$n_W = \cos\theta_C \, n + \sin\theta_C \, \Lambda^0, \tag{2.5.5a}$$

$$\Lambda^0_W = -\sin\theta_C \, n + \cos\theta_C \, \Lambda^0, \tag{2.5.5b}$$

where n_W and Λ^0_W are weak eigenstates[26], n and Λ^0 being the usual mass eigenstates; θ_C is known as the *Cabibbo angle*. The transitions must now be described by matrix elements involving these new states, that is, so-called *weak currents* must be employed:

$$\mathcal{M} \propto G_F \int d^3x \, j_h^{W\,\dagger}(x) \cdot j_l^W(x), \tag{2.5.6}$$

where the $j_{h,\,l}^W$ describe weak hadronic and leptonic currents respectively and are thus of the form

$$j_{h,\,l}^{W,\mu} := \bar{\psi}_f^{W}(g_V\gamma^\mu + g_A\gamma^\mu\gamma_5)\psi_i^W. \tag{2.5.7}$$

In this theory the leptonic states do not suffer such mixing. As we shall see later, this is to do with the fact that neutrinos are considered massless. However, in recent

---

[24] The Λ^0 was discovered by Hopper and Biswas (1950) as a neutral so-called '*V-particle*' (see section 4.1.5), decaying either purely hadronically into a pion–nucleon pair or β^- into a proton, electron and antineutrino.

[25] Note that both the experimental and theoretical identification of a particle is essentially made by way of its mass (and when necessary charge).

[26] Use of the same symbols n and Λ will be justified *a posteriori* by the smallness of the mixing angle; i.e. n_W is mostly n, likewise Λ^0_W and Λ^0.

years we have learnt that neutrinos are not, in fact, massless and so there can indeed be the same type of mixing in the leptonic sector too; we shall discuss this later.

With the mixing structure describe above, it is evident that matrix elements between the real physical states will pick up a factor $\sin \theta_C$ or $\cos \theta_C$ depending on whether the initial hadronic state is strange or not. That is, the hadronic current is now

$$
\begin{aligned}
j_h^{\mathrm{W},\mu} &= \bar{\psi}_p^{\mathrm{W}} \left(V^\mu - A^\mu \right) \psi_n^{\mathrm{W}} \\
&= \cos \theta_C \, \bar{\psi}_p \left(V^\mu - A^\mu \right) \psi_n + \sin \theta_C \, \bar{\psi}_p \left(V^\mu - A^\mu \right) \psi_\Lambda.
\end{aligned}
\tag{2.5.8}
$$

Taking $\sin \theta_C \simeq 0.22$, both decays are then well described by one and the same coupling G_{F}. This picture is well corroborated by the description it provides of other transitions, for example, the analogous pair of decays $\pi^- \rightarrow \pi^0 e^- \bar{\nu}$ and $K^- \rightarrow \pi^0 e^- \bar{\nu}$. In fact, there are many similar β-like transitions to which the theory may be applied. Note, however, that quite why the *same* Cabibbo angle should work for both baryons and mesons is evidently a mystery until we move over to the quark description of hadrons.

2.6 The Glashow–Iliopoulos–Maiani (GIM) mechanism

2.6.1 A brief introduction to the quark model

Around the same time the *quark model* of Murray Gell-Mann (1962) was starting to take shape; for a more complete review, see for example Gell-Mann (1961) or section 3.2 in the present notes. Moreover, although originally only intended as a mathematical expression of the underlying *flavour* symmetry, many started to consider the possibility that quarks were, in fact, real physical entities. Among those who saw early on the possibility that this could explain other puzzling experimental observations were Glashow *et al* (1970). The problem they addressed was that of certain weak-decay channels, which were *not* observed, despite there being no apparent reason for suppression. As we shall now see, once again quantum mechanics plays an important rôle, in this case via the phenomenon of *interference*. The particular processes under consideration were possible purely leptonic decays of the neutral kaon and its antiparticle.

Let us first rephrase the Cabibbo picture in terms of quarks. We now know that baryons (among which we find the spin-half nucleons, the proton and the neutron, together with the Λ^0 already mentioned) are all composed of three quarks in particular spin configurations, which for the moment are inessential, whereas mesons (e.g., $\pi^{0,\pm}$, $K^{0,\pm}$ etc) are simply quark–antiquark pairs. The neutron and proton are thus

$$
|p\rangle = |uud\rangle \quad \text{and} \quad |n\rangle = |udd\rangle,
\tag{2.6.1}
$$

where the two quarks u and d are known as *up* and *down*, respectively. The Λ^0 hyperon is similar but contains a strange quark s:

$$
|\Lambda^0\rangle = |uds\rangle.
\tag{2.6.2}
$$

The β-decay process is then seen to be the transformation of either a d or an s into a u, accompanied by the emission of an e^-–$\bar{\nu}_e$ pair. The remaining quarks are considered to be mere spectators and play no rôle. With no change in the reasoning, the Cabibbo mixing (2.5.5) can then be simply re-expressed as

$$|d\rangle_W = \cos \theta_C |d\rangle + \sin \theta_C |s\rangle, \tag{2.6.3a}$$

$$|s\rangle_W = -\sin \theta_C |d\rangle + \cos \theta_C |s\rangle. \tag{2.6.3b}$$

The mixing may now be re-expressed in a more compact and suggestive form via a 2×2 rotation-matrix representation:

$$\begin{pmatrix} d_W \\ s_W \end{pmatrix} = \begin{pmatrix} \cos \theta_C & \sin \theta_C \\ -\sin \theta_C & \cos \theta_C \end{pmatrix} \begin{pmatrix} d \\ s \end{pmatrix}. \tag{2.6.4}$$

This is then just the Cabibbo mixing matrix. In the quark model description, the case of charged-pion and -kaon β-decay is similar:

$$|\pi^-\rangle = |d\bar{u}\rangle \text{ and } |K^-\rangle = |s\bar{u}\rangle. \tag{2.6.5}$$

And again, it is just a d or an s that decays into a u, accompanied by an electron–neutrino pair.

2.6.2 Feynman diagrams

By way of an introduction to the concept, let us represent these decays by means of so-called Feynman diagrams (Feynman 1949)[27]. According to the rules derived by Feynman, each particle participating in a given process is assigned a *line* (internal lines represent propagation, external lines the initial- and final-state wave-functions), while the interactions between particles are indicated by *vertices*. Thus, the diagram describing neutron β-decay is as shown in figure 2.7. In the diagram shown, time flows from left to right, while the vertical axis represents generic spatial position. The arrows on the fermion lines do *not* represent direction of mechanical motion,

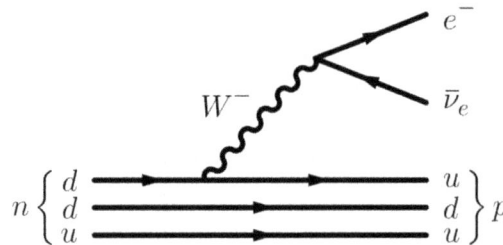

Figure 2.7. The Feynman diagram describing the neutron β-decay process in terms of the weak interaction at the quark-level.

---

[27] The 1965 Nobel Prize in Physics was awarded equally to Sin-Itiro Tomonaga, Julian Schwinger and Richard P Feynman 'for their fundamental work in quantum electrodynamics, with deep-ploughing consequences for the physics of elementary particles'.

but the flow of particle quantum numbers; thus, an antiparticle has an arrow directed backwards (in time).

The real importance of such a representation is that each element (line or vertex) corresponds to a mathematical object, which then goes to make up the matrix element or scattering amplitude. We shall not labour this point here, but merely note that an important aspect of the Feynman rules is that all possible (different) diagrams connecting the same initial and final states must be *added* together to provide the total amplitude, which is then squared and integrated to give the final rate. This opens the way to quantum-interference effects: if two similar diagrams contribute with similar magnitude but *opposite sign*, then they may cancel each other in the sum (at least partially) and can thus lead to vanishing (or suppressed) rates even though there may be no real obstacle or selection rule forbidding the process. A particular example arises owing to a further rule: namely, that, for two diagrams that differ by the exchange of two fermion lines, then there is a relative minus sign. The case we shall now describe is another example of diagram or amplitude cancellation.

2.6.3 Unobserved neutral-kaon decays

Consider then the two neutral pseudoscalar strange mesons K^0 and $\bar{K}^0$, the quark content of these two hadrons is, recall, $K^0 = d\bar{s}$ and $\bar{K}^0 = \bar{d}s$. The assignment *particle–antiparticle* in the case of neutral mesons is dictated by the up–down quark content. The K^0 thus contains a d (or matter), whereas $\bar{K}^0$ contains a $\bar{d}$ (antimatter).

As we shall see later, this pair of particles has a rather complicated dynamics and CP is violated in their decays. However, for the present purposes they may considered simply as the quark composites just defined. Of course, these states are meant to represent the mass eigenstates, or rather the *quark* mass eigenstates. Therefore, to describe possible weak decays, we shall need the rotated states unveiled in the previous section. Since the weak interaction then couples both d and s quarks to the u quarks, we can envisage and indeed calculate the rate for the decay $K^0 \rightarrow e^+e^-$, depicted in figure 2.8. This process is, however, *not* observed at the estimated rate (see e.g. Clark *et al* 1971). Note that by exchanging the rôles of the electron and the neutrino, we have a process in which a kaon apparently disappears into thin air; this too is not seen.

The way out of this *impasse* was provided by Glashow *et al* (1970). Since the weak state d^W couples to the u quarks, they were tempted to hypothesise that the orthogonal combination of mass eigenstates in s^W couples to some new state, the

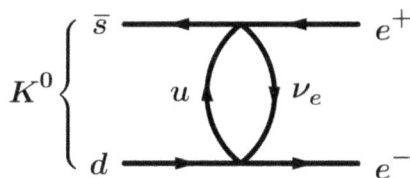

Figure 2.8. A possible process in which a K^0 might decay via internal annihilation into an electron–positron pair (recall that time runs from left to right).

Figure 2.9. The *destructively* interfering diagrams contributing to K^0 decay into an electron–positron pair. Note the minus sign of the lower Cabibbo factor in the right-hand diagram.

charm or c quarks. If this is the case, then there is a second diagram, similar to that of figure 2.8, in which the u quarks are replaced by c. Let us now draw the diagrams also including the $\cos\theta_C$ and $\sin\theta_C$ vertex factors with their relative *signs*, as given in equation (2.6.3) and shown in figure 2.9, we then immediately see that there is a cancellation (or destructive interference) owing to the sign difference. Of course, the u and c quarks in the internal propagators do not have the same mass ($m_c \gg m_u$) and so the cancellation is not perfect. This simply means that the non-observation of such a channel at some level provides an *upper limit* on the mass of the c quarks[28]. This decay was eventually observed much later (Ambrose *et al* 1998), with a rate compatible with theoretical calculations.

Exercise 2.6.1. *Naïvely, one might imagine that a similar GIM-like cancellation should also apply to the decay $K^0 \to \pi^+\pi^-$. By considering the possible quark diagrams responsible for such a channel, demonstrate that this is not the case.*

Note that the possibility of a fourth, charm, quark had already been proposed by Bjorken and Glashow (1964). Their idea was to construct a symmetry between the four known leptons (e, μ and the possibly two neutrino types) and necessarily then four quarks: the original three of Gell-Mann (u, d and s) plus a new, c quarks. The GIM mechanism was proposed in (Glashow *et al* 1970) and the predicted new quark (with a mass that could not be much more than about 1.5 GeV) was discovered in e^+e^- collisions four years later by three independent groups (Aubert *et al* 1974, Augustin *et al* 1974 and Bacci *et al* 1974); see section 4.2.2. This discovery then gave rise to the concept of quark (and lepton) *family*. At the quark level, the states are placed into pairs of up- and down-type, paralleling the first known pair u–d; the corresponding leptons are the e–ν_e and μ–ν_μ pairs. Although it must be recognised that Kobayashi and Maskawa (1973) had already contemplated such a structure (and more) and its consequences, as we shall now see.

---

[28] In reality, Glashow *et al* (1970) considered the process $K^0 \to \mu^+\mu^-$, but it is essentially equivalent to the e^+e^- channel.

2.7 The Cabibbo–Kobayashi–Maskawa (CKM) matrix and CP violation

The following questions now naturally arise:

- Why have we taken the rather one-sided position of rotating (or labelling as specifically weak) the down-type states and not those of the up-type?
- Since the state functions are naturally complex in quantum mechanics, the rotation matrix, in general, could contain complex phases—why then is the parametrisation in terms of only a single real parameter (an Euler angle)?
- What would happen if there were more quark states or families in the game?
- Since this happens for the weak interaction in the quark or hadronic sector, should there not be an analogous mixing among leptons?

All but the last (with which we shall deal immediately) will be treated in the following sections, the answer to the first will be answered at the end of the next subsection, while we shall tackle the other two shortly. With regard to the first, let us just note here that Cabibbo, not knowing yet of the c, had only the possibility he actually adopted.

The case of leptons is conceptually rather simple: in general, mixing occurs because there are two well-defined, unambiguous and distinct bases: the mass eigenstates (i.e. those that propagate and which therefore correspond to experimentally detected particles) and those of the weak interaction (i.e. those that are produced, decay or interact in any way). However, even assuming that the weak-interaction basis also remains well-defined for the leptons, the mass eigenstates of the neutrinos are ambiguous in the Standard Model (SM) since they are all considered *massless*. That is, they are *only* distinguishable by virtue of their weak interactions. Of course, in the wake of the recent experimental and theoretical developments concerning solar and atmospheric neutrinos, their masses are now believed to be non-zero and, indeed, work is already under way on mapping out the leptonic mixing matrix.

2.7.1 CPT

In order to motivate the following discussion, we need to understand a little more of the discrete symmetries C, P and T, mentioned earlier, and their relationship through the so-called CPT theorem. A rather self-contained and sufficiently comprehensive discussion is provided for the unfamiliar reader in appendix B.3; it is highly recommended to consult this before continuing.

2.7.2 The Kobayashi–Maskawa extension

We now turn to the other two intimately related questions posed: what happens in the case of more families and why is there no complex phase in the Cabibbo description? The *naïve* answer to the first of these is that the mixing matrix simply grows to be $n \times n$ for n families, but, as we shall see, there is more to it than this. The second is a loaded question: a complex phase would allow for the violation of time-reversal invariance, which, if CPT is conserved, would be equivalent to and would imply CP non-conservation, as discussed in the appendix.

Let us attack the phase problem first. Once it was known that CP was *not* an exact symmetry of Nature (Christenson *et al* 1964, see section 2.7.6), the quest began for its origins and the possibility of introducing a complex phase, via the weak-interaction mixing matrix, thus became highly relevant. However, Cabibbo (1963) quite rightly described the two-component mixing in terms of a single, real, Euler angle. The reason is that there are the various constraints that must be imposed and also a certain phase freedom.

Firstly, the 2×2 matrix, let us call it V_C, must be unitary ($V_C^\dagger V_C = 1$) in order to respect the ortho-normality of the bases involved. This implies four constraints on the possible four real amplitudes and four complex phases. Matrix theory tells us that the constraints are divided up into three on the amplitudes and one on the phases. The matrix is thus already reduced to just the one real amplitude (or Euler angle) of Cabibbo. Secondly, the number of free phases is further reduced by considering that each of the four quantum-mechanical quark states multiplied by the matrix possesses an arbitrary (unphysical and unmeasurable) intrinsic phase. All except one, which must serve as a reference point, may therefore be rotated so as to absorb phases in the matrix. The three remaining phases may thus all be absorbed into redefinitions of the quark states. Finally then, we are left with a real, one-parameter matrix, as proposed by Cabibbo.

This exercise naturally leads to investigation of the matrix in the case of more than two families (Kobayashi and Maskawa 1973)[29]. A general, complex, $n \times n$, matrix contains n^2 real amplitudes and n^2 complex phases. Again, matrix theory reveals that unitarity imposes n^2 constraints: $n(n + 1)/2$ on the real amplitudes and $n(n - 1)/2$ on the imaginary phases. The $2n$ 'external' quark states allow arbitrary phase rotations and thus absorption of a further $2n - 1$ phases. A little arithmetic then reveals that the most general rotation matrix for n families may be described in terms of

$$\tfrac{1}{2}n(n - 1) \ \ real \ \ \text{Euler angles} \tag{2.7.1a}$$

and

$$\tfrac{1}{2}(n - 1)(n - 2) \ \ complex \ \ \text{phases.} \tag{2.7.1b}$$

The case $n = 2$ just confirms what we have already described, whereas $n = 3$ leads to the possibility of three Euler angles and precisely one complex phase. Thus, the important finding of Kobayashi and Maskawa (1973) is that the mixing induced by

---

[29] The 2008 Nobel Prize in Physics was divided, one half awarded to Yoichiro Nambu 'for the discovery of the mechanism of spontaneous broken symmetry in subatomic physics', the other half jointly to Makoto Kobayashi and Toshihide Maskawa 'for the discovery of the origin of the broken symmetry which predicts the existence of at least three families of quarks in nature'.

the weak interaction in the case of three or more families is sufficient to accommodate a T- or CP-violating phase.

Note that, at that time, a year before the discovery of charm, only the second family of leptons was already known and complete; it would not be until 1975 that evidence of a third family would emerge via the discovery of the τ lepton by Martin Perl et al[30] (see section 4.2.4), with the b quark being discovered by Herb et al (1977).

We still have the first question to answer: namely, the justification to limit mixing to only the down-type quarks. Let us examine more carefully how the currents should be defined in the presence of the most general possible mixing. *A priori, both* up-type ψ_U^W and down-type ψ_D^W weak eigenstates should be considered, as being potentially distinct from their mass-eigenstate counterparts. Thus, *two* independent $n \times n$ mixing matrices should be introduced, say $\mathcal{U}$ and $\mathcal{D}$:

$$\psi_U^W = \mathcal{U}\,\psi_U \text{ and } \psi_D^W = \mathcal{D}\,\psi_D, \tag{2.7.2}$$

where ψ_U and ψ_D now represent n-component spinors in *flavour* space:

$$\psi_U = \begin{pmatrix} u \\ c \\ t \\ \vdots \end{pmatrix} \text{ and } \psi_D = \begin{pmatrix} d \\ s \\ b \\ \vdots \end{pmatrix}. \tag{2.7.3}$$

The natural extension of equation (2.5.8) is then

$$\begin{aligned}
j_W^\mu &= \bar{\psi}_U^W (V^\mu - A^\mu)\,\psi_D^W \\
&= \overline{\mathcal{U}\psi}_U\,(V^\mu - A^\mu)\,\mathcal{D}\psi_D \\
&= \bar{\psi}_U\,\mathcal{U}^\dagger\,(V^\mu - A^\mu)\,\mathcal{D}\,\psi_D \\
&= \bar{\psi}_U\,(V^\mu - A^\mu)\,\mathcal{U}^\dagger\mathcal{D}\,\psi_D =: \bar{\psi}_U\,(V^\mu - A^\mu)\,V_{\text{CKM}}\,\psi_D,
\end{aligned} \tag{2.7.4}$$

where the various steps are made possible owing to the fact that, for example, the Dirac matrices (γ) and the flavour matrices ($\mathcal{U}$ and $\mathcal{D}$) commute since they act on different spaces. We thus find that only the combination (the CKM matrix) $V_{\text{CKM}} = \mathcal{U}^\dagger\mathcal{D}$ is physically important and, moreover, that this can be considered as acting *either* to the right and mixing the down-type quarks, *or* to the left, mixing the up-type quarks. Both interpretations are physically equivalent. It thus becomes a matter of mere convention that the situation is described in terms of the mixing of down-type quarks.

The earlier discussion on the most general parametrisation and the possible introduction of a complex phase must now naturally be applied unaltered to the CKM matrix V_{CKM} with $n = 3$. This means that the matrix can, as anticipated, admit just one complex phase, with the consequent possible violation of time-reversal invariance. Given the CPT theorem, this translates into the parallel and consequent violation of CP, to which we shall now turn our attention.

---

[30] For the associated Nobel prize, see footnote 20 in chapter 4.

2.7.3 The neutral-kaon system

We now come to what is possibly one of the richest systems in particle physics: the K^0–$\bar{K}^0$ pair. Recall that these two particles have the following quark composition:

$$K^0 = d\bar{s} \text{ and } \bar{K}^0 = \bar{d}s \tag{2.7.5}$$

and thus, although neutral, they are *distinct* particles. However, they do have in common several decay modes: as with the charged kaons they can decay into two or three pions (with total charge zero). This fact has the consequence that they can *oscillate*, i.e. each may transform spontaneously, via a virtual two- or three-pion intermediate state, into the opposite particle or antiparticle state. Historically, however, the first problem arose with regard to their CP assignment. Let us stress that the phenomena described here are not to be confused with the case of the τ–θ puzzle described earlier, which involves the *charged* kaons and does *not* imply CP violation.

The immediate question is then: what are the CP signatures of the two- and three-pion final states in neutral-kaon decays? First of all, recall that it is always true that, for a C eigenstate, $C = \pm 1$ (since $C^2 = 1$). Recall too that both neutral and charged pions have negative intrinsic parity.

CP of the two-pion final state
As already discussed, the kaon and pion have spin zero and therefore angular momentum conservation forces the two-pion final state to have $L = 0$. The overall parity of this state is then simply

$$P_{2\pi} = P_\pi^2 = +1. \tag{2.7.6}$$

To discuss the signature under charge conjugation, we must distinguish between the two possibilities $\pi^0\pi^0$ and $\pi^+\pi^-$. In the case of a charged-pion pair, the operation of C interchanges the two and therefore introduces a factor $(-1)^L$ owing to the spatial wave-function, for an s-wave we thus have $+1$. The properties of a fermion–antifermion pair (such as go to make up a neutral pion) under C are such that $C_{\pi^0} = +1$. This is indeed confirmed experimentally by the observation of the principal decay mode

$$\pi^0 \rightarrow \gamma + \gamma \tag{2.7.7a}$$

and the *non*-observation of

$$\pi^0 \rightarrow \gamma + \gamma + \gamma. \tag{2.7.7b}$$

Putting all this together, we find that the two-pion final state in neutral-kaon decay must have $CP = +1$.

CP of the three-pion final state
There are again two possibilities to consider: $\pi^0\pi^0\pi^0$ or $\pi^0\pi^+\pi^-$. The presence of a third particle complicates the discussion of both the spatial-inversion and charge-conjugation properties. Any pair may now have non-zero orbital motion with

respect to the remaining pion, taken as a reference point. However, since the total must still be zero, the two must have identical L, with equal and opposite L_z. Thus, the final state with three neutral pions has spatial-parity signature $P = (-1)^{2L} = +1$ and therefore $CP = (+1)(-1)^3 = -1$. The $\pi^0\pi^+\pi^-$ case is a little more complex as the charge-conjugation signature depends on the relative orbital angular momentum of the charged pair, which may be odd. However, studies of the decay angular distribution indicate $L_{\pi^+\pi^-} = 0$, as might be deduced from the very low Q-value of this decay. The three-pion final state thus always has $CP = -1$.

Although the problem of parity violation was already understood and it was accepted that the weak two- and three-pion decays of the kaons violate P, the product symmetry CP was still believed to hold. Indeed, for example, the neutral pion, which is its own antiparticle, displays no evidence of CP violation. The problem was elegantly solved by Gell-Mann and Pais (1955). Whereas the neutral pion is its own antiparticle, the same is not true for the neutral kaons: neither K^0 nor $\bar{K}^0$ is an eigenstate of $\mathcal{C}$ and therefore certainly not of $\mathcal{CP}$. However, since $\mathcal{C}$ transforms K^0 into $\bar{K}^0$ and *vice versa*, the following linear combinations are easily seen to be eigenstates not only of $\mathcal{C}$ but also of $\mathcal{CP}$:

$$|K_1^0\rangle := \frac{1}{\sqrt{2}}\left(|K^0\rangle + |\bar{K}^0\rangle\right) \tag{2.7.8a}$$

and

$$|K_2^0\rangle := \frac{1}{\sqrt{2}}\left(|K^0\rangle - |\bar{K}^0\rangle\right). \tag{2.7.8b}$$

Note that the standard phase convention for the action of $\mathcal{C}$ sets

$$\mathcal{C}|K^0\rangle = -|\bar{K}^0\rangle \quad \text{and} \quad \mathcal{C}|\bar{K}^0\rangle = -|K^0\rangle. \tag{2.7.9}$$

As always, the intrinsic phase of a transformation such as $\mathcal{C}$ may be altered by redefining (or rotating) the phases of one or other of the states involved. The minus sign is thus merely conventional and has no effect on any physical results. This is the PDG choice (Navas *et al* 2024), but the opposite may also be found in the literature (in which case the rôles of the K_1^0 and K_2^0 combinations defined here are inverted). With this definition and, of course,

$$\mathcal{P}|K^0\rangle = -|K^0\rangle \quad \text{and} \quad \mathcal{P}|\bar{K}^0\rangle = -|\bar{K}^0\rangle \tag{2.7.10}$$

we have

$$\mathcal{P}|K^0\rangle = +|\bar{K}^0\rangle \quad \text{and} \quad \mathcal{P}|\bar{K}^0\rangle = +|K^0\rangle, \tag{2.7.11}$$

which for our new superposition states $|K_{1,2}^0\rangle$ implies

$$\mathcal{P}|K_1^0\rangle = +|K_1^0\rangle \quad \text{and} \quad \mathcal{P}|K_2^0\rangle = -|K_2^0\rangle. \tag{2.7.12}$$

It should now be obvious that the decays are explained by associating the initial state K_1^0 (K_2^0) with the final state containing two (three) pions. In fact, since the two-pion decay mode has a shorter lifetime (by a factor of order 600) the two states are then identified as 'K-short' ($K_S^0 \to 2\pi$) and 'K-long' ($K_L^0 \to 3\pi$):

$$|K_S^0\rangle := |K_1^0\rangle = \tfrac{1}{\sqrt{2}}\Big(|K^0\rangle + |\bar{K}^0\rangle\Big) \qquad (2.7.13a)$$

and

$$|K_L^0\rangle := |K_2^0\rangle = \tfrac{1}{\sqrt{2}}\Big(|K^0\rangle - |\bar{K}^0\rangle\Big). \qquad (2.7.13b)$$

The phenomenology is thus perfectly well explained and CP is *not* yet violated here: in some given production process we can imagine that an s is created and encounters a $\bar{d}$ (e.g. from a virtual $d\bar{d}$ pair) to form a $\bar{K}^0$. Now, such a state may be rewritten, by inverting the above relations, as an equal mixture of K_S^0 and K_L^0, which will then decay according to their natural probabilities into two or three pions. The mean lifetimes are (Navas *et al* 2024):

$$\tau_S = (0.8954 \pm 0.0004) \times 10^{-10}\,\text{s} \qquad (2.7.14a)$$

and

$$\tau_L = (0.5116 \pm 0.0021) \times 10^{-7}\,\text{s}, \qquad (2.7.14b)$$

with K_S^0 decaying predominantly into two pions (99.9%), whereas K_L^0 has a 68% semi-leptonic branching ratio and only 32% into three pions.

2.7.4 Regeneration

Owing to the peculiar form of the neutral-kaon eigenstates, various interesting quantum-mechanical phenomena become imaginable. One such is that known as *regeneration*, suggested by Abraham Pais and Oreste Piccioni (1955). For the purposes of this discussion, we shall ignore any possible effects of CP violation (which we shall discuss shortly); the phenomenon we are about to describe will reveal itself to be a potential source of background for CP-violation measurements. The K_S^0 and K_L^0 may thus be taken as being pure K_1^0 and K_2^0 states, respectively. As already noted, in the generation of strange particles either a strange quark or antiquark is typically produced and therefore either a pure $\bar{K}^0$ or K^0, respectively. For definiteness, let us assume that predominantly K^0 is being produced (as is typically the case since matter contains many more d than $\bar{d}$). That is, the beam created is initially an equal mixture or *superposition* of K_S^0 and K_L^0. Likewise, a beam of pure K_L^0, say (as will always be the case after a period of time that is long with respect to the K_S^0 lifetime), can be viewed as an equal mixture of $\bar{K}^0$ and K^0. The question now arises as to how such states evolve in time or on passage through matter.

Let us first recall that the two states $\bar{K}^0$ and K^0 are very different with respect to their content in terms of ordinary matter (by which we mean up and down quarks): the first contains $\bar{d}$, whereas the second contains d. This means that the first may undergo strong interactions in which the $\bar{d}$ annihilates with a d found in matter, whereas, since ordinary matter does not contain $\bar{d}$, the second may not. Therefore, whereas the $\bar{K}^0$ is very likely to decay or effectively *disappear* on contact with matter, K^0 is not as it may only interact weakly or electromagnetically. The *background* to this disappearance is just their normal weak decays, which are no competition for the strong interaction. It is thus expected that on passage through matter a $\bar{K}^0$ beam should be subject to severe attenuation, whereas a K^0 beam should survive much longer.

More formally, after some time we may say that fractions f and $\bar{f}$ of initially pure K^0 and $\bar{K}^0$ beams will survive, with $f \gg \bar{f}$. Thus, an initially pure $\bar{K}^0$ state will evolve (in the vacuum) into a pure state of K_L^0 (that is, an equal mixture of K^0 and $\bar{K}^0$), which on passing through ordinary matter will become

$$\tfrac{1}{\sqrt{2}}\left(f\,|K^0\rangle - \bar{f}\,|\bar{K}^0\rangle \right) = \tfrac{1}{2}\left((f - \bar{f})|K_S^0\rangle + (f + \bar{f})|K_L^0\rangle \right). \qquad (2.7.15a)$$

Now, since f and $\bar{f}$ are different or rather $(f - \bar{f}) \neq 0$, we have the *reappearance* of K_S^0. In fact, since $f \gg \bar{f}$, then to a good approximation the new state may be expressed as

$$\simeq \tfrac{1}{2} f \left(|K_S^0\rangle + |K_L^0\rangle \right), \qquad (2.7.15b)$$

or roughly equal populations. This can, of course, be easily tested by, e.g., the observation of two-pion decays immediately after passage through matter, where immediately prior there were none. This phenomenon is known as *regeneration*. The first experimental demonstrations were performed by Good *et al* (1961).

As hinted above and will shortly become obvious, any attempts at measuring CP violation through detection of two-pion decays must avoid regeneration, which would swamp the tiny CP-violating effect.

2.7.5 Quantum oscillation

A related, but more subtle, effect is that known as *oscillation*, in which states effectively transform back and forth between K^0 and $\bar{K}^0$. The phenomenon is mathematically the same as the effect known as *beating* in wave mechanics, or more simply, as is the case here, that seen in a system of two *weakly* coupled oscillators. The central point in such phenomena is the presence of two slightly different natural frequencies in the system; in the case of weakly coupled oscillators, if the individual natural frequencies are identical, then the weak coupling induces a splitting between the two lowest possible coupled modes (typically *in* and *out of phase*); the *in-phase* mode usually has the lowest fundamental frequency, whereas the *out-of-phase* mode is slightly higher. This difference results in *beats*: namely, if the starting condition has only one of the two oscillators in motion, then the subsequent evolution will see the other begin to move while the first comes to a stop and vice versa. The frequency of these beats is just the frequency difference between the two lowest modes. In quantum mechanics a simple example is provided by the double potential well, which is illustrated in appendix B.6 for the reader unfamiliar with the problem.

Our oscillating system is just that of the particle states themselves: according to quantum mechanics, for an energy eigenstate we have

$$\phi(t, \boldsymbol{x}) = \phi_0(\boldsymbol{x})e^{-\frac{i}{\hbar}Et}, \tag{2.7.16}$$

where, for our purposes here, the spatial part $\phi_0(\boldsymbol{x})$ is irrelevant. Taking into account special relativity, the energy of a physical particle state must include its rest mass, $E = mc^2$. Thus, neglecting the kinetic energy of the particles involved, E above may be substituted with mc^2. If then the particle (or antiparticle) state under consideration is described as an equal superposition of two states of different masses $m_{1,2}$ (with $m_2 > m_1$ say), at time $t = 0$ we have

$$|a\rangle = \tfrac{1}{\sqrt{2}}\Big(|1\rangle + |2\rangle\Big) \tag{2.7.17a}$$

and

$$|\bar{a}\rangle = \tfrac{1}{\sqrt{2}}\Big(|1\rangle - |2\rangle\Big), \tag{2.7.17b}$$

where, for clarity, any (irrelevant) spatial dependence has been suppressed. The inverse relations are

$$|1\rangle = \tfrac{1}{\sqrt{2}}\Big(|a\rangle + |\bar{a}\rangle\Big) \tag{2.7.18a}$$

and

$$|2\rangle = \tfrac{1}{\sqrt{2}}\Big(|a\rangle - |\bar{a}\rangle\Big). \tag{2.7.18b}$$

At a later time t, an initially pure $|a\rangle$ state becomes (also suppressing now the factors c^2 and $\hbar$)

$$|a, t\rangle = \tfrac{1}{\sqrt{2}}[e^{-im_1 t}|1\rangle + e^{-im_2 t}|2\rangle], \tag{2.7.19}$$

We may then re-express this in terms of the particle–antiparticle states:

$$\begin{aligned}|a, t\rangle &= \tfrac{1}{2}\Big[e^{-im_1 t}(|a\rangle + |\bar{a}\rangle) + e^{-im_2 t}(|a\rangle - |\bar{a}\rangle)\Big] \\ &= \tfrac{1}{2}\Big[(e^{-im_1 t} + e^{-im_2 t})|a\rangle + (e^{-im_1 t} - e^{-im_2 t})|\bar{a}\rangle\Big] \\ &= e^{-i\langle m\rangle t}\Big[\cos\big(\tfrac{1}{2}\Delta m\, t\big)|a\rangle + i\sin\big(\tfrac{1}{2}\Delta m\, t\big)|\bar{a}\rangle\Big],\end{aligned} \tag{2.7.20}$$

where $\langle m\rangle := \tfrac{1}{2}(m_1 + m_2)$ and $\Delta m := m_2 - m_1$.

The particle–antiparticle content is thus seen to oscillate: the sine and cosine coefficients giving particle and antiparticle content

$$\cos^2\left(\tfrac{1}{2}\Delta m\, t\right) = \tfrac{1}{2}(1 + \cos \Delta m\, t) \text{ and } \sin^2\left(\tfrac{1}{2}\Delta m\, t\right) = \tfrac{1}{2}(1 - \cos \Delta m\, t)$$

respectively. As already noted in the classical case, the oscillation frequency is given by the energy (or mass) difference:

$$\omega_{\mathrm{osc}} = \frac{|m_2 - m_1|c^2}{\hbar}. \tag{2.7.21}$$

The physical effect should be evident: a beam initially containing only K^0, say, will at a later time contain some (oscillating) fraction of $\bar{K}^0$. This can be verified experimentally by examining the decays: K^0 (containing $d\bar{s}$) can decay into $\pi^- e^+ \nu$, whereas $\bar{K}^0$ (containing $\bar{d}s$) decays into $\pi^+ e^- \nu$. The measured lepton charge asymmetry (e^+/e^-) as a function of time (or distance travelled) should thus oscillate too. So what can be measured experimentally is the time dependence of the following asymmetry:

$$A_{\Delta m}(\tau) := \frac{[R_+(\tau) + \bar{R}_-(\tau)] - [\bar{R}_+(\tau) + R_-(\tau)]}{[R_+(\tau) + \bar{R}_-(\tau)] + [\bar{R}_+(\tau) + R_-(\tau)]}, \tag{2.7.22}$$

where

$$R_+(\tau) := R(K^0_{t=0} \to e^+ \pi^- \nu_{t=\tau}), \tag{2.7.23a}$$

$$\bar{R}_-(\tau) := R(\bar{K}^0_{t=0} \to e^- \pi^+ \bar{\nu}_{t=\tau}), \tag{2.7.23b}$$

$$\bar{R}_+(\tau) := R(\bar{K}^0_{t=0} \to e^+ \pi^- \nu_{t=\tau}), \tag{2.7.23c}$$

$$R_-(\tau) := R(K^0 t = 0 \to e^- \pi^+ \bar{\nu}_{t=\tau}), \tag{2.7.23d}$$

i.e. the rates of $e^\pm$ detection at time τ after the initial production of either K^0 or $\bar{K}^0$.

The real experimental situation is more complicated owing to the finite and, indeed, rather short lifetimes of the particles involved. So, the effects of decay must now be included in the above description. A state with a finite lifetime may be described by:

$$\phi(t) = \phi_0 e^{-iEt} e^{-\frac{1}{2}\Gamma t}, \tag{2.7.24}$$

where Γ is just the decay rate. This can be seen by considering the number density:

$$|\phi(t)|^2 = |\phi_0|^2 e^{-\Gamma t}, \tag{2.7.25}$$

which satisfies the standard exponential decay-law equation

$$\frac{\mathrm{d}}{\mathrm{d}t}|\phi(t)|^2 = -\Gamma |\phi(t)|^2. \tag{2.7.26}$$

The previous temporal-evolution equations are then modified as follows

$$|a, t\rangle = \frac{1}{2}\left[e^{-im_1 t}e^{-\frac{1}{2}\Gamma_1 t}(|a\rangle + |\bar{a}\rangle) + e^{-im_2 t}e^{-\frac{1}{2}\Gamma_1 t}(|a\rangle - |\bar{a}\rangle) \right]$$

$$= \frac{1}{2}\left[\left(e^{-im_1 t}e^{-\frac{1}{2}\Gamma_1 t} + e^{-im_2 t}e^{-\frac{1}{2}\Gamma_2 t}\right)|a\rangle + \left(e^{-im_1 t}e^{-\frac{1}{2}\Gamma_1 t} - e^{-im_2 t}e^{-\frac{1}{2}\Gamma_2 t}\right)|\bar{a}\rangle \right],$$

which may be rewritten as

$$= f(t)|a\rangle + \bar{f}(t)|\bar{a}\rangle. \tag{2.7.27}$$

Thus, the particle (as opposed to antiparticle) content of the beam is given by

$$|f(t)|^2 = \frac{1}{4}\left| e^{-im_1 t}e^{-\frac{1}{2}\Gamma_1 t} + e^{-im_2 t}e^{-\frac{1}{2}\Gamma_2 t} \right|^2$$

$$= \frac{1}{4}\left[e^{-\Gamma_1 t} + e^{-\Gamma_2 t} + 2e^{-\bar{\Gamma} t}\cos(\Delta m t) \right], \tag{2.7.28}$$

where $\bar{\Gamma} := \frac{1}{2}(\Gamma_1 + \Gamma_2)$. Similarly we can calculate the antiparticle fraction.

Now, since $\Gamma_2 = \Gamma_L \ll \Gamma_S = \Gamma_1$, the only term that survives in the large-t limit is $\frac{1}{4}e^{-\Gamma_2 t}$, which merely implies the expected relative survival (up to its own decay) of the part of the initial beam corresponding to the K_L^0 state. A graphical representation of the fractional intensities (i.e. normalised to the overall $e^{-\Gamma_L t}$ decay behaviour) is displayed in figure 2.10.

Experimentally, as seen in figure 2.10, $\Delta m\, \tau_S \approx \frac{1}{2}$. This is fortunate as it allows just *one* oscillation before the decay process kills the signal. In figure 2.11 an example plot of the experimentally measured asymmetry $A_{\Delta m}$ is shown (Angelopoulos *et al* 1998a). Had the value turned out much smaller, *no* oscillations could have been

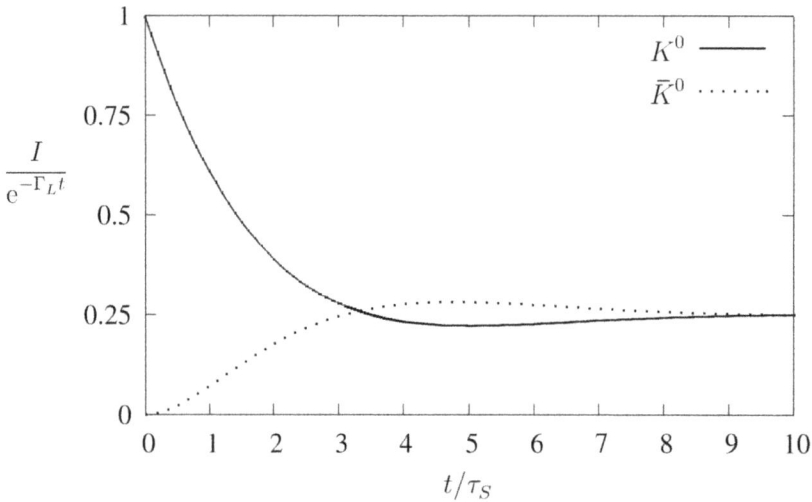

Figure 2.10. The temporal evolution of the K^0 and $\bar{K}^0$ intensities (normalised to the overall $e^{-\Gamma_L t}$ decay behaviour) for a beam of initially pure K^0.

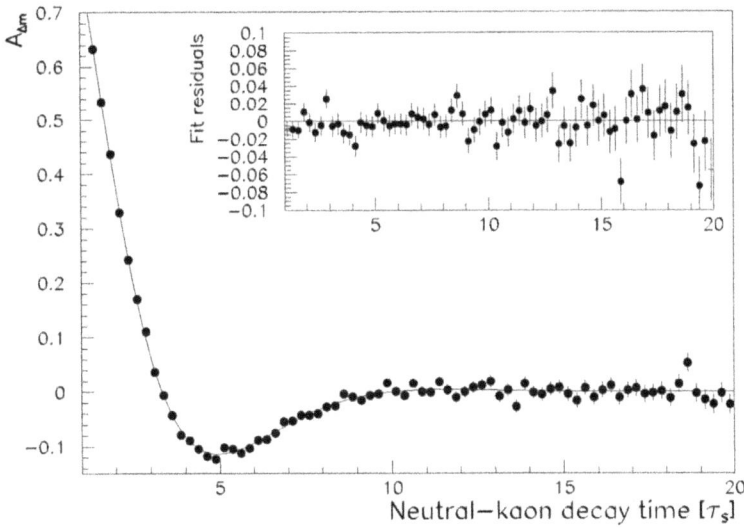

Figure 2.11. The asymmetry $A_{\Delta m}$ versus the neutral-kaon decay time in units of τ_S. The solid line represents the result of a fit (reprinted from Angelopoulos *et al* (1998a), copyright (1998) with permission from Elsevier.

observed. Equally, had it turned out much larger then the risk would have been that the rapid oscillations might have washed themselves out. The measured value is (Navas *et al* 2024)

$$\Delta m = (0.5289 \pm 0.0010) \times 10^{10} \hbar/s \qquad (2.7.29\text{a})$$

$$\hat{=} (3.481 \pm 0.007) \times 10^{-6} \text{ eV}, \qquad (2.7.29\text{b})$$

which, given the value of τ_S, implies

$$\Delta m \, \tau_S = 0.4736 \pm 0.0009 \qquad (2.7.29\text{c})$$

and finally

$$\frac{\Delta m}{m} \sim 0.7 \times 10^{-14}. \qquad (2.7.29\text{d})$$

Experimentally, it is found that $m_L > m_S$ (see e.g. Abouzaid *et al* 2011). The first such mass differences were determined by Good *et al* (1961) and Fitch *et al* (1961).

As remarked earlier, this phenomenon is rather general and more specifically may also occur in the leptonic sector once non-zero neutrino masses are established. Indeed, there is already some understanding of the form of the lepton mixing matrix, which, in order to be non-trivial, requires the neutrino masses to be different. We might, however, remark here that the situation for oscillation is technically slightly different. The neutral-kaon mass difference is exceedingly small as compared to their rather large masses. In the neutrino case the masses are very small (especially with respect to any kinetic energy they might possess), whereas the differences are comparable. This will be dealt with in detail in section 5.4.

2.7.6 CP violation

The enormous difference in decay rates suggests a method to search for possible CP-violating effects. In short, by waiting for long enough (but not too long) all the K_S^0 in an initially purely K^0 or $\bar{K}^0$ sample will have decayed and only the K_L^0 component will have survived. Since the initial populations are equal, the ratio at some later instant t will be

$$\frac{N_S}{N_L} = \frac{e^{-t/\tau_S}}{e^{-t/\tau_L}}, \tag{2.7.30}$$

where, recall, $\tau_S/\tau_L \sim 1/600$. For $\tau_S \ll t \lesssim \tau_L$ this ratio is very small indeed: that is, $N_S/N_L \sim O(e^{-600}) \sim O(10^{-260})$. We should therefore no longer see any two-pion decays *at all*.

Exercise 2.7.1. *Ignoring relativistic time-dilation effects, calculate the mean distances that K_S^0 and a K_L^0 mesons moving at nearly the speed of light will travel before decaying.*

For their experiment, Christenson *et al* (1964)[31] used the Brookhaven National Laboratory (BNL) alternating gradient synchrotron (AGS) 30 GeV [32] proton beam, incident on a beryllium target, to produce a *secondary* beam of neutral kaons. The detector (shown in figure 2.12) was placed a little over 17 m away. A lead collimator and suitable magnetic fields ensured a relatively pure kaon beam while a final collimator guaranteed the direction of motion (important for reconstructing the kinematics). Final pion pairs were selected by requiring their invariant mass to be near that of the K^0 (about 498 MeV); thus excluding three-pion events in which one pion goes undetected.

Put simply, the measured branching ratio was

$$\frac{K_L^0 \to \pi^+ \pi^-}{K_L^0 \to \text{all charged modes}} = (2.0 \pm 0.4) \times 10^{-3}, \tag{2.7.31}$$

based on a two-pion sample of 45 ± 10 events out of a total of 22 700 decays.

After further studies, the interpretation is that the K_L^0 state is not purely K_2^0, but instead contains a small admixture of K_1^0:

$$|K_L^0\rangle := \frac{1}{\sqrt{1+|\varepsilon|^2}}\left(|K_2^0\rangle + \varepsilon|K_1^0\rangle\right), \tag{2.7.32a}$$

---

[31] The 1980 Nobel Prize in Physics was awarded jointly to James Watson Cronin and Val Logsdon Fitch 'for the discovery of violations of fundamental symmetry principles in the decay of neutral K-mesons'.

[32] In their paper Christenson *et al* (1964) employ the notation 'BeV', standing for billion (or 10^9) electron-volts; the present-day accepted form is, of course, 'GeV'.

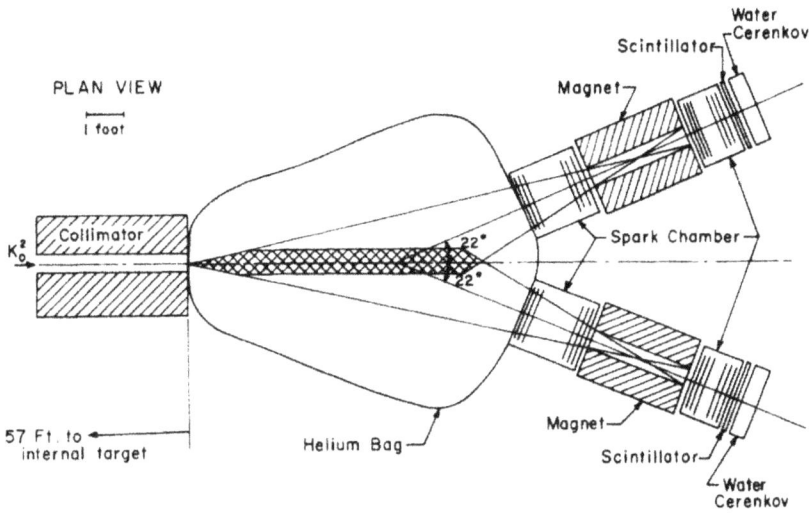

Figure 2.12. The apparatus used to detect two-pion decays of the K_L^0 at a little over 17 m from the production point; figure reproduced with permission from Christenson *et al* (1964), copyright (1964) by the American Physical Society.

where the measured branching ratio implies $|\varepsilon| \simeq 2.3 \times 10^{-3}$. The corresponding K_S^0 state is

$$|K_S^0\rangle := \frac{1}{\sqrt{1+|\varepsilon|^2}}\left(|K_1^0\rangle + \varepsilon|K_2^0\rangle\right). \qquad (2.7.32b)$$

We shall justify the above forms shortly; although a comment is immediately in order: the two states so defined are explicitly no longer orthogonal, whereas the original $K_{1,2}^0$ were; recall equation (2.7.8). This was, however, to be expected since, owing precisely to CP violation, they are now connected via their common two- and three-pion decay final states; i.e. $\langle K_S^0|K_L^0\rangle \neq 0$.

Firstly though, care must be taken over the explanation of the *observed* CP violation we have just described, since there are, in fact, two distinct possible underlying physical mechanisms: namely, *direct* and *indirect* violation. A rather complete and clear pedagogical review of CP violation may be found in Grinstein (2016).

Direct CP violation: The introduction of a complex phase into the CKM matrix allows for CP-violation at an elementary *interaction* level. For example, the decays $K_1^0 \rightarrow 3\pi$ and $K_2^0 \rightarrow 2\pi$ are thus possible. Such *direct* effects are even smaller but have nevertheless also been observed.

Indirect CP violation: It turns out that the K^0–$\bar{K}^0$ case just illustrated is an *indirect* consequence of CP violation inasmuch as it derives from the mixing of CP eigenstates, which in turn is, of course, due to CP violation in the interaction. By the same token, parity violation is observed in atomic physics, due to the effect of mixing of parity eigenstates; in this case such indirect effects are very small indeed.

Further measurements must be made in order to disentangle the two phenomena. Suffice it to note here that experimental results demonstrate that it is the *indirect* effect that dominates in neutral-kaon decays. We shall discuss this more in detail later (see section 5.3).

CP-violation formalism
The K^0–$\bar{K}^0$ system, viewed as a two-state quantum system at rest, is described by a temporal-evolution equation of the form

$$i\frac{d}{dt}\begin{pmatrix} K^0 \\ \bar{K}^0 \end{pmatrix} = \begin{pmatrix} M_{aa} - \frac{i}{2}\Gamma_{aa} & M_{ab} - \frac{i}{2}\Gamma_{ab} \\ M_{ba} - \frac{i}{2}\Gamma_{ba} & M_{bb} - \frac{i}{2}\Gamma_{bb} \end{pmatrix}\begin{pmatrix} K^0 \\ \bar{K}^0 \end{pmatrix}, \tag{2.7.33}$$

where both the M_{ij} and Γ_{ij} represent Hermitian[33] matrices. Note though that this Hamiltonian matrix is itself certainly not Hermitian since the decay process does not conserve probability; it is precisely the anti-Hermitian part $\frac{i}{2}\Gamma$ that describes the kaon decays, while not itself violating CP. The determining constraints here are hermiticity and CPT:

$$\begin{aligned} M_{aa} = M_{bb}, & \qquad \Gamma_{aa} = \Gamma_{bb}, \\ M_{ab} = M_{ba}^*, & \qquad \Gamma_{ab} = \Gamma_{ba}^*. \end{aligned} \tag{2.7.34}$$

The eigenstates are those given in equation (2.7.32) with

$$\frac{1 - \tilde{\varepsilon}}{1 + \tilde{\varepsilon}} = \frac{M_{ab}^* - \frac{i}{2}\Gamma_{ab}^*}{M_{ab} - \frac{i}{2}\Gamma_{ab}}. \tag{2.7.35}$$

Thus, if M_{ab} and Γ_{ab} were both purely real, $\tilde{\varepsilon}$ would vanish and the states K_S^0 (K_L^0) would correspond to the original purely CP-even (-odd) states K_1^0 (K_2^0). On the other hand, if either has an imaginary part, then CP is violated and the two states are not orthogonal, since then

$$\langle K_L^0 | K_S^0 \rangle \approx \mathfrak{Re}\,\tilde{\varepsilon}. \tag{2.7.36}$$

This just reflects the fact that they then share decay channels (both two- and three-pion).

Now, the parameter $\tilde{\varepsilon}$ depends on the phase convention adopted for the states K^0 and $\bar{K}^0$ and may not therefore be taken as a true physical measure of CP violation. However, the real part is phase-convention independent and may be measured in semi-leptonic decays, e.g. via the asymmetry

$$\delta_\ell = \frac{\Gamma(K_L^0 \to \pi^-\ell^+\nu_\ell) - \Gamma(K_L^0 \to \pi^+\ell^-\bar{\nu}_\ell)}{\Gamma(K_L^0 \to \pi^-\ell^+\nu_\ell) + \Gamma(K_L^0 \to \pi^+\ell^-\bar{\nu}_\ell)} = \frac{2\mathfrak{Re}\,\tilde{\varepsilon}}{1 + |\tilde{\varepsilon}|^2} \tag{2.7.37}$$

---

[33] The matrices M_{ij} and Γ_{ij} must both be Hermitian, as their eigenvalues correspond to physical quantities, which can obviously only be real.

The experimental measurement gives $\delta_\ell^{\text{expt}} = (0.330 \pm 0.012)\%$, from which we obtain $\mathfrak{Re}\,\tilde{\varepsilon} = 1.65 \times 10^{-3}$.

The two-pion decay modes allow us to distinguish between direct and indirect CP violation. Two new parameters are conventionally defined: ϵ and ε', which are related to ratios of CP-violating to CP-conserving decay amplitudes for $K^0 \to \pi^+\pi^-$ and $\pi^0\pi^0$:

$$\eta_{+-} := \frac{\mathcal{M}(K_L^0 \to \pi^+\pi^-)}{\mathcal{M}(K_S^0 \to \pi^+\pi^-)} \approx \varepsilon + \varepsilon', \qquad (2.7.38a)$$

$$\eta_{00} := \frac{\mathcal{M}(K_L^0 \to \pi^0\pi^0)}{\mathcal{M}(K_S^0 \to \pi^0\pi^0)} \approx \varepsilon - 2\varepsilon'. \qquad (2.7.38b)$$

The parameter ϵ is a measure of indirect CP violation, which is common to all decay modes. Indirect CP violation in the generic $K \to \pi\pi$ and $K \to \pi\ell\nu$ decays and also in $K_L^0 \to \pi^+\pi^-e^+e^-$ determines

$$|\varepsilon| = (2.228 \pm 0.011) \times 10^{-3}; \qquad (2.7.39)$$

while direct CP violation in $K \to \pi\pi$ decays determines

$$\mathfrak{Re}\,(\varepsilon'/\varepsilon) = (1.66 \pm 0.23) \times 10^{-3}. \qquad (2.7.40)$$

There are many other related experimentally measured CP-violating quantities (see e.g. Navas *et al* 2024); we shall though postpone further study until we have examined the full structure of the CKM matrix in section 5.3.

2.7.7 T violation

At this point, it is natural to ask whether it is possible to observe T-violating effects directly, rather than via the consequent CP violation. That is, without reliance on the CPT theorem. As with parity violation, the physical reversal of time is not possible experimentally. There are however physical variables that change sign on time reversal (e.g. a velocity or three-momentum). It might also be hoped to find processes that can be observed forwards and backwards in time. This is a little more difficult, but we shall discuss such a case shortly. A general discussion of T-violating phenomena may be found in e.g. Schubert (2015).

Now, experimentally measurable quantities are determined in terms of event numbers or detector counts. This requires constructing a *scalar* (or *pseudoscalar*) quantity from the variables to be studied. A single velocity or momentum will thus not be sufficient, a product of two would not suffer a sign change and so we should look for a triple vector product of such quantities. An obvious candidate would be the product $\boldsymbol{p}_n \cdot (\boldsymbol{p}_e \wedge \boldsymbol{p}_\nu)$ in neutron β-decay. Unfortunately, only upper limits on such T-odd effects have been obtained.

Decay processes are difficult, if not impossible, to arrange in the time-inverted direction. However, the K^0–$\bar{K}^0$ pair, with its oscillatory nature implicitly involves a process observable in both temporal directions. The CPLEAR Collaboration performed studies on precisely this system (Angelopoulos *et al* 1998b), obtaining the first direct observation of T violation through a comparison of the time-dependent probabilities of K^0 transforming into $\bar{K}^0$ and of the opposite process. The results were based on the analysis of the neutral-kaon semi-leptonic decays by tagging by the initial-state neutral kaon via the kaon charge in the production reaction $p\bar{p} \to K^\pm \pi^\mp K^0(\bar{K}^0)$ at rest and tagging the strangeness of the neutral kaon at time τ in the decay by the final-state lepton charge.

That is, the collaboration sought to measure the T-violating asymmetry

$$\frac{\mathcal{P}(\bar{K}^0 \to K^0) - \mathcal{P}(K^0 \to \bar{K}^0)}{\mathcal{P}(\bar{K}^0 \to K^0) + \mathcal{P}(K^0 \to \bar{K}^0)} \tag{2.7.41}$$

by measuring the following experimental asymmetry:

$$\mathcal{A}(\tau) := \frac{\mathcal{P}(\bar{K}^0_{t=0} \to e^+ \pi^- \nu_{t=\tau}) - \mathcal{P}(K^0_{t=0} \to e^- \pi^+ \bar{\nu}_{t=\tau})}{\mathcal{P}(\bar{K}^0_{t=0} \to e^+ \pi^- \nu_{t=\tau}) + \mathcal{P}(K^0_{t=0} \to e^- \pi^+ \bar{\nu}_{t=\tau})}. \tag{2.7.42}$$

The result obtained was $\mathcal{A}(\tau) = (6.6 \pm 1.3_{\text{stat}} \pm 1.0_{\text{syst}}) \times 10^{-3}$, measured over the interval $\tau_S < t < 20\tau_S$ (where τ_S is the K^0_S lifetime, see equation (2.7.14)). This is compatible with, but also entirely independent of the corresponding measurements of CP-violation already discussed.

2.7.8 CP-violation and the baryon asymmetry

We shall close this section on CP violation by very briefly recalling its rôle in cosmology. In order to arrive at the observed present-day baryon asymmetry in the Universe (see e.g. Cyburt *et al* 2003),

$$\eta = (n_B - n_{\bar{B}})/n_\gamma = (6.14 \pm 0.25) \times 10^{-10}, \tag{2.7.43}$$

from an initially particle–antiparticle symmetric state, a necessary condition, *inter alia*, is that CP be violated (Sakharov 1967)[34]. However, the precise way in which CP violation contributes to the baryon asymmetry is a rather complex issue, not yet fully understood. We should also note that the value of baryon asymmetry quoted above is not well explained by CP violation generated purely within the strict confines of the SM (see e.g. Bernreuther 2002). The difficulties may be alleviated by the additional possibility of leptonic CP violation; we shall discuss such a phenomenon in the final chapter.

References

Abouzaid E *et al* (KTeV Collab) 2011 *Phys. Rev.* **D83** 092001
Aghanim N *et al* (Planck Collab) 2021 *Astron. Astrophys.* **641** A6 *erratum* ibid **652** (2021) C4
Aker M *et al* (KATRIN Collab) 2022 *Nat. Phys.* **18** 160
Amaldi E 1984 *Phys. Rep.* **111** 1

---

[34] The 1975 Nobel Peace Prize was awarded to Andrei Dmitrievich Sakharov 'for his struggle for human rights in the Soviet Union, for disarmament and cooperation between all nations'.

Ambrose D *et al* (BNL E871 Collab) 1998 *Phys. Rev. Lett.* **81** 4309

Angelopoulos A *et al* (CPLEAR Collab) 1998a *Phys. Lett.* **B444** 38

Angelopoulos A *et al* (CPLEAR Collab) 1998b *Phys. Lett.* **B444** 43

Aubert J J *et al* 1974 *Phys. Rev. Lett.* **33** 1404

Augustin J E *et al* 1974 *Phys. Rev. Lett.* **33** 1406

Bacci C *et al* 1974 *Phys. Rev. Lett.* **33** 1408 *erratum* ibid 1649

Bahcall J N and Glashow S L 1987 *Nature* **326** 476

Bernreuther W 2002 CP violation and baryogenesis *CP Violation in Particle, Nuclear, and Astrophysics* (Berlin: Springer) pp 237–93

Bethe H A and Bacher R F 1936 *Rev. Mod. Phys.* **8** 82

Bjorken J D and Glashow S L 1964 *Phys. Lett.* **11** 255

Cabibbo N 1963 *Phys. Rev. Lett.* **10** 531

Chadwick J 1914 *Verh. Dtsch. Phys. Ges.* **16** 383

Chase C T 1929 *Phys. Rev.* **34** 1069; **36** 984; **36** 1060

Chinowsky W and Steinberger J 1954 *Phys. Rev.* **95** 1561

Christenson J H, Cronin J W, Fitch V L and Turlay R 1964 *Phys. Rev. Lett.* **13** 138

Clark A R *et al* 1971 *Phys. Rev. Lett.* **26** 1667

Cox R T, McIlwraith C G and Kurrelmeyer B 1928 *Proc. Nat. Acad. Sci.* **14** 544

Cyburt R H, Fields B D and Olive K A 2003 *Phys. Lett.* **B567** 227

Dalitz R H 1953 *Phil. Mag. Ser. 7* **44** 1068

Dirac P A M 1928 *Proc. R. Soc. Lond.* **A117** 610

Dirac P A M 1930 *Proc. R. Soc. Lond.* **A126** 360

Fermi E 1933 *Ric. Sci.* **4** 491 ; *see also Nuovo Cim.*, 11 (1934) 1; *transl.*, Gallavotti, G. *ibid.* 1; *Z. Phys.*, 88 (1934) 161; *transl.*, Wilson, F.L. *Am. J. Phys.*, 36, 1150

Fermi E 1934a *Nuovo Cim.* **11** 1 *transl.*, Gallavotti, G. *ibid.* 1

Fermi E 1934b *Z. Phys.* **88** 161 ; *transl.*, Wilson, F.L. *Am. J. Phys.*, 36, 1150

Feynman R P 1949 *Phys. Rev.* **76** 769

Feynman R P and Gell-Mann M 1958 *Phys. Rev.* **109** 193

Fitch V L, Piroué P A and Perkins R B 1961 *Nuovo Cim.* **22** 1160

Focardi S and Ricci R A 1983 *Riv. Nuovo Cim.* **6N11** 1

Frauenfelder H *et al* 1957 *Phys. Rev.* **106** 386

Friedman J I and Telegdi V L 1957 *Phys. Rev.* **105** 1681; **106**, 1290

Gamow G and Teller E 1936 *Phys. Rev.* **49** 895

Garwin R L, Lederman L M and Weinrich M 1957 *Phys. Rev.* **105** 1415

Gell-Mann M 1961 Caltech preprint CTSL-20 *reprinted in The Eightfold Way*, ed M Gell-Mann and Y Ne'eman (Perseus Pub., 2000)

Gell-Mann M 1962 *Phys. Rev.* **125** 1067

Gell-Mann M and Pais A 1955 *Phys. Rev.* **97** 1387

Glashow S L, Iliopoulos J and Maiani L 1970 *Phys. Rev.* **D2** 1285

Goeppert-Mayer M 1935 *Phys. Rev.* **48** 512

Goldhaber M, Grodzins L and Sunyar A W 1958 *Phys. Rev.* **109** 1015

Good R H *et al* 1961 *Phys. Rev.* **124** 1223

Grinstein B 2016 *Proc. of the 8th. CERN–Latin-American School of High-Energy Physics— CLASHEP2015 CERN (Ibarra, March 2015)*; M Mulders and G Zanderighi (CERN) 43 *CERN Yellow Rep. School Proc.*, p 43

Hardy J C and Towner I S 2020 *Phys. Rev.* **C102** 045501

Herb S W *et al* (E288 Collab) 1977 *Phys. Rev. Lett.* **39** 252

Hopper V D and Biswas S 1950 *Phys. Rev.* **80** 1099

Ioffe B L, Okun L B and Rudik A P 1957 *JETP Lett.* **5** 328

Kobayashi M and Maskawa T 1973 *Prog. Theor. Phys.* **49** 652

Kramers H A 1930 *Kon. Ned. Akad. Wetensch. Proc.* **33** 959

Kurie F N D, Richardson J R and Paxton H C 1936 *Phys. Rev.* **49** 368

Lee T-D, Oehme R and Yang C-N 1957 *Phys. Rev.* **106** 340

Lee T-D and Yang C N 1956 *Phys. Rev.* **104** 254 (*erratum ibid* **106** 1371)

Loredo T J and Lamb D Q 2002 *Phys. Rev.* **D65** 063002

Navas S *et al* (Particle Data Group) 2024 *Phys. Rev.* **D110** 030001

Noether E 1918 *Gott. Nachr.* **1918** 235 *transl.*; Tavel M A 1971 *Transp. Theory Statist. Phys.* 1 186

Okubo S, Marshak R E and Sudarshan G 1959 *Phys. Rev.* **113** 944

Pais A and Piccioni O 1955 *Phys. Rev.* **100** 1487

Pauli W 1930 Dear Radioactive Ladies and Gentlemen; *reprinted in Phys. Today* **31N9** 27

Perl M L *et al* 1975 *Phys. Rev. Lett.* **35** 1489

Puppi G 1948 *Nuovo Cim.* **5** 587 *Nuovo Cim.* **6** 194

Sakharov A D 1967 *Pis'ma Zh. Eksp. Teor. Fiz.* **5** 322 *reprinted in Sov. Phys. Usp.* **34** 417

Sargent B W 1933 *Proc. R. Soc. Lond.* **A139** 659

Schubert K R 2015 *Prog. Part. Nucl. Phys.* **81** 1

Sudarshan E C G and Marshak R E 1958 *Proc. of the Int. Conf. on Mesons and Recently Discovered Particles* (Padova, September 1957) Zanichelli, p V-14

Uhlenbeck G E and Goudsmit S A 1925 *Naturwiss.* **13** 953

von Weizsäcker C F 1935 *Z. Phys.* **96** 431

Wigner E 1927 *Gott. Nachr.* **1927** 375 ; *see also Gruppentheorie und ihre An wendung Auf Die Quantenmechanik Der Atomspektren (in German)* (Friedrich Vieweg und Sohn, 1931); *transl. Group Theory and its Application to the Quantum Mechanics of Atomic Spectra* (Academic Press, 1959)

Wu C-S, Ambler E, Hayward R W, Hoppes D D and Hudson R P 1957 *Phys. Rev.* **105** 1413

IOP Publishing

An Introduction to Elementary Particle Phenomenology
(Second Edition)

Philip G Ratcliffe

Chapter 3

Hadronic physics (the quark–parton model)

Three quarks for Muster Mark!
Sure he has not got much of a bark
And sure any he has it's all beside the mark.

James Joyce—*Finnegans Wake*

In this chapter we shall discuss various aspects of the physics of the strongly interacting particles or *hadrons*. Hadrons are not elementary particles but are made up of combinations of quarks and/or antiquarks. The present-day description of interactions between quarks is constructed on the basis of a *non-Abelian gauge theory* known as quantum chromodynamics (QCD). In particular, we shall examine the theoretical and experimental foundations, from the early developments of the quark–parton model, due to Gell-Mann and Feynman, up to the modern version of the theory of strong interactions: namely, QCD.

3.1 Pre-history

3.1.1 Yukawa theory

An early attempt to construct a theory of the strong nuclear interaction is that of Hideki Yukawa (1935). Realising that neither the electromagnetic nor the weak interaction could be responsible for the binding of nucleons inside the nucleus, Yukawa hypothesised an exchange force similar to that of quantum electro-dynamics (QED), but with the photon replaced by a massive scalar particle, the so-called *mesotron*, with obviously a new coupling constant. He noted that a wave

doi:10.1088/978-0-7503-5759-3ch3

equation of the same type as for the electromagnetic scalar potential, derived from Maxwell's equations (1865) and describing the *massless* photon,

$$\left[\boldsymbol{\nabla}^2 - \frac{\partial^2}{\partial t^2}\right]\phi = \rho, \tag{3.1.1}$$

with $\rho = Q\,\delta(\boldsymbol{x})$, leads to a static potential with a purely $1/r$ Coulomb-like dependence, whereas (as already noted) what is needed to describe the finite-range nuclear interaction might be described by something like the following:

$$\frac{e^{-\mu r}}{r} \equiv \frac{e^{-r/a}}{r}. \tag{3.1.2}$$

Now, such a potential would result from a field equation of the form

$$\left[\boldsymbol{\nabla}^2 - \frac{\partial^2}{\partial t^2} - \mu^2\right]\phi = \rho. \tag{3.1.3}$$

The above describes a *massive* particle, of mass μ. For a more detailed derivation, see appendix B.5. The effective range a of such a potential is simply the reciprocal of the mass and thus, exploiting $\mu c^2 \times a = \hbar c$, we therefore see that, for the observed range of order 1 to 2 fm, a new particle with a mass of order 100 to 200 MeV would be required.

There are, however, serious problems that preclude this as a fundamental theory of the nucleon–nucleon interaction. First and foremost, if we wish to go ahead and use such a theory to calculate simple scattering cross-sections, the first step is to extract a value for the effective coupling constant (analogous to α_{QED}). Unfortunately this turns out to be of order 10 (cf $\alpha_{\text{QED}} = \frac{1}{137}$), which immediately renders a perturbative approach inapplicable in general. Although it may be used as a low-energy approximation, where the higher-order corrections are suppressed by powers of collision energy divided by the nucleon mass, there is no possibility of it providing reliable and precise results for high-energy phenomena.

Moreover, while the pion is the clear (and correct) candidate for the exchange field, other similar particles with higher masses were soon discovered. Not only is there a tower of spin-zero resonances with the same quantum numbers as the pion, but increasing masses, there are also (presumably infinitely) many towers of higher-spin resonances. All these new particles (now known to be natural excited states of the basic $q\bar{q}$ system) interact with nucleons in a similar manner and so should be included in any calculation. Again, moving to higher and higher energies, their contributions will become more and more important.

Finally, there is one more rather obvious objection: none of the particles of the theory (proton, neutron and pion) can be considered fundamental. They all have a spatial extensions of the order of femtometres and clearly have non-trivial internal structure. The large anomalous magnetic moments of the two nucleon states indicates that they cannot be simply described by an equation of motion of the Dirac type, unless, that is, we are willing to invoke radiative corrections of order 100% (cf the parts per mille of QED). In any case, to correctly describe their interactions, rather non-fundamental form factors would need to be introduced.

In conclusion then, while the basic idea of Yukawa certainly has some physical relevance and the theory does permit the construction of a reasonable low-energy description with some phenomenological success, it is destined to fail in the high-energy region, where we really wish to have a fundamental and predictive theory.

3.1.2 The bootstrap model

So it was that until as recently as the early 1970s there was no truly fundamental theory of the strong interaction, but rather a model based mainly on ideas of mathematical self-consistency. This somewhat self-generating view of strong-interaction phenomenology led to the name *'bootstrap.'*[1]

Although QED provided a substantially complete quantum field theory for the electromagnetic forces, no such theory could be constructed for the strong interaction. First of all, this is not surprising since we know that the hadrons are not elementary particles and therefore cannot be expected to have point-like interactions. Moreover, as noted, if we insist on constructing a model or effective theory in which, for example, the pion acts as the exchange field for the strong force, simple fits to, say, nucleon–nucleon scattering data indicate a strong fine-structure constant of order 10. This would render a perturbative approach nonsensical and such a model is thus almost useless.

Theorists appealed instead to the general mathematical structure that emerges from quantum field theory approaches and attempted to directly construct the so-called *S*-matrix (or scattering matrix). The natural requirements to be satisfied were then:

- analyticity,
- crossing,
- symmetry.

While the meaning of the last should be obvious, the other two (borrowed, so-to-speak, from field-theoretical descriptions) need a little explanation. At any rate, the idea was simply that, in the absence of a truly fundamental theory from which, in principle, the scattering matrix or *S*-matrix could have been *calculated*, the *S*-matrix should be *constructed* or rather *parametrised* from general principles and experimental data. The above requirements would then become constraints on such a construction.

3.1.2.1 Analyticity

A general mathematical property of scattering amplitudes is that they should be analytic functions of the energies, momenta and other variables involved, which is a rather stringent constraint. It implies, in particular, that Cauchy's theorem is

---

[1] The origins of the expression 'bootstrap', as employed here, are usually ascribed to the fictional character of Baron Münchhausen (Raspe 1786). A particular story attributes him with the fanciful claim that, finding himself once stuck in a swamp and unable to escape, he pulled himself out of the mud by his own bootstraps. The same expression has also long been adopted for the process by which a computer starts itself up or *'boots'*.

applicable and therefore the pole structure or spectrum of the theory determines to a large extent the general nature of the S-matrix.

3.1.2.2 Crossing

Crossing symmetry requires that the amplitudes for processes differing only by the interchange of initial- and final-state particles should be related simply via exchange of the relevant variables (i.e. four-momenta and possible spin variables). Thus, for example, the processes $\pi^0 p \to \pi^+ n$ and $\pi^- p \to \pi^0 n$ should be described by one and the same amplitude with the incoming and outgoing pion momenta exchanged. And by exchanging instead, say, the proton and π^+, the amplitude for $\pi^0 \pi^- \to \bar{p} n$ may be obtained etc.

3.1.2.3 Symmetry

The general concept of symmetry simply refers to the fact that the model must include or respect all known discrete and continuous symmetries of the strong interaction (such as C, P, T etc) and all conservation laws (such as electric charge, strangeness, energy and momentum etc). Moreover, as we shall now see, there are many more observed (at least approximate) symmetries, for which explicit account should be made.

Put all together, these requirements place very tight boundary conditions on the construction of physical scattering amplitudes. Such a picture was partially justified by the development of a string theory[2] of hadronic interactions, which in turn was supported by Regge theory (1959a). However, as far as the *structure* of hadrons was concerned, the vision was one in which *everything was made of everything*, so to speak, and nothing was fundamental. While, to some extent, this permitted a self-consistent description and even some useful predictions, the overall agreement and predictive power were not acceptable for a complete fundamental theory of hadronic physics. Moreover, there was a growing phenomenology, especially where spin dependence was involved, that could not be described correctly.

Before moving on, we should mention that while such an approach could never provide a fundamental description of hadronic interactions, it nevertheless still has some use today. In particular, it does not rely on perturbative techniques and can thus provide important information in those situations where standard perturbative methods fail. For a detailed discussion of the S-matrix, the reader is referred to the classic text by Eden *et al* (1966) although the book by Collins and Martin (1984) provides more insight to modern applications.

3.1.3 Regge theory

We shall now attempt to briefly explain the basic ideas behind the phenomenological description that arises out of the theory proposed by Tullio Regge (1959b). Although the approach is based on rather abstract mathematical notions, it turns out to make

---

[2] A forerunner of, but not to be confused with, modern *superstring* theory, considered to represent a possible theory of *all* known particle interactions including gravity.

almost immediate contact with more phenomenological approaches, for which it thus provides some justification. The ensuing phenomenological approach not only allows the description of hadron physics in kinematic regions not accessible to perturbative QCD, but also finds further justification in the high-energy regime precisely from perturbative QCD calculations. For more detailed presentations of the approach, see e.g. Collins (1977) or Irving and Worden (1977); a more concise, but nevertheless complete introduction, may be found in Collins and Martin (1984).

Central to Regge theory (1959b) is the property of scattering amplitudes that they should naturally be analytic functions of their arguments (such as energies, momenta, masses etc) and that this may be extended to include angular momenta. This allows the amplitudes to be continued analytically into the complex-l plane. With a certain amount of non-trivial mathematical manipulation, which we cannot present here, this leads to the idea of Regge trajectories (1959b). These describe the masses of the known hadronic states as linear functions of l (now the spins of the given states).

Let us take the example of the well-known bound-state solutions to the Schrödinger equation for a particle of (reduced) mass m in a Coulomb-like potential. The energy eigenvalues are give by

$$E_n = -\frac{m\alpha^2}{2n^2}, \tag{3.1.4}$$

where n is the principal quantum number. We have moreover

$$n = n' + l + 1, \tag{3.1.5}$$

where n' and l are, respectively, the radial and orbital-angular-momentum quantum numbers. The above can be inverted to obtain

$$l = -n + 1 + f(E), \tag{3.1.6}$$

where

$$f(E) := -1 + i\pi\alpha\sqrt{2m/E}. \tag{3.1.7}$$

This then gives us the so-called Regge trajectory[3]; i.e. l as a complex function of energy. Now, the Coulomb potential leads to a scattering amplitude

$$A_l(E) = \frac{\Gamma(l - f(E))}{\Gamma(l + f(E))} e^{-i\pi l}, \tag{3.1.8}$$

where, recall, the Euler gamma function $\Gamma(x)$ has poles for values of its argument equal to $0, -1, -2, -3, \dots$. That is, the Regge *poles* (in the numerator here) are given precisely by the Regge trajectory above. And, of course, the poles should represent real particle states in the physical spectrum of the theory. It is found experimentally that the known states of increasing spin with the same quantum numbers do indeed

---

[3] A term also often used is '*reggeon*' (symbol $\mathbb{R}$).

lie on straight lines in the m^2–l plane. This is nicely demonstrated in figure 3.1, showing a very schematic plot Chew and Frautschi (1962).

In the Yukawa picture of section 3.1.1 at lowest order, the scattering amplitude is described in terms of one-pion exchange and thus, for the process described in figure 3.2(a), the amplitude has the following simple form:

$$\mathcal{A}(s,\, t) \propto \frac{1}{t - m_\pi^2}.$$ (3.1.9)

As noted above, t is negative and therefore the apparent pole here causes no problems. However, crossing symmetry implies that the amplitude shown in figure 3.2(b) is equivalent to that of figure 3.2(a) via the interchange of particles 2 and 3 (with their relative momenta and quantum numbers), which in turn implies simply exchanging s and t. That is, these two channels are described by one and the same amplitude function, but in different and disjoint kinematical regions of the variables s and t.

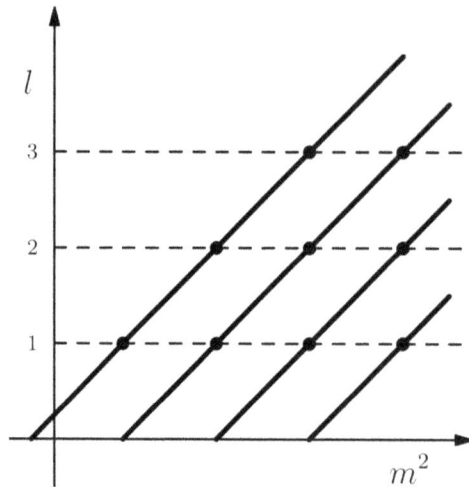

Figure 3.1. A schematic Chew–Frautschi plot showing example Regge trajectories in the l–m^2 plane. The dots represent real particle states located at integer values of l.

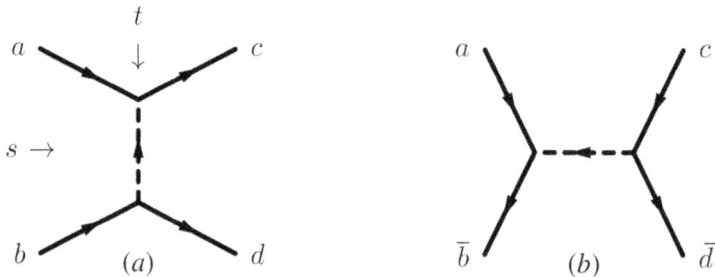

Figure 3.2. The one-pion-exchange Feynman diagrams for (a) the s-channel $ab \rightarrow cd$ and (b) the t-channel $a\bar{c} \rightarrow \bar{b}d$ processes.

Now, from equation (C.1.8), the crossed s-channel process has

$$\cos \theta_t = 1 + \frac{2s}{t - 4m^2}. \qquad (3.1.10)$$

This gives a hint as to the first difficulty with this approach: the amplitude in equation (3.1.9) does not depend on s and thus does not depend on the scattering angle; it can therefore only be an approximation to the s-wave partial amplitude, but we really need the full partial-wave series. Moreover, the pion is only the lightest in a series of particles with the same quantum numbers: ρ, ω, etc, which should all be included if we wish to fully describe such processes.

This is where the concept of a **Regge trajectory** comes to the rescue. The t-channel partial-wave expansion, see e.g. equation (C.5.14), for the amplitude is

$$\mathcal{A}(s, t) = \sum_{l=0}^{\infty} (2l + 1)\mathcal{A}_l(t) \, P_l(\cos \theta_t). \qquad (3.1.11)$$

As viewed in the s-channel, each partial-wave amplitude $\mathcal{A}_l(t)$ refers to the contribution of a *resonance* of the given spin l. Therefore, if these states lie on Regge trajectories, we have all we need to sum the full series. In figure 3.3 two such trajectories are shown. Very similar trajectories, with the same slope parameter, are also observed for the baryons and for strange hadrons. The trajectories may thus be parametrised with

$$\alpha(t) = \alpha_0 + \alpha' t. \qquad (3.1.12)$$

The slope of the physical trajectories α' is found to be almost universally $\sim 0.9 \, \text{GeV}^{-2}$ and, as we shall now show, the intercept α_0 determines the dominant trajectory in any given physical process. That is, the trajectory corresponding to the

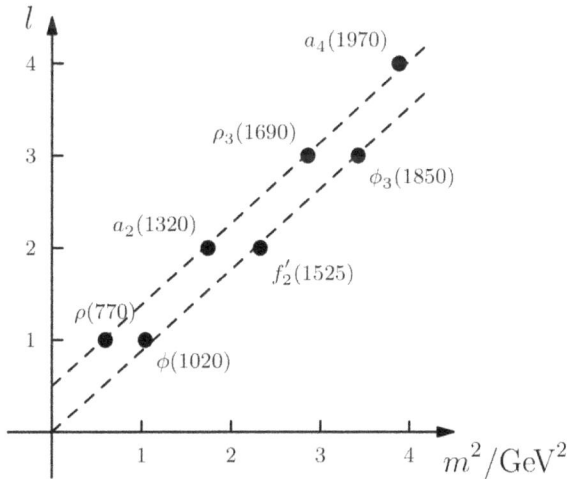

Figure 3.3. A Chew–Frautschi plot displaying Regge trajectories for the beginning of the $\rho(770)$ and $\phi(1020)$ series.

required quantum number exchange with the highest intercept will dominate a given scattering cross-section. The idea that mesons should lie on linear trajectories is corroborated by the predictions of hadronic-string models, where a long-distance phenomenological string-like potential is invoked to explain confinement, and the resulting particle spectrum is just that outlined (see section 3.4.2).

The amplitude for the l-th partial wave then takes the form

$$A_l(t) = \frac{\beta(t)}{1 - \alpha(t)} = \frac{\beta(t)}{\alpha'(m_t^2 - t)}, \tag{3.1.13}$$

where the residue at the pole $\beta(t)$ depends on the coupling of the pole to the scattering particles. Substituting this into the partial-wave expansion above, we have

$$A(s, t) = \sum_{l=0}^{\infty} (2l + 1) \frac{\beta(t)}{1 - \alpha(t)} P_l(\cos \theta_t). \tag{3.1.14}$$

The kinematical region here is such that large s implies $\cos \theta_t \to \infty$. The asymptotic behaviour of the Legendre polynomials for $x \to \infty$ is $P_l(x) \to x^l$ and so, for t fixed

$$A(s, t) \underset{s \to \infty}{\sim} \sum_{l=0}^{\infty} \frac{\beta(t)}{1 - \alpha(t)} \cos^l \theta_t \sim \beta(t) (\cos \theta_t)^{\alpha(t)} \sim \beta(t) s^{\alpha(t)}. \tag{3.1.15}$$

This then is the predicted asymptotic high-energy behaviour of the amplitude for two-body scattering as a function of s for fixed t, on the assumption of the exchange of an entire trajectory (Regge 1959b, 1960). It leads to the following prediction for the differential cross-section:

$$\frac{d\sigma(s, t)}{dt} \sim \frac{1}{s^2} \left| A(s, t) \right|^2 \underset{s \to \infty}{\sim} F(t) \left(\frac{s}{s_0} \right)^{2(\alpha(t) - 1)}, \tag{3.1.16}$$

where $\alpha(t)$ is the leading trajectory with the required quantum numbers and s_0 is presumably some hadronic mass scale, usually taken to be ~ 1 GeV^2. This behaviour has been confirmed by numerous experimental studies (see e.g. Collins 1977, Irving and Worden 1977). As promised, the positive power in α indicates that indeed the dominant trajectory will be that with highest intercept α_0.

By exploiting the parametrisation in equation (3.1.12), the above differential cross-section may be rewritten as

$$\frac{d\sigma(s, t)}{dt} \sim \frac{1}{s^2} \left| A(s, t) \right|^2 \underset{s \to \infty}{\sim} F(t) \left(\frac{s}{s_0} \right)^{2(\alpha_0 - 1)} e^{2\alpha' \ln(s/s_0)t}. \tag{3.1.17}$$

Recalling that t is negative and grows in magnitude with increasing angle, this gives a forward scattering ($\theta = 0$) peak that decreases in width (or shrinks) with increasing s. This is a non-trivial theoretical prediction of the Regge (1959b) model with rising (positive-slope) trajectories that is well borne out by experimental evidence. Note that, in the early 1960s, a very popular model for high-energy scattering was the so-called 'black-disk' approach with a radius that did not depend on energy and which therefore did not predict such a behaviour.

The pomeron

There is one exception to the general rule that the trajectories correspond to physical particle states: the slowly rising cross-section observed in hadronic collisions at high energies suggests the existence of a trajectory with the quantum numbers of the vacuum and intercept around $l = 1$; i.e. higher than any other. The early data for hadronic total cross-sections appeared to be approximately flat for centre-of-mass energies around 10–20 GeV. The known particle trajectories, all having $\alpha_0 \leqslant \frac{1}{2}$, could only lead to decreasing cross-sections.

The prevalent belief in the early 1960s was that high-energy cross-sections would tend asymptotically to constant values, which would require an intercept $\alpha_0 = 1$. The first to postulate the existence of such trajectory was Vladimir Gribov (1961). He also coined the name *pomeron* (symbol $\mathbb{P}$) after Isaak Pomeranchuk. However, it was later found that total cross-sections rise with energy and the actual pomeron parametrisation, as extracted from data, is (Donnachie *et al* 2004)

$$\alpha_{\mathbb{P}}(t) = 1.08 + 0.25t. \tag{3.1.18}$$

That is, the pomeron has a smaller slope that the other reggeons and the intercept is, in fact, slightly greater than 1, leading to rising cross-sections at high energies. Moreover, this will always be the dominant contribution whenever an exchange with the vacuum quantum numbers is permitted.

The question now arises as to the origin of the pomeron: if it does not correspond to any known particle states, what is it? As it has the quantum numbers of the vacuum, it must correspond to a neutral, flavourless, colourless object, which could effectively be some form of two-gluon exchange, as proposed by Francis Low (1975) and Shmuel Nussinov (1975). We shall examine this idea in more detail later, when we discuss QCD. However, note that a reasonable trajectory for the pomeron appears to include a so-called *glueball* candidate (a two-gluon bound state), the

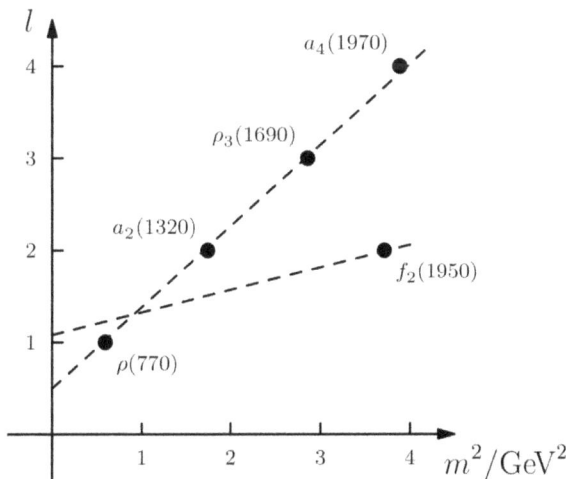

Figure 3.4. A Chew–Frautschi plot displaying the pomeron trajectory, on which lies a glueball candidate, the $f_2(1950)$; the ρ trajectory is also shown for comparison.

$f_2(1950)$, see figure 3.4. The $f_2(1950)$ has been observed in a number of experiments (see e.g. Godizov 2016), being identified as a glueball and the ground state of the pomeron (Donnachie *et al* 2004, Godizov 2016).

Now, while Regge theory (1959b) certainly enjoyed much success, it also had important shortcomings. In many cases, a suitable complication of the theory or weakening of the assumptions made allowed reconciliation, but at the cost of losing predictive power. Certainly though, one of the key issues concerned polarisation effects. For example, a marked polarisation dependence was found in $\pi^- p \to \pi^0 n$ (Bonamy *et al* 1966). That is, a left–right asymmetry in the outgoing neutron with respect to the target proton vertical polarisation of up to 20% was measured. Now, such effects require at least two interfering amplitudes, whereas this process should be described by just the single ρ-trajectory exchange, which would however thus give zero polarisation. Therefore, to explain the effects, a new trajectory with $\alpha_0 > 0$ would be required, but no corresponding particles have been observed.

3.1.4 The birth of quarks and partons

Now, it is almost implicit in the Regge approach that hadrons be considered as composite objects, i.e. bound states of some more elementary fields. This together with the generally unsatisfactory situation with respect to hadronic physics spurred physicists to seek a more fundamental description. Among these were Gell-Mann $(1961)^4$ with the *quark*[5] theory of the observed hadronic symmetries (see also Sakata 1956 and Zweig 1964, for other early work in this direction) and Feynman (1969) with his interacting point-like *parton* constituents of the proton. A concise but illuminating account of early theoretical work on the parton model may be found in Yan and Drell (2014).

While the two approaches were initially followed somewhat independently and were indeed essentially orthogonal in inspiration, it soon became evident that the two resulting pictures coincided, merely describing two different aspects of the same fundamental objects: quark–partons. Mention should also be made of the rôle played by James D Bjorken (1969) in uniting the known symmetries with a high-energy (and therefore short-distance) view of particle interactions. This chapter describes then the unfolding of these two paths to their eventual unification and the successive development of the theory of the strong interaction now known as QCD.

3.1.5 Early attempts to construct bound states

Before moving to the highly successful quark–parton model, it is interesting to examine one of the early attempts to see hadrons as bound states. As the number of new strongly interacting particles grew in the 1940s, it became increasingly evident that they could not all be considered as elementary. However, there was little or no

---

[4] The 1969 Nobel Prize in Physics was awarded to Murray Gell-Mann 'for his contributions and discoveries concerning the classification of elementary particles and their interactions'.
[5] Note that, according to Gell-Mann, the word 'quark' rhymes with 'walk' (*not* 'park') and was inspired by James Joyce's *Finnegan's Wake* (see Gell-Mann 1995, p. 180).

indication as to which, if any, might be truly elementary and which composite. Among those who sought to construct models of bound states were Fermi and Yang (1949).

> *We propose to discuss the hypothesis that the π-meson may not be elementary, but may be a composite particle formed by the associations of a nucleon and an anti-nucleon. The first assumption will be, therefore, that both an anti-proton and an anti-neutron exist[6], having the same relationship to the proton and the neutron, as the electron to the positron. Although this is an assumption that goes beyond what is known experimentally, we do not view it as a very revolutionary one. We must assume, further, that between a nucleon and an anti-nucleon strong attractive forces exist, capable of binding the two particles together.*
>
> Fermi and Yang (1949)

The idea was that the known spin-zero mesons, such as the pions, might actually be nucleon–antinucleon bound states. They considered a non-relativistic quantum mechanical model and found, for example, that a potential-well depth of around 25 GeV would give satisfactory masses to the bound states representing the pions, which would have a size of order $\hbar/m_p$. Moreover, such a model is clearly not in contradiction with Yukawa's approach. The authors note though that the similarity between their model and that of Yukawa theory must break down for phenomena in which sufficiently high energies are involved to break up the meson.

3.2 Gell-Mann's flavour SU(3)

3.2.1 The eightfold way

The reader should already be familiar with the symmetry associated with what is known as *isotopic spin* or, more simply, *isospin* (for a brief introduction see appendix B.4). We shall now describe how such a picture is extended to include strangeness. Let us first remark that the necessity for introducing this new quantum number arose from observations such as the relatively long lifetime of the Λ^0 hyperon (and other so-called *strange* particles). Despite having a large enough mass to decay comfortably into $p\pi^-$ or $n\pi^0$ (i.e. with sufficient phase-space or Q-value so as not to be suppressed), the Λ^0 lifetime is 2.6×10^{-10} s, far from that of a strong decay, for which typical lifetimes are of order 10^{-23} s. In addition, we notice that in *strong*-interaction processes certain particles (such as kaons and hyperons) are only ever produced in pairs, which on examination may be consistently assigned labels of either strange or anti-strange.

Note that, for historical reasons, the strangeness associated with the strange quark (and hence with baryons that contain one) is -1, whereas a strange antiquark then has $S = +1$[7]. Now, defining also a new quantum number B or baryon number,

---

[6] Note that, at the time, both the antiproton and antineutron were yet to be discovered (see section 4.1.6).

[7] Quite simply, the positive kaon was naturally associated with positive strangeness, before it was realised that it contains a strange antiquark.

which is $+1$ for qqq baryons (-1 for antibaryons) and 0 for $q\bar{q}$ mesons, we find by inspection the Gell-Mann–Nishijima relation (Gell-Mann 1953; Nakano and Nishijima 1953):

$$Q = I_3 + \tfrac{1}{2}(B + S) = I_3 + \tfrac{1}{2}Y, \tag{3.2.1}$$

where Q is the electric charge of the baryon in units of the proton charge and where we have also taken the liberty of introducing yet another quantum number,

$$Y := B + S, \tag{3.2.2}$$

the hypercharge. This formula correctly reproduces the charges of *all* known hadrons (i.e. *both* baryons and *mesons*).

Combining isospin with hypercharge leads to a natural set of *periodic* tables for the baryons and mesons (see table 3.1). The fact that both *baryons* and *mesons* fall into octets is an accidental property of SU(3). That they should also be the lowest mass states is, however, a property of the strong interaction itself. In table 3.1 the masses of the particles are approximately the same along the rows (separately for each group, of course), with the exception of the isospin singlets: Λ^0 and η^0. The third component of isospin varies horizontally while the vertical axis represents the hypercharge (or strangeness since B is constant in any given diagram). The early identification of the pseudoscalar meson octet led Ohnuki already in 1960 to predict the existence of the (then unknown) η^0 (mass 548 MeV and $J^{PC} = 0^{-+}$), discovered shortly after by Pevsner *et al* (1961).

The rightmost group in table 3.1 actually contains *nine* particles: together with the octet, there is an SU(3) would-be singlet (of larger mass). In fact, also in the pseudoscalar meson case a ninth (singlet) particle may be identified: namely, the $\eta'(958)$. In the case of the vector mesons the similarity of the masses favours strong mixing and thus the distinction between singlet and octet member loses any clear meaning. The baryon case is rather more complex: the colour (see later) and flavour wave-functions should be antisymmetric and thus zero orbital angular momentum and spin-½ are not possible if the wave-functions are to be overall antisymmetric as required by Fermi–Dirac statistics.

Table 3.1. From left to right: the arrangements of the lowest-mass baryons, pseudoscalar mesons and vector mesons into three octets of *flavour* SU(3).

	n	p		K^0	K^+			K^{*0}		K^{*+}	
Σ^-	$\Sigma^0\Lambda^0$	Σ^+	π^-	$\pi^0\eta^0$	π^+	ρ^-	$\rho^0\omega^0$ ϕ^0	ρ^+			
	Ξ^-	Ξ^0		K^-	$\bar{K}^0$			K^{*-}		$\bar{K}^{*0}$	

At any rate, such an arrangement in octets is readily explained via an underlying *flavour* SU(3) symmetry. In mathematical terms, baryons are products of three fundamental representations, whereas mesons are constructed from one fundamental representation and one anti-fundamental representation. Such composite objects can then be decomposed into the following irreducible representations[8]:

$$3 \otimes 3 \otimes 3 = 1_A \oplus 8_M \oplus 8_M \oplus 10_S \tag{3.2.3a}$$

and

$$3 \otimes \bar{3} = 1 \oplus 8, \tag{3.2.3b}$$

where we see that both representation products contain at least one octet. The suffixes A, M and S above indicate the symmetry with respect to exchange of any pair of quarks (antisymmetric, mixed and symmetric, respectively). The pseudoscalar meson SU(3) singlet is naturally associated with the $\eta'(958)$, subsequently discovered in 1964 independently by Kalbfleisch *et al* (1964) and Goldberg *et al* (1964).

As noted, the baryon case is, however, rather more complex. The baryon wavefunctions should be antisymmetric (i.e. singlets) in both flavour and colour[9]; zero orbital angular momentum and spin-½ are thus not simultaneously possible if the wave-functions are to be antisymmetric overall, as required by Fermi–Dirac statistics (Fermi 1926, Dirac 1926). The lightest such state must therefore have $L = 1$ and will lie much above the other ground-state spin-half baryons. Consultation of the PDG 2024 tables reveals that the $\Lambda(1405)$, with $J^P = \frac{1}{2}^-$ and assumed $L = 1$, is generally ascribed the rôle of an SU(3) baryon singlet. However, the real physical object is considered to be rather more complex and is quite probably a superposition of the state considered and a sort of K–N molecule (i.e. a kaon–nucleon bound state) see e.g. the review 'Pole Structure of the $\Lambda(1405)$ Region' in Navas *et al* (2024), p. 973.

The extra baryon octet is observed as a set of so-called N^* resonances, with similar properties but heavier. The decuplet nicely accommodates the set of spin three-halves, isospin three-halves baryon resonances shown in table 3.2. The question mark as the lowest entry represents a particle that was unknown when the table was first laid down: namely, the Ω^- (an sss state), but which was discovered shortly after by Barnes *et al* (1964) in bubble-chamber[10] experiments (see figure 3.5). It is interesting to note that in 1973, by reanalysing earlier cosmic-ray photographic-emulsion data, Álvarez[11] demonstrated that it had actually been unwittingly 'seen'

---

[8] Compare these with the more familiar spin compositions: e.g. $2 \otimes 2 = 1 \oplus 3$ for two spin-1/2 objects.

[9] We shall soon see that antisymmetry in colour is required for a three-quark state to be colour neutral (for colour see later).

[10] The 1960 Nobel Prize in Physics was awarded to Donald Arthur Glaser 'for the invention of the bubble chamber.'

[11] The 1968 Nobel Prize in Physics was awarded to Luis Walter 'Alvarez 'for his decisive contributions to elementary particle physics, in particular the discovery of a large number of resonance states, made possible through his development of the technique of using hydrogen bubble chamber and data analysis.'

Table 3.2. The arrangement of the lowest-mass, spin-3/2, isospin-3/2, baryons into a decuplet of *flavour* SU(3).

Δ^-	Δ^0	Δ^+	Δ^{++}
	Σ^{*-}	Σ^{*0}	Σ^{*+}
		Ξ^{*-}	Ξ^{*0}
		?	

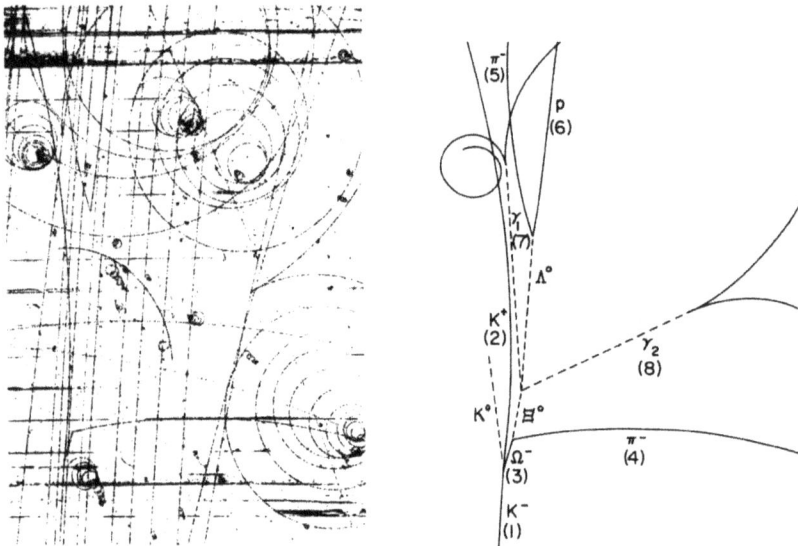

Figure 3.5. The bubble-chamber image in which the Ω^- was discovered. A K^- strikes a proton, producing $\Omega^- K^0 K^+$. These unstable particles then all decay further. The dashed lines indicate neutrals, which do not produce tracks; figure reproduced with permission from Barnes *et al* (1964), copyright (1964) by the American Physical Society.

as early as 1954. Now, not only does the table evidently predict the existence of such a particle, but also its mass. Indeed, SU(3) symmetry in this case leads to the simple prediction that the mass spacing between the table rows is constant, as is verified experimentally.

3.2.2 SU(3) mass relations

Now, the approximate SU(3) symmetry of the Gell-Mann model also leads to a surprisingly good description of the baryon-octet masses, in this case via just three parameters. Such detailed agreement is highly non-trivial: whereas the decuplet mass spacing is even, in the case of the octet it is not and, for example, the Λ^0–Σ^0 mass difference can neither be ignored nor simply ascribed to electromagnetic effects. The celebrated Gell-Mann–Okubo mass formulæ (Gell-Mann 1962, Okubo 1962) may thus be used, for example, to very successfully 'predict' the mass of the Λ^0 hyperon, given the mean masses of the three isospin multiplets: n–p, $\Sigma^{\pm,\,0}$ and $\Xi^{-,\,0}$ (see also Coleman and Glashow 1961).

To derive these formulæ, we need to examine the representations of SU(3). Recall that SU(2) is locally isomorphic to SO(3) (technically, it provides a double covering for the rotation group) and has just three generators, which are conveniently represented by the three Pauli σ-matrices:

$$\begin{pmatrix} 0 & 1 \\ 1 & 0 \end{pmatrix}, \qquad \begin{pmatrix} 0 & -i \\ i & 0 \end{pmatrix}, \qquad \begin{pmatrix} 1 & 0 \\ 0 & -1 \end{pmatrix}. \tag{3.2.4}$$

The SU(3) group, which incidentally is *not* related to any SO(N) group, has eight generators (in general, the group SU(N) has $N^2 - 1$ generators forming an adjoint representation, while the fundamental representation is N-dimensional), which can be constructed in a very similar fashion

$$\lambda_1 = \begin{pmatrix} 0 & 1 & 0 \\ 1 & 0 & 0 \\ 0 & 0 & 0 \end{pmatrix}, \quad \lambda_2 = \begin{pmatrix} 0 & -i & 0 \\ i & 0 & 0 \\ 0 & 0 & 0 \end{pmatrix}, \quad \lambda_3 = \begin{pmatrix} 1 & 0 & 0 \\ 0 & -1 & 0 \\ 0 & 0 & 0 \end{pmatrix},$$

$$\lambda_4 = \begin{pmatrix} 0 & 0 & 1 \\ 0 & 0 & 0 \\ 1 & 0 & 0 \end{pmatrix}, \quad \lambda_5 = \begin{pmatrix} 0 & 0 & -i \\ 0 & 0 & 0 \\ i & 0 & 0 \end{pmatrix}, \tag{3.2.5}$$

$$\lambda_6 = \begin{pmatrix} 0 & 0 & 0 \\ 0 & 0 & 1 \\ 0 & 1 & 0 \end{pmatrix}, \quad \lambda_7 = \begin{pmatrix} 0 & 0 & 0 \\ 0 & 0 & i \\ 0 & -i & 0 \end{pmatrix}, \quad \lambda_8 = \frac{1}{\sqrt{3}} \begin{pmatrix} 1 & 0 & 0 \\ 0 & 1 & 0 \\ 0 & 0 & -2 \end{pmatrix}.$$

The matrices in the first row are evidently a direct extension to 3×3 of the Pauli matrices (i.e. they simply rotate between u and d quarks). On close inspection, we also see that the pairs (λ_4, λ_5) and (λ_6, λ_7) play a rôle similar to that of (λ_1, λ_2), connecting u to s (a so-called V-spin pair) and d to s (a U-spin pair), respectively. Indeed, the usual ladder operators raising or lowering the third component of isospin (or I-spin) and their extension to V-spin and U-spin may be constructed, respectively, as follows:

$$\tfrac{1}{2}(\lambda_1 \pm i\lambda_2), \qquad \tfrac{1}{2}(\lambda_4 \pm i\lambda_5), \qquad \tfrac{1}{2}(\lambda_6 \pm i\lambda_7). \tag{3.2.6}$$

Finally, the eigenvalues of the matrix λ_8 correspond to hypercharge.

To describe (though not truly explain) the baryon mass spectrum, we would naturally wish to write a formula of the form $m_B = \langle B|M|B \rangle$, where M represents the (unknown) mass operator. We thus seek an SU(3) representation of the baryons themselves. Without formally deriving such, let us simply state that the following does the job:

$$
\begin{pmatrix}
\frac{1}{\sqrt{2}}\Sigma^0 - \frac{1}{\sqrt{6}}\Lambda^0 & \Sigma^+ & p \\
\Sigma^- & -\frac{1}{\sqrt{2}}\Sigma^0 - \frac{1}{\sqrt{6}}\Lambda^0 & n \\
\Xi^- & \Xi^0 & \frac{2}{\sqrt{6}}\Lambda^0
\end{pmatrix}. \tag{3.2.7}
$$

The interpretation of equation (3.2.7) is that the matrix used to represent any given baryon will have entries corresponding to the coefficients of that baryon in the above matrix. Thus, for example,

$$
p = \begin{pmatrix} 0 & 0 & 1 \\ 0 & 0 & 0 \\ 0 & 0 & 0 \end{pmatrix}, \qquad \Lambda^0 = \frac{1}{\sqrt{6}}\begin{pmatrix} -1 & 0 & 0 \\ 0 & -1 & 0 \\ 0 & 0 & 2 \end{pmatrix} \qquad \text{etc.} \tag{3.2.8}
$$

As far as the SU(3) dependence is concerned, the interaction is then constructed by simply multiplying the tensors representing the physical states together with the relevant interaction matrix and saturating the indices. For the two-dimensional representations adopted here, this simply means matrix multiplication and an overall trace.

If SU(3) were exact, the interaction would have the unit-matrix form shown below as M_0 while a term that violates SU(3) via the strange-quark mass (a reasonable though not proven hypothesis) should take the form of δM below:

$$
M_0 = m_0 \mathbb{1} \quad \text{and} \quad \delta M \sim \delta m\, \lambda_8. \tag{3.2.9}
$$

We now need simply evaluate traces of the various possible products of the relevant matrices. Recall that traces are group invariants and therefore depend only on the structure of the symmetry-group and not on the particular representation adopted. The SU(3)-symmetric piece leads to something like $\mathrm{Tr}[\bar{B}M_0 B]$, where $\bar{B}$ implies the Hermitian conjugate of the corresponding matrix. Since M_0 is proportional to the unit matrix this reduces trivially to $\mathrm{Tr}[\bar{B}B] = 1$.

The SU(3)-breaking term is a little more involved since λ_8 does not generally commute with B and thus *a priori* there are two possible inequivalent orderings. The best we can do is associate each with a new parameter, as follows:

$$
m_B = \langle B|M|B \rangle = m_0 + \delta m_1 \mathrm{Tr}[\bar{B}B\lambda_8] + \delta m_2 \mathrm{Tr}[B\bar{B}\lambda_8]. \tag{3.2.10}
$$

The final form is thus a three-parameter expression for the masses of the eight baryons. It is not difficult to show that any choice for the matrix δM that does not violate SU(2) (i.e. that treats u and d quarks equally) would lead to an equivalent formula; that is, ultimately giving the same predictions. The term m_0 corresponds to the value all the baryon masses would have for exact SU(3) symmetry while the two terms in $\delta m_{1,2}$ describe the symmetry breaking.

Exercise 3.1. *Using the representation given in equation (3.2.7), evaluate the two coefficients of $\delta m_{1,2}$ in equation (3.2.10) for each of the four independent cases: $B = N, \Lambda, \Sigma$ and Ξ. N.B. The expressions for individual isospin multiplets (e.g. p and n) will turn out to be identical, as isospin is not broken in this construction. Thus, e.g. for N, we should use the average of the proton and neutron masses.*

Now, since we have not considered isospin breaking, we have already effectively set $m_p = m_n$, the masses of the three Σs equal and also the pair of Ξs to have equal masses. Therefore, there are actually only four independent quantities to consider. However, this still leaves room for a prediction (or rather 'post-diction'): for example, the Λ^0 mass is completely determined by the others. The precise result depends on how the individual contributions from the three separate isospin multiplets are weighted and is also affected by the presence of electromagnetic corrections; nevertheless, the mass so obtained is in excellent agreement with the experimental value. Alternatively, the four equations obtained from equation (3.2.10) may be used to eliminate the three unknown mass parameters; this leads to

$$3m_\Lambda + m_\Sigma = 2m_N + 2m_\Xi. \tag{3.2.11}$$

This is just one of many mass formulæ obtainable assuming an approximate, but nevertheless broken, SU(3) flavour symmetry.

Exercise 3.2. *Use the coefficients derived in the previous exercise to verify the above mass relation.*

Finally, note that we may also consider taking into account the small SU(2) or isospin breaking (leading to the proton–neutron mass difference). This certainly has at least two origins: the differing charges of the u and d quarks and the differing masses of the same. In such a simple picture (there are no dynamics here) the two effects cannot be separated, but may both be included in the formulæ via the inclusion of another breaking term:

$$\delta M' \sim \delta m \, \lambda_3. \tag{3.2.12}$$

Two more parameters are necessary, thus leading to a five-parameter formula to describe though the *eight* independent baryon masses. Indeed, another relation, due to Sidney Coleman and Sheldon Glashow (1961), deals precisely with these so-called electromagnetic corrections:

$$(m_p - m_n) - (m_{\Sigma^+} - m_{\Sigma^-}) + (m_{\Xi^0} - m_{\Xi^-}) = 0. \tag{3.2.13}$$

Exercise 3.3. *Introduce SU(2) breaking as described in equation (3.2.12) and thus add two new coefficients, say $\delta m_{3,4}$. Calculate the four coefficients of δm_{1-4} and determine the parameters using the five independent cases of say p, n, Λ^0 and $\Xi^{0,-}$. Using the values thus found, derive equation (3.2.13) and 'predict' the $\Sigma^{0,\pm}$ masses.*

Exercise 3.4. *As a final check, insert the known values into the previous mass formulæ (3.2.13) and examine how closely they are actually satisfied.*

As noted above, similar relations may also be applied to the resonance decuplet and are equally successful. Moreover, similar formulæ may be obtained for the meson octets. There, however, the non-relativistic approximation implicit in our approach fails and the relations deduced apply better to the squared masses.

3.2.3 The nature of quarks

At this point it is perhaps relevant to note that although it is often stated that Gell-Mann himself did not believe in quarks as real physical objects inside hadrons, he has more recently claimed:

> *I always believed they were real—I just said that they had such strange properties that they were better stuck away where they can't be seen. But I didn't know that one could find them inside particles.*

At any rate, many did begin to believe in the physical reality of quarks and indeed, independently, Feynman (1972) was already working towards a description of the possible constituents of hadrons or, more precisely, of the way they might reveal their presence through interaction with an external probe at very high energies (we should not forget here the important contributions of Bjorken 1969).

3.2.3.1 Quark spins
Taking the physical reality of quarks seriously and assuming the lowest-mass baryons (the spin-$\frac{1}{2}$ octet and the spin-$\frac{3}{2}$ decuplet already mentioned) to all be composed of three quarks in an s-wave state, the quarks themselves are evidently required to be spin-$\frac{1}{2}$. That is, they should be fermions and this turns out to be very important for further developments.

3.2.3.2 Quark charges
Their electric charges are easily determined from, say, the p–n system:

$$2Q_u + Q_d = Q_p = 1 \quad \text{and} \quad Q_u + 2Q_d = Q_n = 0, \tag{3.2.14}$$

leading to

$$Q_u = \tfrac{2}{3} \quad \text{and} \quad Q_d = -\tfrac{1}{3}. \tag{3.2.15}$$

The same conclusions are reached by considering the Δ quadruplet. Indeed, these charge and spin assignments, together with $Q_s = -\frac{1}{3}$ correctly reproduce the charges, spins and parities (taking into account the orbital angular momentum assignments) of all known hadrons, both baryons and mesons (not yet including, of course, charm or bottom). Finally, taking the individual quark magnetic moments as free parameters, as we shall soon see, it is possible to obtain similar formulæ for the baryon magnetic moments; the agreement here is not quite so striking, but is nevertheless another success for the theory.

3.2.4 Basic properties of the quarks

We now list the basic properties of the six known quarks. As noted, they are all spin-½ fermions and are subject to all three fundamental interactions.

The quark quantum numbers

The quarks (antiquarks) possess the additive baryon number ⅓ (−⅓) and, *by convention*, have positive (negative) parity (table 3.3). The various quantum numbers they possess are related to the charge Q (in units of the elementary charge $|e|$) via a generalised Gell-Mann–Nishijima formula

$$Q = I_3 + \frac{\mathcal{B} + S + C + B + T}{2}, \tag{3.2.16}$$

where $\mathcal{B}$ is the baryon number, I_3 indicates the possible effective up/down flavour while S, C, B and T are respectively the strangeness, charm, beauty and top quantum numbers.

The convention adopted is that the generalised quark flavour (I_3, S, C, B, or T) has the same sign as its charge Q. According to this convention, all flavours carried by a charged meson have the same sign as the charge; e.g. the K^+ ($u\bar{s}$) strangeness is $+1$, the B^+ ($u\bar{b}$) beauty is $+1$, and both the charm and strangeness of the D_s^- ($s\bar{c}$) are -1; antiquarks, of course, have the opposite flavours. To be clear, whereas the u and c quarks naturally have *positive* quantum numbers with respect to their respective flavours, the s and b quarks unfortunately have *negative* quantum numbers. Finally, hypercharge is defined as

$$Y = \mathcal{B} + S - \frac{C - B + T}{3}. \tag{3.2.17}$$

Table 3.3. The additive quantum numbers of the quarks.

	d	u	s	c	b	t
Q - electric charge	$-\frac{1}{3}$	$+\frac{2}{3}$	$-\frac{1}{3}$	$+\frac{2}{3}$	$-\frac{1}{3}$	$+\frac{2}{3}$
I - isospin	$\frac{1}{2}$	$\frac{1}{2}$	0	0	0	0
I_3 - isospin, z-component	$-\frac{1}{2}$	$+\frac{1}{2}$	0	0	0	0
S - strangeness	0	0	-1	0	0	0
C - charm	0	0	0	$+1$	0	0
B - beauty/bottomness	0	0	0	0	-1	0
T - topness (truth)	0	0	0	0	0	$+1$

The u and d quarks thus have Y equal to ⅓, for the s quark it is $-⅔$ and for all other quarks it is zero.

The quark masses

The question of quark masses is rather delicate; the fact that they are confined and are never observed outside a hadronic bound state means that their masses are not directly accessible, being always inextricably entangled with their binding energies. For the heavier quarks (c, b and t), this is less of a problem and in any case they may be determined via e.g. the cross-sections and thresholds for production in e^+e^- annihilation. For the lighter quarks (u, d and s), various theoretical techniques have been applied, each with limitations. For completeness, the values as quoted by the Particle Data Group (Navas *et al* 2024) are listed in table 3.4. For a more in-depth discussion, see the review on *Quark Masses* in Navas *et al* (2024), p 180.

3.2.5 Mesons as quark–antiquark bound states

The simplest objects in the quark model are the lowest-lying pseudoscalar meson states: $\pi^{0,\pm}$, $K^{0,\pm}$, $\overline{K}^0$ and η^0. The standard quark-model representations are shown in table 3.5. The unusual signs in the quark assignments for π^0 and η^0 are a result of the particular standard choice of isospin representation for the antiquark doublet (see appendix B.4.3 for an explanation). The spin assignment here is simple: all these states are $S=0$ spin-singlet states and are therefore $\frac{1}{\sqrt{2}}[|\uparrow\downarrow\rangle - |\downarrow\uparrow\rangle]$. As has been discussed earlier, assuming the lowest-lying meson states to have $L=0$ and considering the effect of the symmetry operations of $\mathcal{P}$ and $\mathcal{C}$, we see that they must have $J^{PC} = 0^{-+}$.

Table 3.4. The quark masses as quoted by the PDG (Navas *et al* 2024).

u	2.16 ± 0.07 MeV	c	1.2730 ± 0.0046 GeV
d	4.70 ± 0.07 MeV	b	4.183 ± 0.007 GeV
s	93.5 ± 0.8 MeV	t	172.57 ± 0.29 GeV

Table 3.5. The quark composition of the lowest-lying pseudoscalar $q\bar{q}'$ meson octet.

$$K^0 = |d\bar{s}\rangle \qquad\qquad K^+ = |u\bar{s}\rangle$$

$$\pi^- = |u\bar{d}\rangle \qquad \begin{aligned} \pi^0 &= \tfrac{1}{\sqrt{2}}\left[|u\bar{u}\rangle - |d\bar{d}\rangle\right] \\ \eta^0 &= \tfrac{1}{\sqrt{2}}\left[|u\bar{u}\rangle + |d\bar{d}\rangle\right] \end{aligned} \qquad \pi^+ = |d\bar{u}\rangle$$

$$K^+ = |s\bar{u}\rangle \qquad\qquad \overline{K}^0 = |s\bar{d}\rangle$$

The natural interpretation of the spin-one nonet is then $L = 0$, $S = 1$ spin-triplet $q\bar{q}$ states in the same combinations as above. It turns out, however, that the two isoscalar states ω^0 and ϕ^0 are particular superpositions of the corresponding pseudoscalar states; we shall discuss this further in section 4.2.2. The conclusion is though that these must all have $J^{PC} = 1^{--}$.

Exercise 3.5. *By considering the effect of the symmetry operations $\mathcal{P}$ and $\mathcal{C}$, show that the $L = 0$, $S = 0$ mesons must have $J^{PC} = 0^{-+}$, whereas the vector mesons with $L = 0$, $S = 1$ have $J^{PC} = 1^{--}$.*

Once we understand the nature of the mesons as $q\bar{q}'$ bound states, then we also have a prediction for the natural sequence of J^{PC} assignments. The orbital angular momentum may be $L = 0, 1, \ldots$, while the spin part can only be $S = 0$ or 1; the total spin J must satisfy $|L - S| \leqslant J \leqslant L + S$. Recalling equation (2.2.1), we have that $P_{q\bar{q}'} = (-1)^{L+1}$. Only neutral particles such as the π^0 can be eigenstates of charge conjugation. It is not difficult to see in this case that $P_{q\bar{q}} = (-1)^{L+S}$ since it interchanges, say, a u and $\bar{u}$ and thus too the spins. Finally then, we find the following natural sequence: 0^{-+}, 0^{++}, 1^{--}, 1^{+-}, 1^{--}, 2^{--}, 2^{-+}, 2^{++} etc (see Exercise B.3.1). The missing states, 0^{--}, 0^{+-}, 1^{-+}, 2^{+-} etc, are not allowed for a $q\bar{q}$ pair and are thus considered *exotic*.

3.2.6 Baryon magnetic moments

As a final example of the quark-model predictions, we now examine the baryon-octet magnetic moments. The construction of a baryon wave-function is more complicated. First of all it is necessary to anticipate the inclusion of colour (see section 3.4.1): it turns out that the antisymmetry of the three-quark wave-function, necessary to satisfy Fermi–Dirac statistics is provided by this new degree of freedom. The important point here is that the space–spin part of the wave-function is then *symmetric*.

The simplest way to construct, say, the proton wave-function is to consider first the two u quarks. these may form either a spin-0 or spin-1 state:

$$\text{spin-0} = \frac{1}{\sqrt{2}}\left[|u^\uparrow u^\downarrow\rangle + |u^\downarrow u^\uparrow\rangle\right] \quad \text{and} \quad \text{spin-1} = |u^\uparrow u^\uparrow\rangle. \tag{3.2.18}$$

These two states must then be combined with the remaining spin-½ d quark. The Clebsch–Gordan coefficients for combining spin-0 and spin-1 with spin-½ to produce an overall spin-½ state are $-\sqrt{1/3}$ and $\sqrt{2/3}$, respectively. The final form is therefore

$$|p^\uparrow\rangle = \sqrt{2/3}\,|u^\uparrow u^\uparrow d^\downarrow\rangle - \sqrt{1/6}\left[|u^\uparrow u^\downarrow d^\uparrow\rangle + |u^\downarrow u^\uparrow d^\uparrow\rangle\right]. \tag{3.2.19}$$

Strictly speaking, the d quark should also be placed in each of the three possible positions and so the full wave-function takes the form

$$|p^\uparrow\rangle = \sqrt{\tfrac{1}{18}}\big[2|u^\uparrow u^\uparrow d^\downarrow\rangle - |u^\uparrow u^\downarrow d^\uparrow\rangle - |u^\downarrow u^\uparrow d^\uparrow\rangle$$
$$+ 2|u^\uparrow d^\downarrow u^\uparrow\rangle - |u^\uparrow d^\uparrow u^\downarrow\rangle - |u^\downarrow d^\uparrow u^\uparrow\rangle + 2|d^\downarrow u^\uparrow u^\uparrow\rangle - |d^\uparrow u^\uparrow u^\downarrow\rangle - |d^\uparrow u^\downarrow u^\uparrow\rangle\big]. \tag{3.2.20}$$

The wave-functions of all the other octet baryons may be generated by suitably applying the I-, U- and V-spin raising and lowering operators defined earlier (for more details see, e.g., Close 1979).

Exercise 3.6. *Derive the wave-functions for the entire baryon octet by applying the I-, U- and V-spin raising and lowering operators to the proton wave-function derived above.*

Armed with these and defining the quark magnetic moments $\mu_{u,d,s}$, for the proton magnetic moment we immediately find

$$\mu_p = \tfrac{2}{3}(2\mu_u - \mu_d) + \tfrac{1}{3}\mu_d = \tfrac{4}{3}\mu_u - \tfrac{1}{3}\mu_d, \tag{3.2.21}$$

where $\mu_{u,d}$ are the light quark magnetic moments, which are defined in the standard (Dirac) manner:

$$\mu_{u,d} = Q_{u,d}\,\frac{e\hbar}{2m_{u,d}} \qquad (Q_u = \tfrac{2}{3},\ Q_d = -\tfrac{1}{3}). \tag{3.2.22}$$

The neutron magnetic moment may be obtained by exchanging u and d in equation (3.2.21). Assuming the u and d to be endowed with elementary magnetic moments and to have the same effective mass, then $\mu_d/\mu_u = Q_d/Q_u = -\tfrac{1}{2}$. This leads immediately to the prediction $\mu_n/\mu_p = -\tfrac{2}{3}$, which is remarkably close to the experimental value of -0.68 (Bég *et al* 1964, Sakita 1964).

Exercise 3.7. *Derive the magnetic moments of the entire baryon octet in terms of quark magnetic moments from the wave-functions obtained in the previous exercise.*

The quark magnetic moments may then be determined using, say, the proton, neutron and Λ^0 values:

$$\mu_p = 2.79\ \mu_N = \tfrac{4}{3}\mu_u - \tfrac{1}{3}\mu_d, \tag{3.2.23a}$$

$$\mu_n = -1.91\ \mu_N = \tfrac{4}{3}\mu_d - \tfrac{1}{3}\mu_u, \tag{3.2.23b}$$

$$\mu_\Lambda = -0.61\ \mu_N = \mu_s. \tag{3.2.23c}$$

Inverting instead the above expressions for μ_p and μ_n, leads to

$$\mu_u = \tfrac{4}{5}\mu_p + \tfrac{1}{5}\mu_n = 1.85\mu N \qquad (3.2.24a)$$

and

$$\mu_d = \tfrac{4}{5}\mu_n + \tfrac{1}{5}\mu_p = -0.97\mu N \qquad (3.2.24b)$$

The d quark thus has a magnetic moment opposite to and approximately half that of the u quark, which is as might have been expected, given their charge ratio (assuming they have similar masses).

In other words, assuming the same form for quarks as for elementary Dirac (1928) particles and using the definition of the nuclear magneton as $\mu_N := e\hbar/2m_p$, the following general formula for the quark *effective*[12] masses is obtained:

$$m_q = \frac{1}{\mu_q}\frac{Q_q}{Q_p}\, m_p. \qquad (3.2.25)$$

Then, inverting the above relations for the magnetic moments, we obtain

$$m_u = 338\text{ MeV}, \quad m_d = 322\text{ MeV and } m_s = 513\text{ MeV}. \qquad (3.2.26)$$

These are strikingly close to what might be expected if the masses of the nucleons were essentially due to the masses of the quarks constituting them. The magnetic moments for the rest of the baryon octet may now be predicted. The agreement here is rather poorer than the earlier mass formulæ. It should, however, be noted that we have been treating the quarks as real physical entities without any knowledge of their dynamics, which could induce important quantum corrections; these masses are governed by non-perturbative effects, which such a simple model does not and cannot hope to address. Indeed, since $m_n > m_p$, the d quark might have been expected to be slightly the heavier of the two, which does not appear to be the case.

Exercise 3.8. *Using the result of the previous exercise, calculate the numerical values of the magnetic moments of the baryon octet and compare them with their experimental values.*

3.3 Feynman's parton model

3.3.1 High-energy electron–proton scattering

At this point in history (the late 1960s) the experimental capabilities became the determining factor in progress. At the Stanford Linear Accelerator Center (SLAC) the machine then in operation was capable of delivering an intense electron beam with an

---

[12] As we shall explain later, these are not the elementary particle masses appearing in, say, the QCD Lagrangian.

energy of around 2 GeV. It was being used to study the internal structure of the proton in much the same way that in 1911 Ernest Rutherford[13] and his collaborators, Hans Geiger and Ernest Marsden (1909), had used α-particles to study the internal structure of the atom. The correct energy to use is, of course, a question of the length scale to be resolved. An α-particle with a kinetic energy of 5 MeV has momentum

$$p_\alpha = \sqrt{2\, m_\alpha c^2\, E_\alpha} \simeq \sqrt{2 \times 4000 \times 5}\ \text{MeV c}^{-1} = 200\ \text{MeV c}^{-1}. \tag{3.3.1}$$

Using the scale set by $\hbar c \sim 200$ MeV fm, we immediately see that the best resolution attainable is of the order of 1 fm (in practice it will always be rather poorer), which is what would be necessary to approach the nuclear size. Rutherford and collaborators were indeed able to provide an estimate for the size of a nucleus based on the observed deviations from the simple Rutherford formula.

If we now wish to look *inside* the proton, we evidently need a resolution roughly an order of magnitude better, which would mean a factor of 100 in the α-particle energy. However, in order to avoid confusing signals, we also require a probe that does *not* partake in the strong interaction and which thus avoids a convolution of the non-trivial structure of the projectile and that of the target. The obvious choice then, as indeed already much used in nuclear physics, is the electron. At such energies it is already highly relativistic, which fortunately simplifies the calculations since then $pc \approx E$ and so, e.g., an electron of energy 2 GeV has a resolving power of order 0.1 fm.

3.3.2 The parton model

In order to appreciate how and why Feynman (1972) was led to develop a picture of hard, point-like constituents inside the proton (for an interesting account of the model, see Feynman 1972), we need to take a few steps back to the work done in nuclear physics. We wish to describe e–p interactions at *very high* energies, where the possibility arises to 'shatter' the proton into numerous (hadronic) fragments. Such a process is known as *deeply inelastic scattering* (DIS), see figure 3.6. The initial and final electron four-momenta are $\ell^\mu = (E, \ell)$ and $\ell'^\mu = (E', \ell')$ while θ denotes the angle between ℓ and ℓ' in the laboratory frame. The initial proton momentum is $p^\mu = (M, \mathbf{0})$ in the laboratory frame. In the energy regime of interest the energy transferred by the photon is

$$\nu := E - E' \gg M, \tag{3.3.2}$$

---

[13] Much to his surprise as a physicist, Ernest Rutherford was awarded the 1908 Nobel Prize in Chemistry 'for his investigations into the disintegration of the elements, and the chemistry of radioactive substances'. We might note that a similar surprise was reserved for Marie Curie, who was awarded the 1911 Nobel Prize in Chemistry 'in recognition of her services to the advancement of chemistry by the discovery of the elements radium and polonium, by the isolation of radium and the study of the nature and compounds of this remarkable element'. She was, however, also awarded the1903 Nobel Prize in Physics, which was divided, one half awarded to Antoine Henri Becquerel 'in recognition of the extraordinary services he has rendered by his discovery of spontaneous radioactivity', the other half jointly to Pierre Curie and Marie Curie, *née* Skłodowska 'in recognition of the extraordinary services they have rendered by their joint researches on the radiation phenomena discovered by Professor Henri Becquerel.'

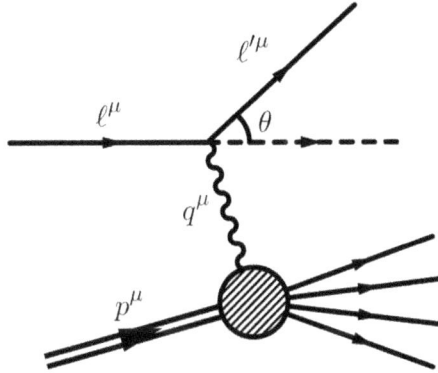

Figure 3.6. The Feynman diagram describing the electron–proton deeply inelastic scattering process.

where M is the nucleon mass. What interests us here is that this is more-or-less the same as the condition that the effective wavelength of the photon be much smaller than the radius of the proton. We thus require a formalism capable of dealing with *inelastic* processes, which, by virtue of the short wavelengths (high energies) involved, take place in very small regions of space and thus probe the small-scale structure of the targets. Let us once again stress that, particularly in the field of particle physics, *high-energy* and *short-distance* are considered synonymous, as too are *low-energy* and *long-distance*; the dimensional translation factor is, as always, $\hbar c \sim 200$ MeV fm.

Exercise 3.9. *Show that the variable ν, defined in equation (3.3.2), can be defined in a Lorentz-invariant manner.*
Hint: consider the scalar four-product $p \cdot q$ in the laboratory frame.

3.3.3 High-energy elastic ep scattering

The starting point will be the simplest form of e–p interaction: namely, elastic scattering. The low-energy case is dealt with in appendix C.2.1. The extension to relativistic energies is then given in appendix C.2.2, where we present the effective Mott formula, applicable when nuclear recoil may be neglected. At higher energies the recoil of the target nucleon is no longer negligible and the reduced Mott formula equation (C.2.6) must be replaced by the full form (C.2.8):

$$\frac{d\sigma}{d\Omega}^{\text{Mott}} = \frac{E'}{E} \frac{d\tilde{\sigma}}{d\Omega}^{\text{Mott}} = \frac{E'}{E}\left(1 - \beta \sin^2 \tfrac{\theta}{2}\right)\frac{d\sigma}{d\Omega}^{\text{Ruther}}, \qquad (3.3.3)$$

where the factor $\frac{E'}{E}$ accounts for the (now non-negligible) recoil effect of the target nucleon. The four-momentum squared of the exchange photon assumes a certain importance and in the high-energy limit, where the electron mass may be safely neglected, we have

$$q^2 = (\ell - \ell')^2 \simeq -4EE' \sin^2 \tfrac{\theta}{2}. \qquad (3.3.4)$$

Since this expression is evidently always negative, it is traditional to introduce the positive variable $Q^2 := -q^2$. This quantity appears in the propagator for the photon and therefore naturally sets the scale for the process. We shall see that in the energy regime of interest *both* ν and Q^2 become large.

As the electron energy increases and its motion becomes ultra-relativistic, it also becomes necessary to include the magnetic interaction (normally suppressed by a factor v/c). For a point-like particle having gyromagnetic ratio exactly two, the full elastic cross-section takes on the form

$$\frac{d\sigma}{d\Omega}^{\text{Dirac}} = \left[1 + 2\tau \tan^2 \tfrac{\theta}{2} \right] \frac{d\sigma}{d\Omega}^{\text{Mott}}, \tag{3.3.5}$$

where the suffix 'Dirac' indicates a point-like cross-section for a spin-half object and the variable $\tau := \frac{Q^2}{4M^2}$. Note that the new term, proportional to $\tan^2(\tfrac{\theta}{2})$, disappears for $\theta = 0°$, reflecting the spin-*flip* nature of the magnetic interaction together with the usual requirement of angular-momentum conservation, coupled to electron-helicity conservation due to the vector nature of the interaction.

However, neither the proton nor the neutron is point-like. Indeed, even the neutron has an appreciable magnetic moment and can therefore scatter high-energy electrons with a cross-section comparable to that of the proton. Furthermore, since we are now moving into a regime where the substructure becomes apparent, we must also take into account both the charge and magnetic-moment *distributions* inside the nucleons. As discussed in appendix C.3, this simply requires the inclusion of form factors, which are nothing other than Fourier transforms of the distributions in question. The cross-section for elastic electron–nucleon scattering thus takes on the Rosenbluth form (Rosenbluth 1950):

$$\frac{d\sigma}{d\Omega}^{\text{Rosen}} = \left[\frac{G_E^2(Q^2) + \tau\, G_M^2(Q^2)}{1 + \tau} + 2\tau\, G_M^2(Q^2)\tan^2 \tfrac{\theta}{2} \right] \frac{d\sigma}{d\Omega}^{\text{Mott}}. \tag{3.3.6}$$

In the limit $Q^2 \to 0$, where the photon wavelength becomes infinite, the scattering becomes effectively point-like and the electric and magnetic form factors $G_E(Q^2)$ and $G_M(Q^2)$ take on their so-called 'static' values:

$$G_E^p(0) = 1, \qquad G_M^p(0) = 2.79, \tag{3.3.7a}$$

$$G_E^n(0) = 0, \qquad G_M^n(0) = -1.91, \tag{3.3.7b}$$

that is, at zero momentum transfer the electric form factor measures the total charge, whereas the magnetic form factor measures the magnetic moment (in units of the *nuclear* magneton). In the absence of a theory for these form factors, the Q^2 dependence must simply be measured experimentally. The dependence on θ allows a two-dimensional plot (Q^2 also depends on E and so is an independent variable), from which the functions $G_{E,M}(Q^2)$ may be extracted separately. For example, note that for Q^2 constant, the ratio $\frac{d\sigma}{d\Omega}^{\text{Rosen}} / \frac{d\sigma}{d\Omega}^{\text{Mott}}$ is linear in $\tan^2 \tfrac{\theta}{2}$, with slope $2\tau G_M^2(Q^2)$; see figure 3.7.

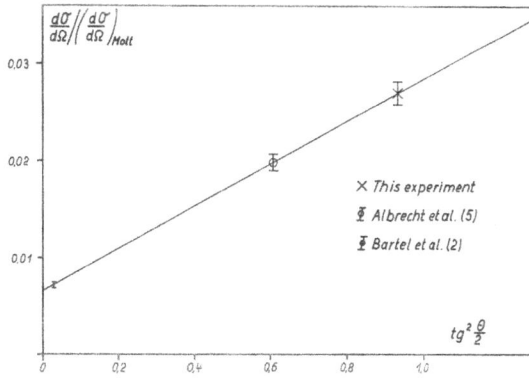

Figure 3.7. An example experimental Rosenbluth plot of $\frac{d\sigma}{d\Omega}^{\text{expt}} / \frac{d\sigma}{d\Omega}^{\text{Mott}}$ as a function of $\tan^2 \frac{\theta}{2}$ for $Q^2 = 2.9\text{GeV}^2$ (reprinted from Bartel et al (1967), copyright (1967) with permission from Elsevier) .

Performing such measurements, the three form factors with a non-zero limiting value are all found to have a dipole-like behaviour (of the form mentioned in appendix C.3), whereas the neutron electric form factor is more difficult both to measure and categorise:

$$G_E^p(Q^2) = \frac{G_M^p(Q^2)}{2.79} = \frac{G_M^n(Q^2)}{-1.91} = \left[1 + \frac{Q^2}{M_V^2}\right]^{-2}, \qquad (3.3.8)$$

where the single, phenomenological, mass parameter is $M_V \simeq 0.84$ GeV. It can be shown that such a dipole form corresponds to an exponentially decaying charge density:

$$\rho(r) = \rho(0)e^{-ar} \text{ with } a = 4.2 \text{ fm}^{-1}. \qquad (3.3.9)$$

Taking the $Q^2 \to 0$ limit of the experimentally measured slope, we deduce a typical root-mean-square radius for the nucleon of approximately 0.8 fm.

3.3.4 Deeply inelastic scattering

As the energy transfer increases, processes other than elastic scattering become possible and inelastic scattering sets in. Since the strict one-to-one constraint of the relation between the outgoing electron energy and scattering angle is then lost, for a fixed detector (or spectrometer) angle, a broad spectrum of energies will be observed. The upper limit is obviously the standard elastic-scattering final-state energy, but many events are seen for energies well below this. In such cases there has evidently been a very large energy transfer. Now, as mentioned in appendix C.3, for Q^2 much larger than the inverse size of the nucleon, the cross-section should fall off very rapidly; i.e. it will have an extra factor $\sim Q^{-4}$ with respect to the natural point-like behaviour. So the natural question now arises: how will the true measured cross-section behave?

The first important structure observed is due to the quadruplet of Δ resonances, with masses around 1230 MeV. On absorbing a photon, the proton is excited into a Δ^+ and the generic subprocess observed is then just $\gamma^* N \to \Delta$; recall that the photon has $I(J^{PC}) = 0, 1(1^{--})$. It is convenient to introduce a new variable W, the final hadronic-state invariant mass. The four-momentum of the hadronic state emerging after photon absorbtion (assuming that nothing else is emitted) is simply $p^\mu + q^\mu$ and the invariant mass sought is thus

$$W^2 := (p + q)^2 = M^2 + 2M\nu - Q^2, \qquad (3.3.10)$$

which is linear in E', recalling that for fixed beam energy and scattering angle with

$$q^2 = -4EE' \sin^2 \frac{\theta}{2}, \qquad (3.3.11)$$

Recall too that, as defined (that is, not specifically as the energy transfer in the rest frame, but as $M\nu = p \cdot q$), ν is actually implicitly a Lorentz invariant, as is of course Q^2. Figure 3.8 displays a typical cross-section or spectrum for $ep \rightarrow eX$ near to the Δ resonance mass peak. Allowing for the underlying inelastic events (e.g. pion production), the resonance has a classic Breit–Wigner shape (for a detailed discussion, the reader is referred to appendix C.6), from which we deduce a mass of a little over 1200 MeV and a width of around 100 MeV. Up to around 2 GeV in W it is possible to identify other resonances, after which the continuum production of multiparticle states starts to dominate.

Exercise 3.10. *Calculate the maximum possible resolution (i.e. the shortest photon wavelength) obtainable using an electron beam of energy 4.879 GeV. For the same beam energy, calculate the maximum value of W.*

The question now is what will happen on moving to higher energies? The situation is very similar to that of Rutherford in the early 1900s: here the proton

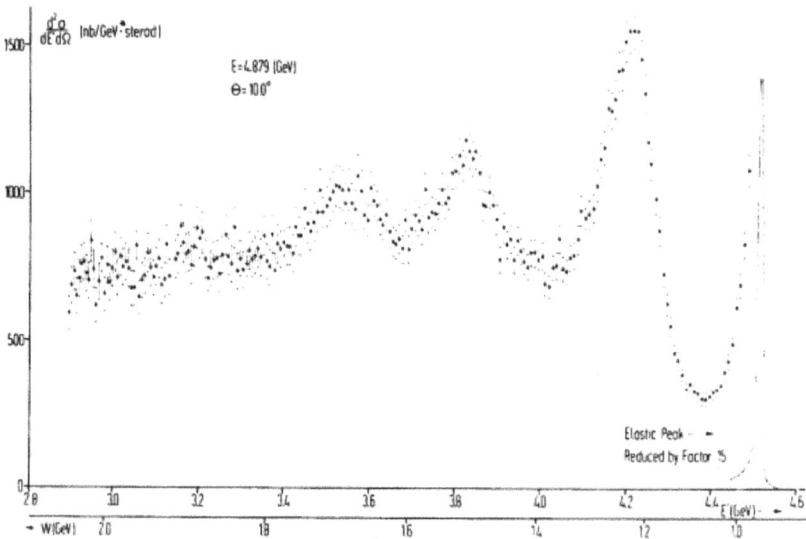

Figure 3.8. The cross-section for $ep \rightarrow eX$ for an electron beam energy $E = 4.879$GeV and scattering angle $\theta = 10°$ (reproduced from Bartel *et al* (1968), copyright (1968) with permission from Elsevier).

has taken over the role of the atom (and the electron that of the α-particle). So, if the proton is just a diffuse sphere of charge, again the large-angle cross-section is expected to fall off very rapidly. Instead the remarkable behaviour found in deeply inelastic scattering (DIS) at the end of the sixties (Bloom *et al* 1969, Breidenbach *et al* 1969) was that, far from dying away as Q^2 increased, as the above form factors would predict, the cross-sections remained large and (up to an overall dimensional scale factor) were independent of Q^2 for fixed ν/Q^2 ratio (see figure 3.9)[14]. Note that such *scaling* behaviour was already observed from about $W = 2$ GeV onwards and is

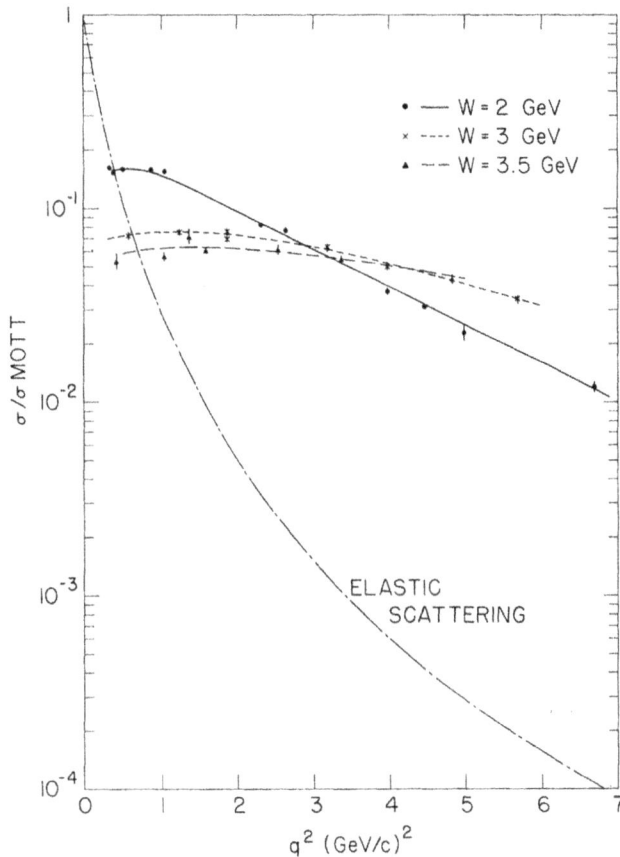

Figure 3.9. The ratio $(d^2\sigma/d\Omega dE')/\sigma^{Mott}$ as a function of q^2 for $W = 2$, 3 and 3.5 GeV in units of GeV^{-1}. Also shown is the cross-section for elastic e–p scattering divided by σ^{Mott}, $(d\sigma/d\Omega)/\sigma^{Mott}$ calculated for $\theta = 10°$ using the dipole form factor. The relatively slow variation of the inelastic cross-section with q^2, as compared to the elastic cross-section, is clearly seen; figure reproduced with permission from Breidenbach *et al* (1969), copyright (1969) by the American Physical Society.

---

[14] The 1990 Nobel Prize in Physics was awarded equally to Jerome I Friedman, Henry W Kendall and Richard E Taylor 'for their pioneering investigations concerning deep inelastic scattering of electrons on protons and bound neutrons, which have been of essential importance for the development of the quark model in particle physics'.

therefore often termed 'precocious'. An early review may be found in Friedman and Kendall (1972).

Before continuing, let us try to understand the *expected* decrease in physical terms and thus better realise the implications of these findings. As the energy of the probe increases, the wavelength of the exchange photon decreases. At the instant it interacts the photon effectively coherently 'sees' only a small volume inside the nucleon, given roughly by the photon wavelength. Therefore, if the charge of the nucleon is distributed more-or-less uniformly throughout its volume, the interaction strength is expected to decrease rapidly with increasing energy. Since this does not happen, we are forced (just as was Rutherford) to entertain the idea that there are small, dense (or rather point-like) objects inside the nucleon, in which its charge is concentrated, thus avoiding the decreasing cross-section.

3.3.5 Bjorken scaling

For inelastic scattering there are two independent variables, which historically were taken as Q^2 and ν, and in place of the Rosenbluth formula (3.3.6) we write

$$
\begin{aligned}
\frac{d^2\sigma}{d\Omega dE'} &= \left[W_2(Q^2, \nu) + 2W_1(Q^2, \nu)\tan^2 \frac{\theta}{2} \right] \frac{d\tilde{\sigma}}{d\Omega}^{\text{Mott}} \\
&= \frac{4\alpha^2 E'^2}{(Q^2)^2} \left[W_2(Q^2, \nu)\cos^2 \frac{\theta}{2} + 2W_1(Q^2, \nu)\sin^2 \frac{\theta}{2} \right].
\end{aligned}
\tag{3.3.12}
$$

That is, in place of the form factors $G_{1,2}$ we now have the structure functions $W_{1,2}$. However, as noted, the data are well described as a function of a single variable. This had already been largely foreseen by Bjorken (1969) essentially via dimensional analysis. He argued that since at high energies (and it must be admitted that he had rather more the just a few GeV in mind) all relatively *small* masses and/or energy-like parameters could be neglected. Therefore, once the *naïve* dimensions of, say, a cross-section (e.g. the E'^2/Q^4 pre-factor in the above formula) had been factored out, the remaining *dimensionless* form factors could only depend on *dimensionless* variables.

In DIS at high energies, for example, unless there is a new scale due to some new physics or dynamics, only two large quantities with dimensions of energy or mass remain important: namely, Q^2 and ν. We can only construct the adimensional so-called Bjorken scaling variable x_B, which satisfies the following kinematical constraints:

$$
0 \leqslant x_B := \frac{Q^2}{2M\nu} \equiv \frac{Q^2}{2p \cdot q} \leqslant 1.
\tag{3.3.13}
$$

It is thus an ideal candidate as the variable against which to plot the data. It might be mentioned here that Bjorken (1969) himself used $x := Q^2/M\nu$, which differs by a factor two. Indeed, in the early literature the inverse $\omega := M\nu/Q^2$ is also found while many authors (including Bjorken himself) still often simply used ν as the independent kinematic variable. We shall soon see that in the Feynman picture (1969) x_B also has a very special meaning. In 1969 Bjorken showed that the correct scaling behaviour is then obtained via the following substitutions:

$$MW_1(Q^2, \nu) \rightarrow F_1(x_B), \tag{3.3.14a}$$

$$\nu W_2(Q^2, \nu) \rightarrow F_2(x_B). \tag{3.3.14b}$$

So there are now two apparently *different* approaches (Bjorken and Feynman), which nevertheless lead to exactly the same result. It is perhaps easier to understand the connection between Bjorken scaling and the Feynman parton picture by considering space–time arguments. As always, the resolving power of the probe is determined by the corresponding de Broglie wavelength, which is to be compared with the other length scales typical of the problem. In the case of electron–proton scattering, there is only the size of the proton (or its inverse mass) and thus, if the scattering is coherent and constructive over only a small fractional volume of distributed charge inside the proton, we should expect the typical rapid fall-off of the old Thomson plum-pudding model of the atom. If the scattering were instead off effectively point-like objects (partons), then, on dimensional grounds, there could be no such effect; i.e. a point has no length scale. In terms of energy, this translates into no possible high-energy fall-off, if there is no other relevant energy scale.

For completeness, we shall now present the DIS cross-section formulæ in terms of the form factors $F_{1,\,2}(x_B)$ or *structure functions* (as they are now known). First of all though, one final useful Lorentz-invariant kinematical variable must be introduced:

$$y := p \cdot q / p \cdot k \quad (0 \leqslant y \leqslant 1), \tag{3.3.15}$$

which, in the nucleon rest frame, is just the fractional energy lost by the incoming electron, ν/E. We also introduce the Mandelstam variable $s := (p + \ell)^2 \simeq 2ME$ (see section C.1.1). The cross-section for charged-lepton ($\ell = e, \mu$) scattering off a nucleon ($N = p, n$) target is

$$\frac{\partial \sigma^{\ell N}}{\partial x \partial y} = \frac{4\pi\alpha^2}{sx^2y^2} \left[xy^2 \, F_1^{\ell N}(x, Q^2) + (1 - y - x^2y^2M^2/Q^2) \, F_2^{\ell N}(x, Q^2) \right]. \tag{3.3.16}$$

This form for the cross-section is dictated by consistency with Lorentz invariance and parity conservation, but assumes that possible terms proportional to quark masses (actually m_q^2/Q^2) may be neglected[15]. In the case of neutrino scattering via parity-violating charged-current ($W^\pm$-boson) exchange, a further structure function, F_3, appears and we thus have

$$\frac{\partial \sigma^{\nu N}}{\partial x \partial y} = \frac{G_F^2 \, s}{2\pi(1 + Q^2/M_W^2)^2} \left[xy^2 \, F_1^{\nu N}(x, Q^2) + (1 - y - x^2y^2M^2/Q^2) \, F_2^{\nu N}(x, Q^2) \right.$$
$$\left. + xy(1 - y/2) \, F_3^{\nu N}(x, Q^2) \right]. \tag{3.3.17}$$

---

[15] This approximation is well justified by the very small masses of the up, down, and strange quarks ($m_u \sim 2.2$ MeV, $m_d \sim 4.8$ MeV, and $m_s \sim 93.4$ MeV, see Navas *et al* 2024). Note that these are the masses of what are known as *current quarks*, *vis à vis* the *constituent quarks* of Gell-Mann; that is, they refer to 'bare' quarks not yet 'dressed' by the gluon field.

For $\bar{\nu}N$ scattering, the sign of the last term (F_3) is inverted. The two preceding formulæ hold for unpolarised charged leptons and target nucleons, whereas the neutrino is, of course, always polarised owing to parity violation.

If both the charged-lepton beam and nucleon target are polarised along the beam direction, a measurable asymmetry may be defined:

$$A(x, Q^2) := \frac{\mathrm{d}\sigma^{\ell N}_{++} - \mathrm{d}\sigma^{\ell N}_{+-}}{\mathrm{d}\sigma^{\ell N}_{++} + \mathrm{d}\sigma^{\ell N}_{+-}}, \tag{3.3.18}$$

where the suffixes $++$ and $+-$ represent the lepton polarisation fixed parallel to the beam $(+)$ and the nucleon polarised parallel $(+)$ or antiparallel $(-)$. The cross-section difference in the numerator[16] is then

$$\frac{\partial\sigma^{\ell N}_{++}}{\partial x \partial y} - \frac{\partial\sigma^{\ell N}_{+-}}{\partial x \partial y} = \frac{8\pi\alpha^2 y}{MQ^2} \Big[(1 - 2/y^2 + 2x^2y^2M^2/Q^2)\, G^{\ell N}_1(x, Q^2)$$
$$+ 4x^2(M^2/Q^2)\, G^{\ell N}_2(x, Q^2) \Big], \tag{3.3.19}$$

where two new spin-dependent structure functions $G^{\ell N}_1$ and $G^{\ell N}_2$ [17] have been introduced.

3.3.6 The Feynman picture

It was, however, Feynman (1969) who gave more specific meaning to x_B and the *structure functions* $F_{1,2}$. Assuming that there were point-like spin-half objects (which he called *partons*) inside the nucleon and that it was with these that the high-energy electromagnetic probe interacted, Feynman calculated the resulting cross-section, much in the same fashion as in appendix C.4 for *quasi-elastic* scattering (see figure 3.10). The parton approach also leads to a simplification: the two *a priori* unrelated structure functions $F_{1,2}$ turn out to be a single function (because the

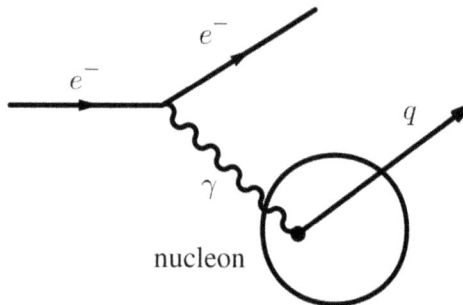

Figure 3.10. The electron–nucleon deeply inelastic scattering process according to the Feynman parton-model picture (the blob inside the circle represents the struck parton inside the nucleon).

---

[16] The denominator in the asymmetry is just equation (3.3.16).

[17] These are not to be confused with the original Rosenbluth form factors of equation (3.3.6).

partons are treated as elementary Dirac particles with gyromagnetic ratio 2)[18]. Indeed, Curtis Callan and David Gross (1969) quickly showed the following relation to hold in the high-energy so-called *scaling limit*:

$$F_2(x) = 2xF_1(x). \tag{3.3.20}$$

It is important to realise that this and indeed the entire picture is to be taken as an approximation that would be precise only in the limit $Q^2 \to \infty$, where so-called higher-power corrections, behaving as $(M^2/Q^2)^n$ with $n \geqslant 1$, die out. In fact, it turns out that higher-order quantum corrections ($\propto \alpha_s^n$) also spoil the simple picture, although they may be incorporated in a systematic way (which will be explained later) to provide precise and experimentally verifiable numerical predictions.

Let us now try to understand the physical meaning of x_B and the structure functions that depend on it. It is natural (following the quasi-elastic scattering analogy) to attempt to describe the process as a collision between a virtual photon and a parton bound inside a nucleon. Indeed, the lepton–photon vertex may be considered as so thoroughly understood as to be of no interest here. The only obstacle is that the partons are considered massless (or as having negligibly small masses) and thus their rest-frame is ill defined. It is therefore necessary to work in a relativistically boosted frame. There are many possible choices of so-called *infinite-momentum frames*, but that which lends itself best to the present purpose is the Breit (or so-called 'brick-wall') frame, in which the struck parton turns through 180° while retaining the same energy and absolute momentum, i.e. it simply reverses its direction of motion, as if it had indeed collided with a brick wall (see figure 3.11). In this frame the photon evidently carries zero energy and q^μ is therefore purely space-like. We shall also take the z-axis as the direction of the proton in this frame (the photon momentum is thus negative). Now, since $x_B := Q^2/2p \cdot q$ and $q^0 = 0$, then $q_z = -2x_B p_z$. To reverse the direction of the incoming parton (with momentum k, say) evidently requires $q_z = -2k_z$. We thus finally obtain

$$x_B = k_z/p_z. \tag{3.3.21}$$

This is then the famous parton-model relation: x_B is just the fraction of the proton momentum carried by the parton as seen in an infinite-momentum frame.

Figure 3.11. A *pseudo*-Feynman diagram depicting a head-on photon–quark collision in the Breit (or brick-wall) frame.

---

[18] The justification for treating the (unknown, unseen) partons as elementary and point-like is not clear at this stage; nor can the impulse approximation be justified yet. These are thus to be taken as merely working hypotheses.

Exercise 3.11. *Derive the above relation explicitly and thus demonstrate that the adimensional Bjorken scaling variable x_B, as defined in equation (3.3.13), is bounded to lie in the range* [0, 1].

Exercise 3.12. *Repeat the above exercise (again explicitly) for a general infinite-momentum frame (i.e. one in which the proton is not at rest and has large momentum) and thus demonstrate that the relation is of more general validity.*

We now have a picture in which a parton carrying a fraction x_B of the parent-hadron momentum collides with an electron at very high energy and with very large momentum transfer. It may thus be reasonable (though more on this later) to avail ourselves of the impulse approximation already used in describing quasi-elastic scattering. In such an approach the deeply inelastic scattering e–p cross-section for a given x_B can be calculated as simply the product of the probability of finding a parton with that momentum fraction and the cross-section for its *elastic* scattering with the incoming electron.

In other words, the photon–proton part of the DIS cross-section may be rewritten as

$$\hat{\sigma}^{\gamma^* p}(q, p) = \sum_i \hat{\sigma}^{\gamma^* q_i}(q, x_B p) \times f_i(x_B), \tag{3.3.22}$$

where i labels the ith quark type or flavour (u, d, s, $\bar{u}$, ...) and $f_i(x_B)$ is thus defined to be the probability of finding that quark with momentum fraction x_B. The left-hand side is parametrised in terms of the structure functions $F_{1,2}(x_B)$, while on the right-hand side the parton cross-section is just the relevant Mott expression for point-like massless elastic scattering[19].

Comparison of the two expressions immediately leads to the identification of the structure functions with probability distributions or densities for partons:

$$\frac{1}{2x} F_2(x) = F_1(x) = \frac{1}{2}\sum_i Q_i^2 \left[f_i(x) + \bar{f}_i(x) \right], \tag{3.3.23}$$

where the sum runs over the different types or flavours of partons that might be found inside a hadron, Q_i is the charge (in units of the proton charge) of the ith. parton type, f_i its probability distribution or density with respect to the momentum fraction x and $\bar{f}_i$ that of the corresponding antiparton. From here on, for simplicity of notation and according to accepted convention, we shall usually drop the suffix B on the Bjorken variable x_B. Although obvious, it should perhaps also be stressed that x here has nothing whatsoever to do with spatial position or configuration space.

Antiquarks are included in the sum over parton types since the spontaneous quark–antiquark production, as predicted by quantum field theory, implies that at

---

[19] Note that such a simple factorisation into a product of a point-like high-energy cross-section and a non-perturbative parton distribution is highly non-trivial; the theoretical justification is a very important proof in the application of perturbative QCD.

any given instant in time a hadron will also contain some (albeit small) fraction of antiquarks. Note that the cross-section is only sensitive to the charge squared Q_i^2 and therefore the contributions of partons and antipartons are indistinguishable. At this point we may as well start calling Feynman's partons *quarks*[20]. We shall at times still continue to use the term *parton* since it may be taken to refer to *any* constituent of the proton, neutron or other hadrons; as we shall see shortly, there are also the gluons (the carriers of the strong interaction) to consider.

In the case of polarised DIS, as described in equation (3.3.19).

$$G_1(x) = \frac{1}{2}\sum_i Q_i^2 [\Delta f_i(x) + \Delta \bar{f}_i(x)], \qquad (3.3.24)$$

where the spin-dependent quark densities $\Delta f_i(x)$ are given by the difference of the density for a quark polarised parallel and antiparallel to the parent polarisation (see e.g. Anselmino *et al* 1995). The structure function $G_2(x)$ is related to transverse-polarisation effects and has no simple parton interpretation (see e.g. Barone *et al* 2002).

3.3.7 Difficulties with the Feynman approach

A number of (actually rather deep) questions now arise, which absolutely beg clarification:

1. parton transverse-momentum effects have been ignored,
2. no theory of the distribution $f(x)$ has been given,
3. gluons have not been included,
4. the destiny of the struck quark is not specified,
5. binding-energy effects have been ignored.

We shall deal with the second in some detail later. We shall also try to provide some understanding of the fourth and fifth shortly; they are related but are also somewhat more complex and profound issues. Let us begin, however, with a few brief comments.

3.3.7.1 *Parton transverse-momentum effects*

Let us comment first on the momentum components of the quark in the plane transverse the z-axis (defined by the proton–photon directions in their centre-of-mass frame). In the above treatment they were totally ignored, this may be justified by assuming that they are due to Fermi motion of the quarks inside the nucleon, which is restricted to low momenta. According to the familiar *uncertainty principle* due to Werner Heisenberg (1927)[21], the mean (internal) momentum will be of the order of

---

[20] 'In that work [Feynman] referred to quarks, antiquarks, and gluons, of which they were made, but he didn't call them quarks, antiquarks, and gluons. He called them 'partons', which is a half-Latin, half-Greek, stupid word. Partons. He said he didn't care what they were, so he made up a name for them. But that's what they were: quarks, antiquarks, and gluons, and he could have said that.' (Excerpt from an interview with Murray Gell-Mann, Kruglinski 2009.)

[21] The 1932 Nobel Prize in Physics was awarded to Werner Heisenberg 'for the creation of quantum mechanics, the application of which has, *inter alia*, led to the discovery of the allotropic forms of hydrogen'.

the inverse size of the nucleon in which they are bound and thus presumably of order 200 MeV. Moreover, transverse components are unaffected by longitudinal boosts. As a first approximation this is just fine although there are certain circumstances where the rôle of transverse momenta is non-negligible.

3.3.7.2 Parton distributions

As to a theory for the quark–parton distributions $f(x)$, it simply does *not exist* as yet. The bound-state problem in QCD is still far from being solved and although various theoretical techniques and models have been developed, none provides truly satisfactory solutions for the bound states of three quarks or quark–antiquark pairs and certainly no approach is sufficiently advanced to provide complete *ab initio* calculations of the parton densities. We shall though discuss this question later and explain what can be deduced *a priori* and what can be obtained through experimental measurement.

3.3.7.3 Gluons

As far as gluons are concerned, being electrically neutral, they cannot contribute to DIS at leading order in QCD. However, at higher order in perturbation theory, a gluon inside the proton may spontaneously split into a quark–antiquark pair, of which one or other may then interact with the photon. Such effects can be calculated and included systematically in phenomenological descriptions. In fact, it is found that approximately only half the proton momentum is carried by quarks and antiquarks (these last carry only some 6% or so), the other half being associated with gluons.

Moreover, as we shall soon see, gluons carry the *colour* charge of QCD and therefore interact not only with quarks but also with each other. This means that in hadron–hadron collisions there may be a parton-level process in which two gluons collide and fuse to produce, e.g. a quark–antiquark pair. Indeed, at high energies such gluon–gluon fusion processes can even dominate the scattering cross-section. Yet another possibility is a gluon–quark Compton-like process.

3.3.7.4 The struck-quark destiny and binding

The remaining two points are conceptually much tougher. Especially in view of the fact that, no matter how high we go in energy experimentally, it has (so far) proved *impossible* to liberate a quark from inside its host hadron.

That is, the Feynman description implicitly assumes that the struck quark is expelled from the parent nucleon; the *naïve* kinematics of the process would suggest the picture shown in figure 3.12. This immediately raises the question of the fate of the struck quark, as it apparently never effectively emerges in the laboratory.

Somehow then, despite the large energy deposited on the quark, it is so strongly bound inside the hadron that it cannot in fact escape. The problem of binding leads us to two concepts that are central to the theory of strong interactions: namely, *confinement* and *asymptotic freedom*. Indeed, so important are they that they deserve a dedicated section. However, this will be postponed until we have described at least a little of the nature of the interaction involved, namely QCD.

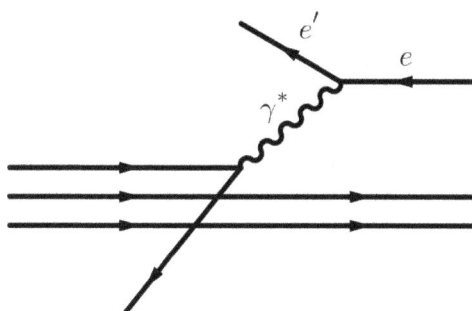

Figure 3.12. The full picture of the electron–nucleon deeply inelastic scattering process in which the struck quark moves off in a different direction with respect to the remnant spectator quarks, thus determining the breakup of the nucleon.

3.3.7.5 Feynman's picture and Bjorken scaling

Before moving on to the fundamental theory of the strong interaction, there is one final issue we should address. The more inquisitive reader may now be raising the natural question of the relationship between Bjorken scaling and the parton-model description of Feynman: that is, how is it that Bjorken's (*naïve*) dimensional analysis arrived at the same phenomenological description (i.e. the Q^2-scaling behaviour of the structure functions) as Feynman's detailed dynamical construction, despite making no reference to any possible underlying particles or dynamics in the high-energy deeply inelastic processes considered?

The crux of the matter is that Feynman's partons (as distinct from Gell-Mann's quarks) are specifically posited as being point-like (i.e. they are true Dirac fermions) and, insofar as they are assumed to have at most negligible masses, they therefore do not set any scale (length, energy or otherwise). Such a lack of any intrinsic scale is precisely the hypothesis from which Bjorken arrived at his scaling conclusions. Had, for example, Feynman's partons not been point-like, then their size (d say) would have set an energy scale ($\hbar/d$), which would have allowed a non-trivial energy dependence to enter the structure functions. In other words, the Feynman picture would not have described the observed scaling behaviour. Of course, had they been very small (e.g. $\ll 1$ fm), then at the energy scales of the early experiments they could still have been considered effectively point-like.

3.4 Quantum chromodynamics

The attempt to construct a fundamental theory of the strong interaction took as its starting point the already well-established QED. This is the field-quantised version of the classical field theory of electrodynamics, a gauge theory (the photon is rigorously massless) and as such possessing an important symmetry, which guarantees suitable high-energy behaviour under quantisation. However, QCD differs from QED in that the gauge structure has a much richer non-Abelian symmetry, i.e. it is of the Yang–Mills type (Yang and Mills 1954), an already existing 'toy model'. Apart from various theoretical complications, this leads to a very important difference: there is not just one single charge but three. By analogy with the primary colours in optics, these are traditionally identified as red, blue and green. A quark may thus carry any one of these three charges, whereas an antiquark carries an anti-colour charge and, as we shall explain, the (massless) gluon is also coloured.

Let us briefly explain what is meant by a Yang–Mills (Yang and Mills 1954) or *non-Abelian* gauge theory. Such a theory is an extension of the standard *Abelian* gauge symmetry defined as invariance under the following transformation:

$$\psi(x) \rightarrow e^{-ig\Lambda(x)}\psi(x) \tag{3.4.1a}$$

and

$$A^{\mu} \rightarrow A^{\mu} + \partial^{\mu}\Lambda(x), \tag{3.4.1b}$$

for any scalar function $\Lambda(x)$. If the fermion field ψ possesses a global (internal) symmetry associated with some new charge that is multivalued, then the local version of such a symmetry is defined by

$$\psi_i(x) \rightarrow e^{-ig\,\Lambda^a(x)\,T^a_{ij}}\psi_j(x), \tag{3.4.2a}$$

and

$$A^{a\mu} \rightarrow A^{a\mu} + \partial^{\mu}\Lambda^a(x) + gf^{abc}\Lambda^b(x)A^{c\mu}(x), \tag{3.4.2b}$$

where there are now N arbitrary functions $\Lambda^a(x)$. The f^{abc} are the structure constants of the symmetry group, here SU(N), and the T^a_{ij} ($i, j = 1, \cdots, N$ and $a = 1, \cdots, N^2 - 1$) belong to the fundamental representation, satisfying the defining algebra

$$[T^a, T^b] = if^{abc}T^c. \tag{3.4.3}$$

As in the Abelian case, the $N^2 - 1$ mediating gauge fields $A^{a\mu}$ are strictly massless, but are now carriers of the associated charge (colour) and may thus interact among themselves. This self-interaction leads to a highly non-trivial structure of the QCD vacuum and the non-linearity of theory renders it impervious, with present techniques, to complete solution. In particular, the bound-state problem is as yet unsolved.

For completeness, we now present the QCD Lagrangian. QCD is a local gauge theory based on the symmetry group SU(3)$_{\text{col}}$ with colour-triplet quark matter fields, this fully determines the QCD Lagrangian density to be

$$\mathcal{L} = -\frac{1}{4}\sum_{a=1}^{8}F^{a\mu\nu}F^a_{\mu\nu} + \sum_{j=1}^{N_f}\bar{q}_j(i\not{D} - m_j)q_j, \tag{3.4.4}$$

where q_j are quark fields, with N_f different flavours and masses m_j; the index a runs over the eight gluon 'colours'; $\not{D} = \gamma^{\mu}D_{\mu}$, where γ^{μ} are the Dirac matrices and D_{μ} is the covariant derivative (its form being dictated by the *principle of minimal coupling*):

$$D_{\mu} := \partial_{\mu}1 - ig\mathbf{A}_{\mu}, \tag{3.4.5}$$

and g is the gauge-field–matter coupling constant. The gauge vector-potential matrix is defined as $\mathbf{A}_\mu := \sum_a t^a A_\mu^a$, where A_μ^a ($a = 1, 8$) are the gluon fields and t^a are the SU(3)$_{\text{col}}$ group generators in the triplet representation of quarks (i.e. t_a are 3×3 matrices acting on the q flavour index). As above, the generators obey the commutation relations $[t^a, t^b] = if^{abc} t^c$, where f^{abc} are the completely antisymmetric structure constants defining SU(3)$_{\text{col}}$; the normalisation of f^{abc} and of g is fixed by the relation $\text{Tr}[t^a t^b] = \frac{1}{2}\delta^{ab}$; and the gluon field-strength tensor is

$$F_{\mu\nu}^a = \partial_\mu A_\nu^a - \partial_\nu A_\mu^a - gf^{abc}A_\mu^b A_\nu^c. \qquad (3.4.6)$$

In analogy with QED, it is convenient to define a strong-interaction fine-structure constant:

$$\alpha_s := \frac{g^2}{4\pi}. \qquad (3.4.7)$$

3.4.1 Motivation for colour SU(3)

Let us now examine how such a theory came to be considered. First of all though, there must be no confusion between SU(3)$_{\text{flav}}$ and SU(3)$_{\text{col}}$; the first refers to an approximate *global* symmetry of the u, d, s etc quark fields, irrespective of their interactions, and contains *no* real dynamics, whereas the second is the exact *local* gauge symmetry of the strong interaction (i.e. the quark–gluon dynamics).

The choice of SU(3)$_{\text{col}}$ as the strong-interaction gauge group (Greenberg 1964, Han and Nambu 1965) is uniquely determined by a number of phenomenological and theoretical observations (see Muta 1998, for example). Note that the following do not constitute a requirement regarding the interaction, but merely indicate the nature of the necessary *global* symmetry group. However, it is natural, following in the footsteps of the highly successful theory of QED, to extend it to a *local gauge* symmetry and thus introduce a very desirable interaction. For a variety of reasons (which we shall now discuss), it is thus necessary to enlarge the symmetry group beyond the simple single-parameter space of the QED U(1) to SU(3)$_{\text{col}}$.

(a) The group must admit a totally antisymmetric colour-singlet ('white') baryon composed of three quarks, qqq. Note that states with, e.g., four quarks have never been observed. From the study of hadron spectroscopy it is known that the lowest-mass baryons, the spin-½ octet and the spin-$3/2$ decuplet of SU(3)$_{\text{flav}}$ (the approximate *flavour* symmetry that rotates the three light quarks u, d and s), are composed of three quarks in what are assumed to be colour-singlet states. Indeed, the qqq wave-function must be antisymmetric in colour, in order to satisfy Fermi–Dirac statistics. Consider, for example, a Δ^{++} with spin-z component $+3/2$: this has the form $|u^\Uparrow u^\Uparrow u^\Uparrow\rangle$ in an s-wave (likewise, Ω^- should be $|s^\Uparrow s^\Uparrow s^\Uparrow\rangle$ with $L = 0$), i.e. three identical fermions in the same state. In space, spin and flavour the wave-function is thus totally symmetric and hence antisymmetry in colour is required for overall antisymmetry. This requirement is neatly satisfied by

$SU(3)_{col}$ and the natural construct $\epsilon_{abc}\, q^a q^b q^c$, where a, b and c are $SU(3)_{col}$ indices.

(b) The group structure must admit complex representations in order to distinguish between quarks and antiquarks. In fact, there exist $q\bar{q}$ mesonic states, whereas no analogous qq bound states are known. Among the simple groups, this restricts the choice to E_6, $SU(N)$ with $N \geqslant 2$ and $SO(4N+2)$ with $N \geqslant 2$, taking into account that $SO(6)$ has the same algebra as $SU(4)$.

(c) The choice of the gauge group $SU(N_c = 3)_{col}$ is also confirmed *a posteriori* by many processes that directly or indirectly *measure* N_c. We shall now present some important examples.

3.4.1.1 The hadron-production rate in e^+e^- annihilation

The e^+e^- annihilation process proceeds via the production of a virtual intermediate boson (γ or Z^0), which then 'decays' into a fermion–antifermion pair (see figure 3.13). These may be charged leptons or quarks (neutrinos are also possible in the case of the Z^0). The overall rate for any given channel is proportional to the charge squared of the final-state fermion. Final states containing only hadrons are assumed to have their origins in a $q\bar{q}$ pair. Counting a separate contribution for each quark colour, the total rate for hadron production in e^+e^- annihilation is thus proportional to N_c:

$$R_{e^+e^-} := \frac{\sigma(e^+e^- \to \text{hadrons})}{\sigma_{\text{point}}(e^+e^- \to \mu^+\mu^-)} = N_c \sum_f Q_f^2 \qquad (\text{for} \quad 2m_f < E_{CM}),$$

where the sum runs over individual contributions (weighted by Q_f^2, the quark electric charge squared) from accessible $q_f \bar{q}_f$ final states. Important, known, quantum corrections have been neglected here (they will be discussed later).

Above the $b\bar{b}$ threshold but well below m_Z we have $q_f = u$, c, d, s and b (t is, of course, too heavy):

$$R_{e^+e^-} \approx \left[2 \times \left(\tfrac{2}{3}\right)^2 + 3 \times \left(-\tfrac{1}{3}\right)^2 \right] N_c = \tfrac{11}{9} N_c. \tag{3.4.8}$$

The data nicely indicate $N_c = 3$, as seen from figure 3.14 (Navas *et al* 2024). Note that the cross-section excess in the data of a few percent with respect to the value $^{11}/_3$ can be accounted for by the QCD radiative (or quantum) corrections mentioned above.

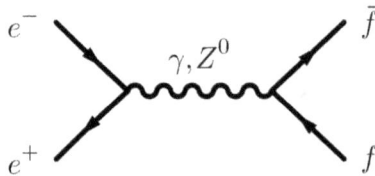

Figure 3.13. The e^+e^- annihilation process into a fermion–antifermion pair, via the intermediate production of a virtual photon or Z^0 boson.

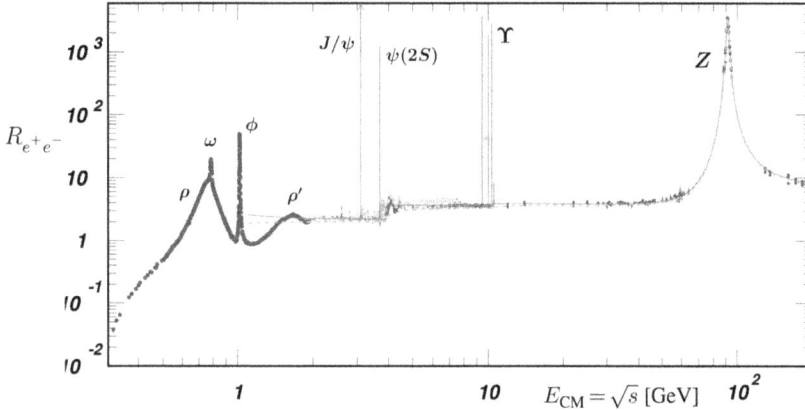

Figure 3.14. $R_{e^+e^-}$ as a function of total centre-of-mass energy $E_{CM} = \sqrt{s}$. The rises in the baseline just above the charm and bottom thresholds are clearly visible; figure reproduced with permission from Navas *et al* (2024). CC BY 4.0.

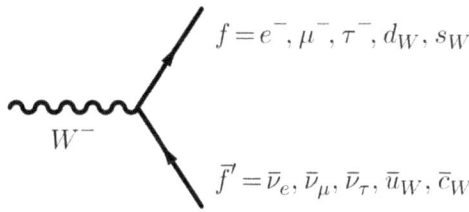

Figure 3.15. The general $W^- \rightarrow f \bar{f}'$ decay process (the charge-conjugate channels naturally also exist for the W^+).

3.4.1.2 The branching ratio $B(W^- \rightarrow e^- \bar{\nu}_e)$

A similar example is provided by the W^- decay rate (see figure 3.15). Again, in the Born approximation, the allowed fermion–antifermion ($f\bar{f}'$) final states in W^- decay are $e^-\bar{\nu}_e$, $\mu^-\bar{\nu}_\mu$, $\tau^-\bar{\nu}_\tau$, $(d\bar{u})_W$ and $(s\bar{c})_W$; the final state $(b\bar{t})_W$ is not possible since the top quark is too heavy to be produced. Quark and lepton mass effects (m_c and m_τ) may be more-or-less neglected here and so each channel type (lepton or quark) contributes equally, except that for quarks there are N_c colours and an extra weight factor $N_c = 3$ must therefore be included for the quark channels:

$$B(W^- \rightarrow e^-\bar{\nu}_e) \equiv \frac{\Gamma(W^- \rightarrow e^-\bar{\nu}_e)}{\Gamma(W^- \rightarrow \text{all})} \approx \frac{1}{3 + 2N_c}. \tag{3.4.9}$$

For $N_c = 3$, $B = 11\%$ (it would be 20% for $N_c = 1$); to be compared with the experimental value $B = 10.75 \pm 0.13\%$.

3.4.1.3 The branching ratio $B(\tau^- \rightarrow e^- \bar{\nu}_e \nu_\tau)$

The τ lepton, having a mass of very nearly 1777 MeV, may decay into a number of final states, both leptonic and hadronic. The basic process $\tau \rightarrow f \bar{f}' \nu_\tau$, obviously an analogue of β-decay, is depicted in figure 3.16. Considering the energetically available final-state channels, the $f\bar{f}'$ pair may be $e^-\bar{\nu}_e$, $\mu^-\bar{\nu}_\mu$, or $(d\bar{u})_W$. In principle, neglecting the small mass effects, each should contribute with equal weight. However, if the quarks are coloured, the number of $(d\bar{u})_W$ states available becomes $N_c = 3$. The branching ratio $\mathcal{B}(\tau^- \to e^-\bar{\nu}_e\nu_\tau)$ is then

$$\mathcal{B}(\tau^- \to e^-\bar{\nu}_e\nu_\tau) \equiv \frac{\Gamma(\tau^- \to e^-\bar{\nu}_e\nu_\tau)}{\Gamma(\tau^- \to \text{all})} \approx \frac{1}{2 + N_c}. \tag{3.4.10}$$

For $N_c = 3$, $\mathcal{B} = 20\%$ (it would be 33% for $N_c = 1$), whereas the experimental number is $\mathcal{B} = 17.84 \pm 0.05\%$ (the poorer agreement in this case is explained by the larger QCD radiative corrections since the mass of the τ^- is small and thus α_s large, see later).

3.4.1.4 The Drell–Yan process

At leading order in QCD, the rate for Drell–Yan processes (1970), e.g. $pp \to \mu^+\mu^-X$, is *inversely* proportional to N_c. Such a process proceeds via $q\bar{q}$ annihilation into a virtual (massive) photon, which subsequently decays into a $\mu^+\mu^-$ pair (see figure 3.28). Thus, for example, a quark of any given colour in one hadron must find an antiquark of the *same* colour in the other and hence only $1/N_c$ of the cases may actually proceed.

3.4.1.5 The decay rate $\Gamma(\pi^0 \to 2\gamma)$

The decay rate $\Gamma(\pi^0 \to 2\gamma)$ is quadratic in N_c; the amplitude is depicted in figure 3.17. It may be reliably calculated via so-called soft-pion theorems and is related to the so-called *chiral anomaly*:

$$\Gamma(\pi^0 \to 2\gamma) \approx \left(\frac{N_c}{3}\right)^2 \frac{\alpha^2 m_{\pi^0}^3}{32\pi^3 f_\pi^2} = (7.73 \pm 0.04) \times \left(\frac{N_c}{3}\right)^2 \text{eV}, \tag{3.4.11}$$

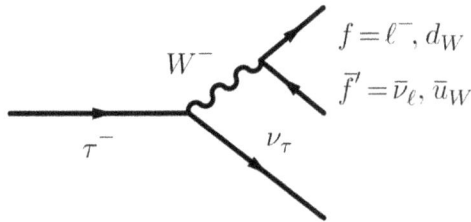

Figure 3.16. The general $\tau^- \to \nu_\tau f\bar{f}'$ decay process; ℓ^- stands for either of the two lighter charged leptons e^- or μ^- (the natural charge-conjugate channels also exist for the τ^+).

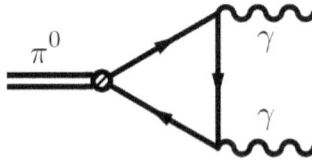

Figure 3.17. The quark description of decay $\pi^0 \to 2\gamma$. Given the natural mass scale involved (m_π), the internal fermion loop only involves the light quarks u, d and s.

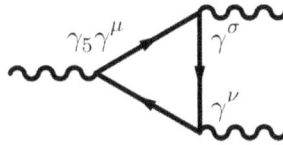

Figure 3.18. The triangle diagram contributing to the ABJ anomaly; the loop contains all the leptons and all the quarks.

where $f_\pi = (130.7 \pm 0.37)$ MeV is the charged-pion decay constant. The measured experimental value is $\Gamma = (7.7 \pm 0.5)$ eV, in good agreement with $N_c = 3$.

3.4.1.6 Cancellation of the Adler–Bell–Jackiw (ABJ) triangle anomaly
Another rather more technical problem, but related to the previous case, is that of the ABJ (Adler 1969; Bell and Jackiw 1969) so-called *triangle anomaly*. At a classical level electrodynamics possesses a U(1) vector symmetry, but the quantum field theory also has a U(1) axial-vector symmetry; i.e. an additional invariance with respect to transformations under γ_5. The associated axial-vector current is analogous to that which contributes to the Gamow–Teller transitions in nuclear β-decay. However, the one-loop triangle graph with two vector vertices and one axial-vector vertex breaks this symmetry; the axial-vector current is therefore not conserved (see figure 3.18). The anomaly contribution is proportional to the charge of the fermion circulating in the loop and is thus: -1 for the charged leptons, zero for neutrinos, $+2/3$ for up-type quarks and $-1/3$ for down-type quarks. Therefore, if and only if there are precisely three colours of quark, each quark–lepton family contributes with an overall coefficient

$$-1 + 0 + 3 \times \left(\tfrac{2}{3} - \tfrac{1}{3}\right) = 0. \tag{3.4.12}$$

3.4.2 QCD at high and low energies

That quarks (and gluons) appear to be inextricably bound inside hadrons (baryons and mesons), i.e. that they are not allowed to propagate in free space, is ascribed to the property (presumably of QCD) known as *confinement*. The behaviour observed in high-energy collisions, where the scattering processes between quarks, gluons and other particles occur as though the partons themselves were instead free, despite being bound, is referred to as *asymptotic freedom* (again, a property of QCD). While the latter can be demonstrated rigorously in the perturbative approach, the former is so far only a reasonable conjecture that may be partially understood and justified through arguments of plausibility.

The confinement problem is a serious obstacle to Feynman's picture: the impulse approximation is applicable to quasi-elastic scattering in nuclear physics because the interaction time for the probe is much shorter than that of the nuclear motion; i.e. nucleons bound inside the nucleus do not actually *feel* the potential until they *touch* the boundaries (internally the potential is approximately constant). It is also true that the binding energy is much less than the potential-well depth; i.e. the binding is loose. Finally, nucleons do actually emerge (and the energy difference due to the well depth is manifest in their spectrum); quarks, on the other hand, do not and would thus appear to have *infinite* binding energy. The potential might therefore be presumed to be far from flat inside the nucleon. How then is it that they behave as though free?

Asymptotic freedom

The answer to this question was provided by David Gross and Frank Wilczek (1973b) and, independently, by David Politzer (1973)[22,23]. These three theorists examined the behaviour of coupling constants in general quantum field theories. An important theoretical aspect of all quantum field theories, with observable and significant experimental consequences, is *renormalisation*. Put simply, owing to quantum (or so-called *radiative*) corrections, various physical parameters of the theory are shifted from their *naïve* (or bare) values. An important and well-studied example is the anomalous magnetic moment of the electron: in Dirac's theory (see appendix B.1) the gyromagnetic ratio for the electron is exactly 2, but calculable corrections increase this by about one part in a thousand. The experimental measurement confirms this value. In QED it had long been known that another important and observable consequence of renormalisation is to *transmute* the coupling constant α into a function that varies with energy scale (generically indicated as Q^2).

To understand this, let us imagine trying to measure the charge of an isolated electron, in a vacuum, by using another (infinitesimal) charge as a probe at some large distance r. Now, in quantum field theory the vacuum is *not* strictly empty; it is rather a sort of bubbling soup of virtual particle–antiparticle pairs being continually created and subsequently annihilating spontaneously. When such a pair is formed, the particle with positive charge is attracted to the electron that is the object of our measurement, whereas the other is repelled. Consider then the Gaussian sphere at radius r: there will be a net movement of *neutralising* charge towards the electron under study. According to Gauss' theorem, this reduces the effective charge as measured by the probe, leading to a so-called 'screening' effect. However, as the probe approaches the object, the screening diminishes and the measured charge thus increases. To derive this behaviour, the full power of perturbation theory must be applied to the quantum field theory under study.

In quantum field theory the description of such an effect is given by the Feynman diagram displayed in figure 3.19. Such a diagram provides (part of) the first-order perturbative correction to the (relativistic) Coulomb potential between two electrons

---

[22] The 2004 Nobel Prize in Physics was awarded to David J Gross, H David Politzer and Frank Wilczek 'for the discovery of asymptotic freedom in the theory of the strong interaction'. The work of Gross and Wilczek (1973b) and, independently, of Politzer (1973) marks the effective birth of QCD as the theory of strong interactions.

[23] It should be mentioned that the same result had already been announced by Gerardus 't Hooft (1999) in a remark from the floor to a talk by Kurt Symanzik during a small meeting in Marseilles in 1972 (see e.g. Hooft (1999)). However, despite Symanzik urging him to publish, as his PhD advisor Martinus Veltman did not consider the findings so important, the work remained unpublished.

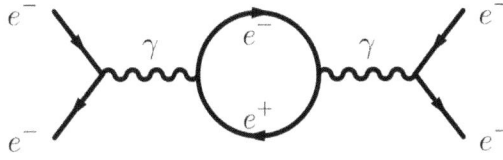

Figure 3.19. Time runs from top to bottom and thus the diagram represents the exchange of a virtual photon between two electrons. However, while it is propagating, the photon may transform temporarily into a virtual e^+e^- pair.

(or any pair of charged particles). It describes precisely the vacuum-polarisation effect of the virtual pair-production process mentioned above. The full calculation in QED, first performed by Landau *et al* (1954)[24] reveals that not only does it lead to a nett screening of the Coulomb interaction, but that this reduction depends on the energy characterising the interaction (e.g. the centre-of-mass energy of the interacting particles). The dependence found is such that the screening effect decreases with increasing energy, or equivalently at shorter distances; i.e. the QED coupling increases in strength at high energy. In fact, there comes a point, at high enough energies, where the charge becomes effectively infinite, whereas the large-distance (or $Q^2 \approx 0$) limit of α remains well-defined and finite; its value is just the oft-quoted $\frac{1}{137}$.

Fortunately, it turns out that the scale or Q^2 for which the physically measured charge would apparently diverge (the so-called Landau pole) is far beyond the Planck mass or energy[25] and is thus of little physical relevance. That is, until the rôle of gravity in quantum field theory is understood and/or becomes important phenomenologically, we need not and indeed cannot sensibly address such a problem. Moreover, long before such a catastrophe is encountered, perturbation theory necessarily breaks down, owing to the growing magnitude of α. That is, perturbative techniques can provide no idea of what happens in such an energy region. *Hic sunt leones.*

The Q^2 *variation* of α_{QED}, in contrast, is very physical and is indeed an observable effect. The **TOPAZ** detector at the Transposable Ring Intersecting Storage Accelerator in Nippon (**TRISTAN**) accumulation ring (**KEK**) measured α_{QED} at a centre-of-mass energy of 57.77 GeV, using two entirely independent methods. Exploiting a purely electromagnetic interaction, via comparison of the cross-section for $e^+e^- \rightarrow \mu^+\mu^-$ and $e^+e^- \rightarrow e^+e^-\mu^+\mu^-$, the value found was (Levine *et al* 1997)

$$\alpha_{\text{QED}}^{-1} = 128.5 \pm 1.8 \text{ (stat)} \pm 0.7 \text{ (syst)} \tag{3.4.13a}$$

Through the γ–Z^0 interference contribution to the total hadronic cross-section, they also deduced (Miyabayashi *et al* 1995)

$$\alpha_{\text{QED}}^{-1} = 128.6^{+0.9}_{-0.8} \text{ (stat)} {}^{+2.7}_{-2.5} \text{ (syst)} \tag{3.4.13b}$$

---

[24] The 1962 Nobel Prize in Physics was awarded to Lev Davidovich Landau 'for his pioneering theories for condensed matter, especially liquid helium'.
[25] The Planck mass is defined as $m_\text{P} := \sqrt{\hbar c / G} \approx 10^{19} \text{GeV}$ (or 22 μg).

The two results, displayed in figure 3.20, are in good agreement with the theoretical prediction (Hagiwara *et al* 1994):

$$\alpha_{QED}^{-1} = 129.6 \pm 1.6. \tag{3.4.14}$$

The above behaviour is typical of Abelian gauge theories (such as QED) and indeed most simple quantum field theories, with the exception of *non*-Abelian theories (such as QCD and, by the way, the weak sector of the electroweak theory). In such theories also the gauge fields themselves carry the charge of the interaction and can thus interact with each other, *even in the absence of matter fields*. In QCD, besides a diagram similar to that of figure 3.19 (with now a quark loop and a gluon instead of the photon), there is thus a further loop correction in the case of virtual gluon exchange, in which a virtual gluon–gluon pair is temporarily produced, see the Feynman diagram shown in figure 3.21. It turns out that their contribution to the

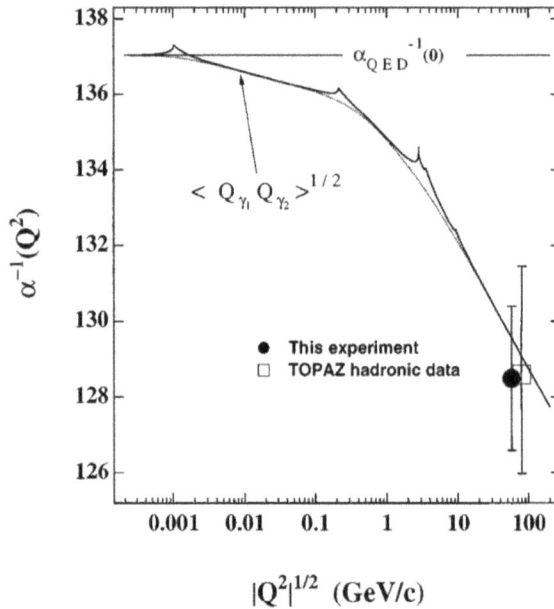

Figure 3.20. The measured and theoretical electromagnetic coupling as a function of momentum transfer Q. The solid and dotted lines correspond to the predictions for positive and negative Q^2, respectively. The hadronic data point has been shifted for clarity of the display. Reprinted with permission from Levine *et al* (1997), copyright (1997) by the American Physical Society.

Figure 3.21. Time is taken as before in the previous case, but here the diagram represents the exchange of a virtual gluon between two coloured quarks and while the gluon is propagating, it may transform temporarily into a virtual gg pair.

vacuum *colour* polarisation has the *opposite* effect to that of the fermions; i.e. it *antiscreens* an isolated colour charge. Moreover, it is stronger than the screening effect of fermions loops, but it too becomes weaker as we approach the colour charge source of the chromo-Coulomb field. Therefore, for QCD the overall effect is the opposite; i.e. the measured charge decreases at short distance or high energies. This has profound implications for the use of perturbation theory in hadronic physics.

The results of the calculations performed by Politzer (1973), Gross and Wilczek (1973b) may be summarised as follows. For convenience, as all variation is logarithmic in energy scale (Q^2), we introduce the so-called β-function as the logarithmic derivative of $\alpha(Q^2)$ with respect to the scale:

$$Q^2 \frac{\partial \alpha(Q^2)}{\partial Q^2} = \frac{\partial \alpha(t)}{\partial t} = \beta(\alpha(t)), \qquad (3.4.15)$$

where we have defined

$$t := \ln \frac{Q^2}{\mu^2}, \qquad (3.4.16)$$

with μ an arbitrary parameter; varying μ merely translates the t-axis, leaving derivatives unaffected. The β-function may be calculated perturbatively in quantum field theory and depends in an essential way on the type of theory. We thus make a power (or perturbative) expansion in α:

$$\beta(\alpha) = -\alpha^2(b_0 + b_1\alpha + b_2\alpha^2 + ...). \qquad (3.4.17)$$

Note that the first term turns out to be already order α^2. The sign of the first coefficient is crucial in determining whether the coupling constant increases or decreases with growing energy scale (the overall minus sign is conventional).

In QED we find:

$$b_0^{\text{QED}} = -\frac{1}{3\pi}\sum_f N_{cf}\, Q_f^2, \qquad (3.4.18a)$$

where $N_{cf} = 3$ for quarks and 1 for charged leptons. The sum runs over all fermions of charge Q_f that are active at the chosen energy scale. By active we mean energetically accessible at the scale determined by Q^2; i.e. having $m^2 \lesssim Q^2$. The importance of the result here is that the derivative is positive and therefore α_{QED} grows with increasing intereaction energy.

In QCD, however,

$$b_0^{\text{QCD}} = \frac{11N_c - 2N_f}{12\pi}, \qquad (3.4.18b)$$

where, as usual, N_f is the number of active flavours of quarks. Therefore, provided $N_f < 17$, the β-function in QCD is *negative*. In fact, b_1 through to b_4 are known for QCD (for b_4, see Baikov *et al* (2017) and references therein for the lower-order calculations) and all have the same sign as b_0 for N_f not too large. An important proven result is that, in four space–time dimensions, only non-Abelian gauge theories display such behaviour (Gross and Wilczek 1973a, 1973b, 1974, Politzer 1973, 1974).

If α is *small* enough for perturbation theory to be valid, defining $\alpha_0 := \alpha(t)|_{t=0}$, the leading-order solutions to the differential equations are simply

$$\text{QED:} \qquad \alpha(t) \simeq \frac{\alpha_0}{1 - |b_0|\alpha_0 t} \qquad\qquad (3.4.19a)$$

and

$$\text{QCD:} \qquad \alpha(t) \simeq \frac{\alpha_0}{1 + |b_0|\alpha_0 t}. \qquad\qquad (3.4.19b)$$

The solution for QED makes explicit the nature of the Landau pole for $t = 1/|b_0|\alpha_0$

Exercise 3.13. *Invert equation (3.4.19) to find the value of Q^2 at the Landau pole in QED. How does this compare, e.g. with the Planck mass?*

A different and more transparent form may be adopted for QCD. By defining

$$\frac{1}{\alpha_0} =: b_0 \ln(\mu^2/\Lambda_{\text{QCD}}^2), \quad \text{for } b_0 > 0, \qquad\qquad (3.4.20)$$

it may be rewritten as

$$\alpha(Q^2) \simeq \frac{1}{\dfrac{1}{\alpha_0} + b_0 t} = \frac{1}{b_0 \ln \dfrac{\mu^2}{\Lambda_{\text{QCD}}^2} + b_0 \ln \dfrac{Q^2}{\mu^2}} = \frac{1}{b_0 \ln \dfrac{Q^2}{\Lambda_{\text{QCD}}^2}}, \qquad (3.4.21)$$

where a dimensional parameter Λ_{QCD} has been introduced to replace μ (and α_0). The logarithmic decrease of $\alpha(Q^2)$ with Q^2 is thus made manifest.

At this point Λ_{QCD} may be considered as an independent physical parameter, substituting the non-physical α_0 (the value of α for $Q^2 = \mu^2$). Of course, Λ_{QCD}^2 is none other than the value of Q^2 at the Landau pole in QCD. In quantum field theory jargon this parameter exchange is known as *dimensional transmutation*. The exact value extracted experimentally depends on the order of perturbation theory used and the energy scale (through the number of active flavours), but generally lies between 200 and 300 MeV.

In other words, so effective is the antiscreening of gluons that, unless there are more than 17 different quark types, the behaviour of the QCD coupling is the opposite of that in QED and the charge *decreases* with increasing Q^2. This leads to the notion of *asymptotic freedom*. With this concept in hand, the apparent freedom of the quarks inside the proton can be justified: as long as they are probed at high enough energies, the effective interaction strength with the surrounding nucleon is small. To have some idea of this, for $Q^2 \simeq M_Z^2$ we find (experimentally) $\alpha_s \simeq 0.1180 \pm 0.0009$ (Navas *et al* 2024), where we have used the standard notation

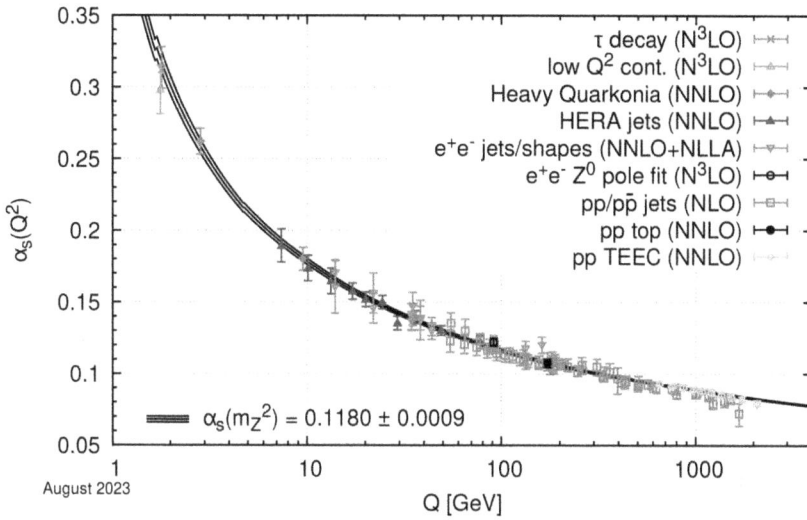

Figure 3.22. The theoretical and experimental running of $\alpha(Q^2)$: the curves correspond to the QCD calculation (with uncertainties) and a *one-parameter* fit to the data; the data points correspond (in more-or-less increasing order of energy) to: τ-lepton decay rate, Υ decay rate, DIS scaling violation, event shapes in e^+e^- annihilation to hadrons, Z^0 production, jets in pp and $p\bar{p}$ interactions and top production (reproduced with permission from Navas *et al* (2024) CC BY 4.0.

of α_s to indicate the *strong* coupling constant (i.e. that of QCD). To be honest, at the 1969 SLAC energies it was larger by about a factor 3 or 4; in other words, it was not really very small. In any case, there are now many independent measurements of $\alpha(Q^2)$ and the agreement with perturbative QCD calculations is excellent (see figure 3.22).

As is often the case, the viewpoint of the physicists involved is enlightening: the following is the motivation given by Gross for pursuing this line of research.

> *I decided, quite deliberately, to prove that local field theory could not explain the experimental fact of scaling and thus was not an appropriate framework for the description of the strong interactions. Thus, deep inelastic scattering would finally settle the issue as to the validity of quantum field theory. The plan of the attack was twofold. First, I would prove that 'UV Stability,' the vanishing of the effective coupling at short distances, later called asymptotic freedom, was necessary to explain scaling. Second, I would show that there existed no asymptotically free field theories. The latter was to be expected.*

David J Gross (2005)

3.4.2.1 *Running masses*
As we have already noted, the effect of renormalisation is not limited to the running of the gauge couplings, but also generates scale dependence in the all other

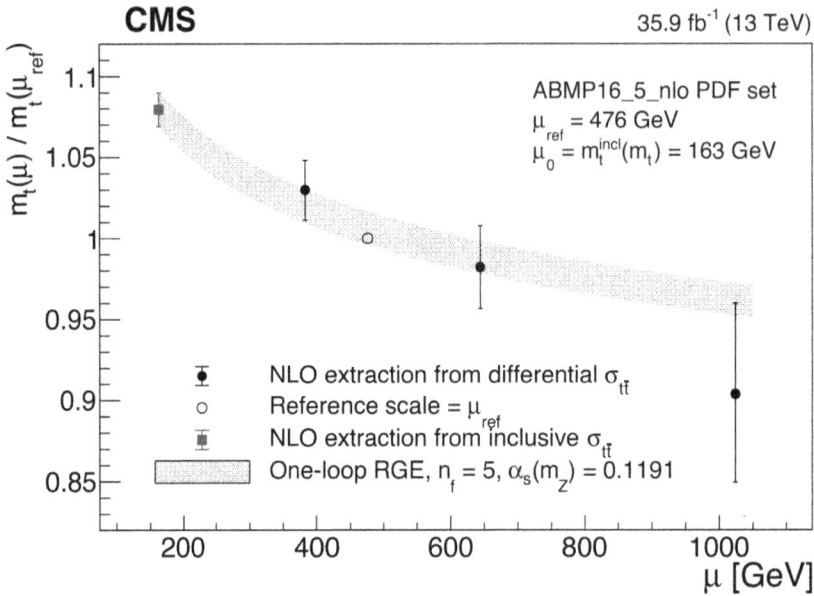

Figure 3.23. The experimentally extracted running of the top quark mass $m_t(\mu)/m_t(\mu_{\text{ref}})$ compared to the theory prediction at one-loop precision, with $n_f = 5$, evolved from an initial scale $\mu_0 = 163\,\text{GeV}$ (reprinted from Sirunyan *et al* 2020, CC BY 4.0).

parameters of the theory. In particular, even the quark masses can no longer be considered constant; not only has their scale dependence been calculated in QCD to five-loop order (Baikov *et al* 2014), but it has also been measured experimentally for the top quark (see e.g. Sirunyan *et al* 2020), as shown in figure 3.23[26]. The scale variation is determined in a similar manner to the running couplings:

$$\mu^2 \frac{\partial m(\mu^2)}{\partial \mu^2} = -\gamma\big(\alpha_s(\mu^2)\big)\, m(\mu^2), \tag{3.4.22}$$

where $\gamma\big(\alpha_s(\mu^2)\big)$ is the calculable so-called mass anomalous dimension; to first order in α_s, we find $\gamma = \alpha_s/\pi$. A review of studies performed at HERA on charm, beauty and top quarks by Behnke *et al* (2015) contains discussions of the running masses of the charm and beauty quarks.

Confinement

How then, on the other hand, can such a picture be reconciled with the phenomenon of confinement? Here the discussion necessarily becomes less rigorous as there is presently no way of performing complete and reliable *ab initio* calculations in the low-energy regime, where the coupling is strong and the non-trivial vacuum structure of QCD comes into play. Some very plausible arguments can, however,

---

[26] Note that the initial value of the top mass used for evolution is $\mu_0 := m_t(m_t) = 163$ GeV, which is the so-called running mass (as deduced from the inclusive $t\bar{t}$ production cross-section, a definition that differs from the so-called pole mass, currently quoted in e.g. the Navas *et al* (2024) tables as 172.4 ± 0.7 GeV.

be made. Note first that the $1/r$ behaviour of the Coulomb potential is a consequence of living in *three* spatial dimensions: the flux lines are distributed over the surface of sphere and therefore the force decreases as $1/r^2$ (see figure 3.24 (a)). In a spatially *one*-dimensional world the flux-line 'density' is constant (as in figure 3.24(b)), so that the force is constant and therefore the potential is proportional to r; evidently a *confining* potential. Theoretical work has been performed on such theories and much is known about them.

However, we do live in three dimensions. How then might the *effective* number of dimensions be reduced? Recall that the gluons themselves carry the colour charge and thus interact strongly with one another. It can be shown that a plausible effect of this is to *squeeze* the flux lines. If they are really forced into a tube-like one-dimensional structure, then the force law indeed will become string-like and the potential will become linear in r (see figure 3.25).

We can imagine that such squeezing is only operative for large separations of two 'isolated' colour charges and that at short distances the configuration returns to a three-dimensional Coulomb-type behaviour. The potential might then be described (approximately) by the form

$$V_{QCD} \sim -\frac{a}{r} + b\,r. \tag{3.4.23}$$

The more usual Coulomb behaviour will only dominate in the short-distance regime, whereas the string-like, linear, second term will naturally take over at sufficiently large distances. Indeed, a close inspection of the flux-tube picture shown in figure 3.25, particularly of the region very near to one of the charges, reveals that it depicts just such a situation. The precise scale at which the switch occurs depends on the values of the two coefficients: a (the coupling constant) and b (the *string*

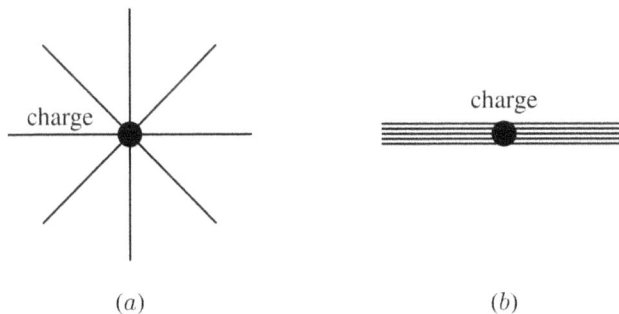

(a) (b)

Figure 3.24. The force-field flux lines around an isolated charge for spatially (a) three-dimensional and (b) one-dimensional field theories.

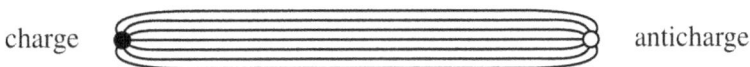

Figure 3.25. The force-field flux lines between a colour charge and anti-charge, assuming a one-dimensional tube-like behaviour.

tension). As mentioned in the section on Regge theory, this type of potential naturally leads to a $q\bar{q}$ meson spectrum of the form $m(J) = m_0 + J\delta m$ (where J is the meson spin), which is precisely what is observed experimentally (the Regge trajectories discussed earlier).

In the light of the foregoing discussion, it might be tempting to consider confinement and asymptotic freedom as being opposite sides of the same coin (the term *infrared slavery* is often used). However, this over-simplification is *misleading* and should be avoided. As we have seen, the same behaviour that leads to a diminishing coupling constant with growing energy also leads to an apparent divergence (the Landau pole) for low energies. It would be wrong, however, to associate confinement with this effect. At a formal level, at large distances the strong-coupling regime takes over and perturbative arguments no longer apply; *here be dragons*. More physically, as we have just discussed, confinement must be a result of the peculiar vacuum structure of QCD and *may* arise owing to effective string-like forces between colour charges at large distances (which can have nothing to do with perturbation theory). Consider attempting a power-series expansion in α of e.g. $e^{1/\alpha}$; i.e. such behaviour may be 'invisible' to perturbation theory.

3.5 A brief survey of quark and gluon densities

Having, hopefully, convinced the reader of the validity of the picture presented by combining the ideas of Gell-Mann and Feynman, we must now demonstrate something of its utility. More to the point, as we shall see, there are various predictions of the model that, given their success, on the one hand lend strong support to the model and on the other provide useful information for both experimental analysis and planning. For more detailed and in-depth discussions on the topic of parton densities, the reader is referred, e.g., to the book by Roberts (1990).

The predictive power of the model, coupled with QCD, is twofold: firstly, the structure functions are universal and may be used to calculate cross-sections for processes other than DIS and, secondly, the scale variation is calculable, which means that relatively low-energy information gathered early in history may be exploited to make predictions for future high-energy experiments.

The intuitive picture we have derived for the DIS process is essentially that of a convolution of two basic ingredients: parton distributions or densities (which may be thought of as fluxes) and partonic cross-sections. It can be shown that for high-energy processes, where α_s is small, many hadronic processes may described in a similar manner. From process to process, the partonic (hard or high-energy) scattering cross-sections will vary, but are *calculable*, whereas the (incalculable) parton densities are assumed to be the same, i.e. they are *universal* up to calculable scale variations. We may therefore exploit DIS, say, to *measure* them and then employ the functions thus measured to make *predictions* for other processes.

3.5.1 Quark densities from electron scattering

Still today we have no reliable way to calculate the densities $f(x)$ from first principles and therefore do indeed need to measure them. Here we shall briefly review what is known about their general behaviour. In DIS at moderate energies (e.g. for $Q^2 \lesssim m_c^2$) only the three lightest quarks (u, d and s) contribute appreciably and so we may write

$$x^{-1}F_2^{ep}(x) = \tfrac{4}{9}[u_p(x) + \bar{u}_p(x)] + \tfrac{1}{9}[d_p(x) + \bar{d}_p(x)] + \tfrac{1}{9}[s_p(x) + \bar{s}_p(x)] \quad (3.5.1a)$$

and

$$x^{-1}F_2^{en}(x) = \tfrac{4}{9}[u_n(x) + \bar{u}_n(x)] + \tfrac{1}{9}[d_n(x) + \bar{d}_n(x)] + \tfrac{1}{9}[s_n(x) + \bar{s}_n(x)], \quad (3.5.1b)$$

where the suffixes p and n on the quark densities indicate that they refer to a parent proton or neutron respectively. Now, assuming isospin to be a good symmetry, we expect $u_p = d_n$, $u_n = d_p$, $s_p = s_n$ etc. Exploiting this symmetry, the accepted convention is to drop the suffixes p and n and employ densities that refer to the *proton*. We thus write

$$x^{-1}F_2^{ep}(x) = \tfrac{4}{9}[u(x) + \bar{u}(x)] + \tfrac{1}{9}[d(x) + \bar{d}(x)] + \tfrac{1}{9}[s(x) + \bar{s}(x)] \quad (3.5.2a)$$

and

$$x^{-1}F_2^{en}(x) = \tfrac{1}{9}[u(x) + \bar{u}(x)] + \tfrac{4}{9}[d(x) + \bar{d}(x)] + \tfrac{1}{9}[s(x) + \bar{s}(x)]. \quad (3.5.2b)$$

3.5.2 Valence and sea quark separation

At this point there are still too many unknown functions to be able to determine or extract very much, but let us see what may be assumed and/or deduced. A first reasonable assumption is that, at least on average, the u and d antiquark and s quark and antiquark (or so-called *sea*-quark) densities should be suppressed with respect to those of the two *valence* quarks. In fact, we might decompose the u and d densities as follows:

$$q(x) = q_{\text{val}}(x) + q_{\text{sea}}(x) \qquad (q = u, d), \quad (3.5.3)$$

where by *sea* we mean those quarks and antiquarks produced spontaneously and only fleetingly, whereas the *valence* quarks are those of Gell-Mann. We therefore have

$$\int_0^1 dx\, u_{\text{val}}(x) = 2 \quad \text{and} \quad \int_0^1 dx\, d_{\text{val}}(x) = 1. \quad (3.5.4)$$

Moreover, since the sea quarks are always produced as quark–antiquark pairs, they must exist in equal numbers overall:

$$\int_0^1 dx\, q_{\text{sea}}(x) = \int_0^1 dx\, \bar{q}_{\text{sea}}(x) \qquad (q = u, d, s). \quad (3.5.5)$$

It might be hoped that equality holds at each value of x, but there is no guarantee of this; the integrals *must*, however, be equal.

The Nachtmann inequality

So, assuming the isospin independence of the sea quarks, let us subsume all the sea densities into one global function, say $\Sigma(x)$; thus,

$$x^{-1}F_2^{ep}(x) = \tfrac{4}{9}u_{\text{val}}(x) + \tfrac{1}{9}d_{\text{val}}(x) + \Sigma(x) \tag{3.5.6a}$$

and

$$x^{-1}F_2^{en}(x) = \tfrac{1}{9}u_{\text{val}}(x) + \tfrac{4}{9}d_{\text{val}}(x) + \Sigma(x). \tag{3.5.6b}$$

Now, since the individual quantities on the right-hand side are all positive definite, then the ratio $F_2^{en}(x)/F_2^{ep}(x)$ is bounded to lie between $\tfrac{1}{4}$ and 4 (Nachtmann 1972). It can only attain one or other bound if $\Sigma(x)$ is negligible. On the other hand, if the sea should dominate anywhere, then the ratio there will be approximately unity. The data (see figure 3.26) show that for x very small, the ratio does indeed tend to unity, whereas for large x it tends to the value $\tfrac{1}{4}$. This then is interpreted as implying that the valence quarks are important for large values of x, whereas the sea grows as $x \to 0$. The observed large-x limiting behaviour indicates that u dominates over d as $x \to 1$.

Valence-quark isospin dependence

The proton–neutron structure function difference may also be considered:

$$x^{-1}\left[F_2^{ep}(x) - F_2^{en}(x)\right] = \tfrac{1}{3}\left[u_{\text{val}}(x) - d_{\text{val}}(x)\right]. \tag{3.5.7}$$

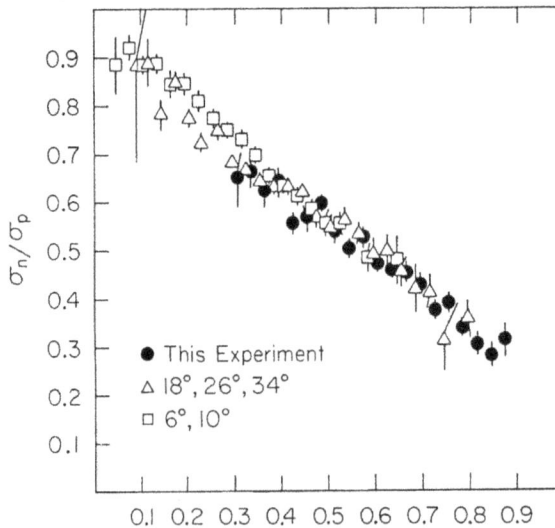

Figure 3.26. The ratio σ_n/σ_p for deeply inelastic scattering as a function of Bjorken x (reproduced from Bodek *et al* (1974), copyright (1974) with permission from Elsevier).

Note that the sea-quark contributions only cancel exactly in the limit of perfect SU(2) or isospin symmetry. The result then is the difference between purely valence quarks. As a first approximation, we might make the assumption

$$u_{val}(x) \simeq 2\, d_{val}(x) \quad \text{and, say,} \quad d_{val}(x) \simeq q_{val}(x), \tag{3.5.8}$$

where $q_{val}(x)$ represents a generic single valence-quark density. The right-hand of equation (3.5.7) is then roughly $\frac{1}{3} q_{val}(x)$. With the availability of more precise and varied data, allowing the separation of $u_{val}(x)$ and $d_{val}(x)$, it was found that this is not a particularly good approximation, especially for large x, where the following behaviour is typically found:

$$\frac{d_{val}(x)}{u_{val}(x)} \sim 1 - x. \tag{3.5.9}$$

That is, the d quarks are relatively suppressed for $x \to 1$. A possible explanation for this may be found in the Pauli[27] exclusion principle (Pauli 1925), which should force the more numerous u quarks to distribute themselves more evenly (i.e. to higher x). Unfortunately, with no real theory of the bound state in QCD, this remains merely a plausible conjecture.

Sea-quark isospin dependence

Some weight is, however, lent to such an argument by the observation of a surprisingly large difference between the antiquark distributions $\bar{u}(x)$ and $\bar{d}(x)$: experimentally we find (Amaudruz *et al* 1991)

$$\bar{d}(x) \simeq 2\, \bar{u}(x). \tag{3.5.10}$$

It thus appears that, as far as the sea is concerned, SU(2) is broken quite strongly. This may be explained by again appealing to the Pauli exclusion principle: the proton contains more valence u quarks than d and thus the u–$\bar{u}$ content of the sea should be suppressed. An experimentally observed consequence is precisely a non-cancellation of the sea in the proton–neutron F_2 difference integral. We shall examine this question in more detail in section 3.5.6.

3.5.3 Quark densities from neutrino scattering

In order to obtain more independent information new probes are needed. That is, currents coupling differently to the various quarks. Neutrino scattering via the weak interaction provides just such a case; the detected final state is no different to standard charged-lepton scattering ($\nu_e N \to e^- X$). However, such a process is sensitive to different combinations of the quark densities. Defining analogous structure functions for neutrino DIS, we have

$$x^{-1} F_2^{\nu p}(x) = 2[d(x) + \bar{u}(x)] \tag{3.5.11a}$$

---

[27] The 1945 Nobel Prize in Physics was awarded to Wolfgang Pauli 'for the discovery of the Exclusion Principle, also called the Pauli Principle'.

and

$$x^{-1}F_2^{\nu n}(x) = 2[u(x) + \bar{d}(x)]. \tag{3.5.11b}$$

Charge conservation requires a negatively charged quark in the neutrino–proton case (and then $d \leftrightarrow u$ for the neutron), whereas the strange and anti-strange contributions are suppressed at low energies owing to the requirement that the final state be either a (heavy) c quark with a cross-section factor $\cos^2 \theta_C$ or a u quark with $\sin^2 \theta_C$. Similar formulæ apply to antineutrino scattering.

Now, consider isoscalar targets (i.e. with equal numbers of protons and neutrons, such as a deuteron or ^{12}C, which just average over the proton and neutron structure functions. In this case the electron-to-neutrino DIS ratio is

$$\frac{F_2^{ep}(x) + F_2^{en}(x)}{F_2^{\nu p}(x) + F_2^{\nu n}(x)} = \frac{\frac{5}{9}(u + \bar{u} + d + \bar{d}) + \frac{2}{9}(s + \bar{s})}{2(u + \bar{u} + d + \bar{d})} \geqslant \frac{5}{18}. \tag{3.5.12}$$

This combination ratio was first studied experimentally by Deden *et al* (1975) and is shown in figure 3.27. The fact that it saturates well for $x \gtrsim 0.2$, but not below (not seen clearly in the figure owing to the large error bars), again suggests that the sea quarks are suppressed in this region, but are present for smaller values.

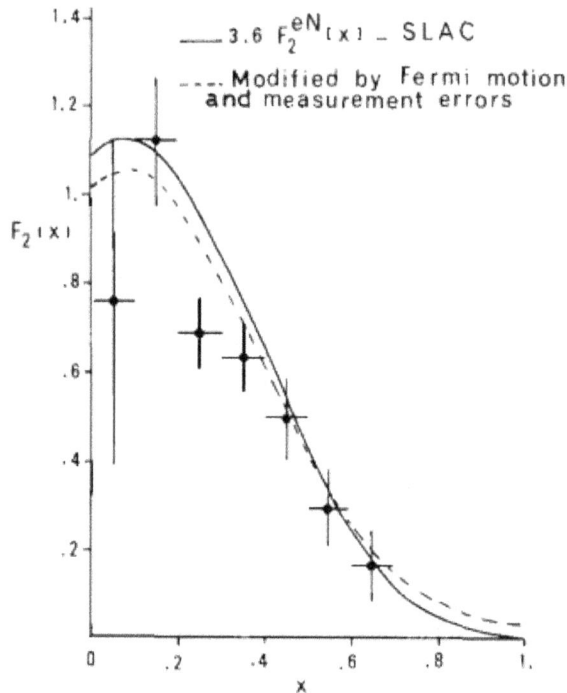

Figure 3.27. Early Gargamelle data (filled circles) for the denominator in equation (3.5.12) compared with $\frac{18}{5}$ times the SLAC data (curves) for the numerator as a function of Bjorken x (reproduced from Deden *et al* (1975), copyright (1975) with permission from Elsevier).

3.5.4 Quark densities from the Drell–Yan process

The process proposed by Sidney Drell and Tung-Mow Yan (1970) has already been mentioned in connection with the justification for the three colour charges; here we shall discuss how it contributes to our knowledge of parton distribution functions. The basic process, $pp \to \ell\bar{\ell}X$, is depicted in figure 3.28. A quark from one of the colliding protons annihilates with an antiquark from the other, producing a virtual photon This then decays into a charged lepton–antilepton pair, which are the objects finally detected. To lowest-order in QCD, the Drell–Yan (DY) cross-section is

$$\frac{\mathrm{d}^2\sigma}{\mathrm{d}x_1\,\mathrm{d}x_2} = \frac{4\pi\alpha}{9sx_1x_2} \sum_{i=u,d,s,\ldots} Q_i^2 \left[f_i^a(x_1)\, \bar{f}_i^b(x_2) + \bar{f}_i^a(x_1)\, f_i^b(x_2) \right], \qquad (3.5.13)$$

where $s := (p_a + p_b)^2$, Q_i is the fractional charge of the ith type quark and $f_i^{a,b}$ ($\bar{f}_i^{a,b}$) represent the quark (antiquark) densities inside protons a and b. In the high-energy limit, defining the four-momentum of the photon q^μ, we have simply $q^2 = x_1 x_2 s$. The initial states may actually be any combination of protons, neutrons and their antiparticles, pion and kaon beams have also been used. While the expression is clearly more complicated than the standard DIS cross-sections, involving as it does a *product* of quark densities, the data for such processes may be used in global fitting programmes and thus contribute to the quark-density determinations. By using pion or kaon beams the relative densities for mesons may also be determined.

Historically, this process played a significant rôle in our understanding of the importance quantum corrections in QCD. Early comparisons of the *naïve* leading-order predictions (exploiting densities extracted from standard ep DIS scattering) found enormous discrepancies. The measured cross-sections were typically a factor of 2–3 larger than the theoretical predictions. The explanation came from Guido Altarelli, Keith Ellis and Guido Martinelli (1978), who calculated the next-to-leading order QCD radiative corrections; i.e. those due both to gluon exchange between the participating quarks and to real gluon emission. The central finding was a very important difference between the DIS case, where the q^2 of the exchanged

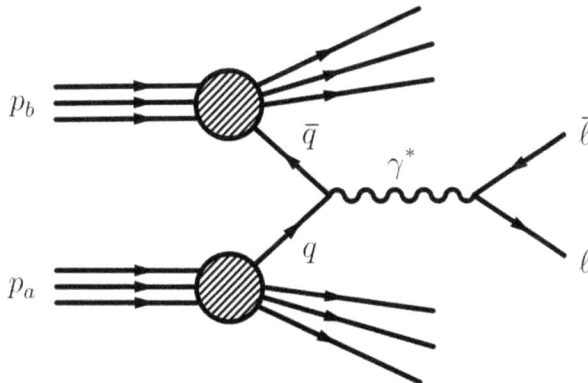

Figure 3.28. The generic DY process; the final lepton pair $\ell\bar{\ell}$ may be any of the three charged leptons.

photon is large and *negative* (space-like), and the DY case, where the q^2 of the produced photon is large and *positive* (time-like).

3.5.5 Quark densities from semi-inclusive scattering processes

Another important process used for the experimental study of parton densities is semi-inclusive deeply inelastic scattering; this is depicted generically in figure 3.29. The struck quark is presumed to *hadronise* or *fragment* into a real hadron, which is then detected experimentally. It is not, of course, possible to trace the origin of the detected hadron back to a specific quark on an event-by-event basis. However, the so-called fragmentation functions, which describe the probability that a given quark will produce a given hadron, are well measured in, for example, electron–positron annihilation experiments (see also section 3.5.10.3). Thus, the data from semi-inclusive DIS may be incorporated into global fits determining the full set of densities.

3.5.6 Sum rules for quark and gluon densities

The quark (and gluon) densities may be integrated with various weights and thus related to known static properties of the nucleons; this leads to so-called sum rules. Evidently, the entire integrals cannot be determined purely from data since the regions $x \sim 0$ and $x \sim 1$ are experimentally inaccessible. Since Q^2 must be kept finite and not small, the limit $x \to 0$ would require an infinitely large beam energy, while for $x \to 1$ we find that the DIS cross-sections tend to zero and therefore data are limited by statistics in this region. However, from theoretical arguments (which we shall discuss shortly) reliable extrapolations can be made to both these limits.

First of all then, the total charges of the proton and neutron are given by the sum over integrated quark densities weighted with their individual charges:

$$1 = \int_0^1 dx \left[\tfrac{2}{3}(u - \bar{u}) - \tfrac{1}{3}(d - \bar{d}) \right] \tag{3.5.14a}$$

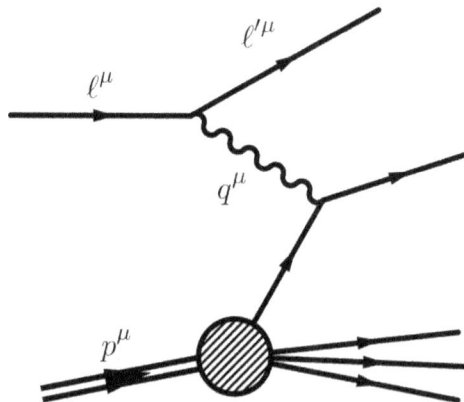

Figure 3.29. The Feynman diagram describing semi-inclusive deeply inelastic scattering.

and

$$0 = \int_0^1 dx \left[\tfrac{2}{3}(d - \bar{d}) - \tfrac{1}{3}(u - \bar{u}) \right]. \tag{3.5.14b}$$

The s-quark contribution evidently vanishes in this sum rule since the overall strangeness must be zero; that is

$$0 = \int_0^1 dx[s - \bar{s}]. \tag{3.5.15}$$

Rearranging, this leads to

$$2 = \int_0^1 dx[u - \bar{u}] = \int_0^1 dx u_{val} \tag{3.5.16a}$$

and

$$1 = \int_0^1 dx[d - \bar{d}] = \int_0^1 dx d_{val}, \tag{3.5.16b}$$

which is, of course, just what would be expected from a simple valence-quark model picture. Since experiments can never cover the complete x interval $[0, 1]$, an important use of these sum rules is to fix the overall normalisation of functional fits to data.

There are many such sum rules, but a last important example should be mentioned: by weighting the integral with x itself, the total fraction of the parent momentum carried by the quarks is calculated. If there were nothing other than quarks inside the nucleon then, summed over all quark types, this would give unity. In contrast, the experimental value is

$$\int_0^1 dx x \, [u + \bar{u} + d + \bar{d} + s + \bar{s}] \simeq 0.5, \tag{3.5.17}$$

for Q^2 in the few-GeV^2 region. From this we may deduce that the gluon density is actually rather important, although invisible to DIS, and that

$$\int_0^1 dx x \, g(x) \simeq 0.5, \tag{3.5.18}$$

where $g(x)$ is the probability of finding a gluon with momentum fraction x. This can be verified in other experiments; i.e. in hadron–hadron interactions (for example the DY process mentioned earlier), where collisions between gluons and quarks (and even gluons and gluons) may contribute. The presence of gluons is also made manifest when considering higher-order corrections: a gluon inside the target hadron may *split* into a quark–antiquark pair, one of which may then interact with the virtual photon.

Exercise 3.14. *Show that the total quark momentum fraction may be obtained directly from the following simple combination of deeply inelastic scattering structure function integrals (as measured below the charm threshold):*

$$\int_0^1 dx \left[\tfrac{9}{2} F_2^{ep+en} - \tfrac{3}{4} F_2^{\nu p + \nu n} \right]$$

There are also a number of sum rules based on the symmetries of the various probing interactions involved (generally electromagnetic or weak). We shall now examine a few of the better known examples; though the following list is by no means exhaustive.

The Adler sum rule

In its original formulation *before* the advent of the parton model, the following sum rule due to Stephen Adler (1966) was expressed in terms of the neutrino structure functions:

$$\int_0^1 \frac{dx}{x}\left[F_2^{\nu p} - F_2^{\bar{\nu} p}\right] = 2. \tag{3.5.19}$$

It follows directly from the relations in equations (3.5.14) and (3.5.15). Note, however, that the original derivation did not depend on the underlying parton structure or the nature of quarks. Note too that, unlike many other sum rules of a similar type, this does not receive perturbative corrections; it is simply twice the difference of the u and d valence-quark numbers and as such is protected. Indeed, it is entirely independent of the underlying QCD theory. Unfortunately, experimental verification is difficult and present precision does not provide significant corroboration.

The Gross–Llewellyn Smith sum rule

This sum rule again involves neutrino scattering and as such is also difficult verify. Moreover, it receives important (but calculable and known) QCD corrections. The rule, due to David Gross and Christopher Llewellyn Smith (1969), states that

$$\int_0^1 \frac{dx}{x}\left[F_3^{\nu p} + F_3^{\bar{\nu} n}\right] = 6. \tag{3.5.20}$$

In quark terms, the left-hand side becomes

$$\int_0^1 dx\left\{2[d(x) - \bar{u}(x)] + 2[u(x) - \bar{d}(x)]\right\} = 2\int_0^1 dx\left[u_{\text{val}}(x) + d_{\text{val}}(x)\right]. \tag{3.5.21}$$

The Gottfried sum rule

A sum rule attributed to Kurt Gottfried (1967), again originally expressed in terms of structure functions is the following:

$$\int_0^1 \frac{dx}{x}\left[F_2^{ep} - F_2^{en}\right] = \tfrac{1}{3}. \tag{3.5.22}$$

The derivation relies on the assumption of isospin invariance and, in particular, that the u and d sea-quark distribution integrals are equal.

$$\int_0^1 \frac{dx}{x}\left[F_2^{ep} - F_2^{en}\right] = \tfrac{1}{3}\int_0^1 dx\left[u(x) + \bar{u}(x) - d(x) - \bar{d}(x)\right]$$

$$= \tfrac{1}{3}\int_0^1 dx\left[u_{\text{val}}(x) - d_{\text{val}}(x) + 2u_{\text{sea}}(x) - 2d_{\text{sea}}(x)\right] \tag{3.5.23}$$

$$= \tfrac{1}{3} - \tfrac{2}{3}\int_0^1 dx\left[u_{\text{sea}}(x) - d_{\text{sea}}(x)\right].$$

If the sea is isospin symmetric, the last remaining integral cancels to yield the Gottfried (1967) sum rule (3.5.22). Here the experimental investigation can achieve meaningful precision and the results of the New Muon Collaboration at CERN evidenced a very significant shortfall (Amaudruz *et al* 1991); the measured value is

$$\int_0^1 \frac{dx}{x} \left[F_2^{ep} - F_2^{en} \right] = 0.235 \pm 0.026. \tag{3.5.24}$$

Recall though that the basic assumption made is normally justified by e.g. the smallness of the neutron–proton mass difference $((m_n - m_p)/m_p \simeq 1.4\permil)$. However, this case is very particular: while the externally visible differences between the neutron and proton may appear very small, their internal structures have specific isospin signatures, which can heavily condition the production of the u and d sea quarks. We may then indeed envisage a sizable isospin asymmetry of the quark sea inside the proton; see e.g. Preparata *et al* (1991). To be precise, the creation of sea-quark pairs inside a proton occurs in a background already containing two u but only one d. It might thus be imagined that a so-called Pauli-blocking effect should suppress the production of $u\bar{u}$ with respect to $d\bar{d}$ pairs, an effect foreseen much earlier by Richard Field and Richard Feynman (1977), thus explaining the experimental results.

Other plausible explanations exist: a much discussed example considers the possible effect of the virtual pion cloud that presumably surrounds all nucleons (see e.g. Eichten *et al* (1992)). For example, a proton may spontaneously transform into a temporary virtual neutron (udd) and π^+ ($u\bar{d}$) pair. The extra quark content of this virtual state is to be considered a contribution to the proton sea; thus, since it contains three d or $\bar{d}$ quarks but only two u, the effect would cause just the u–d sea asymmetry required.

The Bjorken sum rule

The final sum rule that we shall examine is that attributed to James D Bjorken (1966), again originally expressed in terms of structure functions:

$$\int_0^1 dx \left[G_1^{ep} - G_1^{en} \right] = \tfrac{1}{6} |g_A/g_V| \simeq 0.21, \tag{3.5.25}$$

where the ratio g_A/g_V was given in equation (2.5.3). This sum rule also receives important (but calculable and known) QCD corrections. Despite some early confusion, it has now been confirmed to better than the 10% level (see e.g. Deur *et al* 2019).

3.5.7 The shapes of parton distributions

As a first guess, assuming the quarks inside the nucleon to be non-interacting, for just three valence quarks the densities might be expected to be simple sharp spectral lines at precisely $x = \tfrac{1}{3}$ (see figure 3.30(a)). Treating them now as an interacting gas, the energy and momentum may be redistributed and we should expect a rather broader spectrum, still centred around $x = \tfrac{1}{3}$ (see figure 3.30(b)). Finally, allowing for gluon emission (or *bremsstrahlung*) and quark–antiquark pair production we can

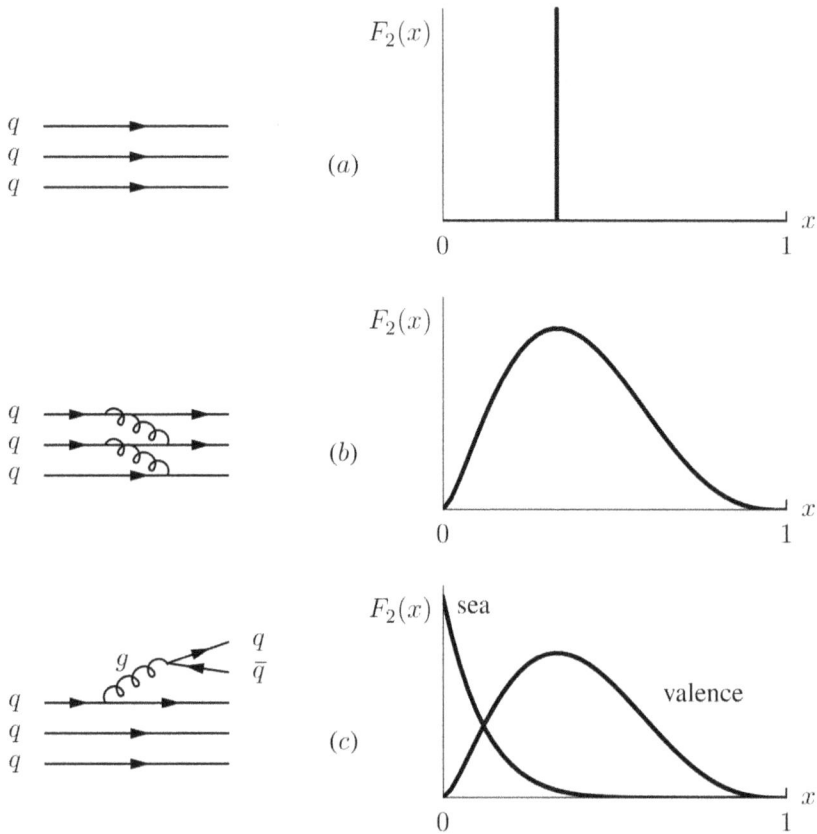

Figure 3.30. The DIS structure function $F_2(x)$ in the *naïve* and QCD-enhanced parton models: (a) the simple narrow spectral line expected for three static, non-interacting, valence quarks; (b) the spectral broadening effect of interactions between the three valence quarks; (c) the effect of spontaneous $q\bar{q}$ pair creation via the gluon field.

imagine that the peaks should move down in x and the new sea and gluon distributions should be important for low energies and momenta (see figure 3.30(c)).

Now, we do not possess theoretical methods for calculating the parton distributions from first principles, as we should need a theoretical approach valid at low energies; i.e. we would essentially be attempting to calculate the quark and gluon wave-functions for, say, the proton. These are fundamentally static or low-energy properties of the hadrons and their interactions. While their are techniques of sorts available, they do not generally solve the full problem.

For example, the technique known as QCD sum rules Shifman *et al* (1979) permits the calculation of certain expectation values, which in turn may be related to integrals of parton distributions. Similarly, lattice gauge theory, which we shall discuss later, can in principle provide information on overall hadron properties. However, as yet, no non-perturbative technique exists capable of producing the full x dependence of the distribution functions *ab initio*. Recall that the perturbative approach is only applicable to QCD for high-energy processes and is therefore not a viable approach to any form of bound-state problem.

3.5.7.1 The large- and small-x behaviour.

The limiting cases of $x \to 0$ and 1 do admit some theoretical investigation and we are able to successfully predict the expected behaviours of the various parton distribution functions in these regions. Beyond the somewhat *naïve* intuition that behaviour is probably well approximated by power-laws of the form x^a and $(1 - x)^b$, respectively, what we can do is obtain some indication of the exponents here, a and b.

The $x \to 0$ behaviour

From the definition of the Bjorken scaling variable $x := Q^2/2M\nu$ (see equation (3.3.13)), where Q^2 is minus the four-momentum transfer squared and $2M\nu = 2p. q \simeq W^2 + Q^2$ (see equation (3.3.10)), we see that there are two ways to consider the small-x limit. The first, and most *naïve*, would be to take $Q^2 \to 0$ keeping W (the photon–hadron invariant mass) constant. This, of course, takes us out of the perturbative regime for QCD and there is little else known about this kinematical region.

The other possibility is to keep Q^2 large and constant while considering the limit W, or rather $\nu \to \infty$. Recall now that Q^2, the momentum transfer, is just what we called t (or rather $-t$) in section 3.1.3 and that this then is just the kinematical regime of applicability of Regge theory. Moreover, as we shall see in section 4.2.2, the photon may convert into a virtual hadronic resonance having the same quantum numbers (i.e. $J^{PC} = 1^{--}$ and charge zero). This permits the application of Regge theory to the deeply inelastic scattering of electrons off hadrons in just the limit $x \to 0$.

With a little work (see e.g. Gilman 1972), the total cross-section for Compton-like virtual-photon–nucleon scattering in this limit ics found to be

$$\sigma_{\gamma*N}^{\text{tot}} \xrightarrow{\nu \to \infty} \sum_{\mathbb{R}} \beta_{\mathbb{R}} \, \nu^{\alpha_0^{\mathbb{R}} - 1},$$

(3.5.26)

where the $\beta_{\mathbb{R}}$ are just coefficients and the sum is over all possible contributing trajectories; i.e. with the appropriate quantum numbers. In terms of parton distributions, this translates into

$$f(x) \xrightarrow{x \to 0} \sum_{\mathbb{R}} \tilde{\beta}_{\mathbb{R}} \, x^{-\alpha_0^{\mathbb{R}}}.$$

(3.5.27)

The leading reggeons are just those mentioned earlier: the pomeron ($\mathbb{P}$), with intercept $\alpha_0 \simeq 1$, and the ρ trajectory, with intercept $\alpha_0 \simeq 1/2$; all others have lower intercepts. The first, having the quantum number of the vacuum, will contribute to the flavour independent functions, i.e. the gluon and the overall sea-quark density, while the second will determine the small-x behaviour of the valence-quarks (and also any eventual variations between the different sea-quark flavours as described in section 3.5.6). We thus predict the following approximate small-x behaviours:

$$q_{\text{val}} \propto 1/\sqrt{x}, \quad q_{\text{sea}} \propto 1/x \text{ and } g \propto 1/x \quad \text{for } x \to 0.$$

(3.5.28)

The $x \to 1$ behaviour

The large-x behaviour too may be predicted, although here we shall exploit QCD and the notion of gluon exchange between the quarks bound inside a nucleon. The limit $x \to 1$ clearly corresponds to a single parton carrying all of the parent hadron energy–momentum. More importantly for us, it also corresponds to the *elastic* limit. That is, scattering in which the proton remains intact and is not broken up. To see this, consider equation (3.3.10): for $x \to 1$, $W^2 \to M^2$. In order that the spectator (valence) quarks should follow the struck parton, they must receive non-negligible energy and momentum via gluon exchange, as shown in figure 3.31. The kinematics is such that the off-shell quark propagators each lead to a power of $1 - x$ at the amplitude level, which, combined with one phase-space negative power, all leads to the so-called quark-counting rules Brodsky and Farrar (1973):

$$f(x) \propto (1 - x)^{2n_s - 1} \quad \text{for} \ x \to 1, \tag{3.5.29}$$

where n_s is the number of spectator quarks that must be realigned with the struck quark via gluon exchange.

In the figure, describing valence-quark scattering, there are just two spectator quarks and therefore two high-energy gluons must be exchanged. In the case of a sea quark or antiquark, produced as part of a quark–antiquark pair, there are four spectator quarks to be realigned (the three valence quarks and the other of the pair). And, although a gluon cannot interact with a photon, considering a similar large-angle scattering process with another hadron, say a pion, in which it could interact with a quark or another gluon, there would be all the three valence quarks to realign. The resulting large-x behaviour is thus predicted to be

$$q_{\text{val}} \propto (1 - x)^3, \quad q_{\text{sea}} \propto (1 - x)^7 \ \text{and} \ g \propto (1 - x)^5, \quad \text{for} \ x \to 1. \tag{3.5.30}$$

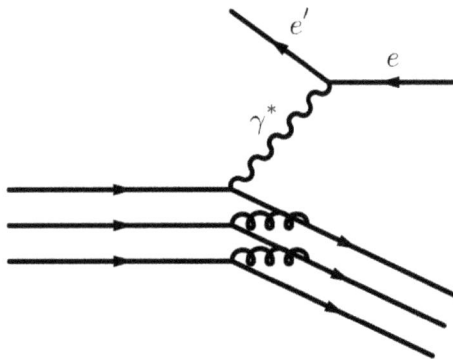

Figure 3.31. The picture of the electron–nucleon deeply inelastic scattering process in the elastic limit: the energy and momentum of the struck parton (here a quark) is shared with the spectator quarks, thus avoiding the breakup of the nucleon.

3.5.7.2 DIS parametrisations and fits to data.
Present-day fits to various data on DIS and other high-p_T hadronic processes can be generically described by the following functional form:

$$q_f(x) = Ax^{-a}(1 - x)^b. \tag{3.5.31}$$

Precision fits employ rather more sophisticated parametrisations, but the basic structure is that given above. A typical set of approximate global-fit values of the parameters for the various partonic distributions is given in table 3.6. Note, however, that these parameters are not constant and change (or evolve) with Q^2, just as the coupling constant; the values given correspond to $Q^2 \simeq 5\text{GeV}^2$. For a more detailed discussion, including QCD effects, see, e.g., Roberts (1990) or Brock *et al* (1995). An example of the fits to the experimentally measured quark distributions is shown in figure 3.32. Most of the above-mentioned features are clearly evident. Moreover, the power behaviour does correspond roughly to that predicted by the quark counting rules and Regge theory just discussed.

3.5.8 The QCD evolution equations

In discussing asymptotic freedom earlier, we introduced the concept of scale variation due to renormalisation effects in quantum field theory. We remarked

Table 3.6. Typical approximate global-fit values of the two powers a and b corresponding to $Q^2 \simeq 5\text{GeV}^2$ for the parametrisations displayed in equation (3.5.31).

	u_{val}	d_{val}	Sea	Gluons
a	0.5	0.5	1	1
b	3	4	8	5

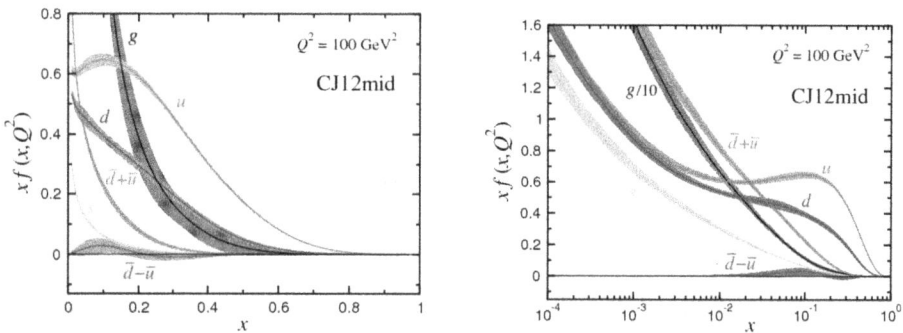

Figure 3.32. The shape of the quark densities for various quark and antiquark flavours, combinations and gluons, shown in the left (right) panel on a linear (logarithmic) x-axis. In the right panel the gluon is scaled by $1/10$. The figures are reproduced with permission from Owens *et al* (2013), copyright (2013) by the American Physical Society.

that the effects are experienced by almost all physical quantities; the parton densities are no exception. Here we may understand such effects in terms of resolution: as we move higher in energy, the partons are resolved more and more finely; that is, we become aware that what appeared as, say, a quark at some distance, on closer inspection reveals itself to be a quark accompanied by a gluon it has emitted. It is immediately obvious that this more finely resolved quark will be carrying slightly less energy, as some has been shared with the emitted gluon. Consequently, the quark and gluon parton densities measured at some high scale will be shifted towards lower x values with respect to those measured at a lower scale.

This is a somewhat technical question, which is addressed in a perturbative manner, much as the Q^2 variation of α_s. Vladimir Gribov and Lev Lipatov (1972), Yuri Dokshitzer (1977), Guido Altarelli and Giorgio Parisi (1977)[28] all contributed more-or-less independently and using a variety of approaches to formulate the calculations, leading to the so-called DGLAP evolution equations. In the simplest case of a flavour non-singlet (i.e. valence-quark) parton density and at first order in perturbation theory, the evolution is governed by the following integro-differential equation:

$$\frac{df(x, Q^2)}{d \ln Q^2} = \frac{\alpha_s(Q^2)}{2\pi} \int_x^1 \frac{dy}{y} \, P\left(\frac{x}{y}\right) f(y, Q^2), \qquad (3.5.32)$$

where $f(x, Q^2)$ represents some non-singlet parton density and $P(z)$ the corresponding so-called DGLAP splitting function. The convolution form does allow exact $\ln Q^2$ integration if transformed to Mellin-moment space (moments in x). Usually though, especially at higher orders, the solution is obtained via numerical integration in x-space.

Let us take a moment to understand this equation. The splitting function $P(z)$ is essentially a measure of the probability that a parton with momentum fraction y will emit a gluon, thus reducing its momentum to some $x < y$ [29]. The splitting clearly has the effect of increasing the population of the lower-x region at the expense of large-x partons. This also implies that to determine $f(x, Q^2)$ using data for $f(y, Q_0^2)$ measured at some $Q_0^2 < Q^2$, we need to know $f(y, Q_0^2)$ for the entire range $x \leqslant y \leqslant 1$, but that smaller values of y are irrelevant. We do need therefore to extrapolate the data to $x \to 1$, but this is usually a rather reliable process.

Now, at leading order, there are four distinct parton splitting processes to be considered, these are depicted in figure 3.33. They describe, from left to right, the splitting functions P_{qq}, P_{qg}, P_{gq}, P_{gg}. Each leads to a different, calculable function P_{ij}. The last three contribute to the evolution of flavour-singlet (sea quark and gluon) densities. The second, P_{qg}, represents the possibility that a quark (or antiquark), via *emission* of a quark (or antiquark), effectively transforms into a gluon, which then scatters with some other object in a high-energy collision. The third, P_{gq}, represents

---

[28] The 2021 Nobel Prize in Physics was awarded to Giorgio Parisi 'for the discovery of the interplay of disorder and fluctuations in physical systems from atomic to planetary scales'.

[29] In a sense it may be considered a sort of diffusion equation.

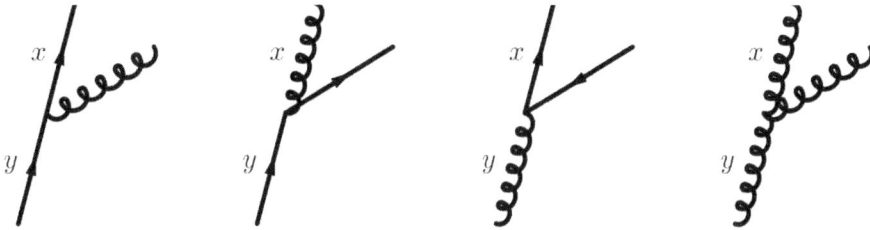

Figure 3.33. The four possible DGLAP splitting processes. The leftmost is that appearing in equation (3.5.32), while the others represent the three possible singlet splittings, one converting a quark into a gluon and the last two stemming from a gluon.

the natural pair-creation process (the final scattering object may be either a quark or antiquark). Finally, the fourth, P_{gg}, is the probability (peculiar to non-Abelian theories) that a gluon may also split into a gluon pair, one of which then interacts with the external probe.

At next-to-leading order, double-parton emission must be considered and the equations immediately become rather more complicated. The perturbative expansion is obtained by considering more and more contemporary such ramifications and the calculations become increasingly arduous to perform. The splitting functions are known to third order in α_s (Moch *et al* 2004). Moreover, Altarelli and Parisi (1977) had also already included the helicity-dependent versions at leading order, although they had been evaluated earlier (Sasaki 1975, Ahmed and Ross 1975). The helicity-dependent splitting functions have also been calculated to third order (Blümlein *et al* 2022).

The evolution is vital to the correct fitting of data taken at differing energy scales in various experimental setups and also to providing accurate predictions for higher-energy processes, such as $W^\pm$ or Higgs boson production in hadron–hadron collisions. For a more thorough explanation of the very technical aspects of higher-order calculations in perturbative QCD and also the methods of implementation in data fitting and cross-section prediction, see e.g. the CTEQ review by Brock *et al* (1995). A clear and concise review of the present situation has been presented by Johannes Blümlein (2023).

3.5.9 The Balitsky–Fadin–Kuraev–Lipatov (BFKL) pomeron

The evolution described above has the particular effect of causing the distributions to rise at small x. How strongly they rise depends on the particular splitting considered. It is natural to ask whether such a rise could have anything to do with the pomeron described earlier. That is, can it have a determining effect on the leading power of $1/x$ for small x? The question is important as the rôle and nature of the pomeron so far described are unclear.

Moreover, Sandy Donnachie and Peter Landshoff (1998) have over the years analysed in great detail the high-energy behaviour of hadronic cross-sections, coming to the conclusion that there must be two distinct pomerons. One (the so-called *soft* pomeron) should have an intercept of around 1.08, as shown above, while the other (the so-called *hard* pomeron) has an intercept around 1.4. This second pomeron is also strongly indicated by the DIS data collected by the DIS experiments H1 and ZEUS at HERA (DESY), where the very high energies available (27.5 GeV electron or positrons and 820-GeV protons collided to provide a centre-of-mass energy of 300 GeV) allowed parton densities to measured down to x below 10^{-5}, with Q^2 still large enough to justify analysis via perturbative QCD (Adloff *et al* 1997, Breitweg *et al* 1997).

Just 20 years earlier Ian Balitsky, Victor Fadin, Eduard Kuraev and Lev Lipatov, (Kuraev *et al* 1977, Balitsky and Lipatov 1978) had improved the perturbative QCD analysis of the evolution, deriving an asymptotic behaviour and calculating the effective leading intercept, finding

$$\alpha_0 = 1 + 12 \ln 2 \, \frac{\alpha_s(Q^2)}{\pi}. \tag{3.5.33}$$

The value of the intercept thus depends on the precise value of Q^2, but taking a representative value of, say, $\alpha_s(Q^2) = 0.2$ (i.e. for Q^2 relatively large), we find an intercept $\alpha_0 \simeq 1.53$. This was clearly very much in line with what was later observed experimentally and led to the term 'BFKL pomeron'. However, the picture is still far from clear and there is continuing debate on the nature and behaviour of the two pomerons (see e.g. Donnachie and Landshoff 1998).

3.5.10 Other quark–parton model topics

In closing this chapter, we should at least make mention of certain topics relating to the quark–parton model not covered here in detail, but which are nonetheless rather important.

3.5.10.1 Meson and other parton densities

First of all, the entire exercise may now be carried out in parallel for mesons; the main difference is merely that there are then only two valence objects: the quark and antiquark. Indeed, any hadron might be considered. Measurements can be performed by colliding, e.g. pion beams with standard hydrogen or other nuclear targets. Experimentally, only very limited information is available for the charged pions and almost nothing for other hadrons. Moreover, at sufficiently high-momentum transfers even a *real photon* may be viewed in the same manner; here, of course, there are *no* valence quarks and the entire partonic content is composed of sea quarks and gluons.

3.5.10.2 Spin correlations

There is another label that may be attached to the parton distributions: namely, spin. That is, we may ask how the spin or helicity of a quark or gluon is correlated to the spin or helicity of the parent hadron. Starting from the mid-seventies (and with ever-increasing interest since the late eighties) a number of experimental groups have been active in this area. Again, we can find constraining sum rules (see e.g. the Bjorken sum rule), in this case related to the axial-vector couplings of nucleon and hyperon β-decays.

3.5.10.3 Fragmentation functions

We must also mention that the, so-to-speak, *inverse* processes may be defined and studied. That is, we may ask the probability that a given quark emerging from the hard-scattering with the DIS photon will materialise in the laboratory as a given hadron with a given fraction of the parent quark's energy. In such a way, it is natural to define so-called *fragmentation* functions, which can also be studied experimentally (especially in e^+e^- collisions). The situation is, however, rather more complex than for the case of distribution functions, from both the theoretical and experimental points of view. We shall comment a little more on the process of fragmentation or *hadronisation* in the next chapter, where we shall present quark-model inspired pictures.

3.5.10.4 Fracture functions

A further development in recent years has been the study of hadrons emerging from the *remnants* of initial hadronic state. That is, those hadrons whose genealogies are not traceable to the struck quark, but rather to so-called *spectator* quarks inside the initial hadron (i.e. those that did not interact directly with the external probe). We can then define *fracture* functions. We shall, however, not delve further into this subject and simply refer the reader to the literature (see, e.g., Trentadue and Veneziano 1994).

3.5.10.5 Exotic quark and gluon states

So far we have only mentioned the simplest and natural states predicted by the quark model: namely, the $q\bar{q}$ mesons and qqq baryons. There is, however, no compelling reason to restrict our attention to such basic combinations and the colour-charge carrying (and not to say mass-generating) gluons immediately provide the possibility of structures known as *glueballs*; i.e. bound states made up purely of gluons without valence-type quarks (although a sea of quark–antiquark pairs will always be present). Moreover, as has already been mentioned in the context of the SU(3) classification and, in particular, the qqq colour-singlet baryons, we might also consider more exotic forms of bound states such as $q\bar{q}q\bar{q}$ and $qqqq\bar{q}$, known, respectively, as tetraquarks and pentaquarks (Fritzsch and Gell-Mann 1973).

The possibility of meson–meson or even meson–nucleon molecules ($q\bar{q}$–$q\bar{q}$ and qqq–$q\bar{q}$, is also actively considered; such states would appear very similar to the tetraquarks and pentaquarks just mentioned. Finally, the hexaquark (or dibaryon), which may be either $qqqqqq$ or $qqq\bar{q}\bar{q}\bar{q}$ (or qqq–qqq and qqq–$\bar{q}\bar{q}\bar{q}$), is a theoretical possibility too (Dyson and Xuong 1964) and there exists an experimental candidate, the so-called $d^*(2380)$ (Bashkanov *et al* 2009, Adlarson *et al* 2011).

Glueballs: the question of bound states consisting only of gluons immediately arises, since the gluons (in contrast to photons) are also carriers of the charge they mediate. It should thus clearly be possible to form a colour-neutral state with two or more gluons bound together. Of course, lacking the ability to reliably calculate bound states, there is no rigorous proof for this. In any case, lattice-QCD calculations, performed for a theory without quarks, predict the lightest glueball to have $J^{PC} = 0^{++}$ and a mass around 1600 to 1700 MeV (which could be lower in the full theory). On the experimental side the situation is rather cloudy. The main problem is that this mass region is rich with states that are already difficult to classify with certainty and which, in any case, should probably mix with any such pure glue state, leaving at best some sort of hybrid superposition of gg and $q\bar{q}$ states, or even possible meson–meson molecules.

Tetraquarks: a tetraquark is a theoretically possible bound state containing two quarks and two antiquarks[30]. For example, the $a_0(980)$ and $f_0(980)$ might be tetraquark states, but they could also be KK molecules, given their large branching ratios into KK, despite the very limited phase space. However, it turns out that the serious candidates experimentally identified appear to consist of one light and one heavy $q\bar{q}$ pair. The first confirmed such tetraquark state was reported independently in 2013 by two experiments: BES III in Beijing (Ablikim *et al* 2013) and Belle at KEK (Liu *et al* 2013). The two groups both published observations of the $Z_c^\pm(3900)$, which is deduced to be a $q\bar{q}'c\bar{c}$ state. Remarkably, this state was discovered as a resonance peak in the decay products of yet another exotic particle, the $Y(4260)$. The Z_c mass and its predominant decay into $\pi^\pm J/\psi$ indicate that it contains a $c\bar{c}$ pair and has a non-zero charge, which then requires the presence of another quark–antiquark pair of different flavours (e.g. $u\bar{d}$ or charge conjugate).

Pentaquarks: in the early 2000s several experiments also observed other resonance peaks in the mass spectra of the final states they were producing. The masses of many of these resonances appeared around 1.54 GeV with narrow widths and were interpreted as evidence for an exotic pentaquark state, denoted Θ^+, composed of the quarks $uudd\bar{s}$, already predicted theoretically by Diakonov *et al* in 1997. An example reaction was $\gamma n \to K^+ K^- n$, using 2.4 GeV photons on a carbon target (Nakano *et al* 2003). There was, however, no compelling evidence for identification as a pentaquark (or alternatively a baryon–meson molecule) and many experiments failed to find such an object.

Then in 2015, the LHCb collaboration (Aaij *et al* 2015) presented evidence for pentaquarks from the final-state invariant-mass distributions in $\Lambda_b^0 \to J/\psi\, K^- p$. This was interpreted as the decay of the Λ_b^0 via an intermediate $K^- P_c^+$ state with $P_c^+ \to J/\psi\, p$, instead of the expected $\Lambda^* J/\psi$ followed by $\Lambda^* \to K^- p$; see figure 3.34. From a detailed analysis of the final state, the experiment excluded known mechanisms and indicated two narrow states close together, $P_c^+(4380)$ and

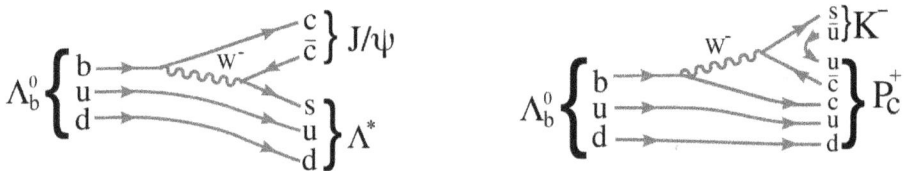

Figure 3.34. The Feynman diagrams for (left) $\Lambda_b^0 \to J/\psi\Lambda^*$ and (right) $\Lambda_b^0 \to K^- P_c^+$ decays; figures taken with permission from Aaij *et al* (2015), CC BY 3.0.

---

[30] Note that, despite the name, it cannot contain four quarks as this could not be a colour-neutral state.

$P_c^+(4450)$. Both were observed to decay strongly into $J/\psi\ p$ and hence should have a valence quark content $uudc\bar{c}$.

Hexaquarks and dibaryons: the last of the exotic multi-quark states we shall examine are those containing either six quarks, or three quarks and three antiquarks. Such states were hypothesised long ago, but have only relatively recently found any genuine experimental support. As with the previously discussed multi-quark states, there are two (theoretically) distinct possibilities: compact six-quark/antiquark clusters, or bound baryon/antibaryon hadronic molecules. Experimentally though, the distinction is not clear and the objects observed cannot easily be classified.

The earliest consideration of such states appears to be by Robert Jaffe (1977). The prediction was made using the bag model of hadrons (Chodos *et al* 1974, DeGrand *et al* 1975). The lightest such object is predicted to be an *s*-wave flavour-singlet dihyperon with $J^P = 0^+$ and mass 2150 MeV. Given the predicted mass ($<2m_{\Lambda^0}$), the new state can only decay weakly. Note that, were it lighter than $m_p + m_{\Lambda^0} = 2055$ MeV, then its decay would be doubly weak and it would thus be very long lived (cf nuclear double β-decay). The model also predicts a slightly heavier *s*-wave dibaryon flavour-octet with $J^P = 1^+$, which would probably decay strongly and therefore be short-lived.

3.5.11 Reconciling the Gell-Mann and Feynman pictures

As a final note, let us observe that, whereas we have suggested here that there is a direct correspondence between the quarks of Gell-Mann and the partons of Feynman, the physical connection is not at all simple or straightforward. The Gell-Mann picture is tied to a very stationary hadron, whereas the Feynman picture is constructed in an infinite-momentum frame.

3.5.11.1 Constituent quarks
The objects described in Gell-Mann's theory have to do with the static properties of hadrons (charge, isospin, strangeness etc) and certainly no account is given of the gluonic field that presumably binds them. Indeed, for such a picture to make sense, the quarks should be non-relativistic; they should, moreover, provide the masses of the baryons in which they reside (and are in fact confined). We thus talk of *constituent* quarks with, for example, a constituent mass that depends on the specific environment. For baryons, this must therefore be of the order of 310 MeV for up and down quarks (this is also roughly consistent with the baryon magnetic moments) and 500 MeV for strange quarks (again in line with the hyperon magnetic moments). These *cannot*, however, be considered *elementary* particles.

3.5.11.2 Current quarks
On the other hand, the infinite-momentum frame used by Feynman and also the kinematics of, e.g. DIS, imply very light partons with masses of just a few MeV[31]. These are known as *current* quarks; they are what should appear in the QCD

---

[31] The particular values are deduced from other considerations.

Lagrangian and which, in the perturbative sense, know nothing of confinement. That is, they are what may be considered *elementary* particles. In this sense, Feynman's model is somewhat closer to the field-theoretical description of quarks and their interactions via QCD, which leads to a picture of the hadrons as being highly complex superpositions of multi-quark and antiquark states bound (non-perturbatively) by the gluon field. It provides a very accurate description of high-energy processes. The major part of the hadron mass is then believed to be generated as just the potential energy of the gluon field.

3.5.11.3 Constituent versus current quarks

We might imagine then constituent quarks as being current quarks dressed up, so to speak, in a cloud of virtual gluons and also quark–antiquark pairs; it is the resulting interaction energy that provides their constituent masses. We thus see that mass is always considered as being generated via the interaction of the gluons; in the Gell-Mann model this is hidden inside the constituent-quark structure and, while the Feynman model does not directly address the question, the assumption would be that it is still the gluon field that generates the hadron masses measured experimentally.

3.6 Further QCD topics

We shall now briefly discuss a few topics of a more theoretical nature that nevertheless have phenomenological repercussions or relevance. It is not intended to provide a detailed working explanation of any of the following, but rather an idea of what is known and what can be done, together with principal results and achievements.

3.6.1 Lattice gauge theory

In any case, we still have no fully reliable *ab initio* method of calculating general (static or low-energy) hadron properties (e.g. the wave-functions or other properties, such as masses, magnetic moments etc) apart from the lattice QCD approach, formulated by Kenneth Wilson[32] in 1974. This is a non-perturbative method whereby space–time is discretised in order to perform numerical Monte-Carlo type calculations (for a brief review, see e.g. *Lattice Quantum Chromodynamics* in Navas *et al* 2024, p 337).

There are unfortunately various technical and also conceptual difficulties. For example, the very large number of degrees of freedom involved in a non-Abelian theory such as QCD requires very powerful (super) computers and present-day calculations can only achieve something of the order of 30 to 40 points in each of the $3 + 1$ space–time dimensions. The limitation is determined by the chosen lattice spacing (the physical distance between lattice points): it should be both small enough to allow an accurate description of the small-scale structure of, say, the proton and,

---

[32] The 1982 Nobel Prize in Physics was awarded to Kenneth G Wilson 'for his theory for critical phenomena in connection with phase transitions'.

at the same time, that the 30 to 40 steps should imply a volume sufficiently large to allow freedom of movement and avoid surface effects. For example, a lattice spacing of order 0.1 fm is still rather too coarse for good sensitivity to the internal structure, but allows only 3 to 4 fm for any dynamics.

Another difficulty lies in the fact that discretisation introduces the concept of Briouillin zones (just as in the description of crystals), which in turn lead to extra zeros in the dispersion relations (or poles in the spectrum) and a doubling of any fermions introduced for each of the space–time dimensions; i.e. the attempted introduction of a single fermion leads automatically to $2^4 = 16$ identical fermions. This incumbrance is an artifact of the discretisation that does not, however, disappear smoothly in the continuum limit. While there are techniques to circumvent such peculiarities, they also lead to extra complexity of the calculations. It should also be mentioned that the continuum limit is never necessarily guaranteed to be smooth and there may be other spurious effects due to the discretisation that do not disappear smoothly.

Of course, computing power is continually improving and more means to overcome the various conceptual problems are being discovered. Indeed, lattice QCD has already provided useful results regarding form factors, important for weak-interaction physics and also, for example, on the parton distribution functions (a compendium of results is given in Aoki *et al* 2020). More in-depth and specific review articles may be found in Hägler (2010) and Capitani (2003).

3.6.2 The strong-CP problem

Having dealt earlier with the question of CP violation in the weak interaction, we should now also mention that there is a long-standing problem regarding CP in the strong interaction or QCD. Here though the surprise is in a sense the opposite: i.e. why is CP violation *not* observed in the strong interaction? The problem, as we shall now discuss, is that there is again a natural way to violate CP in QCD, so natural indeed that there is no reason it should not occur. The theoretically generally preferred explanation of the non-occurrence requires the existence of new particles called *axions*. Experimental searches for such objects have now become routine parts of high-energy experiments and so we should now examine the phenomenology of these particles. First, we shall motivate the theoretical problem and its possible solution.

3.6.2.1 Strong-CP theory

Although the accepted Lagrangian of QCD is as shown in equation (3.4.4), a further two (CP-violating) terms could have been included:

$$\theta \frac{g^2}{32\pi^2} F^{a\mu\nu} \tilde{F}^a_{\mu\nu} \quad \text{and} \quad m\bar{q}e^{i\theta'\gamma_5}q, \tag{3.6.1}$$

where $\tilde{F}^a_{\mu\nu} := \frac{1}{2}\varepsilon_{\mu\nu\rho\sigma}F^{a\rho\sigma}$ is the dual gauge field-strength tensor and θ, θ' are free parameters determining the amounts of the CP violation generated by two terms.

The presence of $\varepsilon_{\mu\nu\rho\sigma}$ in the first term and that of the Dirac matrix γ_5 in the second clearly indicate the violation of both parity and time-reversal invariance or CP.

The second term would just be a standard mass term for the fermion (quark) field if θ' were zero, but violates both P and T if it is non-zero. This phase may, however, be rotated away by applying the following (arbitrary) chiral transformation:

$$q \to e^{-\frac{1}{2}i\theta'\gamma_5}q \text{ and } \bar{q} \to \bar{q}e^{-\frac{1}{2}i\theta'\gamma_5}. \tag{3.6.2}$$

It turns out that such a phase rotation also alters the value of θ and therefore, while either of these two terms may be eliminated individually, it is not possible to eliminate both simultaneously, unless $\theta \equiv \theta'$, an equality for which there would be no physical justification. Therefore, only one of the terms is usually considered: namely, $F \cdot \tilde{F}$. Note that the presence of non-zero (real) mass terms for the quarks also implies that this term cannot be simply rotated away, although it could were the quarks all massless.

Now, the parameter θ (which is also a sort of phase-rotation angle) could, of course, take on any value (just like any other physical parameter such as mass, charge etc). Since it is essentially just a sort of switch for this term, we may reasonably assume though a natural value of order unity. We then obtain a prediction for the neutron electric dipole moment (EDM) that turns out to be very much larger than present experimental limits. In fact, there is no experimentally observed violation of CP in QCD. And, since there is no obvious reason for it to be conserved by QCD, this becomes a so-called *fine-tuning* problem, known as the strong CP-problem. That is, the θ parameter is extremely small (with respect to its naturally expected magnitude) despite there being no known specific symmetry (exact or approximate) that would force it to be precisely zero or so small[33].

3.6.2.2 Neutron electric dipole moment

Let us take a moment to understand one of the important implications of a possible violation of CP in QCD: namely, a non-zero neutron electric dipole moment. So, first of all we should explain why the neutron electric dipole moment should be zero if CP is a perfect symmetry of particle physics.

To generate a dipole moment, a particle should have a non-trivial (internal) charge distribution. In the case of nucleons this is provided by the Gell-Mann quark model: the neutron contains one u and two d quarks, having opposite charges. Their spatial separation might be expected to be of the order of the size of the neutron itself and so a reasonable estimate for the magnitude of the neutron EDM would be

$$|d_n| \approx O(1\,e\,\text{fm}). \tag{3.6.3}$$

The problem remains as to the orientation; it is, of course, a vector-like quantity. The only axis that can be associated with a nucleon is that defined by its spin or

---

[33] The possibility of EDMs for elementary particles had been considered very early on, even before the discovery of CP violation in weak-interaction physics (Purcell and Ramsey 1950). Moreover, experimental limits had already been derived: $|d_n| < 0.5 \times 10^{-8}e$ fm (Smith *et al* 1957).

magnetic moment, but this is an axial-vector. Under spatial inversion (or parity transformation) the EDM changes sign, whereas the spin does not, while under temporal inversion the EDM remains unchanged but the spin inverts. The conclusion is therefore that a non-zero EDM would violate both P and T (and thus too CP)[34].

The present limit on the neutron EDM is (Navas *et al* 2024).

$$|d_n| < 0.18 \times 10^{-12} e \text{ fm} \quad (90\% \text{ CL}) \tag{3.6.4}$$

Note that the CP-violation present in the weak interaction can at best lead to a neutron EDM *smaller* than such a limit. The problem engendered is to explain why the physical value, if not zero, should be more than twelve orders of magnitude smaller than the naturally expected theoretical value. Such cases are normally attributed to the existence of some symmetry (exact or partial) that forces the value to be zero or vanishingly small. We shall now discuss a different idea, which involves the hypothesis of new particles and thus possible new phenomena for which to search experimentally.

3.6.2.3 Axions

Over the years, numerous solutions for the so-called strong-CP problem have been proposed. One of the best-known is the Peccei–Quinn theory (Peccei and Quinn 1977), involving new scalar particles called *axions* (Weinberg 1978, Wilczek 1978). The theory is somewhat involved and cannot be simply explained here, The idea though of Roberto Peccei and Helen Quinn was to invoke an new axial-vector U(1) symmetry to guarantee the smallness (or near vanishing) of the θ parameter introduced above. This symmetry is broken via a Higgs-like mechanism (see later), which thus leads to neutral pseudoscalar Goldstone boson, called an axion. A rather comprehensive introduction to the physics of axions may be found in Kuster *et al* (2008).

According to early calculations (Weinberg 1978, Wilczek 1978), the mass of the predicted particle should lie in the range 100 keV to 1 MeV. Such a light particle could then only decay into two photons and the estimated lifetime would be of order 0.01 to 10 s. There are a variety of experimentally measurable effects that its presence would have. We shall examine some of these and the related experimental searches in the section on dark matter in the final chapter.

References

Aaij R *et al* 2015 LHCb Collab *Phys. Rev. Lett.* **115** 072001
Ablikim M *et al* 2013 BESIII Collab *Phys. Rev. Lett.* **110** 252001
Adlarson P *et al* 2011 WASA-at-COSY Collab *Phys. Lett.* **106** 242302 *Phys. Rev.* **C90** 035 204; *Phys. Rev. Lett.* **112** 202 301; *Phys. Lett.*, **B743** 325
Adler S L 1966 *Phys. Rev.* **143** 1144

---

[34] Note incidentally that the CP-violating term introduced above $\tilde{F} \cdot F \propto \boldsymbol{E} \cdot \boldsymbol{B}$, which again manifestly violates both P and T.

Adler S L 1969 *Phys. Rev.* **177** 2426

Adloff C *et al* 1997 H1 Collab *Nucl. Phys.* **B497** 3

Ahmed M A and Ross G G 1975 *Phys. Lett.* **B56** 385

Altarelli G, Ellis R K and Martinelli G 1978 *Nucl. Phys.* **B143** 521 *erratum* ibid **B146** 544

Altarelli G and Parisi G 1977 *Nucl. Phys.* **B126** 298

Álvarez L W 1973 *Phys. Rev.* **D8** 702

Amaudruz P *et al* 1991 New Muon Collab *Phys. Rev. Lett.* **66** 2712

Anselmino M, Efremov A and Leader E 1995 *Phys. Rep.* **261** 1 *erratum* ibid **281** 399

Aoki S *et al* 2020 Flavour Lattice Averaging Group *Phys. Rev.* **C80** 113 *Phys. Rev.* **D50** R1

Baikov P A, Chetyrkin K G and Kühn J H 2014 *JHEP* **10** 076

Baikov P A, Chetyrkin K G and Kühn J H 2017 *Phys. Rev. Lett.* **118** 082002 *(transl. Sov. J. Nucl. Phys.* **28** 822)

Balitsky I I and Lipatov L N 1978 *Yad. Fiz.* **28** 1597

Barnes V E *et al* 1964 *Phys. Rev. Lett.* **12** 204

Barone V, Drago A and Ratcliffe P G 2002 *Phys. Rep.* **359** 1

Bartel W *et al* 1967 *Phys. Lett.* **B25** 236

Bartel W *et al* 1968 *Phys. Lett.* **B28** 148

Bashkanov M *et al* 2009 *Phys. Rev. Lett.* **102** 052301

Bég M A B, Lee B W and Pais A 1964 *Phys. Rev. Lett.* **13** 514 *erratum* ibid 650

Behnke O, Geiser A and Lisovyi M 2015 *Prog. Part. Nucl. Phys.* **84** 1

Bell J S and Jackiw R 1969 *Nuovo Cim.* **A60** 47

Bjorken J D 1966 *Phys. Rev.* **148** 1467

Bjorken J D 1969 *Phys. Rev.* **179** 1547

Bloom E D *et al* 1969 *Phys. Rev. Lett.* **23** 930

Blümlein J 2023 http://www.arXiv.org/abs/arXiv:2306.01362

Blümlein J, Marquard P, Schneider C and Schönwald K 2022 *JHEP* **11** 156

Bodek A *et al* 1974 *Phys. Lett.* **B51** 417

Bonamy P *et al* 1966 *Phys. Lett.* **23** 501

Breidenbach M *et al* 1969 *Phys. Rev. Lett.* **23** 935

Breitweg J *et al* 1997 ZEUS Collab *Phys. Lett.* **B407** 432

Brock R *et al* 1995 CTEQ Collab *Rev. Mod. Phys.* **67** 157

Brodsky S J and Farrar G R 1973 *Phys. Rev. Lett.* **31** 1153

Callan C G Jr+ and Gross D J 1969 *Phys. Rev. Lett.* **22** 156

Capitani S 2003 *Phys. Rep.* **382** 113

Chew G F and Frautschi S C 1962 *Phys. Rev. Lett.* **8** 41

Chodos A, Jaffe R L, Johnson K, Thorn C B and Weisskopf V F 1974 *Phys. Rev.* **D9** 3471

Close F E 1979 *An Introduction to Quarks and Partons* (New York: Academic)

Coleman S R and Glashow S L 1961 *Phys. Rev. Lett.* **16** 423 *Phys. Rev.* **134** (1964) B671

Collins P D B 1977 *An Introduction to Regge* (Cambridge: Cambridge University Press)

Collins P D B and Martin A D 1984 *Hadron Interactions* (Adam Hilger)

Deden H *et al* 1975 Gargamelle Neutrino Collab *Nucl. Phys.* **B85** 269

DeGrand T, Jaffe R L, Johnson K and Kiskis J 1975 *Phys. Rev.* **D12** 2060

Deur A, Brodsky S J and De Téramond G F 2019 *Rep. Prog. Phys.* **82** 076201

Diakonov D, Petrov V and Polyakov M V 1997 *Z. Phys.* **A359** 305

Dirac P A M 1926 *Proc. R. Soc. Lond.* **A112** 661

Dirac P A M 1928 *Proc. R. Soc. Lond.* **A117** 610

Dokshitzer Y L 1977 *Zh. Eksp. Teor. Fiz.* **73** 1216 *transl. Sov. Phys. JETP* **46** 641

Donnachie A and Landshoff P V 1998 *Phys. Lett.* **B437** 408

Donnachie S, Dosch H G, Landshoff P and Nachtmann O 2004 *Pomeron Physics and QCD* **vol 19** (Cambridge: Cambridge University Press)

Drell S D and Yan T M 1970 *Phys. Rev. Lett.* **25** 316 *erratum* ibid 902

Dyson F and Xuong N-H 1964 *Phys. Rev. Lett.* **13** 815 *erratum* ibid **14**, 339

Eden R J, Landshoff P V, Olive D I and Polkinghorne J C 1966 *The Analytic S-Matrix* (Cambridge: Cambridge University Press)

Eichten E J, Hinchliffe I and Quigg C 1992 *Phys. Rev.* **D45** 2269

Fermi E 1926 *Rend. Acc. Naz. Lincei* **3** 145

Fermi E and Yang C N 1949 *Phys. Rev.* **76** 1739

Feynman R P 1969 *Phys. Rev. Lett.* **23** 1415

Feynman R P 1972 *Photon-Hadron Interactions* (W.A. Benjamin)

Field R D and Feynman R P 1977 *Phys. Rev.* **D15** 2590

Friedman J I and Kendall H W 1972 *Annu. Rev. Nucl. Part. Sci.* **22** 203

Fritzsch H and Gell-Mann M 1973 *Proc. of the XVI Int. Conf. on High-Energy Physics–ICHEP 72 (Batavia, September 1972)* p 135

Geiger H and Marsden E 1909 *Proc. R. Soc. Lond.* **A82** 495

Gell-Mann M 1953 *Phys. Rev.* **92** 833

Gell-Mann M 1961 Caltech preprint CTSL-20

Gell-Mann M 1962 *Phys. Rev.* **125** 1067

Gell-Mann M 1995 *The Quark and the Jaguar: Adventures in the Simple and the Complex* (Henry Holt and Co.) *reprinted in The Eightfold Way* ed M Gell-Mann and Y Ne'eman (Perseus Pub, 2000)

Gilman F J 1972 *Phys. Rep.* **4** 95

Godizov A A 2016 *Eur. Phys. J.* **C76** 361

Goldberg M *et al* 1964 *Phys. Rev. Lett.* **12** 546

Gottfried K 1967 *Phys. Rev. Lett.* **18** 1174

Greenberg O W 1964 *Phys. Rev. Lett.* **13** 598

Gribov V N 1961 *Zh. Eksp. Teor. Fiz.* **41** 67 *transl. Sov. Phys. JETP* **14** 478; *transl. Sov. J. Nucl. Phys.* **15** 438

Gribov V N and Lipatov L N 1972 *Yad. Fiz.* **15** 781 *Int. J. Mod. Phys.* **A20** 5717; *Rev. Mod. Phys.* **77** 837

Gross D J 2005 *Proc. Natl. Acad. Sci.* **102** 9099

Gross D J and Llewellyn Smith C H 1969 *Nucl. Phys.* **B14** 337

Gross D J and Wilczek F 1973a *Phys. Rev.* **D8** 3633

Gross D J and Wilczek F 1973b *Phys. Rev. Lett.* **30** 1343

Gross D J and Wilczek F 1974 *Phys. Rev.* **D9** 980

Hagiwara K, Matsumoto S, Haidt D and Kim C S 1994 *Z. Phys.* **C64** 559 [Erratum: Z. Phys. C 68, 352 (1995)]

Hägler D I 2010 *Phys. Rep.* **490** 49

Han M-Y and Nambu Y 1965 *Phys. Rev.* **139** B1006

Heisenberg W 1927 *Z. Phys.* **43** 172

Irving A C and Worden R P 1977 *Phys. Rep.* **34** 117

Jaffe R 1977 *Phys. Rev. Lett.* **38** 195 *erratum* ibid 617

Kalbfleisch G R *et al* 1964 *Phys. Rev. Lett.* **12** 527

Kruglinski S 2009 *Discover Magazine* 17 March 2009; *transl. Sov. Phys. JETP* **45** 199 https://www.discovermagazine.com/the-sciences/the-man-who-found-quarks-and-made-sense-of-the-universe

Kuraev E A, Lipatov L N and Fadin V S 1977 *Zh. Eksp. Teor. Fiz.* **72** 377

Kuster M, Raffelt G and Beltrán B (ed) 2008 *Axions: Theory, Cosmology, and Experimental Searches* **vol 741** (Berlin: Springer)

Landau L D, Abrikosov A A and Khalatnikov I M 1954 *Dokl. Akad. Nauk. SSSR* **95** 497

Levine I *et al* 1997 TOPAZ Collab *Phys. Rev. Lett.* **78** 424

Liu Z Q *et al* 2013 Belle Collab *Phys. Rev. Lett.* **110** 252002

Low F E 1975 *Phys. Rev.* **D12** 163

Maxwell J C 1865 *Phil. Trans. R. Soc.* **155** 459

Miyabayashi K *et al* 1995 TOPAZ Collab *Phys. Lett.* **B347** 171

Moch S, Vermaseren J A M and Vogt A 2004 *Nucl. Phys.* **B688** 101

Muta T 1998 *Foundations of Quantum Chromodynamics: An Introduction to Perturbative Methods in Gauge Theories* 2nd edn (Singapore: World Scientific)

Nachtmann O 1972 *Nucl. Phys.* **B38** 397

Nakano T and Nishijima K 1953 *Prog. Theor. Phys.* **10** 581

Nakano T *et al* 2003 LEPS Collab *Phys. Rev. Lett.* **91** 012002

Navas S *et al* 2024 Particle Data Group *Phys. Rev.* **D110** 030001

Nussinov S 1975 *Phys. Rev. Lett.* **34** 1286; *Phys. Rev.* **D14** 246

Ohnuki Y 1960 *Proc. of the Int. Conf. on High-Energy Physics (Geneva)* p 843

Okubo S 1962 *Prog. Theor. Phys.* **27** 949

Owens J F, Accardi A and Melnitchouk W 2013 *Phys. Rev.* **D87** 094012

Pauli W 1925 *Z. Phys.* **31** 765

Peccei R D and Quinn H R 1977 *Phys. Rev. Lett.* **38** 1440

Pevsner A *et al* 1961 *Phys. Rev. Lett.* **7** 421

Politzer H D 1973 *Phys. Rev. Lett.* **30** 1346

Politzer H D 1974 *Phys. Rep.* **14** 129

Preparata G, Ratcliffe P G and Soffer J 1991 *Phys. Rev. Lett.* **66** 687

Purcell E M and Ramsey N F 1950 *Phys. Rev.* **78** 807

Raspe R E 1786 *Baron Münchhausen's Narrative of his Marvellous Travels and Campaigns in Russia* (Utrecht: Stichting De Roos, 1952)

Regge T 1959a *Nuovo Cim.* **14** 951

Regge T 1959b *Nuovo Cim.* **14** 951

Regge T 1960 *Nuovo Cim.* **18** 947 ibid **18** (1960) 947

Roberts R G 1990 *The Structure of the Proton: Deep Inelastic Scattering* (Cambridge: Cambridge University Press)

Rosenbluth M N 1950 *Phys. Rev.* **79** 615

Rutherford E 1911 *Phil. Mag.* **21** 669

Sakata S 1956 *Prog. Theor. Phys.* **16** 686

Sakita B 1964 *Phys. Rev. Lett.* **13** 643

Sasaki K 1975 *Prog. Theor. Phys.* **54** 1816

Shifman M A, Vainshtein A I and Zakharov V I 1979 *Nucl. Phys.* **B147** 385

Sirunyan A M *et al* 2020 CMS Collab *Phys. Lett.* **B803** 135263

Smith J H, Purcell E M and Ramsey N F 1957 *Phys. Rev.* **108** 120 't Hooft G 1999 *Proc. of the 3rd. Euroconf on Quantum Chromodynamics—QCD 98* **74**; N Narison 413 (Montpellier, July 1998); *Nucl. Phys. B (Proc. Suppl.)* **74**

Trentadue L and Veneziano G 1994 *Phys. Lett.* **B323** 201

Weinberg S 1978 *Phys. Rev. Lett.* **40** 223

Wilczek F 1978 *Phys. Rev. Lett.* **40** 279

Wilson K G 1974 *Phys. Rev.* **D10** 2445

Yan T-M and Drell S D 2014 *Int. J. Mod. Phys.* **A29** 1430071 *50 Years of Quarks*, ed H Fritzsch and M Gell-Mann (WSP, 2015), p 227

Yang C-N and Mills R L 1954 *Phys. Rev.* **96** 191

Yukawa H 1935 *Proc. Phys. Math. Soc. Jap.* **17** 48

Zweig G 1964 CERN preprint CERN-TH-401; *reprinted in Developments in the Quark Theory of Hadrons,* **vol 1** ed D B Lichtenberg and S P Rosen (Hadronic Press) 1964–78 p 22

IOP Publishing

An Introduction to Elementary Particle Phenomenology
(Second Edition)

Philip G Ratcliffe

Chapter 4

The new particles (discovery and study)

'There are two possible outcomes: if the result confirms the hypothesis, then you've made a measurement. If the result is contrary to the hypothesis, then you've made a discovery.'

Enrico Fermi

In this chapter we shall trace the history of particle discovery, from the early experiments exploiting the first available source of energetic particles: namely, cosmic rays, right up to the most energetic particle colliders available today, in all their various forms: e^+e^-, ep, pp, $p\bar{p}$ etc. Now, although ep colliders do exist even with very high-energy beams, they are not normally considered as suitable for particle discovery[1] and so, after examining the rôle of cosmic rays, we shall concentrate here on e^+e^- and hadron–hadron machines.

Before discussing the various types of experiments in detail, let us briefly examine the basic requirements for the discovery of new particles. Firstly, of the various reasons that at some given point in time a particle has not previously been discovered, the most common is that its mass is larger than the available energies. Obviously, there are also other possible explanations: it interacts too weakly to be detected (e.g. the neutrino), there are conserved quantum numbers that suppress production processes (e.g. the strange particles) etc. And it is usually more a combination of such effects. The top quark is a good example: its large mass requires high energy, but also the conserved top flavour implies production either via the weak interaction or as a $t\bar{t}$ pair, requiring even higher energy. In any case, the

---

[1] An exception is the early discovery of e.g. the so-called Δ resonances, which we shall discuss.

doi:10.1088/978-0-7503-5759-3ch4

question of mass is fundamental: if a particle is too heavy it simply cannot be produced[2].

To make further discoveries then, we need more control over the interacting system. In particular, we need to control (and raise) the energy. We also often need a *clean* initial system so that the details of the final state may emerge clearly and provide unambiguous indications of any new object produced. These requirements immediately suggest e^+e^- colliders as the prime candidate. The initial particles are point-like, well understood, can be produced with very precisely known energies and have no by-products that might pollute the final state. Having said that, hadron–hadron machines can reach much higher energies and therefore in certain circumstances are unavoidable.

In the following sections we shall systematically examine the discoveries made by direct particle production in dedicated electron–positron colliders and hadron–hadron machines. We begin though with a rather synthetic history of the various early particle discoveries and also those made via use of *ad hoc* secondary beams.

4.1 Cosmic rays and the early accelerator discoveries

It might be said that the birth of particle physics lies in cosmic rays, which were discovered by Victor Hess[3] in 1912 with experiments using balloons. Antimatter was first detected in cosmic rays, as were many of the first strange hadrons, not to mention the muon. Until the fifties cosmic rays were the only source of high-energy particles and still today they remain the source of the highest-energy particles that can be studied: events due to single particles up to $O(10^9 \text{ GeV})$ have been observed.

The energies of these nuclei lie mainly between 100 MeV and 10 GeV, with a long very high-energy tail; the highest recorded energy for a cosmic ray or particle to date is approximately 320 EeV[4] (Bird *et al*, 1995). This is, of course, much higher than any energy that may (or ever will) be produced in the laboratory. Unfortunately however, such events are also exceedingly rare[5] and thus of minimal use experimentally. They can though be exploited, for example, to estimate the proton–proton total cross-section for very high (and otherwise inaccessible) energies and thus check the predictive power of models for hadronic interactions.

4.1.1 The positron

In 1930 Dirac proposed the particle–hole interpretation of the solutions to the relativistic wave equation he had himself derived earlier (Dirac 1928). He had found that the equation not only correctly described the electron (including its gyromagnetic ratio of two), but also contained an object that was an exact copy of the

---

[2] In some cases it is possible to detect the existence of a particle indirectly via its contributions as an intermediate virtual state; however, sensitivity is usually very limited (it is typically a higher-order perturbative effect). For a long time this was the case of e.g. the top quark and the Higgs boson.

[3] The 1936 Nobel Prize in Physics was awarded equally to Victor Franz Hess 'for his discovery of cosmic radiation' and to Carl David Anderson 'for his discovery of the positron'.

[4] Recall that EeV = 10^{18} eV.

[5] The flux at the Earth's surface of such 'ultra-high-energy' cosmic particles (i.e. those exceeding 1 EeV) is approximately one per square kilometre per century.

electron with, however, the opposite charge: the antielectron or positron as it then became known.

Dirac's hole theory (1930) suggested that a photon of energy slightly more than 1 MeV could, in principle, produce an electron–positron pair. However, kinematics does not permit direct production or conversion and more energy is required. Such energy is found naturally in cosmic rays. What are commonly called cosmic rays are, of course, only the by-products of extremely high-energy collisions between, typically, protons of cosmic origin and nuclei in the Earth's atmosphere. Such collisions, although totally uncontrollable, give rise to all energetically accessible states. The problem then lies in detecting the particles produced.

In 1932, Carl Anderson (1933b)[6] and, independently a few months later, Blackett and Occhialini (1933)[7] detected the passage of positively charged particles, similar in mass to the electron, using Wilson cloud chambers. While Anderson did not immediately connect his discovery to Dirac's prediction, Blackett and Occhialini clearly recognised these particles as Dirac's positrons.

There was, however, much skepticism.

> 'Dirac has tried to identify holes with antielectrons...
> ...we do not believe that this explanation can be seriously considered.'
>
> Wolfgang Pauli

> '...even if all this [the positron discovery] turns out to be true, of one thing I am certain: that it has nothing to do with Dirac's theory of holes!'
>
> Niels Bohr

The idea behind the cloud chamber, invented by Charles Wilson (1911)[8], is that a charged particle passing through supersaturated water vapour provokes local condensation. The cloud chamber consists of a container, fitted with a piston, into which a saturated air–vapour mixture is injected. When the piston is moved suddenly to lower the pressure, the temperature also drops rapidly and the vapour passes into a supersaturated phase. Any charged particle traversing the chamber in this moment leaves a track of fine condensation droplets, which may be photo-graphed (possibly from two different angles so as to permit a stereo image). The presence of a magnetic field reveals the sign of the charge.

One of Anderson's positron events is displayed in figure 4.1. The track enters from the bottom with very high energy (deduced from the large radius of curvature

---

[6] For the associated Nobel prize, see footnote 3 in this chapter.

[7] The 1948 Nobel Prize in Physics was awarded to Patrick Maynard Stuart Blackett 'for his development of the Wilson cloud chamber method, and his discoveries therewith in the fields of nuclear physics and cosmic radiation'.

[8] The 1927 Nobel Prize in Physics was divided equally between Arthur Holly Compton 'for his discovery of the effect named after him' and Charles Thomson Rees Wilson 'for his method of making the paths of electrically charged particles visible by condensation of vapour'.

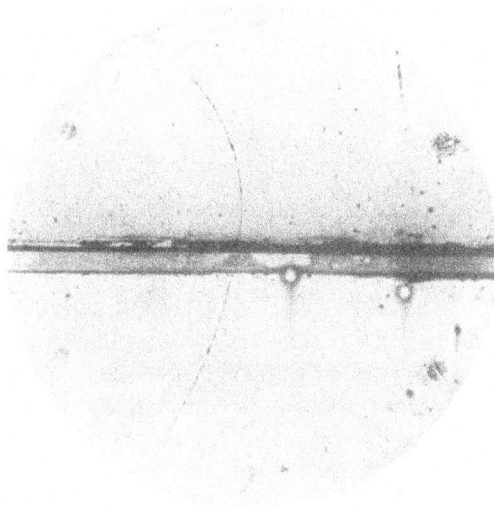

Figure 4.1. A Wilson cloud-chamber photograph showing the passage of a positron. The radius of curvature determines the momentum while the direction indicates the charge sign. The track length for such a relatively low-momentum particle allows the distinction between a positron and a proton; photograph reproduced with permission from Anderson (1933a), copyright (1933) by the American Physical Society.

in the 1.5 T magnetic field used). It then passes through a 6 mm lead strip, which has the purpose of slowing the particles to reveal their direction, and continues for nearly 3 cm before presumably annihilating with an atomic electron. The curvature of the upper track indicates a momentum of approximately 23 MeV. Were it a proton, this would correspond to a very low velocity and it is known that the range would then be only a few millimetres (recall that low-energy cross-sections are typically *inversely* proportional to the velocity).

4.1.2 The neutron

By the late 1920s it was clear that there must be some other particle inside the nucleus besides just the proton: the nuclear mass number is typical around twice the atomic number. Thus, alongside the Z protons, there must be a similar amount of neutral matter. Rutherford (1920) had already hypothesised what he called *'neutral doublets'* (effectively the neutron) formed from closely bound protons and electrons, which could then explain both the masses and charges of different nuclei. That is, in some way, a nucleus of atomic mass number A was supposed to contain A protons together with $A-Z$ electrons, which would then neutralise the charges of $A-Z$ protons leaving an overall charge of just Z.

However, such an explanation immediately runs into serious difficulties. First of all, it would require the intervention of some new attractive force (possibly short range) involving electrons, which should lead to an effective *nuclear* Bohr radius for electrons of O(1 fm). The problem then is due to both the small radius of the nucleus ($r_{nucl}/r_{atom} \approx 1/10^5$) and the small mass of the electron ($m_e/m_p \approx 1/2000$).

The former would imply (albeit *naïvely*) a new coupling constant $O(10^5)$ times larger than the standard electromagnetic coupling α! This would surprisingly and unfortunately be around even 100–1000 times the strong nuclear interaction coupling. It is thus very difficult to imagine that there should be no other manifestation of such a force. The latter, moreover, implies implausibly high kinetic energies. To see this consider the uncertainty principle, which tells us that a particle confined inside a radius R, say, will typically have momentum of order $\hbar c/R$ (recall that $\hbar c \sim 200$ MeV fm); a nuclear size thus implies momenta of order 100 MeV. Whereas a proton with that momentum has a kinetic energy of around 5 MeV (perfectly in line with nuclear-level energies), an electron would be highly relativistic with energy around 100 MeV, which is far too large to be physically tenable.

The second problem lies in the spin of such new particles: pairing up the $(A-Z)$ protons and electrons, we immediately deduce that the extra nuclear matter (i.e. beyond the necessary Z protons) would always be effectively *bosonic*, which would be in direct conflict with the known (half-integer) spins of all $(A-Z)$ odd nuclei. The extra matter must therefore have half-integer spin, i.e. it must be *fermionic*. A spin-half neutron with $m_n \sim m_p$ clearly solves both problems. Recall, that the Bohr radius is inversely proportional to the (reduced) mass of the particle considered. Therefore, substituting the neutron mass for that of the electron relaxes the constraint on the force necessary by a factor $O(10^3)$, which would then be more-or-less in line with the strong nuclear interaction.

Bothe and Becker (1930)[9], found that, on being bombarded with energetic α-particles (emitted by polonium), certain elements (e.g. $^7_3$Li, $^9_4$Be, $^{11}_5$B and $^{18}_9$F) emitted highly penetrating and non-ionising '*radiation*' (cf. β-rays and γ-rays). They observed that electric fields had no effect and therefore deduced (erroneously) that it was γ-radiation, despite it being far more penetrating than any γ-rays ever seen at the time. Following up these experiments, Curie and Joliot (1932a)[10], demonstrated that when this new radiation struck paraffin (or other material containing hydrogen), protons (or 'H-rays' as they were then sometimes known) were often ejected with relatively high energies (up to around 5.3 MeV or velocities of nearly $\frac{1}{10}c$).

Although this last observation itself was not in direct conflict with the γ-ray hypothesis, such an interpretation has severe problems with energy–momentum conservation: a γ-ray (i.e. a photon) would require around 50 MeV of kinetic energy to knock out a proton and this did not seem to be the case (see the next exercise). Indeed, further study (Curie and Joliot 1932b), while not dispelling their belief in the γ-ray description, led them to hypothesise '*a new mode of interaction of radiation with matter.*'

---

[9] The 1954 Nobel Prize in Physics was divided equally between Max Born 'for his fundamental research in quantum mechanics, especially for his statistical interpretation of the wave-function' and Walther Bothe 'for the coincidence method and his discoveries made therewith'.

[10] The 1935 Nobel Prize in Chemistry was awarded jointly to Frédéric Joliot and Irène Joliot-Curie 'in recognition of their synthesis of new radioactive elements'.

The first to interpret these findings correctly was, however, Majorana (1933), who immediately proposed the existence of a new neutral particle of mass similar to the proton, which he called a *'neutral proton'*, or rather the neutron[11].

'Once you eliminate the impossible, whatever remains, no matter how improbable, must be the truth.'

Sherlock Homes
The Sign of the Four (1890)—Sir Arthur Conan Doyle

Moreover, both Rutherford and his assistant James Chadwick were sceptical with regard to the γ-ray hypothesis. In a series of experiments Chadwick (1932)[12] not only showed that from energy–momentum considerations the neutral particles ejected via α irradiation could not be photons, but also that the cross-section was far too large to be an electromagnetic process (it is, of course, a strong interaction). Finally, via the overall energy balance, Chadwick was able to determine the mass of the neutral particles produced in α–beryllium collisions by allowing them to collide with nitrogen and hydrogen and comparing the velocities of the ejected protons. The value he obtained was close (within errors) to the proton mass though a little larger.

Exercise 4.1.1. *Were the γ-ray hypothesis true, the production mechanism via scattering or rather absorption of α-particles by beryllium would be*

$$^{9}_{4}\text{Be} + \alpha \rightarrow {}^{13}_{6}\text{C} + \gamma.$$

Show that, for α-particles of $\sim$5.3 MeV (as emitted by polonium), the γ-rays emitted in the above interaction would have an energy of at most $\sim$14 MeV.

Useful data: $m_\alpha = 4.001\ 506\ u$, $m_{^{9}\text{Be}} = 9.012\ 183\ u$ and $m_{^{13}\text{C}} = 13.003\ 355\ u$.

N.B. For beryllium and carbon, these are atomic masses and so care should be taken to correctly account for the electrons; recall that the α-particle is a fully ionised helium atom, or isolated helium nucleus.

Show that to knock a proton out of the target material with the magnitude of final kinetic energy quoted requires either:

(a) *an extremely energetic photon, or*
(b) *another particle of similar mass (if the projectile kinetic energy is not to be extremely high).*

In both cases, calculate the projectile energy required to produce a proton of 5.3 MeV kinetic energy.

---

[11] Although Fermi encouraged him to publish his ideas immediately, Majorana did not write the paper until rather later (Majorana, 1933).

[12] The 1935 Nobel Prize in Physics was awarded to James Chadwick 'for the discovery of the neutron'.

Hint: non-relativistic kinematics may be used for the proton and neutron; assume a Compton-like process for a photon.
Finally, what is the correct reaction to consider?

4.1.3 The muon

The muon has a (very precisely measured) mass,

$$105.658\ 3755 \pm 0.000\ 0023 \text{ MeV}, \tag{4.1.1}$$

and thus requires much higher energies to be produced in the laboratory than, say, the electron or positron. It was discovered in cosmic-ray experiments independently by Anderson and Neddermeyer (1936) and Street and Stevenson (1937).

Electrons and muons are produced copiously in high-energy collisions and, whereas the electron is stable, the muon is not, decaying with a relatively long lifetime: $\tau_\mu = 2.2 \times 10^{-6}$ s. Ignoring time-dilation effects, a highly relativistic muon therefore travels an average distance of nearly 0.7 km before decaying. On the other hand, the tau lepton is rather heavier and therefore less common, added to which its lifetime is $\tau_\tau = 2.9 \times 10^{-13}$ s. Thus, a τ produced by cosmic-ray interactions never reaches a laboratory on the ground. Moreover, its decay products are either a number of light hadrons or, in the leptonic mode, contain two neutrinos. It was not until 1975 and the availability of high-energy electron–positron colliders that the third of the charged leptons was discovered (see section 4.2.4).

4.1.3.1 Testing the muon's nature

The muon was initially considered a prime candidate as the particle (then known as the *mesotron*) suggested by Yukawa (1935)[13] as the exchange field responsible for the strong interaction. With a mass of 106 MeV, it appeared more similar to the baryons than to the other known charged lepton at that time, the electron. According to Yukawa's theory, such a mass would lead to a range of action around 1 fm or so, which corresponded well to the observed finite range of the strong nuclear force.

The question then arose as to how to ascertain whether or not such an interpretation was correct. Apart from the experimental evidence, which we shall shortly discuss, there are theoretical reasons (not, however, available at that time) for *not* accepting such a rôle for the muon. Conservation of angular momentum requires that the exchange particle have integer spin; the muon is now known to be a fermion. Moreover, the flavour or isospin symmetry (see appendix B.4) of the strong interactions requires that the exchange particle have integer isospin too (the proton and neutron belong to an isospin one-half doublet). A singlet state would not interact (or at best its interactions would be suppressed) and therefore it should have at least one unit of isospin. Finally, the multiplicity of an isospin-one system is three, whereas there exist only two states for the muon: $\mu^\pm$, there being no neutral state.

---

[13] The 1949 Nobel Prize in Physics was awarded to Hideki Yukawa 'for his prediction of the existence of mesons on the basis of theoretical work on nuclear forces'.

Evidently though, it was necessary to examine the strength of its interaction to really determine the nature of the muon. Its decay is seen to be weak ($\tau_\mu \sim 2\ \mu s$), but this alone cannot be interpreted as excluding its strong interaction; the type of interaction through which a particle may decay is also determined by the various conservation laws. In this case conservation of energy is sufficient to exclude a strong decay: the muon is lighter than all known strongly interacting particles and, in fact, decays primarily to $e\bar{\nu}_e\nu_\mu$, none of which are strongly interacting. The behaviour of the muon must then be studied in a strongly interacting environment, e.g. inside the nucleus. The strong interaction has a time scale of the order of 10^{-23} s (the typical decay time for the heavier hadrons such as Δ^{++} etc) and so a muon might be expected to be absorbed on such a time scale by a nucleus (inside which the energy balance may easily be redressed).

Conversi *et al* (1947) set out to measure the lifetime of what had then already been dubbed Yukawa's *mesotron* (the present-day muon) by studying its stopping behaviour in nuclear matter; they had already measured the free lifetime. The experiments they performed turned out to be a disproof of the strong-interaction hypothesis.

In matter (negatively charged) muons lose energy via electromagnetic interactions until they are eventually captured by an atom and become bound, just as an electron. Since the muon is evidently distinguishable from the electrons, it does not suffer Pauli exclusion and, via photon emission, may cascade down to the ground state. At this point, owing to its relatively large mass, it is much nearer to the nucleus than the corresponding K-shell electron would be. Indeed, for a charged particle of mass m, the Bohr radius in an atom with atomic number Z is

$$R^B = \frac{Z}{m\,\alpha}, \tag{4.1.2}$$

which leads to $R_e^B \simeq Z \times 0.6 \times 10^{-10}$ m for an electron. And since then

$$R_\mu^B = \frac{m_e}{m_\mu}\,R_e^B, \tag{4.1.3}$$

for the given mass ratio of approximately 200, this leads to $R_\mu^B \simeq Z \times 3 \times 10^{-13}$ m for a muon.

The strong interaction is evidently negligible at such distances; however, the smaller radius implies that the wave-function for the muon will have a higher density inside the nucleus than would the corresponding electron, by roughly a factor 200^3. As we shall now show, this represents a sufficiently long time spent *inside* the nucleus to test the strong-interaction hypothesis. Conversi *et al* (1947) measured a decay lifetime (i.e. for the disappearance or so-called K-capture of muons) of 0.88 μs (to be compared to the free decay time of 2 μs). Thus, some form of interaction evidently occurs. In order to evaluate the strength of this interaction it is necessary to estimate the mean free path of muons in nuclear matter.

A simple (back-of-the-envelope) estimate may be performed by considering the volume of the nucleus itself as a fraction f of the total volume occupied by a K-shell muon. This is just the ratio $(R_{nucl}/R_\mu^B)^3$. Recall that empirically $R_{nucl} = R_0 A^{\frac{1}{3}}$, where A is just the atomic mass and $R_0 \simeq 1.2$ fm. Using this and equation (4.1.3) leads to

$$f = \left(\frac{R_{nucl}}{R_\mu^B}\right)^3 = 0.27\, A\left(\frac{Z}{137}\right)^3. \tag{4.1.4}$$

For aluminium $Z = 13$ and $A = 27$, giving

$$f \simeq 6 \times 10^{-3}. \tag{4.1.5}$$

This will be roughly the fraction of its lifetime that a muon spends inside an aluminium nucleus. Already, it might be anticipated that, as a fraction of a microsecond, this still leads to a survival time inside the nucleon many orders of magnitude larger that the 10^{-23} s we might have expected. However, let us first estimate the mean free path (for a strongly interacting particle it should not be much larger than about 1 fm). The mean velocity of the muon may be estimated from the Heisenberg uncertainty principle by setting $p_\mu \sim \hbar/R_\mu^B$. In a non-relativistic approximation, this leads to an estimated velocity $v_\mu \sim Z\alpha$, which, given that typically $Z \ll \alpha^{-1}$, justifies the approximation *a posteriori*. Finally, the mean free path is

$$\Lambda = vf\tau, \tag{4.1.6}$$

where τ is the lifetime of such a K-shell state.

Now, since decay rates are additive and inversely proportional to lifetimes (i.e. $\Gamma = \Gamma_d + \Gamma_c$ and $\Gamma \propto 1/\tau$), the rule for combining lifetimes is

$$\frac{1}{\tau} = \frac{1}{\tau_d} + \frac{1}{\tau_c}, \tag{4.1.7}$$

where τ_d and τ_c stand for the free-decay and capture lifetimes, respectively. The measured values are $\tau_d = 2.16$ μs and $\tau = 0.88$ μs in aluminium. We thus obtain

$$\tau_c \sim 1.5\ \mu s. \tag{4.1.8}$$

Inserting this into the formula for the mean free path, equation (4.1.6), leads to

$$\Lambda \sim 20\text{–}30\ cm \tag{4.1.9}$$

In other words, muon survival inside the nucleus far exceeds the expectations for a strongly interacting particle (Fermi *et al* 1947). Indeed, the K-capture time τ_c is more suggestive of a *weak* interaction; this came very much as a surprise:

> 'This result was completely unexpected, and we believed at first that there might be some malfunction in our apparatus.'
>
> Marcello Conversi

Indeed, the revelation that apparently the muon therefore had no particular rôle in the general scheme of particle physics prompted Isidor Rabi[14] to comment publicly, in astonishment:

'Who ordered that?'

Gell-Mann and Rosenbaum (1957) too were rather upset:

'Here we have nature at her most perverse. She has given us a particle for which there is no theoretical justification and no use whatever. The muon was the unwelcome baby on the doorstep, signifying the end of days of innocence.'

Another Nobel Prize winner[15] underlined the importance of the results, expressing the following view much later:

'...modern particle physics began during the last days of the Second World War[16], when a group of young Italians, Conversi, Pancini and Piccioni, who were hiding from the German occupying forces, initiated a remarkable experiment.'

Luis Álvarez—Nobel Lecture (1968)

4.1.4 The pion

In the same year that Conversi *et al* (1947) had discovered that the muon was not Yukawa's mesotron, the charged pion was discovered by Lattes *et al* (1947)[17]. This discovery was made, however, with photographic-emulsion detection methods at high altitude, typically used on mountain tops or even in aeroplanes and here in the Pyrenees.

Shortly after, Burfening, Gardner and Lattes (1948) were the first to produce pions by 380 MeV α-particles incident on a carbon target using the synchro-cyclotron at Berkeley (USA). Despite the recent revelation by Conversi *et al*, muons were still referred to as 'light mesons' in this last paper, while pions were called 'heavy mesons', quoted, respectively, as having masses around 200 and 300 times that of the electron. The present value for the charged-pion mass is (Navas *et al* 2024)

$$m_{\pi^\pm} = 139.570\,39 \pm 0.000\,18 \text{ MeV}. \tag{4.1.10}$$

---

[14] The 1944 Nobel Prize in Physics was awarded to Isidor Rabi 'for his resonance method for recording the magnetic properties of atomic nuclei'.

[15] See footnote 11 in chapter 3.

[16] The series of experiments to measure the muon lifetime actually began in 1943, while the war was still underway and the Allies were bombing Rome, which was occupied by the German forces.

[17] The 1950 Nobel Prize in Physics was awarded to Cecil Frank Powell 'for his development of the photographic method of studying nuclear processes and his discoveries regarding mesons made with this method'.

Mention should be also made here of the somewhat earlier work by Bose and Chowdhry (1941), who had started using similar techniques already in the late 1930s, based in Darjeeling at an altitude of 2000 m. They reported tracks that corresponded to charged particles with an estimated mass, using a new method they had devised, of around 100 to 150 MeV. With hindsight, these were clearly charged pions, although, owing to the large experimental uncertainties, Bose and Chowdhry did not make strong claims at the time.

A word of explanation is clearly in order here. On the basis of available phase-space (or Q-value), we would imagine that the $e\bar{\nu}_e$ final state would be preferred over the heavier $\mu\bar{\nu}_\mu$. However, there is an important question of angular momentum to address. The initial pion state has spin zero, whereas the two final particles are both spin-$1/2$ and so must emerge (back-to-back in the pion rest frame) with *same* helicity. This causes problems on two counts. Firstly, we know that, since this is a weak interaction, the helicity of the electron (or muon) produced should be *negative*, whereas that of the antineutrino should be *positive*. Secondly, as discussed in appendix C.2.2, the vector (and axial-vector) interactions preserve helicity in the massless limit and we know that the $e\bar{\nu}_e$ ($\mu\bar{\nu}_\mu$) pair must originate from a W^-; thus again, they should have opposite helicities. The only way around this selection rule is helicity flip of the charged lepton thanks to its non-zero mass. It can be shown that the amplitude for such a process is proportional to $m_\ell/(m_\pi + m_\ell)$ and the relative factor $O(m_\mu^2/m_e^2)$ then more than compensates the phase-space suppression; as a result the muon decay mode is dominant.

Recall that one reason, *a posteriori*, for not considering the muon as a true mesotron candidate is that, having only two states ($\mu^\pm$), it could only possibly have isospin one half. The two charged-pion states would also suffer the same difficulty and so, of course, the question arises of a third presumably neutral state, allowing an isospin assignment of one.

Now, whereas the charged pions undergo a weak β-type decay principally into $\mu\bar{\nu}_\mu$, the neutral state decays predominantly via an electromagnetic two-photon channel. Moreover, whereas the charged-pion lifetime is $(2.6033 \pm 0.0005) \times 10^{-8}$ s, that of the neutral pion is much shorter, being $(8.43 \pm 0.13) \times 10^{-17}$ s. Neutral pions are thus naturally much harder to detect than the charged versions, as they do not leave tracks in photographic emulsions or cloud chambers. The neutral pion was instead first detected indirectly in cosmic-ray experiments (Chao 1949, Fretter 1949) by observing its decay products: two photons, which produced a low-energy electromagnetic shower. A more reliable observation was again subsequently made at the Berkeley synchro-cyclotron (Bjorklund *et al* 1950); their observations lead to mass of about 300 electron masses. The present value for the neutral-pion mass is (Navas *et al* 2024)

$$m_{\pi^0} = 134.9768 \pm 0.0005 \text{ MeV}. \tag{4.1.11}$$

4.1.5 The strange particles

There were, however, other particles having masses not too dissimilar to the pion and with very similar (weak) decay rates. They were the *kaons*: $K^\pm$ (the $\tau^\pm$ and $\theta^\pm$ discussed earlier), K^0 and $\bar{K}^0$. These then were ideal candidates for discovery in cosmic-ray experiments. Indeed, in the same year as the pion discovery the first 'V' particles were detected too (Rochester and Butler 1947), so-called owing to their distinctive two-pronged decay-state tracks, easily identifiable in cloud-chamber, emulsion and, later on, bubble-chamber experiments (see figures 4.2 and 4.3). Moreover, the strange baryons, Λ^0, $\Sigma^{0,\pm}$ and $\Xi^{-,0}$, having masses a little larger than the nucleons and lifetimes of the order of 10^{-10}s, were also similarly soon discovered.

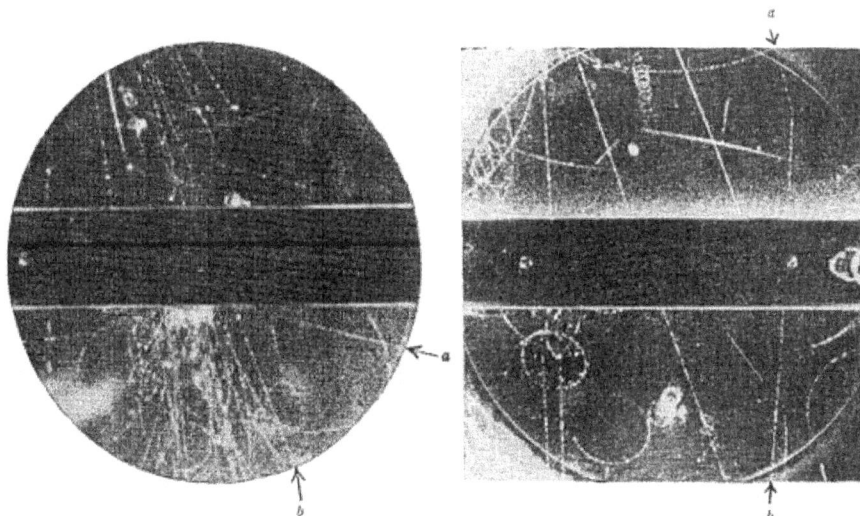

Figure 4.2. In the left-hand cloud-chamber photograph slightly below the lead plate and to the right we see a typical 'V' fork (a–b), probably due to the decay of a neutral (hence no incoming track) kaon into a $\pi^+\pi^-$ pair; the right-hand photograph shows a very open charged 'V' (a–b) top right, probably the signal of a charged kaon decaying into a muon plus neutrino; photographs reproduced from Rochester and Butler (1947) with permission from Springer Nature.

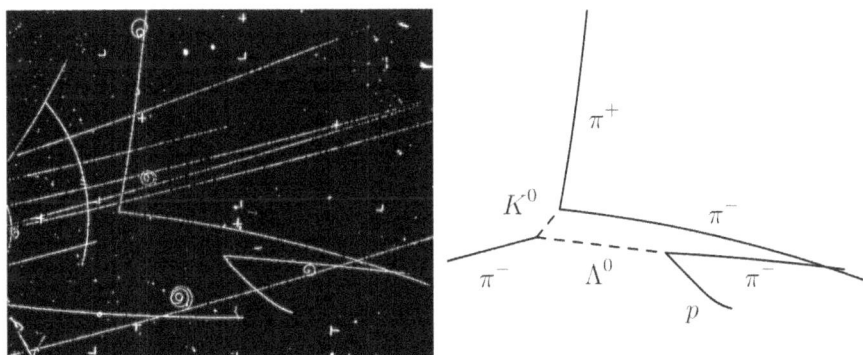

Figure 4.3. A negatively charged pion enters from the left and strikes a proton, producing two uncharged strange particles (a kaon and a lambda baryon, indicated by dashed lines), which leave no tracks until they too decay; photograph with permission from Lawrence Berkeley National Laboratory. CC BY-NC-ND.

'These bubble chambers took pictures on film at the rate of about one per second. Many millions were produced. These had to be scanned, and the events of interest measured and reconstructed. At first we used the simple, manual techniques for scanning and measuring inherited from our cloud chamber predecessors: simple projection tables, protractors for angles, templates for the measurement of the track curvatures and manual computers.'

Jack Steinberger—*Learning About Particles* (Springer, 2005)

Figure 4.3 shows both a photograph and a sketched version of an interesting event: a negatively charged pion enters from the left and strikes a proton, producing two uncharged particles (a neutral kaon and a lambda baryon):

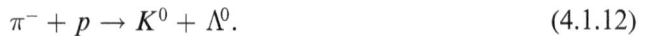

$$\pi^- + p \rightarrow K^0 + \Lambda^0. \tag{4.1.12}$$

Note that this process does not violate any known conservation law (the quark subprocess should be $u\bar{u} \rightarrow s\bar{s}$) and is therefore most likely a strong interaction. The kaon and lambda travel some distance (but, being neutral, leave no tracks) before decaying via the weak processes:

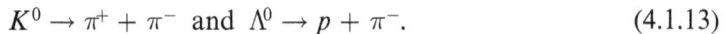

$$K^0 \rightarrow \pi^+ + \pi^- \text{ and } \Lambda^0 \rightarrow p + \pi^-. \tag{4.1.13}$$

4.1.6 The antiproton and antineutron

Following the discovery of the positron and its interpretation via the Dirac equation as the antielectron, it was soon generally believed that all particles should have an antimatter equivalent with opposite charge but exactly equal mass. With this precise idea in mind, the Bevatron (a weak-focusing proton synchrotron) was constructed at Lawrence Berkeley National Laboratory, USA and began operating in 1954, accelerating 6.2 GeV protons onto a fixed target. Particle detection was performed via a bubble chamber. Soon after, the antiproton was discovered by Chamberlain *et al* (1955)[18] and the following year the antineutron was also identified (Cork *et al* 1957) using the same machine.

Exercise 4.1.2. *Calculate the laboratory beam-energy threshold for the production of an antiproton in a fixed-target proton–proton collision.*

4.1.7 The electron neutrino

Postulated in 1930 by Pauli, the neutrino (together with the positron) is an early example of a particle first predicted to exist for theoretical reasons, then found later

---

[18] The 1959 Nobel Prize in Physics was awarded equally to Emilio Gino Segrè and Owen Chamberlain 'for their discovery of the antiproton'.

experimentally. The observation of β-decay with its continuous electron-energy spectrum and apparent non-conservation of angular momentum (Chadwick 1914) prompted Pauli in 1930 to propose the existence of a third light and invisible particle emitted: the neutrino[19]. We might note that indeed none other than Niels Bohr openly opposed this 'improbable' interpretation of β-decay, preferring to accept that energy, momentum and angular momentum were possibly not conserved in such processes.

The electron-neutrino was first detected by Frederick Cowan, Clyde Reines[20] and co-workers (Reines and Cowan 1953). The experiment consisted of two tanks, each containing about 100 litres of water, as detectors, placed near to a nuclear reactor, as a very intense source of electron-antineutrinos (about 5×10^{13} neutrinos s^{-1} cm^{-2}) from β-decay. The antineutrinos were detected via their interaction with protons in the water; the products of which were neutrons and positrons:

$$\bar{\nu}_e + p \rightarrow n + e^+. \tag{4.1.14}$$

The positrons quickly annihilated with electrons in the water, producing two γ-rays. These were detected in further adjacent tanks containing liquid scintillator, which, in response to the γ-rays, emitted light flashes finally collected by photomultiplier tubes. In addition, the neutron produced in the initial interaction provided a second signal: by dissolving cadmium chloride ($CdCl_2$) in the tanks (about 40 kg), a further γ-ray was produced when a cadmium nucleus absorbed the neutron. The reaction induced was

$$n + {}^{108}Cd \rightarrow {}^{109}Cd^* \rightarrow {}^{109}Cd + \gamma. \tag{4.1.15}$$

The neutrino-interaction signal was thus comprised of two γ-rays produced by the positron annihilation, followed several microseconds later by the γ-ray from the cadmium neutron absorption –all three γ-rays being of predetermined energies.

4.1.8 The muon neutrino

It was well known that the muon decayed into an electron and two neutrinos ($\bar{\nu}_e\ \nu$), since the electron spectrum was typical of a three-body decay. The two neutrinos produced also had to be different: one was clearly an electron-antineutrino ($\bar{\nu}_e$); the other, known then as the *neutretto*[21], was clearly a candidate muon-neutrino (ν above).

Let us see why it must be different. If there were just one type of neutrino, then the process $\mu \rightarrow e\gamma$ would become possible (Feinberg 1958) via a so-called *penguin*-like diagram, see figure 4.4. This would lead to a so-called *'flavour-changing neutral current'* (see later). A theory with intermediate charged vector bosons ($W^\pm$) having a

---

[19] Pauli initially called the new particle a neutron. When the neutron (as it is now known) was later discovered, Fermi suggested the name neutrino (being a 'small' neutral particle).

[20] The 1995 Nobel Prize in Physics was awarded 'for pioneering experimental contributions to lepton physics' jointly with one half to Martin L Perl 'for the discovery of the tau lepton' and one half to Frederick Reines 'for the detection of the neutrino'. Note that Cowan unfortunately died in 1974.

[21] Note that the term *neutretto* was also used to indicate other light neutral particles, such as the π^0.

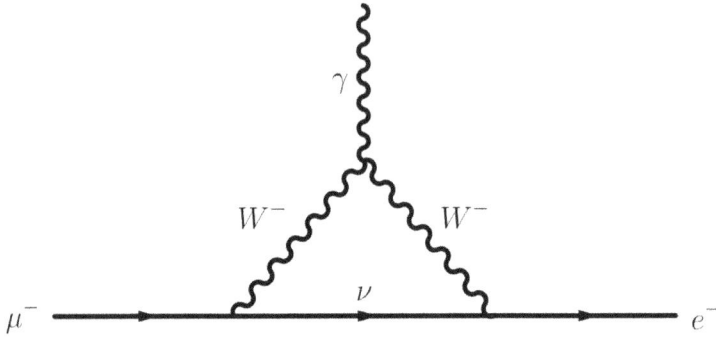

Figure 4.4. A so-called *penguin* diagram leading to the decay $\mu \rightarrow e\gamma$.

mass less than about 300 GeV and just a single type of neutrino would lead to a branching ratio (Feinberg, 1958)

$$\frac{\Gamma(\mu \rightarrow e\gamma)}{\Gamma(\mu \rightarrow e\nu\bar{\nu})} \approx 10^{-4}. \tag{4.1.16}$$

Such a decay is, however, not observed at this level. Indeed, the Mu to E Gamma (MEG) experiment, dedicated to measuring the $\mu \rightarrow e\gamma$ decay, has now placed the following limit on the branching ratio (Baldini *et al* 2016):

$$\mathcal{B}_{\text{expt}} < 4.2 \times 10^{-13} \ (90\% \ \text{CL}). \tag{4.1.17}$$

Note moreover that although the Standard Model does allow such charged-lepton flavour violating processes, the branching ratios are predicted to be extremely small ($\ll 10^{-50}$).

In 1960, Melvin Schwartz had the idea of using high-energy secondary neutrino and antineutrino beams to study the weak interaction. Exploiting such beams, the muon-neutrino (ν_μ) was first detected in 1962 by Leon Lederman, Melvin Schwartz, Jack Steinberger[22] and co-workers (Danby *et al* 1962) using the AGS at BNL. They used a setup conceptually similar to that already described in section 2.7.6 for the study of CP violation. A 15 GeV proton beam incident on a beryllium target produced, *inter alia*, charged pions; see figure 4.5. The proton beam energy was chosen to limit the penetrating power of the muons subsequently produced. The flux of particles so generated then travelled 21 m, with the charged pions decaying in-flight mainly into muons and muon-neutrinos, to a 13.5-m thick iron shield, which blocked all but the neutrinos. These last were predominantly muon-neutrinos and -antineutrinos[23]. A 10-ton aluminium spark chamber was placed immediately

---

[22] The 1988 Nobel Prize in Physics was awarded equally to Leon M Lederman, Melvin Schwartz and Jack Steinberger 'for the neutrino beam method and the demonstration of the doublet structure of the leptons through the discovery of the muon neutrino'.

[23] Recall the dominance of the muon over electron channel in charged-pion decay, due to helicity-flip suppression.

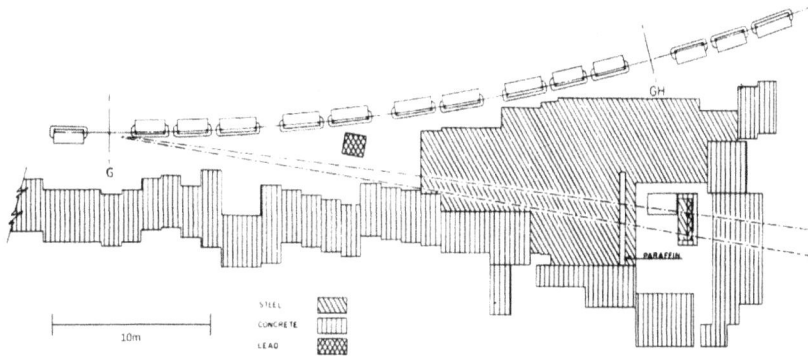

Figure 4.5. A plan view of the alternating gradient synchrotron neutrino experiment at Brook-haven National Laboratory (reprinted with permission from Danby *et al* 1962), copyright (1962) by the American Physical Society.

behind the shield to register neutrino interactions. On interacting with matter, these neutrinos were seen to produce muons and *not* electrons via the process

$$\nu_\mu N \to \mu X, \tag{4.1.18}$$

cf. equation (4.1.14). The muon-neutrino hypothesis was thus verified. The observation that leptons, just as quarks, appear to be arranged in (weak-isospin) doublets then leads naturally to the concept of quark–lepton families.

4.1.9 The tau neutrino

The third and last of the known neutral leptons is the tau-neutrino. The Direct Observation of the Nu Tau (DONuT) experiment was set up at Fermi National Accelerator (FNAL) in the late 1990s, with the specific aim of detecting tau-neutrinos. The experiment ran for a few months during 1997, announcing the results three years later (Kodama *et al* 2001). The claim was based on only four events; however, with an estimated background of less than 0.3 events, the signal was considered valid. The final results from the DONuT experiment were based on data from just nine tau-neutrino events and were published by Kodama *et al* (2008).

A secondary ν_τ beam was generated using 800 GeV protons in the Tevatron at FNAL incident on a tungsten beam dump; a total of 3.54×10^{17} protons were recorded during the live-time of the experiment. It was estimated that only one in 10^{12} of the neutrinos would in fact interact. The collisions produced, *inter alia*, D_s mesons. Among the numerous modes, the purely leptonic decay is predominantly into $\tau \bar{\nu}_\tau$; the τ-lepton subsequently decays into a ν_τ and other lighter charged particles. The above decay mode thus yielded two τ-neutrinos within a distance of a few millimetres; for the purposes of this experiment it was not necessary to distinguish between neutrinos and antineutrinos. Magnet fields, concrete, iron and lead shielding removed all but the neutrinos. The neutrino beam so produced passed through a number of layers of nuclear emulsion. The neutrinos, when interacting in

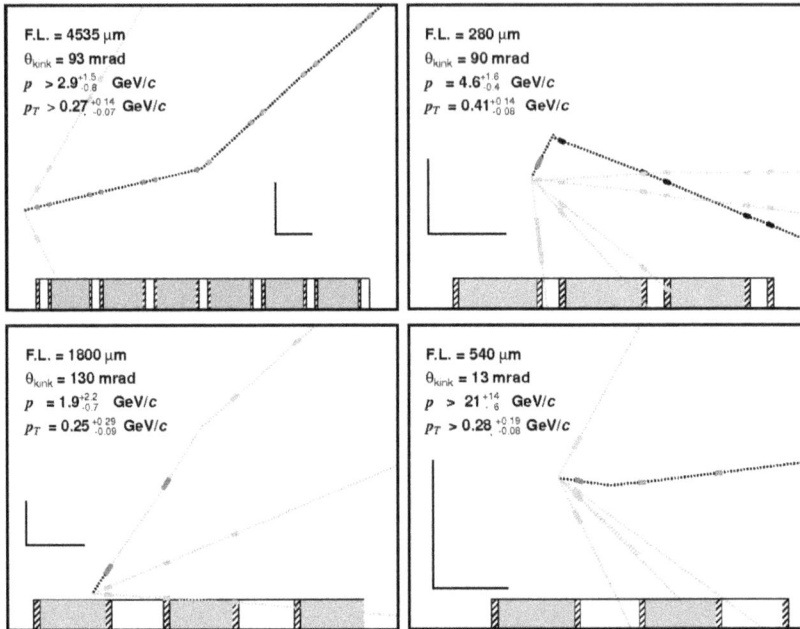

Figure 4.6. The four tau-neutrino events observed by DONuT. The neutrino beam is incident from the left. The scale is given by the perpendicular lines: the vertical line represents 0.1 mm and the horizontal 1.0 mm. At the bottom of each figure the target material is indicated by shaded bars representing steel (grey), emulsion (cross-hatched) and plastic (white). The figures are taken from Kodama *et al* (2001), copyright (2001) with permission from Elsevier.

the detector, produced charged particles leaving visible tracks in the emulsion, which were registered by a system of scintillators and drift chambers. The emulsion signals were then used to reconstruct particle tracks. The interaction of a ν_τ led to a peculiar structure: several tracks would appear from nothing, one of which would then abruptly change direction after a few millimetres (85% of τ-decays produce only a single charged particle, called a *kink*, the rest being neutrals, mainly neutrinos), clearly indicating the decay of the τ-lepton produced; see figure 4.6.

Exercise 4.1.3. *Assuming that the tau neutrino collides with a single nucleon in the detector, draw the Feynman diagram describing the interaction. For an initial low-energy $\bar{\nu}_\tau$, deduce whether the target nucleon must be a proton or neutron; motivate the answer. Finally, calculate the threshold for such a charge-exchange process.*

4.2 Electron–positron colliders

A particular limitation of cloud chambers, emulsions and bubble chambers is that any new particle produced must leave a detectable track. While this obviously severely limits their employment in the case of neutrals, the most important consequence here is that they are also inadequate for very short-lived (intermediate) states such as the π^0 in the example above. We evidently need an indirect method of detecting the presence of such an object. This is where the concept of a resonance *à la* Breit and Wigner (1936) arises (for a detailed discussion, the reader is referred to appendix C.6). Once a particle lifetime becomes too short to be measured directly (e.g. via the mean length of its cloud-chamber, emulsion or bubble-chamber tracks) we must move over to the energy counterpart, the decay rate or *width*, related in the following way:

$$\Gamma = \hbar/\tau. \tag{4.2.1}$$

Thus, for shorter lifetimes we exploit the growing uncertainty in the mass of the state produced, which being (very) unstable is always an intermediate state in any given process. As the term 'width' suggests, the uncertainty principle implies that given only a short time to determine the mass of a particle there will be some natural fluctuation around a central value compatible with the above expression; spectral lines are thus broadened and their width provides an indirect measure of the lifetime.

4.2.1 Resonance production

The concept of resonance in nuclear and particle physics is best approached through the partial-wave expansion, which is briefly outlined in appendix C.5. The earliest applications of the partial-wave and Breit–Wigner approach (see appendix C.6) are found in nuclear physics, where it is used to describe interactions that proceed via the formation of intermediate (possibly excited and/or unstable) nuclear states or resonances.

A first simple and practical example of resonance production in particle physics is the discovery of the Δ baryons in pion–nucleon elastic scattering (Anderson *et al* 1952; see also Hahn *et al* 1952) via the process

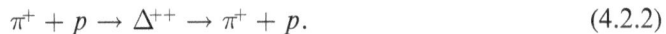

$$\pi^+ + p \rightarrow \Delta^{++} \rightarrow \pi^+ + p. \tag{4.2.2}$$

The nominal masses and widths of the Δ resonances are 1232 MeV and 120 MeV, respectively. In this case the initial (and final) spins are 0 and ½, whereas the intermediate Δ has spin 3/2.

Exercise 4.2.1. *Show that, for a charged-pion projectile incident on a fixed proton target, the resonance peak is attained for a pion kinetic energy a little below* 200 MeV. *Calculate the centre-of-mass pion energy.*

With such energies, although the pion itself is quite relativistic in the laboratory frame, this is no longer true in the centre-of-mass frame and we are justified in using the non-relativistic Breit–Wigner form given in equation (C.6.15). The maximum value of the cross-section is therefore

$$\sigma_R^{\max} = \frac{4\pi\hbar^2}{p^2} \frac{4}{1 \times 2} = \frac{8\pi\hbar^2}{p^2}. \tag{4.2.3}$$

Note that, owing to the rapidly falling underlying cross-section, the experimental peak lies a little below the true resonance energy, corresponding to the point where the experimental value coincides with the maximum-value curve (see figure 4.7). Note also the skew effect due to the variation of Γ over the width of the Δ resonance. This is principally the effect of the variation of Γ in the numerator, which grows with increasing centre-of-mass energy. The high-energy tail is thus rather higher than the low-energy tail. Once out in the tails, the value of Γ used in the denominator usually has less influence.

A further way in which to search for resonances is to examine the invariant mass of, say, a final-state $\pi^-\pi^+$ pair. A possible production mechanism is shown in figure 4.8. Again, the form of the effective propagator for an intermediate resonance, having the same quantum numbers as two pions, leads to the classic Breit–Wigner shape in the invariant-mass distribution (see figure 4.9). In the figure we see three clear peaks with widths of the order of 100–200 MeV. These correspond to:

$$\rho^0(769), \quad \Gamma \simeq 150\,\text{MeV}; \qquad f_2^0(1275), \quad \Gamma \simeq 180\,\text{MeV}$$
$$\text{and} \qquad \rho^0(1700), \quad \Gamma \simeq 200\,\text{MeV}. \tag{4.2.4}$$

Naturally, these peaks are superimposed on a background of non-resonant, continuum production of charged-pion pairs.

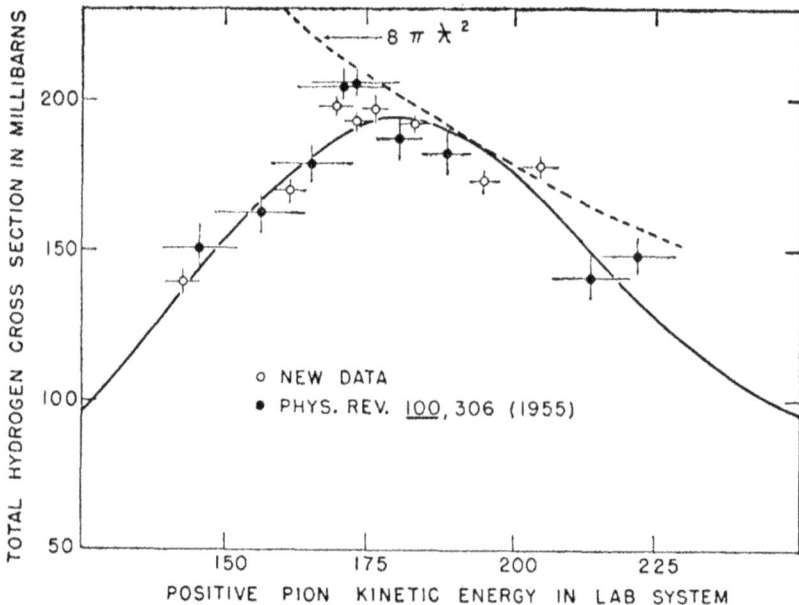

Figure 4.7. The π^+p elastic cross-section as a function of laboratory-frame pion kinetic energy. The dashed line represents the maximum value $8\pi/k^2$, attained at the peak. Note the skew effect due to the variation of Γ over the width of the Δ resonance; figure reproduced with permission from Lindenbaum and Yuan (1958), copyright (1958) by the American Physical Society.

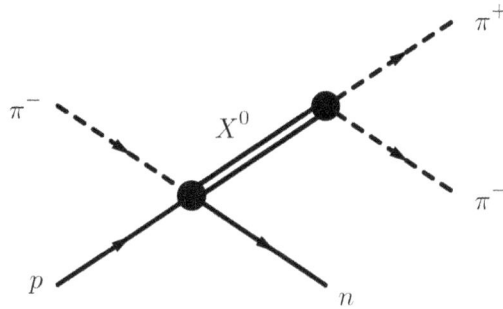

Figure 4.8. The production of an intermediate resonant state X^0 in the $\pi^+\pi^-$ channel of the process $\pi^-p \rightarrow \pi^-\pi^+n$.

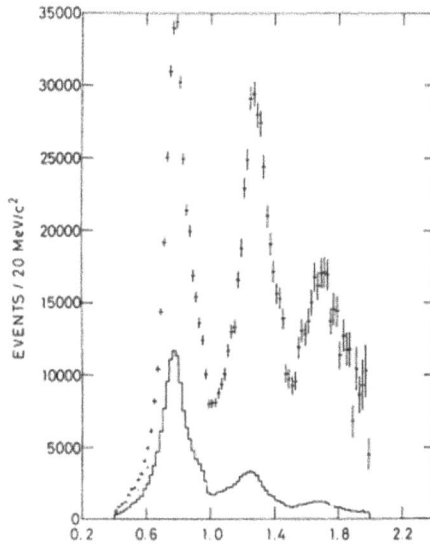

Figure 4.9. The $\pi^-p \rightarrow \pi^-\pi^+n$ cross-section as a function of the final-state $\pi^-\pi^+$ pair invariant mass, for pion-beam momentum 17 GeV c^{-1}. The three peaks correspond to: $\rho^0(769)$, $f_2^0(1275)$ and $\rho^0(1700)$; figure reproduced from Grayer *et al* (1974), copyright (1974) with permission from Elsevier.

In general, by considering a multiparticle state, we see that the invariant-mass distribution of the final composite system measures both the mass and width of any intermediate resonant state having the same quantum numbers as the detected system. To search for strange mesonic resonances, we might thus study the invariant mass of, say, a particular $K\pi$ final state.

One word of caution is in order. As we know from quantum mechanics, to calculate any particular process (scattering or decay), we must sum *all* amplitudes that can contribute. If the peaks are sufficiently narrow and/or well separated (as is the case above), then to a good approximation the contributions may be considered independently. However, it may happen that two or more resonances with the same quantum numbers have very similar masses and thus the Breit–Wigner peaks can overlap. In this case the phase variation and, in particular, the possibility of phase mismatch in the initial production process may mean that across the interval in which the two (or more) resonances contribute there are successive regions in which the two (or more) channels interfere destructively and constructively. These can distort the peaks in an essential way, leaving their appearance very misleading. In fact, for the three-body final state, it turns out that the Dalitz-plot technique (see appendix C.7) allows a complete description in terms of interfering complex resonant amplitudes.

4.2.2 Hadronic resonances in e^+e^- annihilation

As remarked earlier, one of the simplest imaginable situations is the search for and discovery of the neutral resonances that may be produced in e^+e^- annihilation. We have already seen the basic type of (resonance) diagram via which a fermion–antifermion pair may be produced (see figure 4.10). The process is very similar to that of e^+e^- annihilation through a virtual photon. Note that the quantum numbers and energy of the initial state are well determined.

The picture provided here is not yet complete. The initial-state e^+e^- pair cannot couple directly to a hadronic resonance and therefore the annihilation process proceeds via the formation of a virtual photon, which can then transform into any neutral resonance having the same J^{PC} quantum numbers. In such a way, whenever the centre-of-mass energy corresponds to the mass of a neutral, spin-one, flavourless hadronic resonance, we have the picture given in figure 4.11. The total angular

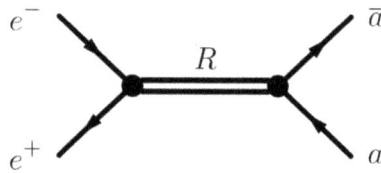

Figure 4.10. The e^+e^- annihilation process into a particle–antiparticle pair ($a\bar{a}$) via a direct intermediate resonant state R.

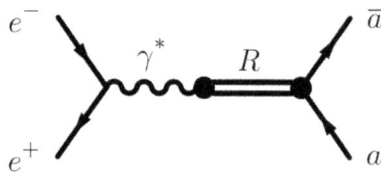

Figure 4.11. The e^+e^- annihilation process into a particle–antiparticle pair ($a\bar{a}$) via a photon coupling to an intermediate resonant state R.

momentum, P and C properties of the resonance so formed are thus predetermined. Indeed, the photon is characterised as having $J^{PC} = 1^{--}$ and isospin $I = 0, 1$; the isospin is not completely determined as the electromagnetic interaction violates isospin (the up quark and down quark have different charges) and therefore the photon need not be a pure eigenstate of isospin.

Now, it turns out to be more useful to examine the cross-section into hadrons rather than the elastic cross-section and we shall need to modify formula (C.6.15) slightly, to take into account that the initial and final states are different (at higher energies many channels do indeed exist). The most general form of the Breit–Wigner approximation for such an *inelastic* process is then

$$\sigma_R \simeq \frac{4\pi\hbar^2}{p^2} \frac{(2J_R + 1)}{(2s_1 + 1)(2s_2 + 1)} \frac{(\Gamma_i/2)(\Gamma_f/2)}{(E - E_0)^2 + (\Gamma_{\text{tot}}/2)^2}, \tag{4.2.5}$$

where the width Γ_{tot} appearing in the numerator is the *total* width of the resonance, whereas in the numerator we have the two *partial* widths Γ_i and Γ_f, corresponding to the decay channels into the particular initial and final states.

To have some intuition as to how this comes about, it is instructive to examine the appropriate Feynman diagram for such a process (see, e.g., figure 4.11). The implicit approximation is equivalent to considering this single amplitude as the product of two factorising sub-amplitudes: namely, $e^+e^- \to R$ and $R \to a\bar{a}$. The process is thus seen in two distinct and independent stages: initial production followed by decay. The first corresponds to the *Hermitian conjugate* of the process $R \to e^+e^-$, whereas the second evidently describes the decay of the resonance R into the state $a\bar{a}$. On taking the square of the modulus, we are thus equivalently calculating the product of the processes $e^+e^- \to R$ and $R \to a\bar{a}$. The spin factors are kept explicitly separate and so are still correct.

From this description, we may also appreciate a little better the nature of the approximation: the resonance R is effectively considered as a real, on-shell particle, which is evidently not true, except precisely at the peak. To some extent, the inclusion of momentum factors, as in equation (C.6.26), can correct for this. However, we are also implicitly treating the couplings at the vertices as point-like, whereas an exact formulation would require form factors. Since we do not have a complete theory of the bound state in the strong interaction, we can only make models and parametrise these.

Inserting the spins of the initial-state electron–positron pair and the intermediate photon, we have

$$\sigma_R \simeq \frac{3\pi\hbar^2}{4p^2} \frac{\Gamma_i \Gamma_f}{(E - E_0)^2 + (\Gamma_{\text{tot}}/2)^2}. \tag{4.2.6}$$

We invoke the photon spin since we are considering a process initiated by electrons, which do not interact strongly and which cannot therefore directly produce strongly interacting particles. Thus, the first vertex in the diagram must be of the $e^+e^-\gamma$ type. However, the photon may convert into any hadronic resonance having the same quantum numbers, in particular having the same spin.

An example case is the production of the $\rho^0(770)$ meson (with $I, J^{PC} = 1, 1^{--}$). As already noted, it has the (strong) decay channel $\pi^+\pi^-$ and will thus be seen as a Breit–Wigner resonance according to figure 4.11 with $R = \rho^0$ and $a\bar{a} = \pi^+\pi^-$.

4.2.2.1 Electron–positron annihilation into hadrons

In the case of e^+e^- annihilation it is easier to simply require that the final state contain only hadronic states (all those allowed). We thus now examine the *spectrum* obtained in e^+e^- annihilation when the final states are restricted to those containing only hadrons. See figure 4.12, in which the total *hadronic* cross-section is plotted as a function of $\sqrt{s} = E_{\rm CM}$. A large number of peaks are evident, from very low energies right up to the maximum so far attained (the 200 GeV of the LEP II run).

The first peak corresponds to resonant production of the $\rho^0(770)$ meson together with the $\omega^0(782)$, the former of width $\Gamma \simeq 146$ MeV and $I, J^{PC} = 1, 1^{--}$, whereas the latter is much narrower with $\Gamma \simeq 8.5$ MeV and $I, J^{PC} = 0, 1^{--}$; these two certainly overlap owing to the broadness of the former; note though that the ρ^0 line shape does not have a true Breit–Wigner form, even with the *p*-wavecorrection (C.6.26). The reason for the large difference in widths is that, whereas $\rho^0(770)$ is part of an isospin triplet (i.e. together with $\rho^\pm$) decaying almost 100% into a charged-pion pair, the $\omega^0(782)$ is a singlet; in quark terms it is almost purely $(u\bar{u} + d\bar{d})/\sqrt{2}$ and its principal decay mode is into three pions[24], with much less phase-space.

A first natural question then regards the suppression of the two-pion decay mode of the ω^0. Pions are bosons and the wave-function must therefore be overall symmetric with respect to pair-wise interchange. A pair of (identical) neutral pions can thus clearly only form a symmetric state. To conserve angular momentum

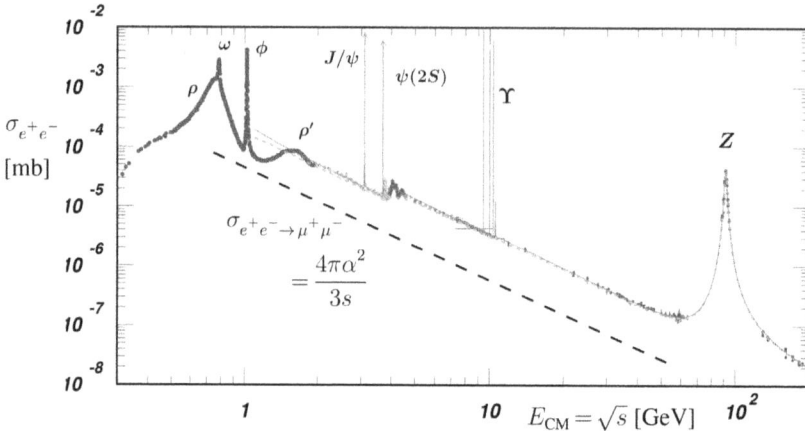

Figure 4.12. The e^+e^- cross-section into hadrons as a function of CM energy. The dashed line indicates the baseline point-like cross-section for a $\mu^+\mu^-$ final state; figure reproduced with permission from Navas *et al* (2024) CC BY 4.0.

---

[24] We should stress though that, in plotting the experimental data in figure 4.12, no specific identification of the final states is made beyond the basic requirement that they be purely hadronic.

($J = 1$), the final-state pion pair here must be in a p-wave, which is antisymmetric and is therefore immediately excluded. We should thus examine the isospin of a neutral pair of charged pions ($\pi^+\pi^-$): the Clebsch–Gordan coefficients for the relevant decomposition of a $\pi^+\pi^-$ state, $|1,+1\rangle \otimes |1,-1\rangle$, are symmetric for $|2, 0\rangle$ and $|0,0\rangle$, but antisymmetric for $|1,0\rangle$. The only accessible state thus has $I = 1$ and is therefore prohibited for ω^0 [25].

A second question regards the suppression of the decay $\rho^0 \to \pi^0\pi^0$.

Exercise 4.2.2. *Explain why the ρ^0 can decay only into $\pi^+\pi^-$ and not into $\pi^0\pi^0$.*

Now, in quark terms, ρ^0 is almost purely $\frac{1}{\sqrt{2}}(u\bar{u} - d\bar{d})$, just as is the π^0, whereas ω^0 is the orthogonal state $\frac{1}{\sqrt{2}}(u\bar{u} + d\bar{d})$. The relative signs in the quark-state representations are the opposite of what might be expected considering e.g. the ordinary angular momentum spin-triplet and -singlet cases[26]. This has to do with the existence of *two* distinct fundamental isospin doublets, see equation (B.4.5). These are chosen this way so as to have the same transformation properties[27]. For a more complete explanation, see appendix B.4.3.

Exercise 4.2.3. *Explain why the ω^0 can decay only into $\pi^+\pi^-\pi^0$ and not into $\pi^0\pi^0\pi^0$.*

The above cases all naturally involve neutral composite final states and it turns out to be very useful to define a new compound symmetry operator and related conserved quantum number: *G-parity*; for a description, see appendix B.4.4. We note finally that the '*prohibited*' decays $\rho^0 \to \pi^+\pi^-\pi^0$ and $\omega^0 \to \pi^+\pi^-$ are in fact observed (albeit with very low rates) and this may be explained as the effect of a very small mixing between the two states ρ^0 and ω^0. Such mixing clearly violates isospin and is therefore severely limited.

Exercise 4.2.4. *Deduce the G-parities of $\rho^0(770)$ and $\omega^0(782)$.*

The successive, again rather narrow, peak corresponds to the $\phi^0(1020)$, also having I, $J^{PC} = 0$, 1^{--} and width $\Gamma \simeq 4.3$ MeV. This is the SU(3) partner to the $\omega^0(782)$ and is almost purely $|s\bar{s}\rangle$. The fact that these two states are arranged in this manner, i.e. *neither* being the natural superpositions of $|u\bar{u}\rangle$, $|d\bar{d}\rangle$ and $|s\bar{s}\rangle$ states,

$$\phi_0 = \frac{1}{\sqrt{3}}\Big[|u\bar{u}\rangle + |d\bar{d}\rangle + |s\bar{s}\rangle\Big] \qquad (4.2.7a)$$

---

[25] Note that this is still just a reflection of the Bose–Einstein statistics (see appendix B.2) that pions must obey, I_3 being again considered merely a label (much as ordinary spin for atomic electrons).

[26] Ordinary spin is different: $|1, 0\rangle = [|1/2, +1/2\rangle + |1/2, -1/2\rangle]/\sqrt{2}$ and $|0, 0\rangle = [|1/2, +1/2\rangle - |1/2, -1/2\rangle]/\sqrt{2}$.

[27] In the case of ordinary spin, there is no such problem since *anti-spin* does not exist.

and

$$\phi_8 = \frac{1}{\sqrt{6}}\left[|u\bar{u}\rangle + |d\bar{d}\rangle - 2|s\bar{s}\rangle\right],$$ (4.2.7b)

is known as *ideal mixing*. Quite why the mixing should be so nearly ideal is not clear and must have to do with the (poorly understood) bound-state dynamics. The determination of these combinations is obtained through comparison with the mass formulæ in SU(3); see, e.g., the section on the *Quark Model* in Navas *et al* (2024), p 312. In any case, these three states belong to the nonet of pseudoscalar mesons, of which the ϕ_0 (N.B. *not* ϕ^0) *would have been* the SU(3) singlet.

4.2.2.2 The Okuba–Zweig–Iizuka (OZI) or Zweig rule

It is interesting to examine in detail and compare the principal decay modes of the two SU(2)-singlet states $\omega^0(782)$ and $\phi^0(1020)$, displayed in table 4.1. Naïvely, phase-space should favour the three-pion decay mode of the ϕ^0 since the Q-value is 600 MeV, as compared to just 24 MeV for the two-kaon decay. Nevertheless, the three-pion mode is relatively suppressed by a factor of nearly 6. This and other similar observations led to the formulation of the so-called OZI or Zweig rule (Okubo 1963, Zweig 1964, Iizuka 1966).

Let us examine the quark diagrams for these decays (see figure 4.13). We see that, whereas at least some of the quark lines are connected between the initial and final state for the three-pion decay of the ω and for the two-kaon decay of the ϕ, the case of $\phi \to 3\pi$ is distinctive in that the initial and final states are completely disconnected with respect to quark lines. The OZI rule then simply states that such decays are suppressed.

This suppression can be understood in terms of gluon exchange. Note though that the rule was derived before the advent of QCD. To connect the initial annihilating quark–antiquark pair to the final state, evidently a number of gluons is needed. Unlike the photon, gluons are carriers of the relevant (colour) charge, a single gluon is therefore not allowed since the intermediate state would then be (colour) charged, whereas the initial and final states are (colour) neutral. A two-gluon state would permit a colour-singlet exchange but would not have the correct J^{PC} quantum numbers; recall that, just as the photon, the gluon has $J^{PC} = 1^{--}$. The minimum number of gluons is therefore three. Since each gluon is associated with an extra factor α_s at the amplitude level (coming from the vertices), the price is high and such a process is evidently suppressed with respect to decays proceeding via connected diagrams. Compare this with the relative decays rates of para- and ortho-positronium, which proceed via two- and three-photon final-state channels, respectively. There are

Table 4.1. The principal decay modes of the two lowest-lying, spin-zero, $SU(2)$-singlet states $\omega^0(782)$ and $\phi^0(1020)$.

$\phi^0(1020) \to$			$\omega^0(782) \to$		
	K^+K^-	49%		$\pi^+\pi^-\pi^0$	89%
	$K^0\bar{K}^0$	34%		$\pi^0\gamma$	9%
	$\pi^+\pi^-\pi^0$	15%		$\pi^+\pi^-$	2%

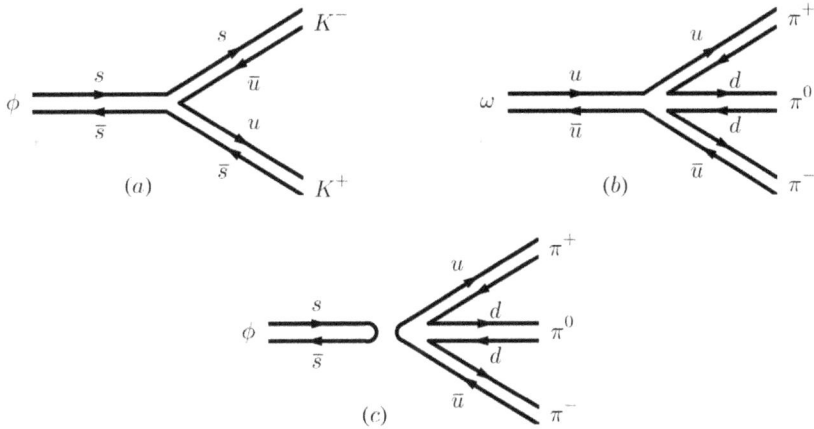

Figure 4.13. Comparison of the quark diagrams for OZI suppressed and unsuppressed decays of quark–antiquark states: (a) $\phi \to K^+K^-$, (b) $\omega \to \pi^+\pi^-\pi^0$, (c) $\phi \to \pi^+\pi^-\pi^0$.

many examples of decays in which the OZI rule is at work and it is now considered both well confirmed and well understood.

Before the next sharp and very distinctive peaks, there is a further broad structure that is little more than a rise to a new baseline. In figure 4.12 this is indicated as the ρ', which up to and including the PDG-1986 publication (Aguilar-Benitez *et al* 1986) was listed as the single $\rho\,(1600)$, but has since been understood to be due to two distinct but broad and heavily overlapping resonances: $\rho\,(1450)$ and $\rho\,(1700)$. The former has a nominal width of ≈ 400 MeV and the latter ≈ 250 MeV. They are evidently radial excitations of the $\rho\,(770)$, but their exact nature has yet to be determined.

4.2.2.3 Discovery of the c quark

The next set of visible peaks corresponds to production of the so-called J/ψ and one of its excited states, ψ'. The observation of this resonance marks the discovery of the c or *charm* quark[28] at BNL by Aubert *et al* (1974) and at SLAC by Augustin *et al* (1974); almost immediately after the first announcements similar evidence was found using the ADONE machine at the National Laboratory of Frascati by Bacci *et al* (1974)[29]. The principal object, the J/ψ, has the following mass and width:

$$m_{J/\psi} = 3097 \text{ MeV} \quad \text{and} \quad \Gamma_{J/\psi} \simeq 93 \text{ keV}, \tag{4.2.8}$$

---

[28] The 1976 Nobel Prize in Physics was awarded equally to Burton Richter and Samuel Chao Chung Ting 'for their pioneering work in the discovery of a heavy elementary particle of a new kind'.

[29] We might mention that shortly before, an experiment employing a broad-band neutrino–antineutrino beam at FNAL had presented evidence for just two dimuon events in neutrino-induced DIS (Rubbia 1974), which could presumably have only been due to c-quark production; see the discussion in section 5.3 of such a process in relation to measuring V_{cd}.

with spin–parity quantum numbers $J^{PC} = 1^{--}$. The second peak is due to the ψ' or $\psi(2S)$ with

$$m_{\psi'} = 3686 \text{ MeV} \quad \text{and} \quad \Gamma_{\psi'} \simeq 294 \text{ keV} \tag{4.2.9}$$

and it too obviously has $J^{PC} = 1^{--}$. These two states are then evidently $c\bar{c}$ in an s-wave and with spins aligned, the ψ' being a radial excitation. Various higher radial-excitation states are now known.

We should immediately remark that the experiment of Aubert *et al* (1974), performed at BNL, did not employ colliding electron and positron beams. Instead, the AGS provided 28 GeV protons, which were made to collide with a beryllium target. The experiment then consisted in the measurement of the invariant-mass spectrum of e^+e^- and $\mu^+\mu^-$ pairs produced in these collisions. The reaction studied was therefore

$$\begin{aligned} p + \text{Be} &\rightarrow J/\psi + \text{anything} \\ &\hookrightarrow e^+e^-,\ \mu^+\mu^-, \end{aligned} \tag{4.2.10}$$

so that typical Breit–Wigner resonance peaks were clearly visible here too.

It is interesting to examine a little more closely the data of the experiment performed at SLAC. The process here is classic e^+e^- annihilation in the Stanford Positron-Electron Accelerating Ring (SPEAR) collider. The possible channels are

$$e^+e^- \rightarrow J/\psi \rightarrow \text{hadrons},\ e^+e^- \text{ and } \mu^+\mu^-. \tag{4.2.11}$$

The three related cross-sections are shown in figure 4.14. The particularly skewed shapes of some of the curves is typical of interference between competing channels. Here both the direct channel (as shown in figure 4.10) and the channel with an intermediate photon (figure 4.11) contribute. It can be shown that the interference pattern corresponds well to interference between these two channels.

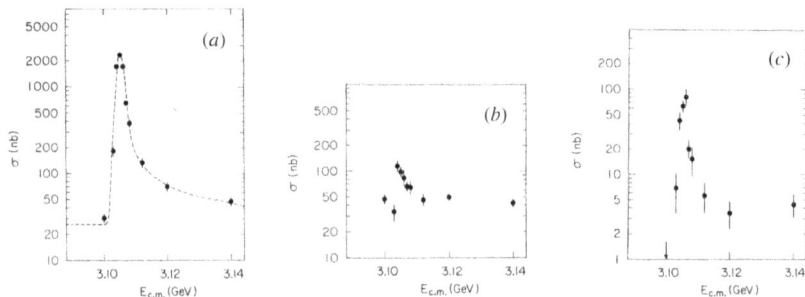

Figure 4.14. The e^+e^- cross-section as a function of centre-of-mass energy around the J/ψ peak for various final states: (a) hadrons, (b) e^+e^- and (c) $\mu^+\mu^-$, $\pi^+\pi^-$ and K^+K^-. The dashed curve in the left-most figure is the theoretical prediction including a Gaussian energy spread of the beams; figure reproduced with permission from Augustin *et al* (1974), copyright (1974) by the American Physical Society.

This is also an example of how too selective a production mechanism can unfortunately hide states. There exists a so-called η_c (for its similarity to the much lower-mass η), also an s-wave $c\bar{c}$ but with spins antiparallel[30]:

$$m_{\eta_c} = 2983.9 \pm 0.4 \text{ MeV} \quad \text{and} \quad \Gamma_{\eta_c} = 32.0 \pm 0.7 \text{ MeV}, \tag{4.2.12}$$

but I^G, $J^{PC} = 0^+$, 0^{-+}. It was not, however, observed in these experiments. Having spin-zero and isospin-zero with also positive C-parity, it is the natural $c\bar{c}$ partner to the original $s\bar{s}$ pseudoscalar η; it cannot be produced directly in e^+e^- annihilation, nor can it decay into e^+e^- or $\mu^+\mu^-$, as it cannot couple directly to a single photon. Its discovery was, however, made indirectly a little later in an e^+e^- machine (SPEAR at SLAC, Partridge *et al* 1980) via the (electromagnetic) radiative decays $J/\psi \to \eta_c\gamma$ and $\psi' \to \eta_c\,\gamma$, with the subsequent strong ($c\bar{c}$-annihilation) decay $\eta_c \to \eta\,\pi^+\pi^-$.

4.2.2.4 Discovery of the b quark

The pattern then repeats itself a little higher in energy: at around 10 GeV in the centre-of-mass, the threshold for Upsilon (Υ) production is reached and several peaks are found very close together, the first of which is the $\Upsilon(1S)$. Historically, the discovery of the b (bottom or beauty) quark (Herb *et al* 1977, Innes *et al* 1977) was actually made at the Fermilab proton machine, where 400 GeV protons were made to collide with copper and platinum targets. The process studied was

$$p + \text{Cu, Pt} \to \Upsilon + \text{anything}$$
$$\quad\quad \hookrightarrow \mu^+\mu^-, \tag{4.2.13}$$

where again the resonance is evident via a final-state spectrum. The various radial excitations of the fundamental Υ state were observed later in the two most energetic e^+e^- machines of the time: the Doppel-Ring-Speicher (DORIS) at the Deutches Elektronische Syncrotron (DESY) in Hamburg and the Cornell Electron Storage Ring (CESR) at Cornell University.

The first Υ state has the following mass and width:

$$m_\Upsilon = 9.46 \text{ GeV} \quad \text{and} \quad \Gamma_\Upsilon \simeq 54 \text{ keV}, \tag{4.2.14}$$

with spin–parity quantum numbers $J^{PC} = 1^{--}$ (just as the J/ψ). On a logarithmic energy scale there are a number of states quite close to each other. In fact, in the first Fermilab experiment the resolution was rather poor (around 0.5 GeV) and the set of peaks thus appeared as one or possibly two broad humps. Nevertheless, since the total width of the distribution was around 1.2 GeV, it was immediately deduced that more than one resonance was present (see figure 4.15). On further investigation, others were indeed found; the first and most prominent are those shown in table 4.2, although there are many others (see too figure 4.16). The evident reduction in the widths with growing radial excitation will be discussed shortly.

As might be expected, there also exists an η_b state (I^G, $J^{PC} = 0^+$, 0^{-+}), which again is not directly accessible via e^+e^- annihilation. In fact, it was not until as

---

[30] Note that this is just as might be expected: all the other ground state (i.e. lowest-mass) $q\bar{q}$ mesons also have their spins antiparallel and thus, with $L = 0$, must be spin-zero states.

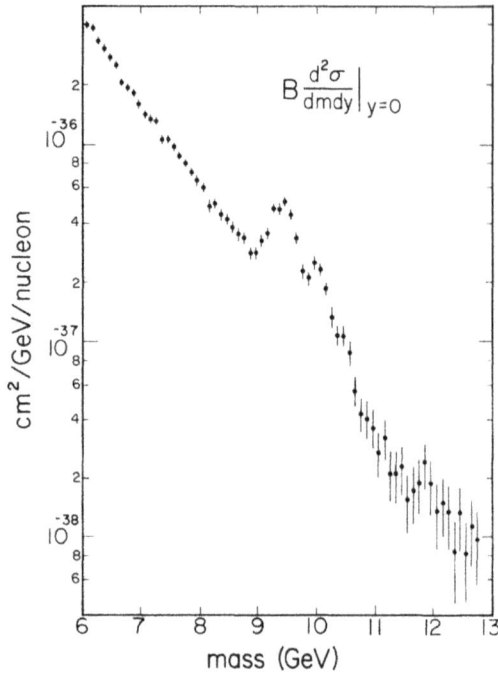

Figure 4.15. The broad overlapping peaks of the lowest-mass Υ states in the Fermilab experiment dimuon spectrum; figure reproduced with permission from Innes *et al* (1977), copyright (1977) by the American Physical Society.

Table 4.2. The $\Upsilon(1S)$ plus the three lowest-mass Υ ($b\bar{b}$, $J^{PC} = 1^{--}$) radial-excitation states.

State	Mass/GeV	Width
$\Upsilon(1S)$	9.460	54 keV
$\Upsilon(2S)$	10.023	32 keV
$\Upsilon(3S)$	10.355	20 keV
$\Upsilon(4S)$	10.579	21 MeV*

*The large jump is due to the opening of the $D\bar{D}$ threshold.

recently as 2008 that this state was observed by the BaBar experiment at SLAC (Aubert *et al* 2009):

$$m_{\eta_b} \simeq 9399 \text{ MeV} \quad \text{and} \quad \Gamma_{\eta_b} \simeq 10 \text{ MeV}. \tag{4.2.15}$$

Once more, the observation was made indirectly in an e^+e^- machine (PEP-II at SLAC) via the radiative decay $\Upsilon(3S) \to \eta_b \, \gamma$.

Again, all of these states are very narrow owing to OZI-type suppression. Indeed, they are seen to be even narrower than the typical J/ψ widths. We might have *naïvely* expected just the opposite on the basis of arguments of both phase-space and

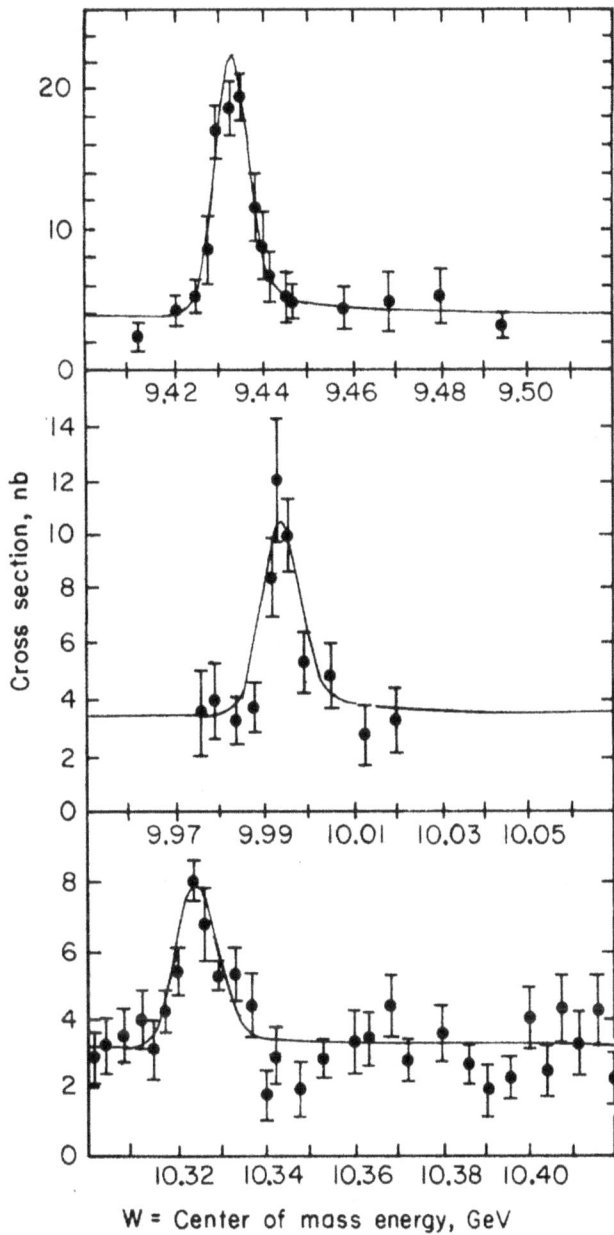

Figure 4.16. The narrow $\Upsilon(1S)$, $\Upsilon(2S)$ and $\Upsilon(3S)$ states observed with the CLEO detector at (CESR); figure reproduced with permission from Andrews *et al* (1980), copyright (1980) by the American Physical Society.

number of channels available. However, since the OZI rule is evidently operative here and the decays must proceed via three-gluon intermediate states, there will be a very strong dependence on α_s. Each gluon implies a factor α_s in the total decay rate and thus, if perturbative calculations can be trusted, the widths are expected to be proportional to α_s^3. The reason this is not $(\alpha_s^3)^2$ is that the final coupling to quarks is not included, as this part of the calculation is considered as a sort of decay of the gluons produced, which must occur with unit probability. Thus, in complete analogy with positronium decay, we expect one power of α_s for each gluon 'emitted'. Note, however, that for exclusive decay-channel rates, where a specific final state must also couple to the gluon, the $(\alpha_s^3)^2$ behaviour would still be expected.

Now, as already explained in the section on the strong interaction, in quantum field theory the couplings are *not* constant, in particular, in QCD the coupling is a decreasing function of energy scale. It can be shown (in quantum field theory) that the correct scale to consider here is of the order of the quark masses involved. For a change of scale from that of the c-quark mass to that of the b quark, the change in α_s is roughly a factor two. The resulting suppression is, of course, offset by various other effects: phase-space, number of channels and, last but not least, the size of the objects involved: the larger mass of the b quark in a non-relativistic approximation leads to a smaller Bohr radius; that is, the wave-function is more concentrated at the origin, thus favouring the $q\bar{q}$ annihilation channel.

It is, of course, impossible to perform complete *ab initio* calculations since we are unable to reliably calculate the bound-state wave-function. However, for a heavy-quark system, such as $c\bar{c}$ or $b\bar{b}$, we might imagine non-relativistic quantum mechanics to be sufficient. And since the appropriate energy scale should be that of the heavy-quark mass, we may assume α_s to be small enough to allow a pseudo-perturbative approach. Moreover, as discussed in the previous chapter, a reasonable assumption for the effective form of the $q\bar{q}$ interaction is given by the so-called Cornell potential (Eichten *et al* 1975, 1978):

$$V(r) = -C_F \frac{\alpha(Q^2)}{r} + b\,r, \qquad (4.2.16)$$

where the first term represents a standard Coulomb-like potential, including the known scale dependence of the coupling ($Q^2 \propto 1/r^2$), and the second term is an empirical string-like potential, held to be dominant at large distances. The coefficient $C_F = 4/3$ is a QCD colour factor, while b (the so-called *string tension*) is just a model parameter, to be fixed by fitting the data.

The Schrödinger equation may be solved for a heavy quark–antiquark pair in such a potential; the resulting 'energy levels' will provide the masses of the various excitations (both radial and rotational). The model may be refined by the inclusion of spin–orbit and spin–spin effects, with the necessary additional parameters. Indeed, such models already have modest success even with lighter systems and, for example, the string tension (the parameter b above) turns out to be approximately universal, with a value of the order of 1 GeV/fm [31].

It is then possible to reliably estimate decay rates. An obvious prediction of such a model would be precisely the observed decrease in the decay widths of the heavy $q\bar{q}$ states with growing radial excitation. The decay annihilation requires the pair to be

---

[31] In more macroscopic units, this is equivalent to 160 kJ m^{-1} or 160 KN.

found at some instant at the origin and thus the rate is proportional to the wave-function at the origin. The radial-excitation states, having greater spatial extension naturally have smaller wave-functions at the origin. Note also that reliability can be improved by taking ratios: for example, we may compare the two-photon and two-gluon decays of the spin-zero quarkonia states, or the decays via a single photon (to two charged leptons) and three gluons in the case of the $J^{PC} = 1^{--}$ states. In this way, to some extent, the uncertainties in the bound-state parameters cancel in the ratio. The measured values for the decay rates agree well with the values of α_s extracted from other sources.

4.2.2.5 Toponium

A few words are in order here on the question of toponium, the *would-be* $t\bar{t}$ bound state. First of all, as we shall see, the t quark is much more massive than the others, having a mass $\sim$173 GeV. This already naturally precludes its discovery in any e^+e^- collider built to-date, or indeed ever proposed. Moreover, the t quark itself is very unstable, with $\Gamma_t = 1.42^{+0.19}_{-0.15}$ GeV, and is likely to have a much shorter intrinsic lifetime than the formation time of any possible bound state. Note that at energies corresponding to the top mass the weak interaction is no longer particularly suppressed by the $W^{\pm}$ and Z^0 boson masses. This would make the $t\bar{t}$ meson a very short-lived and weakly bound object, rendering it highly unlikely that any such resonances will ever be observed.

4.2.3 Discovery of the Z^0 boson

We now turn to the very last of the pronounced peaks in the cross-section plot for $e^+e^- \rightarrow$ hadrons: namely, the Z^0 boson. This is evidently a slightly different situation to those described in the case of $q\bar{q}$ resonances. The Z^0 boson is an elementary field, with a point-like fundamental coupling to an e^+e^- initial (and/or final) state. However, its experimental appearance is very similar: having a finite (large) mass and a finite (but not small) width, a standard Breit–Wigner shape is observed in the cross-section energy dependence for a variety of final states. It was first produced on-shell in large numbers by LEP at Centre Europée de Rechèrche Nucleaire (CERN, Geneva) although incontrovertible evidence for its existence had already been obtained in various other experiments. In particular, the TRISTAN e^+e^- collider at KEK in Japan had already achieved a maximum 32 + 32 GeV. At such energies the presence of the Z^0 Breit–Wigner tail is very evident. Moreover, the earlier CERN experiments aimed at $W^{\pm}$ discovery had also produced some Z^0 events (see section 4.3.1) and the Stanford Linear Collider (SLC) at SLAC was already running.

At intermediate energies (i.e. below the peak), where the Z^0 boson is not completely dominant and where interference with the photon intermediate state is thus non-negligible, we can check (via the presence of interference) certain of the quantum numbers of this resonance. This is important: the spin and parity could, in principle, be different to that of the photon, since the coupling to fermions could be different. In an e^+e^- collision all possible spin configurations of the initial pair are indeed present:

$$|+\hat{z}\rangle|+\hat{z}\rangle, \quad |+\hat{z}\rangle|-\hat{z}\rangle, \quad |-\hat{z}\rangle|+\hat{z}\rangle, \quad |-\hat{z}\rangle|-\hat{z}\rangle, \qquad (4.2.17)$$

where the $\hat{z}$ indicates that these are *not* helicities, but spin projections along the z-axis. They therefore represent the helicities of the particle moving in the positive z direction and *minus* the helicities for the other. These states should then be rearranged into multiplets according to *total spin*:

$$\text{spin}-1: \quad |+\hat{z}\rangle|+\hat{z}\rangle, \quad \frac{1}{\sqrt{2}}\big(|+\hat{z}\rangle|-\hat{z}\rangle + |-\hat{z}\rangle|+\hat{z}\rangle\big), \quad |-\hat{z}\rangle|-\hat{z}\rangle, \quad (4.2.18a)$$

$$\text{spin}-0: \quad \frac{1}{\sqrt{2}}\big(|+\hat{z}\rangle|-\hat{z}\rangle - |-\hat{z}\rangle|+\hat{z}\rangle\big). \quad (4.2.18b)$$

The coupling to a spin-one boson merely selects a particular multiplet; three out of four of the above states are acceptable, reflecting the following factor already explicitly present in the Breit–Wigner formulæ:

$$\frac{(2J_R + 1)}{(2s_1 + 1)(2s_2 + 1)} = \frac{3}{4}. \quad (4.2.19)$$

Note that, for the Z^0 to interfere with the photon, it must have the same J^{PC}. As we shall see in the next chapter, the Z^0 is to be considered a close relative of the photon, as too are the $W^{\pm}$.

As deduced from the Breit–Wigner resonance line-shape, the mass and width parameters of the Z^0 are (Navas *et al* 2024):

$$m_Z = 91.1880 \pm 0.0020 \,\text{GeV} \quad \text{and} \quad \Gamma_Z = 2.4955 \pm 0.0023 \,\text{GeV}. \quad (4.2.20)$$

It is a spin-one boson, just as the photon, but its parity and charge-conjugation properties are *not* defined: the weak coupling of the Z^0 violates both parity and charge-conjugation (just as that of the $W^{\pm}$) and it cannot therefore be an eigenstate of either parity or charge conjugation. This last observation does not preclude its interfering with the photon since it should thus be considered as a superposition of both opposite parity and charge-conjugation eigenstates (just as, e.g. the K^0 and $\bar{K}^0$ are superpositions of CP eigenstates). Note finally that such precise measurements[32] on such a broad object require the inclusion of the phase-space corrections to the width itself as s various across the Breit–Wigner peak, see equation (C.6.26). Note though that, since the Z^0 is an elementary particle, the Breit–Wigner formula is accurate even out into the tails of the distribution.

Now, recall that the width Γ_Z, as deduced from the line-shape, is the *total* width; that is to say, it includes all available channels without discrimination. Remembering that the width is none other than the decay rate, we see that if there are a number of *distinct* (i.e. *non*-interfering) final states, then the total width (or rate) is simply the sum of all the partial widths (or rates). This leads to some very important cross checks.

---

[32] Indeed, such is the precision achieved at LEP, that tidal effects on the length of the ring (variations of order 1 mm), which can contribute up to 40 MeV to the beam energy, must be taken into account (Arnaudon *et al* 1995). Even gravitational effects, owing to the nearby Jura mountains and seasonal changes in the level of water in Lake Geneva, have been considered.

4.2.3.1 Z^0 partial decay widths

First of all, universality of the weak coupling in the neutral-current channel may be checked by comparing the measured Z^0 decay rates or branching ratios into electrons, muons and tau leptons:

$$\Gamma_Z^{e^+e^-} / \Gamma_Z^{\text{tot}} = (3.3632 \pm 0.0042)\%, \tag{4.2.21a}$$

$$\Gamma_Z^{\mu^+\mu^-} / \Gamma_Z^{\text{tot}} = (3.3662 \pm 0.0066)\%, \tag{4.2.21b}$$

$$\Gamma_Z^{\tau^+\tau^-} / \Gamma_Z^{\text{tot}} = (3.3696 \pm 0.0083)\%. \tag{4.2.21c}$$

Note that phase-space differences due to lepton masses are essentially negligible here. We can, of course, also *calculate* the partial decay rates. The generic rate for a real (i.e. on mass-shell) Z^0 decaying into a fermion–antifermion pair is

$$\Gamma_Z^{f\bar{f}} = \frac{C\,G_\text{F}M_Z^3}{6\sqrt{2}\,\pi}\left[g_{Vf}^2 + g_{Af}^2\right], \tag{4.2.22}$$

where G_F is just the usual Fermi weak coupling constant, e.g. as measured in μ-decay. The coefficient C is defined as follows:

$$C := \begin{cases} 1 & \text{for lepton pairs,} \\ N_\text{c}\left[1 + a + 1.409a^2 - 12.77a^3 - 80.0a^4\right] & \text{for quark pairs.} \end{cases} \tag{4.2.23}$$

The second line here includes the corrections due to the strong interaction (or QCD), where $a = \alpha_s(M_Z)/\pi$ (i.e. α_s as evaluated at the energy scale of the Z^0, for which we recall the present world average is ~ 0.118). Finally, the vector- and axial-vector weak coupling constants (or weak charges) are

$$g_{Vf} = I_{3f} - 2Q_f \sin^2\theta_W, \tag{4.2.24a}$$

$$g_{Af} = I_{3f}, \tag{4.2.24b}$$

where I_{3f} is just the third component of *weak* isospin ($+\frac{1}{2}$ for up-type quarks and neutrinos, $-\frac{1}{2}$ for down-type quarks and charged leptons), Q_f is the electric charge of the fermion in units of $|e|$ and θ_W is the weak mixing angle (with $\sin^2\theta_W \simeq 0.2234$), which we shall discuss later. Inserting the values of the various parameters (all of which may, in principle, be measured independently in other processes), leads to (Navas *et al* 2024)

$$\Gamma_Z^{f\bar{f}} = \begin{cases} 299.87 \pm 0.20 \text{ MeV} & (u\bar{u}), \\ 299.81 \pm 0.20 \text{ MeV} & (c\bar{c}), \\ 382.75 \pm 0.14 \text{ MeV} & (d\bar{d}, s\bar{s}), \\ 375.73 \pm 0.18 \text{ MeV} & (b\bar{b}), \\ 167.145 \pm 0.015 \text{ MeV} & (\nu\bar{\nu}), \\ 83.955 \pm 0.009 \text{ MeV} & (e^+e^-, \mu^+\mu^-) \\ 83.772 \pm 0.009 \text{ MeV} & (\tau^+\tau^-). \end{cases} \tag{4.2.25}$$

The predicted total width is then

$$\Gamma_Z^{\text{tot}} = 2494.00 \pm 0.87 \text{ GeV}, \tag{4.2.26}$$

in excellent agreement with the measurements.

4.2.3.2 The number of light neutrinos

Hidden inside the numbers and formulæ just presented is an interesting and perhaps surprising measurement: the number of *light* neutrinos. First of all, we say light neutrinos since there would obviously be no sensitivity to a neutrino with a mass greater than half that of the Z^0. Now, of course, although the statistics gathered at LEP is considerable (many tens of millions of Z^0 bosons have been produced and detected), there is little chance of detecting either the neutrino or antineutrino produced in a Z^0 decay. Note that to be certain, we should need to detect *both* and measure their energies, in order to fully reconstruct the mass of the decaying object. However, since the line-shape provides the total width and the other partial widths may all be measured directly, the 'invisible' width can be deduced from the difference. The following combination is then usually extracted:

$$N_\nu = \left(\frac{\Gamma_{\text{inv}}}{\Gamma_\ell}\right)_{\text{expt}} \left(\frac{\Gamma_\ell}{\Gamma_\nu}\right)_{\text{th}} = \left(\frac{\Gamma_{\text{tot}} - \Gamma_{\text{vis}}}{\Gamma_\ell}\right)_{\text{expt}} \left(\frac{\Gamma_\ell}{\Gamma_\nu}\right)_{\text{th}}. \tag{4.2.27}$$

The double ratios in these equalities are chosen so as to improve the reliability of the calculation by favouring the cancellation of various systematic effects, both theoretical and experimental. Combining the results from all four LEP experiments, finally leads to (Navas *et al* 2024)

$$N_\nu = 3.0025 \pm 0.0061 \tag{4.2.28}$$

That is, the three known neutrinos (the partners to the electron, muon and tau lepton) are confirmed as the only *light* neutrinos. Incidentally, this number also places very stringent limits on the existence of other light particles coupling to the Z^0 and which might hitherto have gone undetected for some reason.

Many further tests of the electroweak theory may be performed at the Z^0 peak. For example, there are various parity-violating angular asymmetries that may be measured and that are directly related to the parameters of the theory (in particular, to $\sin\theta_W$). All measurements so far performed provide *no* evidence of any flaw or shortcoming in the theory. We shall discuss these questions more in detail in the following chapter.

4.2.4 Discovery of the τ lepton

The remaining major discovery made at high-energy e^+e^- colliders is that of the τ lepton ($m = 1777$ MeV, $\tau = 290 \times 10^{-15}$ s), again with SPEAR at SLAC (Perl *et al* 1975)[33]. The experiment obtained just 64 events of the form

$$e^+e^- \rightarrow e^\pm + \mu^\mp + \text{missing energy}, \tag{4.2.29}$$

---

[33] For the associated Nobel Prize in Physics, see footnote 20 in this chapter.

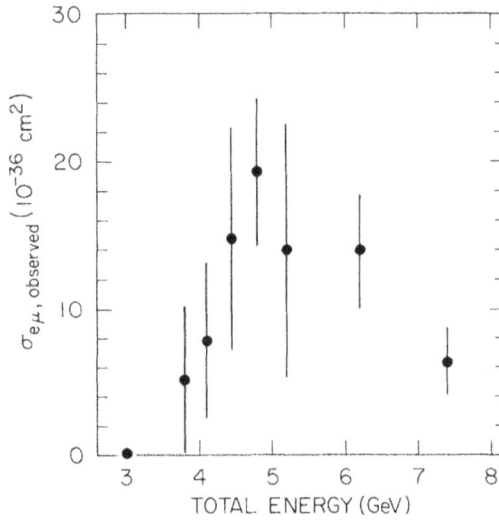

Figure 4.17. The observed cross-section for the $e^{\pm}+\mu^{\mp}+$ missing-energy events as a function of e^+e^- centre-of-mass energy; figure reproduced with permission from Perl *et al* (1975), copyright (1975) by the American Physical Society.

in which no other charged particles or photons were detected. Most of these events were detected at or just above a centre-of-mass energy of 4 GeV (see figure 4.17). The missing-energy and missing-momentum spectra indicated that at least two additional particles had been produced in each event. There being no conventional explanation for such events, they were attributed to the production of a new charged-lepton pair $\tau^+\tau^-$.

The processes being observed here are then presumed to be

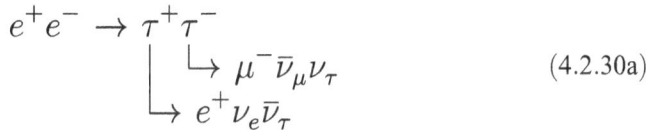

$$e^+e^- \to \tau^+\tau^-$$
$$\hookrightarrow \mu^-\bar{\nu}_\mu\nu_\tau$$
$$\hookrightarrow e^+\nu_e\bar{\nu}_\tau \qquad (4.2.30a)$$

and

$$e^+e^- \to \tau^+\tau^-$$
$$\hookrightarrow e^-\bar{\nu}_e\nu_\tau$$
$$\hookrightarrow \mu^+\nu_\mu\bar{\nu}_\tau. \qquad (4.2.30b)$$

Under normal circumstances, neither of these two (charged) mixed-flavour final states is possible at the level measured since they would represent *individual* (electron and muon) lepton-number violation. However, of course, the undetected neutrinos (missing energy) balance all the relevant quantum numbers.

4.2.5 Open-flavour and other particle production

Shortly above each new quark threshold, which is always first seen as a $q\bar{q}$ resonance, the threshold for the corresponding *open* production is reached. That is, the centre-of-mass energy is sufficient to produce a meson–antimeson pair, each containing the new quark (antiquark), paired up with a light antiquark (quark).

4.2.5.1 D mesons

Immediately above the J/ψ peak we see a *fall* in the cross-section, back to the previous level, owing to the fact that we are now off resonance but have not yet reached the threshold for open charm. As soon as there is sufficient energy available in the centre-of-mass to produce a D-meson pair, we see a rise to the new level. The lowest-lying D mesons have the canonical $J^P = 0^-$ while the masses and decay widths are shown in table 4.3. An extra charged state is now possible by replacing the d with an s quark; it is naturally a little heavier than the others.

Note that here we do not talk of decay widths, but of lifetimes. These mesons do not have access to strong annihilation channels and only decay via the weak decay of the individual c and $\bar{c}$ quarks or via $c\bar{q}$ (or charge conjugate) annihilation into $W^\pm$ (where possible). The lifetimes are therefore considerably longer and the resonance widths too narrow to be determined as such. By using modern silicon-strip tracking detectors with vertex resolution capabilities of the order of a few microns, the lifetimes are thus typically determined via precise measurement of track lengths.

It turns out to be very instructive to examine the lifetimes and possible decay modes of the D mesons. An immediately striking feature is the marked difference between the $D^\pm$ and both the D^0 and $D_s^\pm$ lifetimes, by more than a factor two. Now, the semi-leptonic partial widths are all essentially equivalent and so the origin is to be sought in the hadronic decays. The different types of quark diagrams that may contribute are shown in figure 4.18. A possible partial solution immediately suggests itself: the annihilation process is only important for $D_s^\pm$, being Cabibbo suppressed for $D^\pm$ (where it involves $\sin \theta_C$), but is of course impossible for D^0. On the other hand the exchange diagram may be important for D^0 but is not possible for $D^\pm$.

It is, however, unlikely that such individual contributions could be large enough to create the disparity highlighted above. Note also that the annihilation process suffers a similar helicity-conservation suppression to that described earlier in charged-pion decay (see section 4.1.4). An additional possible solution to the problem (suggested by Guberina *et al* 1979) comes from the recognition that in the case of $D^\pm$ the $\bar{q}$ in figures 4.18(*a*) and (*b*) is actually a $\bar{d}$, which may then interfere destructively with the other $\bar{d}$ already present (see section B.3.4 for a brief explanation of this type of Pauli suppression). Clearly, for neither D^0 nor $D_s^\pm$ is such an effect possible.

Table 4.3. The lowest-lying D-meson masses and decay widths.

State	Quark content	Mass/MeV	$\tau/10^{-12}$ s
$D^\pm$	$(c\bar{d}, d\bar{c})$	1870	1.04
D^0, $\bar{D}^0$	$(c\bar{u}, u\bar{c})$	1865	0.41
$D_s^\pm$	$(c\bar{s}, s\bar{c})$	1968	0.49

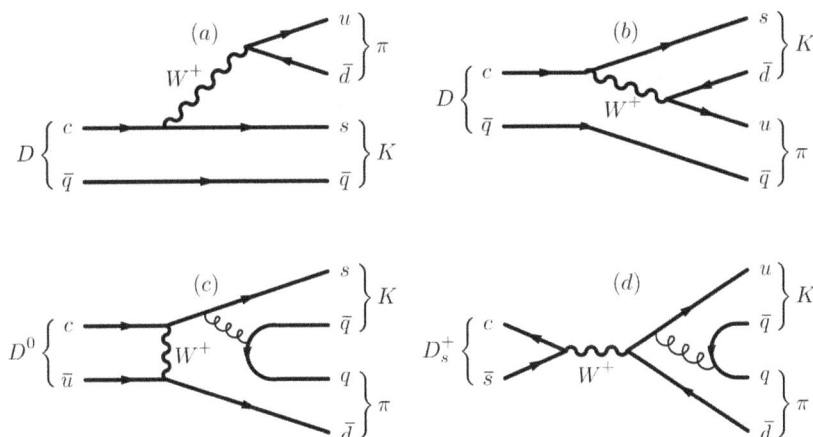

Figure 4.18. The different possible quark-diagram contributions to generic D meson decay into $K\pi$: (a) external spectator, (b) internal spectator, (c) exchange and (d) annihilation.

Table 4.4. The lowest-lying B-meson masses and decay widths.

State	Quark content	Mass/MeV	$\tau/10^{-12}$ s
$B^\pm$	$(u\bar{b}, b\bar{u})$	5279.3	1.64
B^0, $\bar{B}^0$	$(d\bar{b}, b\bar{d})$	5279.6	1.52
B_s^0, $\bar{B}_s^0$	$(s\bar{b}, b\bar{s})$	5366.9	1.51
$B_c^\pm$	$(c\bar{b}, b\bar{c})$	6274.9	0.51

4.2.5.2 B mesons

Slightly above the Υ peak we see a rise in cross-section to a new level corresponding to the threshold for open-b production, which initiates with the production of B mesons. This is in precise analogy with the previous case of D mesons. Again, the lowest-lying B mesons have $J^P = 0^-$; the masses and decay widths are shown in table 4.4. Here there is yet another new possibility: the doubly heavy, charged, meson states D^+ ($c\bar{b}$) and D^- ($b\bar{c}$).

4.2.5.3 W^+W^- pair production

The final structure that emerges from a study of the hadronic cross-section in e^+e^- annihilation is the opening of the threshold for the production of W^+W^- pairs. So far such a study has only been possible in one machine: LEP at Centre Europée de Rechèrche Nucleaire (CERN, Geneva). After having thoroughly studied the Z^0 resonance with centre-of-mass energies around 91 GeV, the machine was slowly pushed to its design limit of about 100 GeV beam energy, thus providing a total centre-of-mass energy of 200 GeV. This permitted the study of the highest-energy process accessible in this machine: namely, W^+W^- production:

$$e^+e^- \to \gamma, Z^0 \to W^+W^-. \tag{4.2.31}$$

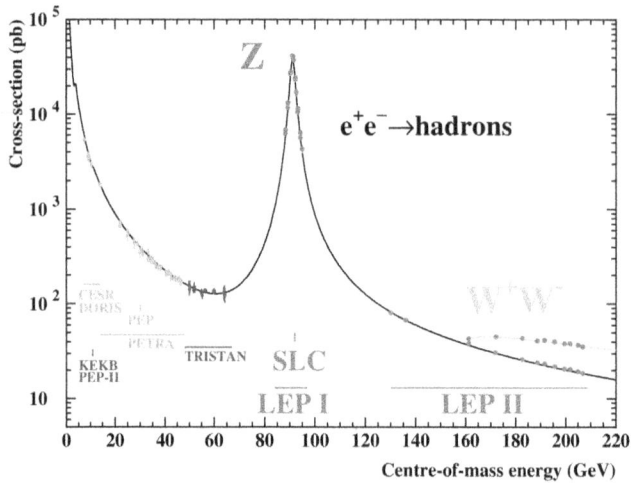

Figure 4.19. A plot of the hadronic cross-section in e^+e^- at LEP, showing the onset of W^+W^- production at around 160 GeV (figure from Schael *et al* 2006), copyright (2006) with permission from Elsevier.

This can be seen as the enhancement of the curve in figure 4.19 starting at around 160 GeV. Note that while we might assume the photon channel to be rather natural, since, after all, the $W^\pm$ is charged, the triple weak-boson coupling $Z^0W^+W^-$ is a precise prediction of the electroweak theory. As such, it has to be tested; once again, the measurements performed revealed absolutely *no* indication of deviations with respect to theoretical predictions.

4.2.6 Jets in e^+e^- annihilation

Before leaving the topic of e^+e^- collisions, there is one final type of process that deserves detailed examination. We have discussed a great deal the production of hadrons in such collisions; before the advent of QCD and the quark model, the description of the process $e^+e^- \to$ hadrons took the form of the intermediate creation of what was typically called a *fireball*. In such a picture the centre-of-mass energy was more-or-less uniformly distributed among final-state particles emerging over the entire 4π solid angle. However, once the idea of point-like objects that could be produced in particle–antiparticle pairs was established, this picture changed radically. We have already seen the relevant quark diagrams many times; the main point to appreciate is that, at leading order, the dominant process sees the production of a quark–antiquark pair, with a large amount of kinetic energy and which are therefore projected out of the interaction region *back-to-back*. Thus, if the *hadronisation* process does little to change the direction of motion, we might expect to see something like two back-to-back *jets* of particles (see figure 4.20).

4.2.6.1 The Lund string model

The final detected hadronic state is born then of just two objects, separating at high velocity. Unfortunately, we do not know how these two fundamental fields convert or *fragment* into the hadrons that eventually materialise in the laboratory and

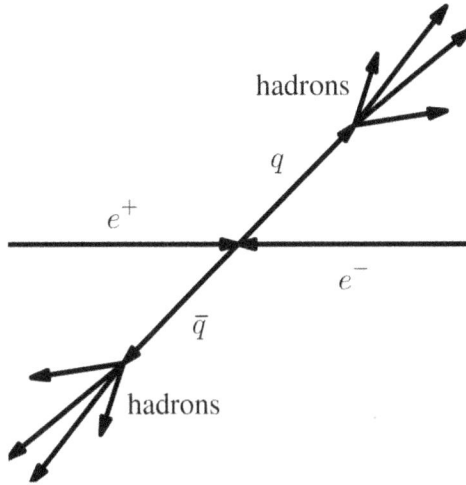

Figure 4.20. A schematic view of the jet-like structure that might be expected, based on the assumption of a process initiated by quark–antiquark pair production.

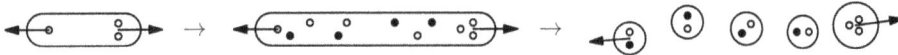

Figure 4.21. The Lund string-model picture of quark fragmentation. The hadrons created at the ends of the string will be the fastest (or so-called *leading*) particles and each will be followed by a jet of slower hadrons.

propagate some finite distance until either entering directly into a detector or decaying into other lighter, more stable hadrons. At any rate, one present conventional picture of the *hadronisation* process (the Lund model; Andersson *et al* 1983) is as follows. As the $q\bar{q}$ separate, the field lines stretch out between them to form a string-like object, in which energy is stored according to an approximately linear potential. As this stored potential energy grows, the threshold for real pair production is approached and at some point, the string may break producing a quark–antiquark pair, which will then form the new ends of the two strings so produced (see figure 4.21). Although a small amount of energy may be converted into transverse momentum, the principal direction of the quarks thus created will be parallel to the initial motion. This process will iterate, with the smaller strings breaking again and again until the energy has been fully dissipated.

The end picture then has a number of final-state particles, over which the energy of the initial quark–antiquark system is shared (in relation to position along the string) and which are moving more-or-less collinearly; this is called a *jet*. At low centre-of-mass energy and consequent low final-state multiplicity, it is hard to distinguish such behaviour. However, at higher energy the two-jet nature of the majority of hard-scattering events clearly emerges. Models based on these ideas, with a simple stochastic choice of the string breaking points provide a good description of the jet-like events observed; an example experimentally observed two-jet event is shown in figure 4.22.

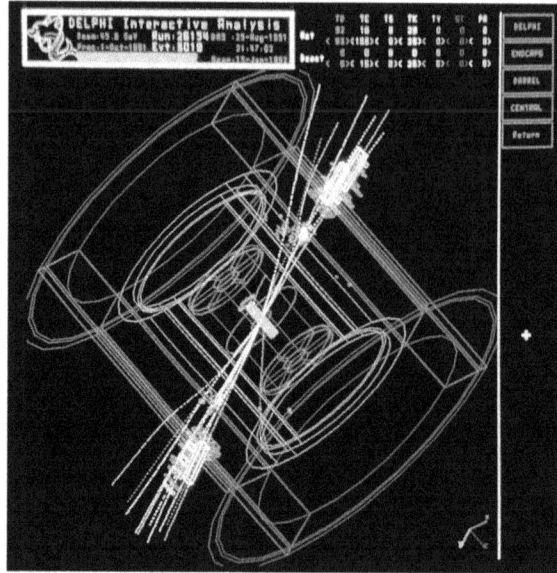

Figure 4.22. The reconstruction of a two-jet event as experimentally observed in the DELPHI detector at LEP; image reproduced courtesy of the CERN Courier, 22nd. Mar. 2002, p. 18. Copyright IOP Publishing. CC BY 4.0.

4.2.6.2 Event shapes in e^+e^-

While the qualitative picture is clear, it is obviously desirable to have a quantitative measure of the jet-like nature of the observed events. A number of variables have been defined to describe or quantify the *shape* of such jet-like events; three prominent examples are: *sphericity* (Bjorken and Brodsky 1970), *spherocity* (Banfi *et al* 2010) and *thrust* (Brandt *et al* 1964). Note that these event-shape variables may all be evaluated both experimentally and via the various theoretical simulations available, thus allowing a clear comparison between the hadronisation models and the data. Note too that, in principle, they may also be applied to the hadronic final states also observed in hadron–hadron scattering processes, although best suited in that case is spherocity.

Sphericity: (Bjorken and Brodsky 1970) is defined on an event-by-event basis as

$$S := \min_{\hat{n}} \frac{3\sum_i \boldsymbol{p}_{\mathrm{T}i}^2}{2\sum_i \boldsymbol{p}_i^2}, \tag{4.2.32}$$

where the sphericity axis $\hat{n}$ is defined as precisely that minimising S, $\boldsymbol{p}_i$ is the three-momentum of the ith particle in the event and $\boldsymbol{p}_{\mathrm{T}i}$ its component perpendicular to $\hat{n}$. A perfect two-jet event, i.e. with all outgoing particles aligned precisely along a single axis, will have $S = 0$, whereas for a completely isotropic event $S = 1$. The graph in figure 4.23 shows the measured mean sphericity in e^+e^- annihilation as a function of centre-of-mass energy in comparison with predictions of the jet model (solid line) and an isotropic phase-space model (dashed curve).

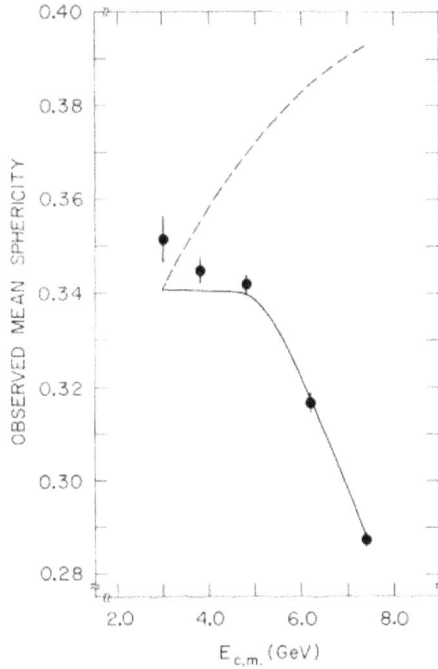

Figure 4.23. The mean experimental sphericity in e^+e^- annihilation as a function of centre-of-mass energy, compared with predictions by the jet model (solid line) and an isotropic phase-space model (dashed curve). The figure is reproduced with permission from Hanson *et al* (1975), copyright (1975) by the American Physical Society.

Spherocity: (or transverse spherocity, Banfi *et al* 2010) is again defined on an event-by-event basis, somewhat similarly to sphericity, as

$$S_0 := \frac{\pi^2}{4}\min_{\hat{n}_T}\left(\frac{\sum_i |\boldsymbol{p}_{T,i} \wedge \hat{\boldsymbol{n}}_T|}{\sum_i |\boldsymbol{p}_{T,i}|}\right)^2, \qquad (4.2.33)$$

where $\hat{n}_T$ is a unit transverse vector, again chosen to minimise the value of S_0, and $\boldsymbol{p}_{T,i}$ is the transverse component of the ith particle. Transverse is meant here with respect to the beam axis; allowing the variable to be used also in hadron–hadron processes, where the unknown momentum fractions of the colliding partons induce an unknown longitudinal boost. Just as for sphericity, a perfect two-jet event, with all outgoing particles aligned precisely along a single axis, will have $S_0 = 0$, whereas for a completely isotropic event $S_0 = 1$.

Exercise 4.2.5. *Show that the extreme values of sphericity and spherocity are attained for perfectly aligned two-jet events* $(S = S_0 = 0)$*, and completely isotropic events* $(S = S_0 = 1)$*.*

Thrust: (Brandt *et al* 1964) is defined too on an event-by-event basis as

$$T := \max_{\hat{n}} \frac{\sum_i |\boldsymbol{p}_i \cdot \hat{n}|}{\sum_i |\boldsymbol{p}_i|}, \tag{4.2.34}$$

where again $\boldsymbol{p}_i$ is the momentum of the ith particle in the event and now the thrust axis $\hat{n}$ is that which maximises the thrust variable T. With this definition, it is easy to show that $\frac{1}{2} \leqslant T \leqslant 1$. The upper limit is attained for precisely aligned two-jet events, whereas $T = \frac{1}{2}$ for completely isotropic events

Exercise 4.2.6. *Demonstrate that the two extreme values of thrust are attained for precisely aligned two-jet events ($T = 1$) and completely isotropic events ($T = \frac{1}{2}$).*

4.2.6.3 Gluon jets

So far we have appealed to non-perturbative aspects of QCD and its string-like long-distance potential. There is also a perturbative aspect to the interaction, in which gluons may be emitted, essentially via *bremsstrahlung*, see figure 4.24(*a*). Given the magnitude of α_s, such a process has a non-negligible probability. Moreover, it is quite likely to produce a gluon with large energy and at a large angle with respect to the direction of the emitting quark. There are then three rapidly moving particles, which, according to the string model, should give rise to three jets. Again, with sufficient centre-of-mass energy, such events are clearly visible. Indeed, since the probability of gluon emission is directly proportional to α_s, the ratio of the numbers of events with three and two jets is a direct measure of α_s. This method of extracting the strong coupling constant agrees well with other results.

We can naturally go further and consider events with four or even more jets, see figure 4.24(*b*). However, there is evidently a limit to how many jets we can sensibly hope to identify experimentally. The available energy is a severe limitation but there is also the question of overlapping of adjacent jets.

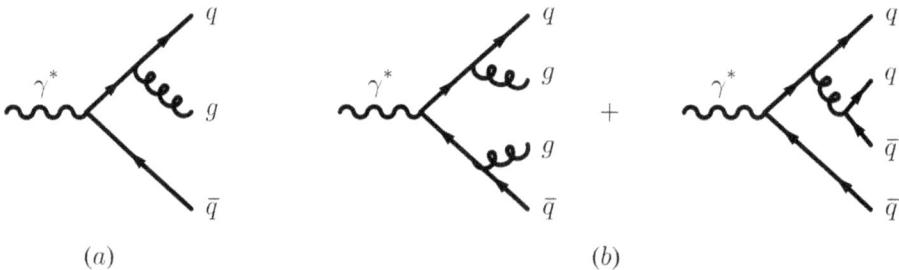

(a)   *(b)*

Figure 4.24. Hard gluon *bremsstrahlung* and $q\bar{q}$ pair-production processes leading to (a) three- and (b) four-jet events.

4.2.6.4 Particle flow

Given the overall colour neutrality of the $q\bar{q}g$ system, if the gluon carries a colour charge, e.g. blue–anti-red, say, the quark will be red and the antiquark, anti-blue. So, going back to the string model, we see that in the case of a three-jet event the string will stretch from the quark to the gluon and on to the antiquark (see figure 4.25). Note how the string is stretched between the quark and gluon and also between the antiquark and gluon, thus generating final-state hadrons mainly in the angular region between those pairs of partons. However, there is no string directly between the quark and antiquark. We should therefore expect this angular region to be depleted with respect to the other two. Now, since the gluon should on average carry less energy, there should be a correlation between the number of particles in intermediate regions and the energy of the away-side jets. In other words, on average, we expect to find fewer particles opposite to the lowest-energy jet as compared with the other two. This prediction of the model is also experimentally well verified (see figure 4.26).

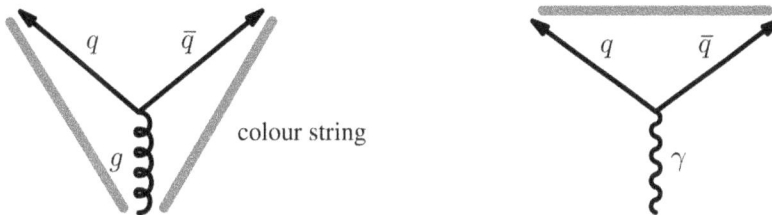

Figure 4.25. The evolution of a three-jet event in the Lund string picture. Of the two possibilities shown, that on the left better describes the data. In particular, it naturally leads to a depletion of particle and energy flow on the opposite side to the gluon jet. The right-hand figure would be as expected for *bremsstrahlung* of, say, a hard (colourless) photon and would lead to a very different final-state particle distribution.

Exercise 4.2.7. *Show that in the case of a perfectly symmetric three-jet event, the thrust value is* $T = \frac{2}{3}$.

4.2.6.5 The Marchesini–Webber model

In a different approach, the idea of parton emission (i.e. gluon *bremsstrahlung* and quark–antiquark pair production) led to the development of a perturbative model of hadronisation by Marchesini and Webber (1984), in which each emission is treated more-or-less independently and thus iterated (see figure 4.27). At the end of such a *parton shower* we are again left with a large number of quarks distributed over a range of momenta. The hadronisation *ansatz* adopted consists of combining each nearby colourless quark–antiquark pair and assuming it will form the meson state nearest in mass and quantum numbers. Again, a good description of the observed phenomenology is obtained. The depletion phenomenon on the opposite side to the gluon jet is seen here as the result of a so-called colour-coherence effect.

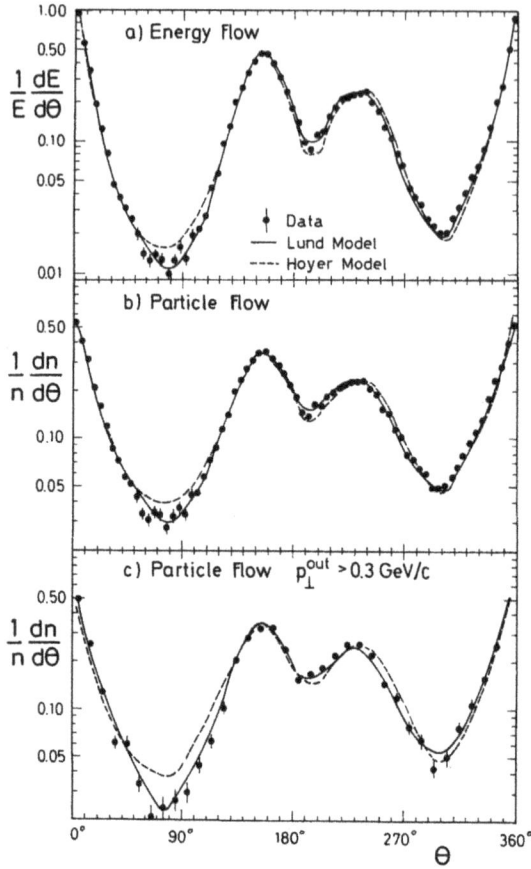

Figure 4.26. Correlations between the energy/particle flow in the angular region between the two jets opposite the lowest-energy jet. Comparisons are presented with a QCD-based model (Hoyer *et al* 1979) and with the Lund string model (Andersson *et al* 1983); figure reproduced from Bartel *et al* (1983) with permission from Springer Nature.

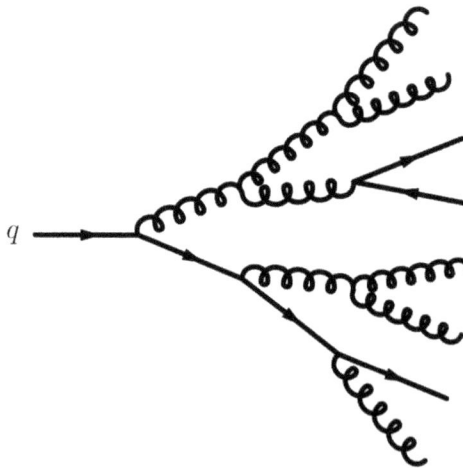

Figure 4.27. An example quark–gluon branching sequence in the perturbative QCD-based parton-shower model for quark fragmentation.

In conclusion, let us remark that in all approaches to the description of the final multi-hadron state there is a great deal of modelling, with necessarily a large number of parameters to be fitted or *tuned* to the data. That these models work as well as they do is a clear indication that our heuristic picture cannot be far from the truth. That said, it must always be borne in mind that we have as yet no first-principles approach to the problem, and then again, with enough parameters, even an incorrect model may appear satisfactory[34].

4.3 Proton–antiproton colliders

In this final section of the chapter on new particles we now turn briefly to the case of proton–antiproton (and proton–proton) colliders. All else being equal, a hadron collider would not normally be chosen to perform new-particle searches owing to its inherently dirty nature. Conceptually, the problems are essentially two-fold, but with a common origin: the actual collisions that take place in such a collider at the elementary level are not between the hadrons themselves but between the partons they contain. The two resulting difficulties can be classified as initial- and final-state.

The difficulty with the initial state is that *a priori* we can have no direct knowledge (i.e. event-by-event) as to the details: parton type, energy–momentum, spin etc. What we do have is a statistical description: we have parton distributions, measured in other processes, that give us the probability of finding a given parton type with a given momentum fraction inside a given hadron. Indeed, event-by-event we can only hope to reconstruct the initial kinematics from measurement of an exclusive final state, which, however, limits what may be measured. Incidentally, for somewhat technical theoretical reasons, the dependence on exclusive final states generally seriously reduces the reliability of perturbative QCD predictions. Even considering only statistical analyses, we nevertheless require accurate prior information on the parton densities and in the case of gluons, for example, or in general for very small values of x_B, this is not yet available.

As far as the final state is concerned, the difficulty here is that a large part of the hadron remnants disappears along the beam pipe and cannot therefore be detected. Moreover, it is possible for remnants of the spectator systems (the unstruck quarks) to fall within the detector acceptance. This all makes complete reconstruction impossible and therefore precludes certain types of analyses.

Naturally though, there are situations where there is no choice: typically this has to do with the available energy. The highest centre-of-mass energy achieved to date in an e^+e^- machine is 200 GeV at LEP, whereas the Tevatron at Fermilab ran for many years at a centre-of-mass energy of slightly less than 2 TeV, while LHC at CERN is designed to achieve 14 TeV centre-of-mass energy; although it should be noted that this is a proton–proton machine. For the sake of completeness, let us mention that in its heavy-ion mode LHC is also intended to provide collisions between, e.g. lead or gold ions with approximately 2.6 TeV per *nucleon*. Finally, we

---

[34] We need only recall the Ptolemaic or geocentric model of the Universe with its necessary epicycles, to understand how complex but *wrong* models can nevertheless produce apparently or rather approximately correct answers.

have already seen cases (such as the discovery of J/ψ) in which discovery is equally as favourable in hadron–hadron as in e^+e^- machines. Indeed, the η_c, for instance, cannot be produced directly in an e^+e^- machine.

4.3.1 Discovery of the $W^\pm$ boson

A clear example of the need for a hadron–hadron machine is the case of the discovery of the $W^\pm$ bosons[35]. Given that the $W^\pm$ couple to a pair of *different* flavour fermions (e.g., $u\bar{d}$, $e^+\nu_e$ etc), a simple e^+e^- collision is no longer useful. Since high-energy collisions between electron or muon and neutrino beams of sufficient intensity are unattainable, the only choice is to collide proton and, preferably, antiproton beams (to have both quarks and antiquarks in large numbers). For discovery of the $W^\pm$ in $p\bar{p}$ collisions, the following are the main channels:

$$u + \bar{d} \rightarrow W^+ \rightarrow e^+ + \nu_e, \ \mu^+ + \nu_\mu, \tag{4.3.1a}$$

and

$$\bar{u} + d \rightarrow W^- \rightarrow e^- + \bar{\nu}_e, \ \mu^- + \bar{\nu}_\mu. \tag{4.3.1b}$$

As a by-product, in the same collisions the Z^0 is also accessible, although a little more difficult, via

$$u + \bar{u}, \ d + \bar{d} \rightarrow Z^0 \rightarrow e^+e^-, \ \mu^+\mu^-. \tag{4.3.1c}$$

The final states indicated above are simply those easiest to detect.

Being a parton–parton interaction, the true centre-of-mass energy available in such a collision is given by

$$E_{CM}^{parton} = \sqrt{x_1 x_2} \ E_{CM}^{hadron}. \tag{4.3.2}$$

Moreover, recall that valence quarks might *naïvely* be expected to carry on average roughly one third of the total hadronic energy or momentum; i.e. the momentum fractions are $\langle x_{1,2} \rangle \approx 1/3$. We thus presumably need of order at least three times the $W^\pm$ mass, or $E_{CM}^{hadron} \gtrsim 240$ GeV in the laboratory collisions.

Exercise 4.3.1. *Show that the true centre-of-mass energy available in a parton–parton collision is given by $E_{CM}^{parton} = \sqrt{x_1 x_2} \ E_{CM}^{hadron}$, where $x_{1,2}$ are the two colliding parton momentum fractions.*

In reality, the situation is rather worse, for two reasons. First of all, interactions, via which the gluons and sea quarks are generated inside the proton, significantly reduce the energy share of the valence quarks: as mentioned earlier, it is found

---

[35] The 1984 Nobel Prize in Physics was awarded equally to Carlo Rubbia and Simon van der Meer 'for their decisive contributions to the large project, which led to the discovery of the field particles W and Z, communicators of weak interaction'.

experimentally that the gluons carry slightly more than half the total energy of the proton. Secondly, this effect grows with energy scale. A higher energy scale is equivalent to finer spatial resolution and, as we look at the proton in ever finer detail, we resolve more and more gluons and sea quarks. The resulting mean effective valence and sea-quark fractions relevant to such an experiment are then

$$\langle x_{\text{val}} \rangle \approx 0.12 \ \text{ and } \ \langle x_{\text{sea}} \rangle \approx 0.04. \tag{4.3.3}$$

In order to produce a $W^{\pm}$ in a $p\bar{p}$ collider we might therefore expect to require total centre-of-mass energies nearer to $80/0.12 \sim 670$ GeV, whereas for pp this rises to around $80/\sqrt{0.12 \times 0.04} \sim 1200$ GeV. In practice, slightly lower energies may be employed by working in the tails of the parton distributions; i.e. accepting lower event rates.

The cross-section for $W^{\pm}$ production in a *parton–parton* collision may be calculated using the standard Breit–Wigner form, see equation (C.6.15):

$$\sigma(u\bar{d} \to W^+ \to e^+\nu_e) \simeq \frac{1}{N_c}\frac{1}{3}\frac{(2J+1)}{(2s_u+1)(2s_d+1)}$$
$$\times \frac{4\pi\hbar^2}{p^2}\frac{\left(\frac{1}{2}\Gamma_{u\bar{d}}\right)\left(\frac{1}{2}\Gamma_{e\nu}\right)}{\left(E_{\text{CM}} - M_W\right)^2 + \left(\frac{1}{2}\Gamma_{\text{tot}}\right)^2}, \tag{4.3.4}$$

where the factor N_c in the denominator accounts for the requirement that for a quark of a given colour, the corresponding antiquark must have precisely that anti-colour out of the three possible; the further factor 3, associated with spin degeneracy, arises owing to the requirement that the quark and antiquark be left and right handed respectively; i.e. the $W^{\pm}$ are produced in only one of the three possible helicity states.

At the peak energy ($E_{\text{CM}} = M_W$) the maximum total cross-section (in natural units) is therefore

$$\sigma_{\text{max}}(u\bar{d} \to W^+ \to e^+\nu_e) \simeq \frac{4\pi}{3M_W^2}\mathcal{B}_{u\bar{d}}\,\mathcal{B}_{e\nu} = \frac{4\pi}{81M_W^2} \simeq 9.2 \text{ nb.} \tag{4.3.5}$$

where $\mathcal{B}_{u\bar{d}} = \Gamma_{u\bar{d}}/\Gamma_{\text{tot}}$ and $\mathcal{B}_{e\nu} = \Gamma_{e\nu}/\Gamma_{\text{tot}}$. The values of the branching ratios ($\mathcal{B}_{u\bar{d}}$ and $\mathcal{B}_{e\nu}$) used are based on the observation that each single decay channel (leptonic or coloured quark) has the same weight (neglecting the tiny phase-space variations). There are thus three leptonic ($e\nu_e$, $\mu\nu_\mu$ and $\tau\nu_\tau$) and two quark channels ($u\bar{d}_W$ and $c\bar{s}_W$) of three colours, giving a total of nine equal-weight channels. We therefore have

$$\mathcal{B}_{u\bar{d}} = N_c/(3 + 2N_c) = 1/3 \ \text{ and } \ \mathcal{B}_{e\nu} = 1/(3 + 2N_c) = 1/9. \tag{4.3.6}$$

To calculate the cross-section for hadronic collisions, we need to integrate (or average) over the width of the resonance and the relevant partonic distributions. That is, schematically,

$$\sigma(p\bar{p} \to W^+ \to e^+\nu_e) = \int \mathrm{d}x_1\,\mathrm{d}x_2 \sum_{i,j} q_i(x_1)\,\bar{q}_j(x_2)\,\sigma(q_i\bar{q}_j \to W^+ \to e^+\nu_e), \tag{4.3.7}$$

where the sum runs over all suitable quark–antiquark combinations i, j (with momentum fractions $x_{1,2}$) that can produce a W^+. The partonic cross-section $\sigma(q_i \bar{q}_j)$ is calculated for quark and antiquark momentum fractions x_1 and x_2, respectively. For the actual experiments (UA1: Arnison *et al* 1983a and UA2: Banner *et al* 1983) performed in the Sp$\bar{\text{p}}$S collider at CERN, with very high-intensity proton and antiproton beams at 270 GeV (in later, so-called *ramped*, runs the CERN machine achieved beam energies of 318 GeV), the cross-sections are

$$\sigma(p\bar{p} \to W^+ \to e^+\nu_e) \sim 1 \text{ nb} \tag{4.3.8}$$

and

$$\sigma(p\bar{p} \to Z^0 \to e^+e^-) \sim 0.1 \text{ nb}. \tag{4.3.9}$$

In the 1982 runs, UA1 recorded a luminosity of 10^{29} cm^{-2} s^{-1} and the W discovery was announced on the basis of an integrated luminosity of 28 nb^{-1}. Over the period 1981–85, UA1 and UA2 gathered an integrated luminosity of ~ 0.9 pb^{-1}. The above cross-sections should be compared to the total $p\bar{p}$ cross-section at such energies, which is measured to be about 40 mb. In other words, the two signals were at a level of 10^{-8} and 10^{-9} of the total, respectively. Such a weak signal required very distinctive events or final states.

Now, since nearly all of the available energy must be used to provide the $W^\pm$ mass and since to maximise the cross-section the quark and antiquark will also tend to have similar velocities, the boson so-produced will be almost at rest in the laboratory. When it decays, there is then a high probability that the final-state lepton–neutrino pair will emerge more-or-less back-to-back (in the laboratory) and also at large angles with respect to the beam direction, i.e. with high transverse-momenta or p_T. Moreover, the detected charged lepton will have large energy $\approx \frac{1}{2}M_W$ while, of course, the neutrino will inevitably go undetected. The signature is thus:

- a single, isolated, high-p_T electron track in the central tracking detector;
- a very localised shower in the electromagnetic calorimeter;
- large missing p_T when all transverse momenta are added vectorially.

This determines to a large extent the detector geometry and topology. Schematically then, the principal components of the detector setup are as follows (see figure 4.28)

- a central tracking detector with a high magnetic field to observe charged particles and measure their momenta;
- electromagnetic shower counters, which detect both electrons and photons;
- hadron calorimeters;
- muon detectors.

figure 4.29 shows an example of the electromagnetic-calorimeter energy deposition due to a $W^\pm$ event. We mention in passing that a few similar, but initially inexplicable, events with instead a single *hadronic* jet were also found (see e.g.

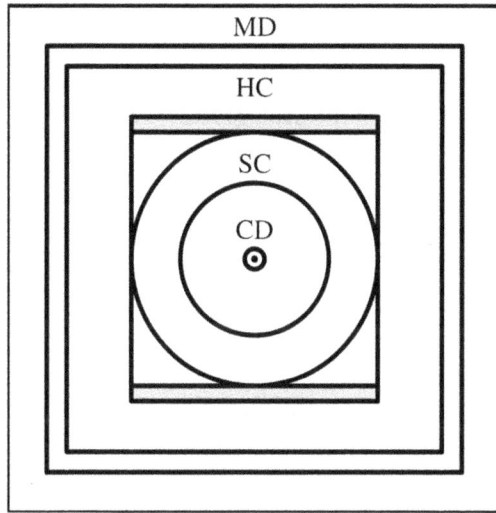

Figure 4.28. A schematic view of a transverse section of the UA1 detector at Centre Europée de Rechèrche Nucleaire (CERN, Geneva). The beams travel along the centre (in and out of the page). CD is the central detector, SC the shower counters, HC the hadron calorimeters and finally MD are the outer muon detectors. The shaded area represents the magnetic coils.

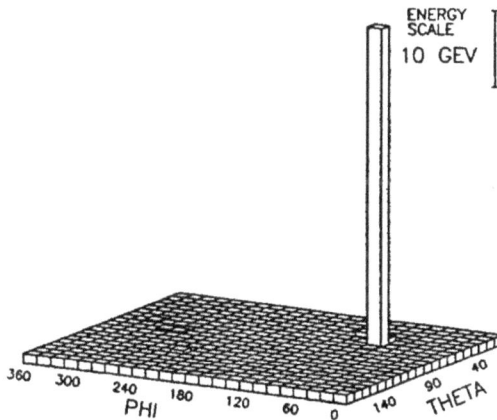

Figure 4.29. A so-called *Lego plot* from the electromagnetic calorimeter due to a $W^{\pm}$ event in the UA2 experiment at Centre Europée de Rechèrche Nucleaire (CERN, Geneva); figure reproduced from Banner *et al* (1983), copyright (1983) with permission from Elsevier.

Rohlf 1984). However, early hopes of a truly new discovery were soon quashed by a full background analysis (see e.g. Ellis *et al* 1986).

Note that the individual longitudinal momenta of the incident partons are not known and therefore nothing is known of the possible Lorentz boost of the decaying system. It is thus impossible to fully reconstruct the event and so nothing can be said about the $W^{\pm}$ mass or width on an event-by-event basis However, the p_T

distribution of the detected electrons may be measured. For the purposes of this explanation, we shall assume that the $W^\pm$ is produced approximately at rest in the laboratory system; a full (numerical) analysis should also include the parton distributions to account for non-zero centre-of-mass motion. The transverse momentum of the outgoing electron is then

$$p_{\mathrm{T}} \approx \tfrac{1}{2} M_W \sin \theta, \qquad (4.3.10)$$

where θ is the electron angle with respect to the beam axis. Therefore,

$$\frac{\mathrm{d}\sigma}{\mathrm{d}p_{\mathrm{T}}} = \frac{\mathrm{d}\sigma}{\mathrm{d}\cos\theta} \frac{\mathrm{d}\cos\theta}{\mathrm{d}p_{\mathrm{T}}} \approx \frac{\mathrm{d}\sigma}{\mathrm{d}\cos\theta}\left[1 - 4p_{\mathrm{T}}^2 / M_W^2\right]^{-1/2}. \qquad (4.3.11)$$

Another often used and related variable is the so-called *transverse mass*:

$$m_{\mathrm{T}} = \sqrt{2p_{\mathrm{T}}^e p_{\mathrm{T}}^\nu (1 - \cos \phi_{e\nu})}, \qquad (4.3.12)$$

where p_{T}^ν is the transverse momentum of the undetected neutrino, reconstructed as the missing transverse momentum.

The so-called *Jacobian* peak (see figure 4.30) thus induced in the p_{T} or m_{T} distribution allows for a fairly precise determination of the mass and width of the $W^\pm$ (via a full parton-model analysis); the present-day world-average values are (Navas *et al* 2024)

$$m_W = 80.3692 \pm 0.0133 \,\text{GeV} \quad \text{and} \quad \Gamma_W = 2.085 \pm 0.042 \,\text{GeV}. \qquad (4.3.13)$$

The precision is seen to be roughly twenty times poorer with respect to the LEP measurements of the Z^0 parameters. Note that the CERN experiments also

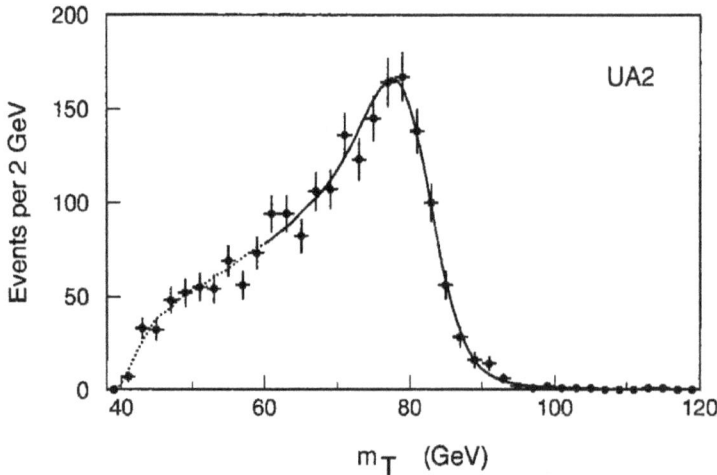

Figure 4.30. The Jacobian peak in the transverse-mass distribution of the electron or positron emitted in the $W^\pm$ decay, as seen in the UA2 experiment at Centre Europée de Rechèrche Nucleaire (CERN, Geneva); figure reproduced from Alitti *et al* (1992), copyright (1992) with permission from Elsevier.

published data on observation of the Z^0 (UA1: Arnison *et al* 1983b and UA2: Bagnaia *et al* 1983).

4.3.2 Discovery of the t quark

The Higgs boson aside (which we shall discuss later), the last major discovery made in a high-energy hadron collider, indeed in any collider, is that of the *top* quark. This was achieved at Fermilab in 1995 by two experiments: CDF (Abe *et al* 1995) and DØ (Abachi *et al* 1995). The basic process observed is

$$p + \bar{p} \to t + \bar{t}$$
$$\quad\quad\quad \hookrightarrow W^- \bar{b}$$
$$\quad\quad \hookrightarrow W^+ b. \tag{4.3.14}$$

Of course, both the W's and b's also decay (either leptonically or hadronically) rather rapidly; the cleaner leptonic channels are used here for the $W^\pm$ while vertex tracking allows full reconstruction of the semi-leptonic b decays. In such a way, for the initial analyses, CDF collected 19 events and DØ 17. Combining all possible production modes (*i.e.* quark–antiquark annihilation and gluon–gluon fusion), the cross-section is estimated to be of order 5 pb at Tevatron energies and for a top-quark mass around 175 GeV, as determined from the reconstructed top-quark mass spectrum (see figure 4.31).

These two experiments continued to collect data over a long period and also performed analyses aimed at detecting the weak bosons, in particular the $W^\pm$. The current values are:

$$m_t = 173.1 \pm 0.9 \text{ GeV} \quad \text{(direct measurement by CDF and DØ)}, \tag{4.3.15a}$$

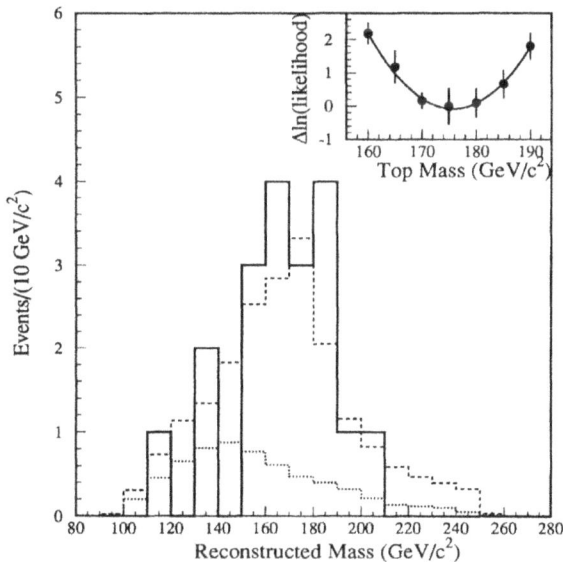

Figure 4.31. The reconstructed top-quark mass spectrum as seen early on in the CDF experiment at Fermilab; figure reproduced with permission from Abe *et al* (1995), copyright (1995) by the American Physical Society.

$$= 178.1 \, {}^{+10.4}_{-8.3} \, \text{GeV} \quad \text{(early extraction from global SM fits)}, \quad (4.3.15b)$$

$$= 172.4 \pm 0.7 \, \text{GeV} \quad \text{(pole mass from the measured cross-section)}, \quad (4.3.15c)$$

$$= 172.57 \pm 0.29 \, \text{GeV} \quad \text{(Navas } et \, al \text{ (2024) world average)}. \quad (4.3.15d)$$

The first number comes from a combined analysis of the direct measurements (ATLAS, CDF, CMS and DØ Collabs. 2014); the second is obtained from the precision fits to high-statistics data of various Standard-Model parameters (e.g. $\sin \theta_W$), in which the top quark plays a rôle via mass-dependent, higher-order corrections; the third is deduced from the measured cross-section (Navas *et al* 2024).

Now, as already commented, the top quark decays so rapidly that there is no likelihood of forming top mesons (Bigi and Dokshitzer 1986). The width is measured to be

$$\Gamma_t = 1.42^{+0.19}_{-0.15} \, \text{GeV}, \quad (4.3.16a)$$

while theoretical calculation in the Standard Model gives (Jeżabek and Kühn 1989)

$$\Gamma_t \simeq \frac{G_F m_t^3}{8\sqrt{2}\,\pi} \left(1 - \frac{M_W^2}{m_t^2}\right)^2 \left(1 + 2\frac{M_W^2}{m_t^2}\right) \left[1 - \frac{2\alpha_s}{3\pi}\left(\frac{2\pi^2}{3} - \frac{5}{2}\right)\right]. \quad (4.3.16b)$$

The theoretical value, obtained using the measured masses (taking $m_t = 172.5$ GeV) and strong coupling constant, is approximately 1.326 GeV, in reasonable agreement with experiment. Note that the first-order QCD correction is about 10%; the second-order correction (Czarnecki and Melnikov 1999, Chetyrkin *et al* 1999) as well as the electroweak NLO corrections (Chen *et al* 2023) are known. Such widths are equivalent to a lifetime of order 10^{-24} s.

4.3.3 The search for quark–gluon plasma

We have seen phenomenologically that the strong interaction confines the quarks and gluons, presumably by means of an effective string-like long-distance potential. However, at short distances (or high energies) the quarks and gluons behave as though they were free and not subject to a confining potential. This picture led Hagedorn (1965) to the idea that at very high energy *and* parton density, QCD should undergo a transition to a sort of *quark–gluon plasma* phase, of which the natural principal consequence or feature would be *deconfinement*. That is, there should be no identifiable hadronic states, but rather effectively free quarks and gluons. For a deeper discussion of the subject, see e.g. Rafelski (2020).

Such a phase presumably existed in the very early moments after the big bang and maybe even occurs inside some very dense neutron stars. We might hope to recreate it in the laboratory by generating very high energy and particle densities in high-energy heavy nucleus–nucleus collisions. To achieve this, Relativistic Heavy-Ion Collider (RHIC) was built at BNL (USA), in which heavy ions (primarily gold) at around 100 GeV per nucleon collide. The LHC programme at CERN also includes a large fraction of heavy-ion physics; the lead-ion energies here will reach approximately 2.7 TeV per nucleon.

In this hypothesised phase of QCD matter, owing to the effective deconfinement, it is believed that, for example, the strange quark should no longer be suppressed and should coexist in roughly equal quantities with the other two light quarks. The critical or Hagedorn temperature for such a transition is estimated to be around 1.7×10^{12} K, which is equivalent to a thermal kinetic energy of around 150 MeV (Hagedorn 1965). This means that strange quark–antiquark pairs should be easily produced in the collisions between quarks and antiquarks occurring naturally in such a highly dense phase. They are then likely to remain, since the annihilation process is relatively rather rare. The signal sought therefore is a sudden and marked enhancement of strange-particle production.

A further commonly sought signal is so-called 'jet quenching' (Bjorken 1982), whereby jets produced in such collisions are expected to loose energy via multiple interactions with the strongly interacting plasma, through which they must travel before emerging. To some extent, this is indeed seen at RHIC by PHENIX and STAR (Adcox *et al* 2002, Adler *et al* 2003) and more recently at LHC by the ALICE collaboration (Acharya *et al* 2020), but there is still some debate as to the significance of the various results obtained (Heinz and Jacob 2000).

References

Abachi S *et al* (D0 Collab) 1995 *Phys. Rev. Lett.* **74** 2632

Abe F *et al* (CDF Collab) 1995 *Phys. Rev. Lett.* **74** 2626

Acharya S *et al* (ALICE Collab) 2020 *Phys. Rev.* **C101** 034911

Adcox K *et al* (PHENIX Collab) 2002 *Phys. Rev. Lett.* **88** 022301

Adler C *et al* (STAR Collab) 2003 *Phys. Rev. Lett.* **90** 082302

Aguilar-Benitez M *et al* (Particle Data Group) 1986 *Phys. Lett.* **B170** 1

Alitti J *et al* (UA2 Collab) 1992 *Phys. Lett.* **B276** 354

Álvarez L W 1968 *Nobel Lectures, Physics 1963–1970* Nobel Foundation (Amsterdam: Elsevier) p 241

Anderson C D 1933a *Phys. Rev.* **43** 491

Anderson C D 1933b *Science* **77** 432; *Phys. Rev.* 491

Anderson C and Neddermeyer S H 1936 *Phys. Rev.* **50** 263
    Neddermeyer S H and Anderson C D *Phys. Rev.* **51** 884

Anderson H L, Fermi E, Long E A and Nagle D E 1952 *Phys. Rev.* **85** 936

Andersson B, Gustafson G, Ingelman G and Sjöstrand T 1983 *Phys. Rep.* **97** 31

Andrews D *et al* 1980 *Phys. Rev. Lett.* **44** 1108

Arnaudon L *et al* 1995 *Nucl. Instrum. Meth.* **A357** 249

Arnison G *et al* (UA1 Collab) 1983a *Phys. Lett.* **B122** 103

Arnison G *et al* (UA1 Collab) 1983b *Phys. Lett.* **B126** 398

ATLAS, CDF, CMS and D0 Collabs 2014 arXiv:1403.4427

Aubert B *et al* (BaBar Collab) 2009 *Phys. Rev. Lett.* **101** 071801 *erratum* ibid **102** (2009) 071 801

Aubert J J *et al* 1974 *Phys. Rev. Lett.* **33** 1404

Augustin J-E *et al* 1974 *Phys. Rev. Lett.* **33** 1406

Bacci C *et al* 1974 *Phys. Rev. Lett.* **33** 1408 *erratum ibid* 1649

Bagnaia P *et al* (UA2 Collab) 1983 *Phys. Lett.* **B129** 130

Baldini A M *et al* (MEG Collab) 2016 *Eur. Phys. J.* **C76** 434

Banfi A, Salam G P and Zanderighi G 2010 *JHEP* **06** 038

Banner M *et al* (UA2 Collab) 1983 *Phys. Lett.* **B122** 476

Bartel W *et al* (JADE Collab) 1983 *Z. Phys.* **C21** 37

Bigi I I Y, Dokshitzer Y L, Khoze V A, Kühn J H and Zerwas P M 1986 *Phys. Lett.* **B181** 157

Bird D J *et al* 1995 *Astrophys. J.* **441** 144

Bjorken J D 1982 Fermilab preprint Fermilab-Pub-82-059-THY Fermilab:Fermilab-Pub-82-059-THY

Bjorken J D and Brodsky S J 1970 *Phys. Rev.* **D1** 1416

Bjorklund R, Crandall W E, Moyer B J and York H F 1950 *Phys. Rev.* **77** 213

Blackett P M S and Occhialini G P S 1933 *Proc. R. Soc. Lond.* **A139** 699

Bose D M and Chowdhry B 1941 *Nature* **147** 240 **148**, 259; *ibid.* **149** (1942) 302

Bothe W and Becker H 1930 *Z. Phys.* **66** 289

Brandt S, Peyrou C, Sosnowski R and Wroblewski A 1964 *Phys. Lett.* **12** 57

Breit G and Wigner E P 1936 *Phys. Rev.* **49** 519

Chadwick J 1914 *Verh. Dtsch. Phys. Ges.* **16** 383

Chadwick J 1932 *Nature* **129** 312

Chamberlain O, Segrè E, Wiegand C and Ypsilantis T 1955 *Phys. Rev.* **100** 947

Chao C Y 1949 *Phys. Rev.* **75** 581

Chen L-B, Li H T, Wang J and Wang Y 2023 *Phys. Rev.* **D108** 054003

Chetyrkin K G, Harlander R, Seidensticker T and Steinhauser M 1999 *Phys. Rev.* **D60** 114015

Conversi M, Pancini E and Piccioni O 1947 *Phys. Rev.* **71** 209; *see also*, Sigurgeirsson, T. and Yamakawa, K.A. *ibid.* 319

Cork B, Lambertson G R, Piccioni O and Wenzel W A 1957 *Phys. Rev.* **104** 1193

Curie I and Joliot J F 1932a *C. R. Acad. Sci. URSS* **194** 273

Curie I and Joliot J F 1932b *C. R. Acad. Sci.* **194** 708; *erratum* ibid 1032

Czarnecki A and Melnikov K 1999 *Nucl. Phys.* **B544** 520

Danby G *et al* 1962 *Phys. Rev. Lett.* **9** 36

Dirac P A M 1928 *Proc. R. Soc. Lond.* **A117** 610

Dirac P A M 1930 *Proc. R. Soc. Lond.* **A126** 360

Eichten E, Gottfried K, Kinoshita T, Lane K D and Yan T-M 1978 *Phys. Rev.* **D17** 3090

Eichten E *et al* 1975 *Phys. Rev. Lett.* **34** 369

Ellis S D, Kleiss R and Stirling W J 1986 *Phys. Lett.* **B167** 464

Feinberg G 1958 *Phys. Rev.* **110** 1482

Fermi E, Teller E and Weisskopf V F 1947 *Phys. Rev.* **71** 314; **72** 399

Fretter W B 1949 *Phys. Rev.* **76** 511

Gardner E and Lattes C M G 1948 *Science* **107** 270 *Phys. Rev.* **75** 382

Gell-Mann M and Rosenbaum E P 1957 *Sci. Am.* **197** 72 *reprinted in* in *Particles and Fields* ed W J Kaufmann (W.H. Freeman, 1980) p 22

Grayer G *et al* 1974 *Nucl. Phys.* **B75** 189

Guberina B, Nussinov S, Peccei R D and Rückl R 1979 *Phys. Lett.* **B89** 111

Hagedorn R 1965 *Nuovo Cim. Suppl.* **3** 147

Hahn T M *et al* 1952 *Phys. Rev.* **85** 934

Hanson G *et al* 1975 *Phys. Rev. Lett.* **35** 1609

Heinz U W and Jacob M 2000 nucl-th/0002042

Herb S W *et al* (E288 Collab) 1977 *Phys. Rev. Lett.* **39** 252

Hess V F 1912 *Phys. Z* **13** 1084

Hoyer P, Osland P, Sander H G, Walsh T F and Zerwas P M 1979 *Nucl. Phys.* **B161** 349

Iizuka J 1966 *Prog. Theor. Phys. Suppl.* **37** 21

Innes W R *et al* 1977 *Phys. Rev. Lett.* **39** 1240; *erratum ibid.* 1640

Jeżabek M and Kühn J H 1989 *Nucl. Phys.* **B314** 1

Kodama K *et al* (DONuT Collab.) 2001 *Phys. Lett.* **B504** 218

Kodama K *et al* (DONuT Collab.) 2008 *Phys. Rev.* **D78** 052002

Lattes C M G, Muirhead H, Occhialini G P S and Powell G F 1947 *Nature* **159** 694

Lindenbaum S J and Yuan L C L 1958 *Phys. Rev.* **111** 1380

Majorana E 1933 *Z. Phys.* **82** 137

Marchesini G and Webber B R 1984 *Nucl. Phys.* **B238** 1

Navas S *et al* (Particle Data Group) 2024 *Phys. Rev.* **D110** 030001

Okubo S 1963 *Phys. Lett.* **5** 165

Partridge R *et al* 1980 *Phys. Rev. Lett.* **45** 1150

Perl M L *et al* 1975 *Phys. Rev. Lett.* **35** 1489

Rafelski J 2020 *Eur. Phys. J. ST* **229** 1

Reines F and Cowan C L 1953 *Phys. Rev.* **92** 830

Rochester G D and Butler C C 1947 *Nature* **160** 855

Rohlf J (UA1 Collab) 1985 *Proc. of the XXII Int. Conf. on High-Energy Physics—ICHEP 84 (Leipzig, July 1984)* **vol 2** ed A Meyer and E Wieczorek (Akad. Wissen.—Inst. Hochenergiephys.) p II.12

Rubbia C 1974 *Proc. of the 17th. Int. Conf. on High Energy Physics—ICHEP 74* (Leipzig, July 1974) J R Smith (Rutherford Lab.) p IV.117

Rutherford E 1920 *Proc. R. Soc. Lond.* **A97** 374

Schael S *et al* (ALEPH, DELPHI, L3, OPAL, SLD, LEP Electroweak Working Group, SLD Electroweak Group, SLD Heavy Flavour Group) 2006 *Phys. Rep.* **427** 257

Schwartz M 1960 *Phys. Rev. Lett.* **4** 306

Street J C and Stevenson E C 1937 *Phys. Rev.* **51** 1005; **52** 1003

Wilson C T R 1911 *Proc. R. Soc. Lond.* **A85** 285; *ibid* **A87** (1912) 277.

Yukawa H 1935 *Proc. Phys. Math. Soc. Jap.* **17** 48

Zweig G 1964 CERN preprint CERN-TH-412

IOP Publishing

An Introduction to Elementary Particle Phenomenology
(Second Edition)

Philip G Ratcliffe

Chapter 5

The Standard Model (where we are now)

'Physicists like to think that all you have to do is say, these are the conditions, now what happens next?'

Richard P Feynman

In this penultimate chapter we shall deal with the so-called *Standard Model* (SM) of particle physics. In particular, we shall turn our attention to more dynamical questions and examine the rôle of interactions. A central issue here, not yet approached, will be the generation of mass. Various theoretical considerations lead us to prefer theories in which no masses are *explicitly* present. The gauge principle (which implies massless exchange fields) is found to be immensely valuable and, indeed, all known phenomena may be described by theories taking the basic form of a QED-like interaction. However, most, if not all, particles have mass and, in particular, the $W^\pm$ and Z^0 are very heavy indeed. As we shall see, the solution is to *induce* effective masses via interaction energies and we shall discover, moreover, that symmetries and their breaking play an important rôle here too.

This then provides the working and now well-tested theory for the complete set of fundamental interactions and elementary particles or fields that have already been introduced and discussed separately in the previous chapters. In this chapter we shall also examine in some detail the very precise predictions of the non-trivial constructions necessary to obtain a self-consistent and functional theoretical description, in particular, of the combined electroweak interaction. This theoretical structure functions very well indeed and leaves little room for variations and there is certainly no purely *phenomenological* necessity for improvement.

doi:10.1088/978-0-7503-5759-3ch5

5.1 Fundamental forces and particles

5.1.1 The table of forces and particles

The building blocks we now have in our hands comprise a number of quarks and leptons, interacting via the exchange of various types of spin-one gauge bosons. Leaving aside gravity and not yet wishing to enter into the special case of the Higgs boson, these are arranged according to particle type and family (or generation) in table 5.1. The individual quark and lepton doublets are ordered from left to right according to mass; there is no other compelling reason to place the particular quark and lepton pairs together in specific families. The bosons mediating the interactions of QED and quantum chromodynamics (QCD) are all massless, as is required of a true gauge theory. It is found experimentally though that the exchange fields of the weak-interaction are massive, being among the heaviest particles known. A major problem in theoretical physics is then precisely how the $W^{\pm}$ and Z^0 might acquire a non-zero mass while remaining true gauge bosons. Let us first of all examine the reasons for requiring them to have a large mass.

Table 5.1. The elementary matter (left) and force (right) fields of the SM; the only fields missing from this table are the spin-zero Higgs boson and the spin-two graviton. The lepton and quark pairs are arranged in their natural weak-isospin doublets.

$$
\begin{array}{l}
\text{leptons} \left\{ \begin{array}{ccc} \nu_e & \nu_\mu & \nu_\tau \\ e^- & \mu^- & \tau^- \end{array} \right. \\[2em]
\text{quarks} \left\{ \begin{array}{ccc} u & c & t \\ d & s & b \end{array} \right\} \begin{array}{c} g \\ \text{(QCD)} \end{array} \left. \begin{array}{c} \gamma \\ \text{(QED)} \end{array} \right\} \begin{array}{c} W^{\pm},\, Z^0 \\ \text{(weak)} \end{array}
\end{array}
$$

$$\underbrace{\qquad\qquad}_{\text{families/generations}}\qquad \underbrace{\qquad\qquad}_{\text{(electroweak)}}$$

$$\underbrace{\qquad\qquad}_{\substack{\text{spin-half fermions}\\ \text{(matter)}}}\qquad \underbrace{\qquad\qquad}_{\substack{\text{spin-one bosons}\\ \text{(gauge fields)}}}$$

5.1.2 The need for massive weak-interaction bosons

5.1.2.1 Unitarity violation in Fermi theory

Fermi theory describes all known low-energy weak-interaction phenomena very well and to high precision. However, once Fermi had introduced his universal four-point interaction, many new processes could be imagined that had yet to be observed or studied experimentally. For example, the following process (improbable as it may appear and difficult to induce experimentally) becomes possible:

$$\nu_\mu + e^- \rightarrow \mu^- + \nu_e. \tag{5.1.1}$$

The theory also determines precisely how to calculate the cross-section. For a sufficiently high energy so that in comparison all particle masses may be neglected, the result is (in natural units, $\hbar = 1 = c$)

$$\sigma_{\text{tot}}(\nu_\mu e^- \rightarrow \mu^- \nu_e) \approx \frac{G_F^2 E_{\text{CM}}^2}{\pi}. \tag{5.1.2}$$

Recall that, since $(\hbar c)^2 = 0.3894$ mb GeV^2, the conversion is performed by multiplying this expression for the cross-section (which has dimension GeV^{-2}) by 0.39 to obtain the answer in mb. A cross-section is thus found that grows with increasing energy, apparently without bounds.

Note that, in fact, such behaviour might have been anticipated on purely dimensional grounds simply by exploiting our knowledge that the Fermi coupling constant, $G_F \simeq 1.166 \times 10^{-6}$ GeV^{-2}, has dimensions E^{-2} (again in natural units). A cross-section has dimensions of an area, which in natural units is also E^{-2}, and therefore, to compensate the two powers of G_F, two powers of E_{CM} are needed. It is obvious that such a behaviour must sooner or later violate unitarity; that is, above some critical energy the interaction probability will exceed unity[1].

This statement may be made more quantitative and even more stringent by considering the partial-wave expansion for the total cross-section:

$$\sigma_{\text{tot}} = \frac{4\pi}{k^2} \sum_\ell (2\ell + 1) \sin^2 \delta_\ell, \tag{5.1.3}$$

where k is the projectile momentum, ℓ is the angular momentum of the individual partial wave and δ_ℓ is the associated phase-shift, which depends on the precise details of the theory. This may be applied to the above case by exploiting one simple property of the Fermi interaction: it is point-like, which implies vanishing impact parameter and therefore zero orbital angular momentum. Only $\ell = 0$ (or s-wave) need be considered, which leads to the following limiting behaviour:

$$\sigma_{\text{tot}} = \sigma_0 \leqslant \frac{4\pi}{k^2} \rightarrow \frac{4\pi}{E_{\text{CM}}^2}. \tag{5.1.4}$$

Equality of the two expressions (5.1.2) and (5.1.4) for σ_{tot} provides an upper limit to E_{CM} for which Fermi theory can be valid: $E_{\text{CM}} \leqslant \sqrt{2\pi/G_F}$; beyond this energy partial-wave unitarity will apparently be violated. Inserting the relevant numbers leads to

$$E_{\text{CM}}^{\text{max}} \approx 730 \text{ GeV}. \tag{5.1.5}$$

---

[1] Note that any eventual variation of the Fermi coupling with energy scale due to renormalisation effects (recall the asymptotic freedom of QCD), being in general only logarithmic, cannot compensate such behaviour.

In other words, well before this energy is reached, some new physics must take over that is evidently not manifest at lower energies but that can tame the unbounded growth of the cross-section. The question is: 'What might it be?'

5.1.2.2 Intermediate vector bosons

Comparison of such behaviour with that of QED provides a hint of how it might be cured (Feynman and Gell-Mann 1958; Sudarshan and Marshak 1958; Sakurai 1958). The equivalent interaction in QED is not a four-fermion interaction but rather a pair of fermion–photon vertices connected via a photon propagator. An indicator of the correct behaviour is that the QED coupling constant is dimensionless. The dimensionality corresponding to G_F is provided by the propagator, which introduces the sought-after high-energy suppression. Diagrammatically then, we might be tempted to make the substitution depicted in figure 5.1.

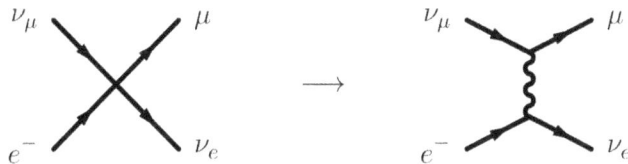

Figure 5.1. A possible cure for the divergent behaviour of Fermi theory via substitution of the four-point coupling with an exchange-boson propagator *à la* QED.

In terms of couplings and propagators, this translates into

$$G_F \to \frac{g^2}{|q^2 - M^2|}, \tag{5.1.6}$$

where g is some new *dimensionless* coupling and we have allowed for a mass M in the new propagator. At very high energies ($|q^2| \gg M^2$), where help is needed, M^2 may be neglected and thus an asymptotic behaviour is attained such as that of QED, i.e. acceptable. For low-energy interactions, where $|q^2| \ll M^2$, q^2 may be neglected in the propagator and thus the standard behaviour of Fermi theory is recovered. Moreover, inverting the relation at low energies gives

$$g \sim M\sqrt{G_F}. \tag{5.1.7}$$

Now, we have already seen that M cannot be more than a few hundred GeV if it is to save unitarity. On the other hand, it cannot be significantly smaller, otherwise the propagator energy dependence would spoil the accurate predictions of Fermi theory. This suggests a value for g that is not so very different from the value of e, the QED coupling. It is then reasonable to speculate that the new theory has indeed a similar coupling to QED and thus M should be around, say, 100 GeV, but we shall justify this a little better shortly.

Charge conservation at the vertices requires that, together with a non-zero mass, the new boson be charged and we should thus introduce something like $W^{\pm}$. Moreover, if we are to reproduce the $V - A$ interaction associated with Fermi theory, then it must be a spin-one particle, just as the photon. Now, whereas photon vertices only transfer energy and momentum, interactions involving $W^\pm$ must also change the nature of the fermion. For example, a u quark is converted into a d quark and *vice versa*. We are thus faced with the choice of either continuing to speak of isospin violation or to admit that the new bosons also carry isospin. This latter is, of course, much more attractive. Indeed, it is more natural to include the leptons in such a picture and thus introduce *weak isospin*. To couple objects of isospin one-half we are forced to introduce bosons of isospin one. This implies a multiplicity three $(2I + 1)$, that is, a triplet. Just as in the pion case, we should thus expect a W^0 (a so-called *weak neutral current*) together with the two $W^\pm$ (the charged currents). At the time, this extra W^0 was clearly not a phenomenological necessity, but opened up a veritable Pandora's box of new possible weak-interaction processes.

5.1.3 Weak neutral currents

Now, the W^0 cannot be the photon since it must both have a large mass and interact with neutrinos. We should remark immediately, moreover, that the particle so introduced is not exactly the Z^0 already mentioned. We shall see later that the Z^0 is in fact a superposition of W^0 and another new field B^0, while γ is the corresponding orthogonal combination; as we shall show, this is all part of the mechanism designed to provide the new (gauge) bosons with non-zero masses.

Without entering yet into the question of gauge invariance, we immediately realise the introduction of a new intermediate neutral boson implies further possible new phenomenology: that of the so-called *weak neutral currents*. An immediate question arises: why had no experimental indication been observed earlier? The answer lies in *both* the similarities *and* the differences with respect to the photon. The neutrality of the W^0 (or Z^0) implies that almost everywhere it may be exchanged a photon may be exchanged too. However, at low energies photon exchange is not suppressed by a large mass. Indeed, the cross-section ratio for W^0 and photon exchange can be estimated in any given low-energy process as follows:

$$\frac{\sigma(W^0)}{\sigma(\gamma)} \sim \left(\frac{1/(q^2 - M^2)}{1/q^2} \right)^2 \approx \left(\frac{q^2}{M^2} \right)^2 \text{ for } |q^2| \ll M^2. \tag{5.1.8}$$

Since we expect $M \sim 100$ GeV, for energies of order 1 GeV the suppression is of order 10^{-8}. While for interference effects this may improve to the square-root and therefore 10^{-4}, to separate the two contributions, absolute cross-section measurements of unprecedented precision would still be required.

If the obvious W^0-exchange processes are swamped by photon exchange, then we should look for something not possible in QED. A possibility might be to exploit neutrino physics, but that is always very difficult. As noted earlier, the photon does not change the nature of matter particles, whereas the $W^\pm$ evidently do. It might then be hoped that some process exists in which the W^0 also provokes a change of flavour. We are thus led back to the well-known question of eigenstates of different interactions. The electron and muon etc. are evidently eigenstates of QED, which is experimentally seen to be perfectly diagonal in these states. Note that the basis states of QED are precisely those we call mass eigenstates and which propagate in the laboratory. However, the natural eigenstates of the weak interaction are not necessarily the same and may be superpositions of these states. In other words, if the charged leptons $e^\pm$, $\mu^\pm$ and $\tau^\pm$ are described via a three-component vector then there should be a unitary transformation U that takes us to the weak basis:

$$\psi_i^{\rm W} = U_{ij}\psi_j. \tag{5.1.9}$$

A similar consideration could naturally be made for the neutrinos, but it is rather irrelevant in this connection since they do not interact with the photon (but see later discussions on neutrino oscillations etc).

Now, interactions are always described in terms of currents; the QED (neutral) current is

$$J_{\rm QED}^\mu = e\bar\psi\,\gamma^\mu\,1_{\rm flav}\,\psi, \tag{5.1.10}$$

where the unit matrix $1_{\rm flav}$ just expresses the fact that the interaction is diagonal in this basis. The equivalent *weak* neutral current (i.e. mediated by the W^0 or rather Z^0 boson) would then be

$$J_{\rm W}^{0\mu} = g\bar\psi^{\rm W}(c_{\rm V}\gamma^\mu - c_{\rm A}\gamma^\mu\gamma_5)\,1_{\rm flav}\,\psi^{\rm W}, \tag{5.1.11}$$

which is expressed in the weak basis, the zero index indicates a neutral current and we also remind the reader that this interaction *probably* violates parity conservation, but for the purposes of this discussion the precise values of $c_{\rm V}$ and $c_{\rm A}$ are quite irrelevant. The fact that the components of $\bar\psi^{\rm W}$ are mixtures of the physical charged-lepton states could, in principle, allow for some flavour changing effect. However, let us rewrite the above current in the physical basis:

$$\begin{aligned} J_{\rm W}^{0\mu} &= g\overline{U\psi}\,(c_{\rm V}\gamma^\mu - c_{\rm A}\gamma^\mu\gamma_5)\,1_{\rm flav}\,U\psi \\ &= g\bar\psi\,U^\dagger(c_{\rm V}\gamma^\mu - c_{\rm A}\gamma^\mu\gamma_5)\,1_{\rm flav}\,U\,\psi \\ &= g\bar\psi\,(c_{\rm V}\gamma^\mu - c_{\rm A}\gamma^\mu\gamma_5)\,U^\dagger\,1_{\rm flav}\,U\,\psi \\ &= g\bar\psi\,(c_{\rm V}\gamma^\mu - c_{\rm A}\gamma^\mu\gamma_5)\,1_{\rm flav}\,\psi. \end{aligned} \tag{5.1.12}$$

It thus remains perfectly diagonal; in effect, a GIM-like mechanism (stemming from the unitarity of U) has protected this construction from any possible effect of *flavour-changing neutral currents*. Indeed, no evidence of any such phenomena is found (but see e.g. Acosta *et al* (2005). We remark, however, for completeness that higher-order quantum (or loop) corrections, *penguin diagrams* (which we shall study shortly), can and do lead to very small but non-vanishing effective flavour-changing neutral currents.

5.1.3.1 The experimental discovery of weak neutral currents
We must then find some process in which the photon cannot participate and yet which has a clear signal. As noted in the section on the muon-neutrino discovery, already in the early sixties, secondary ν_μ and $\bar\nu_\mu$ beams became available. By the late sixties, the production of high-energy (200–300 GeV) and high-intensity proton

beams allowed the generation of high-energy and high-intensity secondary ν_μ and $\bar{\nu}_\mu$ beams. The following interactions thus became possible for study:

$$\nu_\mu + e^- \rightarrow \nu_\mu + e^- \quad \text{and} \quad \bar{\nu}_\mu + e^- \rightarrow \bar{\nu}_\mu + e^-. \qquad (5.1.13)$$

The corresponding experiment was first performed at CERN in the PS and SPS rings by the Gargamelle collaboration (Hasert *et al* 1973a).

A so-called narrow-band neutrino beam was produced via the system shown schematically in figure 5.2. A 400 GeV proton beam impinging on a beryllium target to produce a large number of high-energy particles. By judicious choice of electric and magnetic fields, all but charged pions were filtered out over a length of around 100 m (although the beam did contain some charged-kaon contamination), also selecting an energy of around 200 GeV. Inside a vacuum pipe of about 200 m both the pions and kaons then decayed in-flight into muons and muon neutrinos. A long shielding block of steel and the Earth cleaned up the remaining muons and other stray particles, leaving a pure muon neutrino or antineutrino beam. There is, moreover, a strong correlation between the angle of emission and the energy of the neutrinos so that the spectrum is rather narrow. At the end of all this was Gargamelle, a giant bubble-chamber detector, designed principally for neutrino detection. With a diameter of nearly 2 m and length 4.8 m, it weighed 1000 tonnes and held nearly 12 cubic metres of freon (CF_3Br) at a density of 1.5×10^3 kg m^{-3}.

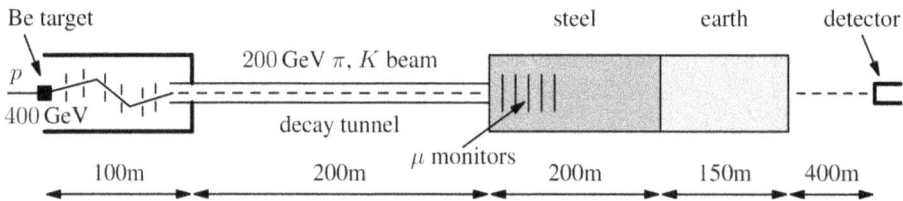

Figure 5.2. A schematic view of the narrow-band neutrino beam at CERN; see the main text for details.

Exercise 5.1.1. *For the given kinematics (i.e. a decaying charged-pion beam of energy 200 GeV) and taking into account the maximum angle subtended by a 2-m-wide detector, calculate the energy spread of the resulting neutrino beam that actually struck Gargamelle (use $m_\pi = 140$ MeV and $m_\mu = 106$ MeV).*

The motivation for using muon neutrinos is that charged currents can thus be excluded in their interactions with electrons (see figure 5.3). By requiring the final state to contain an *electron*, while the initial state contains a *muon* neutrino or antineutrino, the interaction is guaranteed to be via a neutral current. Indeed, the process depicted in figure 5.3(a) is impossible without a weak neutral current. The observation of electrons scattered by the muon-neutrino beam thus constituted a clear indication of the existence of weak neutral currents. Figure 5.4 displays the first such event recorded by Gargamelle.

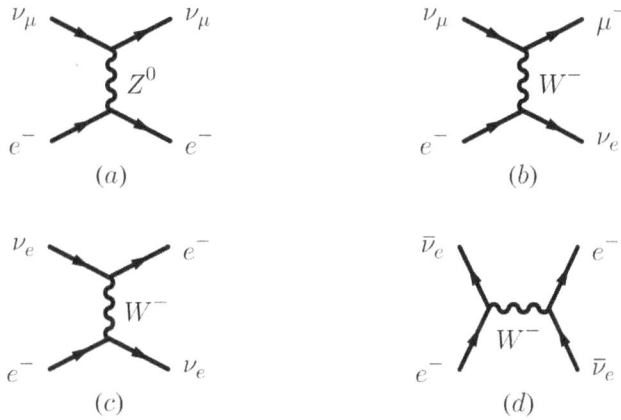

Figure 5.3. A comparison of the possible neutral- and charged-current interactions between muon and electron neutrinos and electrons. (*a*) A neutral-current event with an electron in the final state; (*b*) a charged-current event with instead a muon in the final state; (*c*) and (*d*) electron-neutrino initiated charged-current events with a final-state electron.

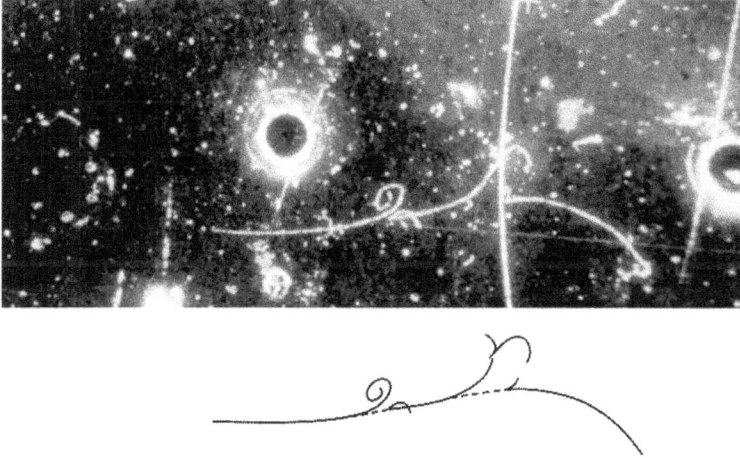

Figure 5.4. The first example of the neutral-current process $\bar{\nu}_\mu + e^- \rightarrow \bar{\nu}_\mu + e^-$. The recoil electron is projected forward with an energy of 400 MeV at an angle of $1.5 \pm 1.5°$ to the beam (which enters from the left), creating a characteristic shower of electron–positron pairs; images reproduced courtesy of the CERN Courier, 19th. Aug. 2013. © IOP Publishing. CC BY 4.0.

We could also look for DIS-type processes where all that is seen to recoil is a purely hadronic system (Hasert *et al* 1973b). The two possible reactions are then

$$\nu_\mu + N \rightarrow \nu_\mu + X \text{ and } \bar{\nu}_\mu + N \rightarrow \bar{\nu}_\mu + X, \tag{5.1.14}$$

where N is a struck nucleon and X the recoiling hadronic system (see figure 5.5). An example event from Gargamelle is shown in figure 5.6.

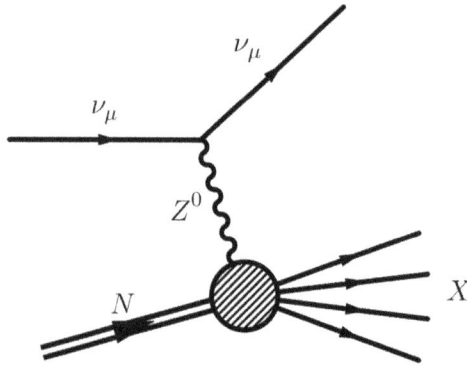

Figure 5.5. Deeply inelastic muon-neutrino–nucleon scattering via neutral-current (Z^0) exchange.

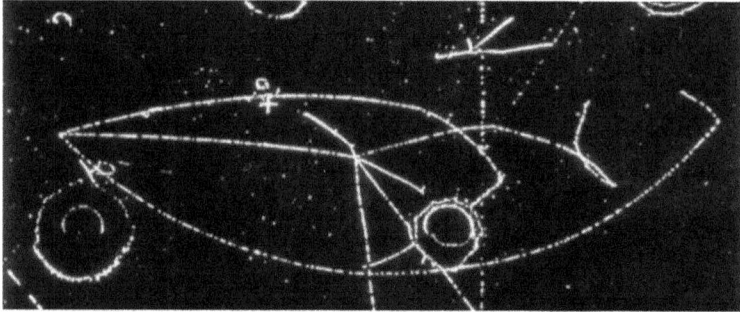

Figure 5.6. An example hadronic neutral-current process, as seen in Gargamelle. The interaction of the neutrino (coming from the left) produces three secondary particles, all clearly identifiable as hadrons as they then interact with other nuclei within the bubble chamber; no associated charged lepton is seen. Image reproduced courtesy of the CERN Courier, 19th. Aug. 2013. © IOP Publishing. CC BY 4.0.

5.1.3.2 Other evidence of weak neutral currents

Nuclear β-decay proceeds via charged-current exchange and is easily studied. On the other hand, since the weak neutral currents do not change flavour they cannot give rise to decays[2]. Thus, it is not possible to detect their *direct* rôle in weak nuclear transitions. However, the presence of a parity-violating interaction in competition with the usual Coulomb potential leads to bound-state nuclear wave-functions that are no longer pure eigenstates of parity. This in turn allows the observation of *indirect* parity violation in, for example, electromagnetic or strong-interaction nuclear transitions (for a more comprehensive discussion see, e.g., Haeberli and Holstein 1995).

We briefly present three such examples in the following. The first is an α-decay:

$$^{16}\text{O}^* \rightarrow {}^{12}\text{C} + \alpha$$
$$J^P \; : \; 2^- \quad\;\; 0^+ \;\; 0^+. \tag{5.1.15}$$

---

[2] In principle, they may contribute to such decays via the annihilation mechanism, but their rôle would always be eclipsed by photon-mediated processes.

This proceeds via the strong and electromagnetic interaction processes but violates parity and should therefore be prohibited. However, neither the initial nor the final nuclear state is a pure parity eigenstate and it is experimentally observed with a width $\Gamma \sim 10^{-10}$ eV (Neubeck $et\ al$ 1974). The next is a γ-emission:

$$^{19}\text{F*} \rightarrow {}^{19}\text{F} + \gamma$$

$$J^P: \quad \tfrac{1}{2}^- \quad \tfrac{1}{2}^- \quad 1^-. \tag{5.1.16}$$

In this case, the transition is electromagnetic, which again cannot directly violate P. Nevertheless, given the spin of the decaying ^{19}F*, the initial state may be polarised in order to examine the up–down photon-momentum asymmetry. An asymmetry of $A_\gamma \sim 10^{-4}$ is found (Elsener $et\ al$ 1984). Lastly, a γ-emission:

$$^{175}\text{Lu*} \rightarrow {}^{175}\text{Lu} + \gamma. \tag{5.1.17}$$

Here the initial state is not polarised, but the photon is experimentally found to emerge with a preference for being left-handed: $P_\gamma \sim 6 \times 10^{-5}$ (Vanderleeden and Boehm 1970). Finally, we might remark that there are analogous, smaller but still measurable, effects in atomic physics (see, e.g., Bouchiat and Bouchiat 1997).

5.1.3.3 Flavour-changing neutral currents and penguin diagrams

Having mentioned the possibility of flavour-changing neutral currents, we should now explain how they may arise. As the expression clearly suggests, the nett effect sought is a change in the flavour of an elementary fermion (usually a quark, but a lepton is also possible) while maintaining the $same$ charge. That is, we seek a transition between either an up- or down-type quark and another of the $same$ up or down type (e.g. $c \leftrightarrow u$ or $s \leftrightarrow d$).

We have already noted that a GIM-like mechanism guarantees the cancellation of any such direct (or tree-level) off-diagonal transitions. However, we also noted in section 2.6 that, in the case of neutral kaon decays and two-step weak transitions, such a cancellation is rendered incomplete owing to the mass difference between the u and c quarks. The question then naturally arises as to whether or not such transitions may effectively occur, perhaps more indirectly.

A clue to the answer is provided by the so-called unitarity triangles (which will be examined in section 5.3). Here it will be sufficient to consider one of the six off-$diagonal$ elements of the unitarity product $V_{\text{CKM}}^\dagger V_{\text{CKM}} = \mathbb{1}$, e.g. the 's–b' product:

$$V_{su} V_{ub}^* + V_{sc} V_{cb}^* + V_{st} V_{tb}^* = 0. \tag{5.1.18}$$

This is just another example of the GIM cancellation; diagrammatically, it should look something like figure 5.7. The crosses represent the weak coupling (i.e. a charged-current interaction involving emission or absorption of a $W^\pm$) and so carry factors of the corresponding CKM matrix elements (as indicated). It should now be clear that the GIM-like cancellation would be complete if and only if the internal propagators (indicated u, c and t in figure 5.7) were identical. They are, of course, not, as the masses are different. Such a mismatch allows transitions between, say, b

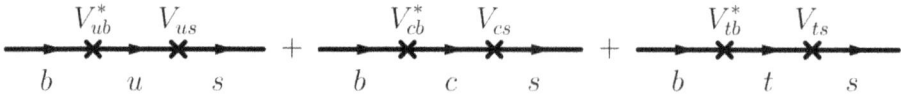

Figure 5.7. A diagrammatic representation of an example GIM cancellation.

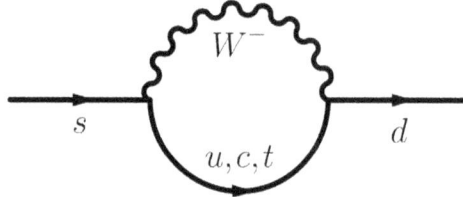

Figure 5.8. An example flavour-changing neutral current penguin sub-diagram allowing the transition of an s into a d quark via repeated intermediate charged-current exchanges.

and s quarks, permitting, for example, B-meson decays into strange mesons that would otherwise be prohibited.

The question was first studied theoretically by Vainshtein *et al* (1975) in the context of non-leptonic neutral-kaon decays. They examined the rôle of more complete sub-diagrams of the type displayed in figure 5.8. Note that, to conserve energy, since the s is heavier than the d quarks, there must be an accompanying real emission (typically a photon or gluon, which may then produce a lepton or quark pair). Note too how the same intermediate boson intervenes in both weak transitions.

The term *penguin* was coined by Ellis *et al* (1977) to describe full diagrams of this type, see figure 5.9. The extra external particles shown in the figure are connected here via gluon exchange and thus this type of diagram is often termed a *gluonic penguin*. Other exchanges are possible and lead to what are then called *electromagnetic* or *electroweak* penguins. The two external particles, quarks here, may both appear in the final (or even initial) state and in the electromagnetic and electroweak cases may be charged leptons or neutrinos.

The phenomenological study of these effects is made particularly interesting by the various peculiar aspects they encompass. First of all, recall that the unitarity of the CKM matrix guarantees that, despite the mixing of the weak states, the Z^0 cannot change flavour at the tree level, see equation (5.1.12). In that case though the vanishing is simply due to $V_{CKM}^\dagger$ *undoing*, so to speak, the basis rotation performed by V_{CKM}. In contrast, the would-be cancellation between diagrams like that of figure 5.9 is due to the particular combinations of CKM matric elements at the weak vertices. To see this, consider all three possible diagrams: together with the internal virtual t quarks, both a u- and a c-quark contribution must be added. The amplitude is then seen to be proportional to

$$V_{su}V_{ub}^* + V_{sc}V_{cb}^* + V_{st}V_{tb}^*, \tag{5.1.19}$$

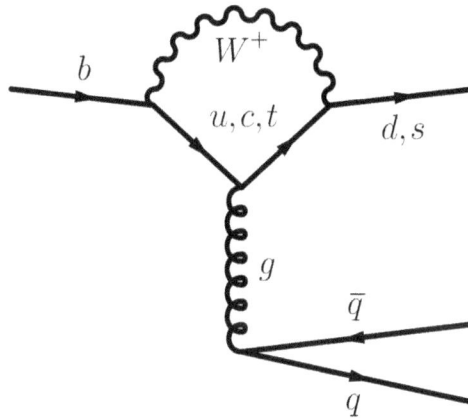

Figure 5.9. An example penguin diagram for the transition from a b into a d or s quarks and thus describing the decay of a B meson via an effective flavour-changing neutral current.

which is just an off-diagonal element of $VV^\dagger$ and thus vanishes. Of course, mathematically this is no different to the previous cancellation, but the way in which it arises is more intimately tied to the form of the weak interaction. Finally, as noted in the discussion of the GIM mechanism, the cancellation is incomplete owing to the differing quark masses and is thus sensitive to their precise values.

An additional, but all important point is that the particles propagating inside the loops are virtual and may therefore be very massive; that is, even at the b-quark mass scale a t quark or heavier object may provide the relevant contribution. This last observation is the key to the importance of such contributions: while the resulting cross-sections may be very small, they give rise to processes that could not otherwise take place; i.e. flavour-changing neutral-current processes. That is, we are not dealing here with small deviations from SM predictions, but rather clear and easily identifiable signals.

Exercise 5.1.2. *Consider the possible decays of the $\bar{B}^0$ (a $b\bar{d}$ meson) via suitable penguin-like diagrams. In particular, construct the most probable diagrams for:*

$$\bar{B}^0 \to \gamma\gamma \quad \text{and} \quad \bar{B}^0 \to \ell^+\ell^-.$$

Exercise 5.1.3. *Considering the penguin diagram shown in figure 5.9, identify the cases (and the corresponding diagrams) for which the following processes may proceed without penguin contributions and those for which they are necessary:*

$$B \to \pi\pi \quad \text{and} \quad B \to K\pi.$$

5.1.4 Further unitarity problems

Unfortunately, however, the problem of growing cross-sections with respect to unitarity limits is not completely solved: the $\nu_\mu e^- \to \mu^- \nu_e$ process previously examined still grows logarithmically but would exceed the bounds only at extremely high energies. More problematically, the new states invoked to tame the Fermi-theory divergences naturally induce new processes, which although often rather exotic are nevertheless theoretically possible and thus too should not violate unitarity.

A particular example is $\nu_e \bar{\nu}_e$ annihilation into a $W^+ W^-$ pair[3] (but there are many others). The first obvious contribution comes from the diagram in figure 5.10(a). This diagram leads to a total cross-section that again grows as s. One solution might be to invoke a sort of GIM-like cancellation with a new (heavy) lepton, as shown in figure 5.10(b), having suitably chosen *ad hoc* couplings (Georgi and Glashow 1972). No such heavy lepton has ever been detected though and there are stringent limits on the mass it may have for the cancellation to still be effective. Of course, there are now also other possible, new, but more natural contributions that might be included, such as the neutral-current annihilation graph of figure 5.10(c). This is evidently a different process, however, and has no specific relationship to the first *unless there is some larger symmetry linking them*.

It turns out that any such attempt to patch up the cross-section growth-rate problem in an *ad hoc* manner still runs foul of the requirement of renormalisability: when quantum corrections are evaluated, this type of theory is not generally renormalisable. Moreover, if we examine the $\nu_e \bar{\nu}_e \to W^+ W^-$ cross-section in detail, we discover that it is precisely the 'extra', longitudinal, component of the $W^\pm$ bosons that is responsible for the unbounded growth.

Now, had the gauge principle somehow been enforced, such contributions would have been automatically absent. Indeed, QED suffers *none* of these problems. That is, the photon has no longitudinal component and QED is renormalisable. However, such highly desirable properties are a direct consequence of the masslessness of the

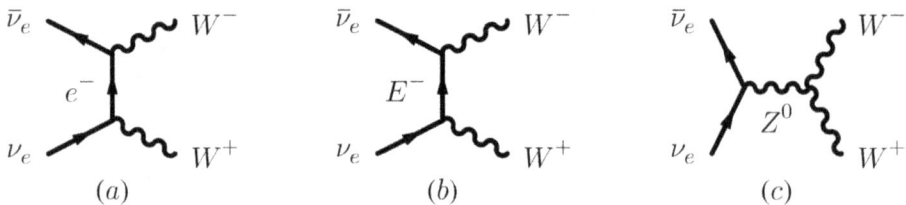

Figure 5.10. Contributions to the annihilation process $\nu_e \bar{\nu}_e \to W^+ W^-$: (*a*) the *t*-channel electron exchange graph, (*a*) a possible heavy-electron exchange graph, (*a*) the *s*-channel neutral-current annihilation graph.

---

[3] This process is very closely related to the $e^+ e^- \to W^+ W^-$ process discussed earlier and which has been observed.

photon; local gauge symmetry is broken by any *explicit* mass term. We recall briefly that gauge symmetry requires that the Lagrangian (and all *physical quantities*) be invariant under the following (local gauge) transformation:

$$A^\mu(x) \to A'^\mu(x) = A^\mu(x) + \partial^\mu \Lambda(x) \tag{5.1.20}$$

and

$$\psi(x) \to \psi'(x) = e^{-ie\Lambda(x)}\psi(x), \tag{5.1.21}$$

where $\Lambda(x)$ is any scalar function of x, $A^\mu(x)$ is the gauge field (for instance, the photon) while $\psi(x)$ is a spinor field representing, say, the electron.

The field equations governing the motion and interaction of such fields are derived from the following Lagrangian:

$$\mathcal{L} = \bar{\psi}(i\gamma^\mu D_\mu - m)\psi - \frac{1}{4}F^{\mu\nu}F_{\mu\nu}, \tag{5.1.22}$$

where the covariant derivative is

$$D^\mu := \partial^\mu + ieA^\mu(x) \tag{5.1.23}$$

and the field-strength tensor $F^{\mu\nu}$ is

$$F^{\mu\nu} := \partial^\mu A^\nu - \partial^\nu A^\mu. \tag{5.1.24}$$

It is straightforward to verify explicitly that the purely gauge term in the above Lagrangian leads to equations Maxwell's (1865) for the photon field. The term $-eA_\mu\bar{\psi}\gamma^\mu\psi$ implicit in equation (5.1.22) leads to the standard fermion–photon interaction. A mass term for the gauge field would take the form $-\frac{1}{2}m^2 A^\mu A_\mu$, which would plainly violate gauge invariance. We thus see that, to render the propagation of the vector fields $W^\pm$ massive, some other means of providing a mass must be found, via an interaction, for example.

5.2 The Higgs mechanism (SSB)

In this section we shall briefly describe the mechanism by which it is possible to retain gauge symmetry (with all the consequent benefits) while allowing the gauge bosons to acquire significant *effective* masses. The concept of a broken symmetry, with the consequent generation of *massless* states was first discussed by Goldstone (1961). That a spontaneously broken symmetry (i.e. a symmetry of the Lagrangian broken by the vacuum) could circumvent the Goldstone theorem and thus avoid the existence of massless states was proposed by Anderson (1963)[4] within the context of superconductivity and successively adapted to particle theory by Nambu and Jona-Lasinio (1961)[5]. Spontaneous symmetry breaking in a gauge theory and the resulting mass generation, known as the Higgs (or more completely Higgs–Brout–Englert–

---

[4] The 1977 Nobel Prize in Physics was awarded jointly to Philip Warren Anderson, Sir Nevill Francis Mott and John Hasbrouck van Vleck 'for their fundamental theoretical investigations of the electronic structure of magnetic and disordered systems'.
[5] For the associated Nobel prize, see footnote 29 in chapter 2.

Guralnik–Hagen–Kibble) mechanism, was first applied to quantum field theory by Higgs (1964), Englert and Brout (1964) and Guralnik *et al* (1964)[6]. A review article by Quigg, (2007) is recommended to the interested reader as a general discussion of the problem of mass generation. A comprehensive historical account of the developments leading to the final construction was given by Kibble (2009).

5.2.1 A real scalar field

The simplest case is a purely scalar theory with a mass term of the '*wrong sign*' (note that it is the term that has the wrong sign and not the mass itself) and a φ^4 self-interaction (which guarantees stability):

$$\mathcal{L} = \tfrac{1}{2}(\partial^\mu \varphi)(\partial_\mu \varphi) + \tfrac{1}{2}m^2\varphi^2 - \tfrac{1}{4!}\lambda\varphi^4. \tag{5.2.1}$$

We remind the reader that in quantum field theory the physical mass of a particle is given (to lowest order in perturbation theory) by the square-root of the second derivative of the potential at the minimum. That is,

$$m^2 := \frac{\partial^2 \mathcal{L}}{\partial \varphi^2}\Big|_{\varphi=\varphi_0} \quad \text{with} \quad 0 = \frac{\partial \mathcal{L}}{\partial \varphi}\Big|_{\varphi=\varphi_0}. \tag{5.2.2}$$

Thus, of course, it normally corresponds to the coefficient of the quadratic term *if and only if* the minimum lies at $\varphi = 0$.

The corresponding potential

$$V(\varphi) = -\tfrac{1}{2}m^2\varphi^2 + \tfrac{1}{4!}\lambda\varphi^4 \tag{5.2.3}$$

has the form shown in figure 5.11, which is evidently symmetric under the discrete transformation $\varphi \rightarrow -\varphi$. However, the point $\varphi = 0$ is no longer the minimum while there are now two equivalent minima at $\varphi = \pm\sqrt{6m^2/\lambda}$. The vacuum or ground state of the system may be one or other, but not both at the same time, and will therefore break the original symmetry.

Since perturbation theory is to be performed around the minimum, we should shift to $\varphi' := \varphi - \upsilon$, where for the purposes of example and without loss of generality we shall take $\upsilon = +\sqrt{6m^2/\lambda}$. The Lagrangian then takes the form

$$\mathcal{L} = \tfrac{1}{2}(\partial^\mu \varphi')(\partial_\mu \varphi') - m^2\varphi'^2 - \tfrac{1}{3!}\lambda\varphi'^3 - \tfrac{1}{4!}\lambda\varphi'^4, \tag{5.2.4}$$

---

[6] The 2013 Nobel Prize in Physics was awarded jointly to François Englert and Peter W Higgs 'for the theoretical discovery of a mechanism that contributes to our understanding of the origin of mass of subatomic particles, and which recently was confirmed through the discovery of the predicted fundamental particle, by the ATLAS and CMS experiments at CERN's Large Hadron Collider'. Although Guralnik, Hagen and Kibble missed out on the Nobel Prize, they did receive many other important recognitions jointly with Higgs, Brout and Englert, such as the 2010 J J Sakurai Prize.

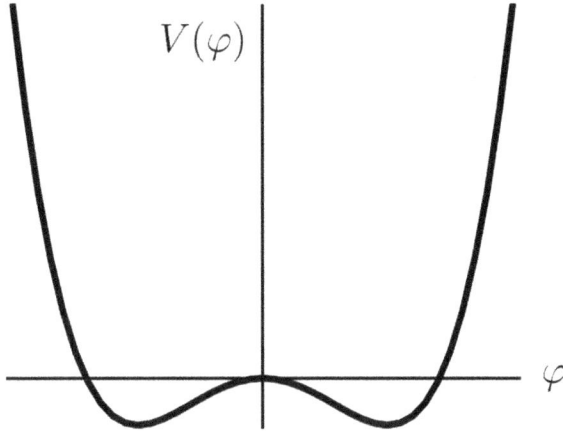

Figure 5.11. The form of the potential for a single, real, scalar field in the case of a mass term with a negative sign.

where an irrelevant constant term has been eliminated. Thus, the true physical state of the theory has a mass $\sqrt{2}\,m$ and a cubic self-interaction has appeared.

5.2.2 A two-component scalar field

The case of a continuous symmetry is rather more interesting. Consider now a *complex* scalar field φ with the following Lagrangian

$$\mathcal{L} = (\partial^\mu \varphi^*)(\partial_\mu \varphi) + m^2 \varphi^* \varphi - \tfrac{1}{2}\lambda(\varphi^*\varphi)^2. \tag{5.2.5}$$

The corresponding potential

$$V(\varphi, \varphi^*) = -m^2 \varphi^* \varphi + \tfrac{1}{2}\lambda(\varphi^*\varphi)^2 \tag{5.2.6}$$

has the so-called Mexican-hat form shown in figure 5.12 and is symmetric under the (global gauge) transformation $\varphi \to e^{i\phi}\varphi$. The possible vacuum states now belong to a continuum, corresponding to the variable $\phi \in [0, 2\pi]$ in the above transformation.

Again, without loss of generality, we take the ground state as $\varphi = v = \sqrt{m^2/\lambda}$, with v real, and thus make the shift to $\varphi' := \varphi - v$. It is convenient to re-express these fields in terms of two *real* scalars $\varphi_{1,2}$:

$$\varphi' = \tfrac{1}{\sqrt{2}}(\varphi_1 + i\varphi_2). \tag{5.2.7}$$

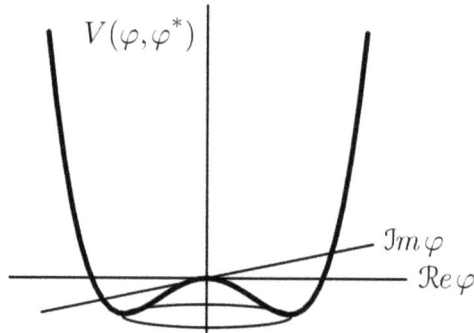

Figure 5.12. The so-called Mexican-hat form of the potential in the case of a complex scalar field with a negative mass-squared term; it possesses a cylindrical symmetry about the vertical axis.

The Lagrangian then takes the form

$$\mathcal{L} = \tfrac{1}{2}(\partial^\mu \varphi_1)(\partial_\mu \varphi_1) + \tfrac{1}{2}(\partial^\mu \varphi_2)(\partial_\mu \varphi_2) - m^2 \varphi_1^2$$
$$- \lambda v \varphi_1 (\varphi_1^2 + \varphi_2^2) - \tfrac{1}{4}\lambda(\varphi_1^2 + \varphi_2^2)^2. \tag{5.2.8}$$

In this case we see that, together with interaction terms of various degrees, there is just a single non-zero mass; that is, one state (φ_2) is left massless. It is easy to see why: the original continuous symmetry implies that there is always a direction (that of the symmetry) in which both the first and second derivatives of the field are zero. This is the essence of the Goldstone theorem (1961), which states that for each broken continuous symmetry there exists a massless field, known as the *Goldstone boson*. We should stress here that the theorem only applies to *continuous* symmetries and is thus only relevant to *multidimensional* symmetry subspaces (i.e. $D \geqslant 2$). We might also point out that Coleman and Weinberg (1973) demonstrated that radiative corrections can create the conditions necessary for spontaneous symmetry breaking, without the need for a 'wrong-sign' mass term.

The Goldstone theorem thus proved to be something of a stumbling block for the field-theoretical description of such phenomena as superconductivity, superfluidity and even ferromagnetism, where the ideas of spontaneous symmetry breaking seemed to be so very useful and even implicit. The problem was simply that no evidence had ever been found to suggest the physical existence of the bosonic *zero-mass* states (the Goldstone bosons) predicted by the Goldstone theorem. The early emphasis was then on the elimination of the undesirable zero-mass bosons generated in the process of spontaneous symmetry breaking in condensed-matter physics (Anderson 1963), rather than on a mechanism to generate the massive gauge-field states so desirable for the weak-interaction theory. This was however exactly what Anderson incidentally achieved.

5.2.3 The Higgs mechanism

We now examine the case of spontaneous symmetry breaking in the presence of a *local* gauge symmetry. We shall find that this provides an exception to the Goldstone theorem, in that the massless bosons do not appear as true individual states of the theory, but are absorbed (or *eaten*) by the gauge bosons to provide the third components for the gauge fields, which in turn become massive. The important point to realise though is that the underlying gauge symmetry survives; the theory thus remains renormalisable and satisfies unitarity in the usual way.

Consider the simplest example of a single charged scalar field minimally coupled to a gauge field (a sort of *scalar* QED). The Lagrangian then takes on the standard form

$$\mathcal{L} = (D^\mu\varphi^*)(D_\mu\varphi) + \mu^2\varphi^*\varphi - \lambda(\varphi^*\varphi)^2 - \tfrac{1}{4}F^{\mu\nu}F_{\mu\nu}, \qquad (5.2.9)$$

where the covariant derivative D^μ is defined as

$$D^\mu\varphi(x) := \left(\partial^\mu - ieA^\mu(x)\right)\varphi(x) \qquad (5.2.10a)$$

and

$$D^\mu\varphi^*(x) := \left(\partial^\mu + ieA^\mu(x)\right)\varphi^*(x). \qquad (5.2.10b)$$

It is again convenient to reparametrise the field φ exponentially in terms of two real fields η and ξ (the 'radial' and 'angular' components, respectively):

$$\varphi(x) = \tfrac{1}{\sqrt{2}}\left(v + \eta(x)\right)\exp\left(i\xi(x)/v\right), \qquad (5.2.11)$$

with the Higgs-boson vacuum expectation value

$$v = \sqrt{\mu^2/\lambda}. \qquad (5.2.12)$$

In the absence of the gauge field and the resultant coupling, the field ξ would have been the massless Goldstone boson associated with the spontaneous breaking of a global U(1) symmetry. However, the presence of a gauge symmetry and corresponding fields induces mixing of the ξ and A^μ fields. To see this, let us rewrite the Lagrangian in terms of ξ and η:

$$\begin{aligned}
\mathcal{L}(\xi, \eta, A^\mu) = &-\tfrac{1}{4}F^{\mu\nu}F_{\mu\nu} + \tfrac{1}{2}\partial^\mu\xi\partial_\mu\xi + \tfrac{1}{2}\partial^\mu\eta\,\partial_\mu\eta \\
&+ \tfrac{1}{2}e^2v^2A^\mu A_\mu - evA_\mu\partial^\mu\xi - \mu^2\eta^2 \\
&+ \text{terms higher than quadratic order.}
\end{aligned} \qquad (5.2.13)$$

The higher-order terms describe various interactions and are of interest for the full phenomenology, but here we are only interested in the question of mass. At first sight, there appear to be the fields of the previous example: η a boson of mass $\sqrt{2}\mu$ and ξ a massless Goldstone boson, together now with a massive gauge field (since a term in $A^\mu A_\mu$ has been generated). However, the presence of the term $A_\mu\partial^\mu\xi$ complicates matters: it directly mixes the ξ and A^μ fields, which cannot then be the true asymptotic or physical states of the theory, and so a little more care is required.

The gauge invariance of the Lagrangian may be exploited to apply the following gauge transformation, which effectively diagonalises the mixed fields:

$$\varphi \to \varphi' = \exp[-i\xi(x)/\upsilon]\,\varphi = \tfrac{1}{\sqrt{2}}(\upsilon + \eta) \tag{5.2.14a}$$

and

$$A_\mu \to A'_\mu = A_\mu - \frac{1}{e\upsilon}\,\partial_\mu \xi. \tag{5.2.14b}$$

With this, the Lagrangian becomes

$$\begin{aligned}
\mathcal{L}(\xi, \eta, A^\mu) = &-\tfrac{1}{4}F'^{\mu\nu}F'_{\mu\nu} + \tfrac{1}{2}e^2\upsilon^2 A'^\mu A'_\mu + \tfrac{1}{2}\partial^\mu\eta\,\partial_\mu\eta - \mu^2\eta^2 \\
&+ \tfrac{1}{2}e^2 A'^\mu A'_\mu\,\eta(2\upsilon + \eta) - \lambda\upsilon\eta^3 - \tfrac{1}{4}\lambda\eta^4.
\end{aligned} \tag{5.2.15}$$

The gauge field has thus acquired an effective mass $e\upsilon$ and the scalar field η has a mass $m = \sqrt{2}\,\mu$ while the field ξ has simply disappeared from the theory.

The physical interpretation should be rather obvious: a massive vector field necessarily has three degrees of freedom, whereas the original massless gauge field only had two; the third is provided by the *would-be* Goldstone boson ξ, which, we say figuratively, has thus been 'eaten' by the gauge field.

As a by-product, the presence of a scalar field with non-vanishing vacuum expectation value also allows the generation of effective mass terms for the matter fields, which may then be initially defined as massless too. It is simply necessary to add a Yukawa-type coupling to each fermion for which a mass is desired (Weinberg 1967):

$$\mathcal{L}_{\text{int}} = -g\,\varphi\,\bar{\psi}\psi. \tag{5.2.16}$$

Shifting the scalar field and rewriting it as above leads to

$$\mathcal{L}_{\text{int}} = -g(\upsilon + \varphi')\bar{\psi}\psi. \tag{5.2.17}$$

The first term in brackets is evidently none other than a mass term for the field ψ and we have thus have

$$m_\psi = g\,\upsilon. \tag{5.2.18}$$

Since the coupling g is arbitrary, the mass is not determined. However, given the measured mass, the relation may be inverted to provide g in terms of the Higgs-field vacuum expectation value. Indeed, this relation tells us that the heaviest fermions will have the strongest couplings.

5.2.4 The Glashow–Salam–Weinberg model

To describe the weak interaction correctly it is found necessary to include the theory of electromagnetism at the same time; although, of course, the photon remains rigorously massless. That is, we are led to the construction of a (*quasi*) *unified* model of the electromagnetic and weak interactions, or *electroweak theory* (Glashow 1961; Salam 1968; Weinberg 1967)[7].

We start then with a weak-isospin triplet of massless spin-one bosons $W_\mu^{(1)}$, $W_\mu^{(2)}$ and $W_\mu^{(3)}$, where now the associated SU(2) symmetry is taken to be a *local gauge* symmetry, thus guaranteeing unitarity and renormalisability but requiring mass-lessness. It turns out that, in addition, we must include a single (isoscalar) neutral gauge boson B_μ^0. This is *not* to be associated directly with the photon and electric charge, but with the weak *hypercharge*. The $W^{\pm,\,0}$ triplet may be more suggestively rewritten as

$$W_\mu^\pm = W_\mu^{(1)} \pm iW_\mu^{(2)} \tag{5.2.19a}$$

and

$$W_\mu^0 = W_\mu^{(3)}. \tag{5.2.19b}$$

The scalar system necessary for spontaneous symmetry breaking here consists of two doublets, which may be expressed as (cf. the $K^\pm$, K^0, $\bar{K}^0$ system)

$$\begin{pmatrix} \varphi_1 \pm i\varphi_2 \\ \varphi_3 \pm i\varphi_4 \end{pmatrix} \quad \text{or} \quad \begin{pmatrix} \varphi^+ \\ \varphi^0 \end{pmatrix} \quad \text{and} \quad \begin{pmatrix} \bar{\varphi}^0 \\ \varphi^- \end{pmatrix}. \tag{5.2.20}$$

The construction of the standard interaction part of the Lagrangian, via general-isation of the electromagnetic case, then leads to the following two terms:

$$\mathcal{L}_{\text{int}} = g\boldsymbol{J}_W^\mu \cdot \boldsymbol{W}_\mu + g'J_Y^\mu B_\mu + \text{Hermitian conjugate}, \tag{5.2.21}$$

to which we must add terms coupling the scalar and gauge fields. The two currents introduced here are $\boldsymbol{J}_W^\mu$ (weak isospin) and J_Y^μ (hypercharge). The SU(2) symmetry of the Higgs potential is spontaneously broken in such a way that the two charged $W^\pm$ acquire a mass just as described in the preceding section while the case of the two neutral vector bosons is a little more complicated. The neutral fields that finally emerge as the physical degrees of freedom of the theory are the mutually orthogonal combinations

$$A_\mu = \frac{g'W_\mu^0 + gB_\mu}{\sqrt{g^2 + g'^2}} \quad \text{(the photon)} \tag{5.2.22a}$$

and

$$Z_\mu = \frac{gW_\mu^0 - g'B_\mu}{\sqrt{g^2 + g'^2}} \quad \text{(the weak neutral boson).} \tag{5.2.22b}$$

---

[7] The 1979 Nobel Prize in Physics was awarded equally to Sheldon Glashow, Abdus Salam and Steven Weinberg 'for their contributions to the theory of the unified weak and electromagnetic interaction between elementary particles, including, *inter alia*, the prediction of the weak neutral current'.

The two independent coupling constants g and g' thus play the rôle of a mixing angle here and it is therefore more convenient to introduce

$$\sin \theta_W := \frac{g'}{\sqrt{g^2 + g'^2}} \qquad \text{and} \qquad \cos \theta_W := \frac{g}{\sqrt{g^2 + g'^2}}. \qquad (5.2.23)$$

In other words,

$$\tan \theta_W := \frac{g'}{g}. \qquad (5.2.24)$$

As to the scalar fields, the two charged scalars $\varphi^\pm$ are absorbed into the $W_\mu^\pm$, respectively, to provide the longitudinal components, as before, while the combination $\frac{1}{\sqrt{2}}(\varphi^0 - \bar{\varphi}^0)$ is absorbed by the combination of W_μ^0 and B_μ corresponding to Z_μ. This all leaves just one physical, massive, scalar field:

$$H^0 := \frac{1}{\sqrt{2}}(\varphi^0 + \bar{\varphi}^0), \qquad (5.2.25)$$

which is precisely the object known as the *Higgs boson*.

Using the fact that hypercharge $Y = Q - I_3$ (a factor 2 has been introduced to avoid numerous spurious factors of ½), so that

$$J_Y^\mu = J_{EM}^\mu - J_{(3)}^\mu, \qquad (5.2.26)$$

we may now rewrite the above in terms of the asymptotic or physical fields. The original interaction Lagrangian

$$\mathcal{L}_{int} = g \left[J_{(1)}^\mu W_\mu^{(1)} + J_{(2)}^\mu W_\mu^{(2)} \right] + J_{(3)}^\mu \left[g W_\mu^{(3)} - g' B_\mu \right] + g' J_{EM}^\mu B_\mu$$
$$+ \text{Hermitian conjugate}$$

then becomes

$$= \frac{g}{\sqrt{2}} \left[J_-^\mu W_\mu^+ + J_+^\mu W_\mu^- \right] + \frac{g}{\cos \theta_W} \left[J_{(3)}^\mu - \sin^2 \theta_W J_{EM}^\mu \right] Z_\mu + g \sin \theta_W \, J_{EM}^\mu A_\mu \qquad (5.2.27)$$
$$+ \text{Hermitian conjugate}.$$

The first bracketed term represents the charged-current interaction, the second the weak neutral current, which we see mixes the two parity-conserving and maximally parity-violating couplings (we shall expand on this later), while the third is identified with the standard electromagnetic interaction and thus immediately leads to the relation

$$e = g \sin \theta_W. \qquad (5.2.28)$$

Leaving aside the fermion mass parameters, we see that the theory is determined by a very small number of physical constants. For example, if we take the Fermi constant G_F and the electromagnetic coupling constant α as known, then both of the heavy-boson masses can be predicted in terms of the same single mixing parameter $\sin \theta_W$:

$$M_W = \left(\frac{\sqrt{2}\, g^2}{8 G_F} \right)^{1/2} = \left(\frac{\sqrt{2}\, e^2}{8 G_F \sin^2 \theta_W} \right)^{1/2} = \frac{37.4}{\sin \theta_W} \text{GeV} \qquad (5.2.29)$$

and

$$M_Z = \frac{M_W}{\cos \theta_W} = \frac{74.8}{\sin 2\theta_W} \text{GeV}. \qquad (5.2.30)$$

As we shall see shortly, $\sin \theta_W$ may also be experimentally determined via various other independent measurable quantities; the present world average value is (Navas *et al* 2024)

$$\sin^2 \theta_W = 0.223\ 48(10), \qquad (5.2.31)$$

although care must be taken in comparison with, e.g., the boson masses, as there are important quantum corrections to be taken into account, which are different for the various physical quantities.

Before moving on to the more phenomenological aspects, we should make one final theoretical comment. It has already been explained that a consequence of field quantisation is the renormalisation of the various parameters of the theory (see the discussion in section 3.4.1). As remarked above, it turns out that gauge invariance protects quantum field theories from certain disastrous consequences of quantum corrections; indeed, it can be proved that gauge theories are systematically renormalisable. The proof is, however, highly non-trivial and is not immediately applicable in the presence of spontaneous symmetry breaking. An import theoretical passage then was the demonstration that the Glashow–Salam–Weinberg theory was renormalisable ('t Hooft, 1971)[8].

5.2.4.1 The Higgs-boson mass

Unfortunately, one important physical quantity is left entirely undetermined: namely, the Higgs-boson mass m_H. This is because, although the Higgs-field vacuum expectation value v is found to be

$$v = 2^{-1/4} G_F^{-1/2} = 246 \text{ GeV}, \qquad (5.2.32)$$

the mass itself is given by $m_H = \lambda v$, but λ is undetermined.

However, the Higgs-boson decay width is only a function of its mass and may thus be calculated; we find $\Gamma_H \sim G_F m_H^3$. Since to have any chance of 'seeing' the Higgs particle (i.e. as a Breit–Wigner resonance peak) its width should be substantially less than its mass, this means that the mass should therefore be less than about $1/\sqrt{G_F}$. A more accurate analysis in terms of partial-wave unitarity in WW scattering places an upper limit of about 1 TeV. Moreover, since the quantum corrections to various physical quantities and processes contain contributions

---

[8] The 1999 Nobel Prize in Physics was awarded equally to Gerardus 't Hooft and Martinus J.G. Veltman 'for elucidating the quantum structure of electroweak interactions in physics'.

depending on the mass, global fits to SM data can provide a window of acceptable masses, which at the 90% CL was (Beringer *et al* 2012)

$$m_H = 99^{+28}_{-23} \text{ GeV}. \tag{5.2.33}$$

Despite the high-precision data available, the limits are not very stringent owing to the weak (logarithmic) dependence. In figure 5.13 the constraints on the Higgs-boson mass m_H as a function of the top mass m_t are displayed as 90% CL allowed regions. The central ellipse marks the 90% CL allowed region combining all data.

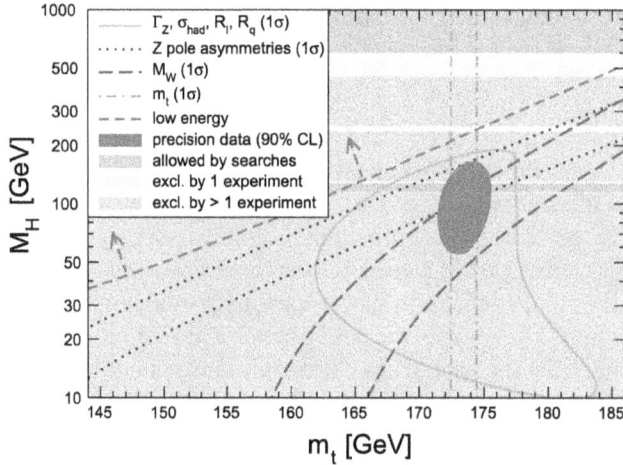

Figure 5.13. The 2012 one-standard-deviation boundaries for the Higgs-boson mass m_H as a function of the top mass m_t from various sources. The ellipse marks the 90% CL allowed region combining all data; figure reproduced from PDG-2012, Beringer *et al* (2012), copyright (2012) by the American Physical Society.

5.2.4.2 Electroweak couplings of the fermions
The charged $W^\pm$ bosons maximally violate P inasmuch as they only interact with left-handed fermions. The extra W^0 necessarily also behaves in the same manner as it belongs to the same multiplet. On the other hand, the photon has equal left- and right-handed couplings and therefore the Z^0, being a mixture of the two, only partially violates P in its interactions with the charged fermions (both quarks and leptons), though still maximally with neutrinos.

The leptonic sector may be characterised as follows ($\ell = e, \mu, \tau$):

$$\psi_L = \begin{pmatrix} \nu_\ell \\ \ell \end{pmatrix}_L \qquad I = \tfrac{1}{2} \qquad I_3 = \begin{cases} +\tfrac{1}{2} \\ -\tfrac{1}{2} \end{cases} \qquad Q = \begin{cases} 0 \\ -1 \end{cases} \qquad Y = -\tfrac{1}{2}, \tag{5.2.34a}$$

$$\psi_R = \ell_R \qquad I = 0 \qquad I_3 = 0 \qquad Q = -1 \qquad Y = -1, \tag{5.2.34b}$$

where we have exploited the definition $Y = Q - I_3$. Now, the Z^0 couples to the neutral current $J^\mu_{(3)} - \sin^2\theta_W J^\mu_{EM}$ and therefore left- and right-handed coupling constants may be defined:

$$g_L := I_3 - Q\sin^2\theta_W \quad \text{and} \quad g_R := -Q\sin^2\theta_W, \quad (5.2.35)$$

where Q is the fermion charge in units of $|e|$. The vector and axial-vector couplings c_V and c_A may also be defined, $g_{L/R} = \frac{1}{2}(c_V \pm c_A)$; they are shown in table 5.2 as functions of $\sin\theta_W$ for the different fermion species. Note that, since $\sin^2\theta_W \sim \frac{1}{4}$, the vector couplings almost vanish for the charged leptons and the up-type quarks and these channels (coupling to the Z^0) thus have almost no parity violation, whereas the neutrino channel remains maximally parity violating.

Table 5.2. The Z^0 vector and axial-vector couplings c_V and c_A, $g_{L/R} = \frac{1}{2}(c_V \pm c_A)$ as functions of $\sin\theta_W$ for the various fermion species; ℓ are the charged leptons while U and D represent the up- and down-type quarks.

	c_V	c_A	g_L	g_R
ν_ℓ	$\frac{1}{2}$	$\frac{1}{2}$	$\frac{1}{2}$	0
ℓ	$-\frac{1}{2} + 2\sin^2\theta_W$	$-\frac{1}{2}$	$-\frac{1}{2} + \sin^2\theta_W$	$\sin^2\theta_W$
U	$\frac{1}{2} - \frac{4}{3}\sin^2\theta_W$	$\frac{1}{2}$	$\frac{1}{2} - \frac{2}{3}\sin^2\theta_W$	$-\frac{2}{3}\sin^2\theta_W$
D	$-\frac{1}{2} + \frac{2}{3}\sin^2\theta_W$	$-\frac{1}{2}$	$-\frac{1}{2} + \frac{1}{3}\sin^2\theta_W$	$\frac{1}{3}\sin^2\theta_W$

5.2.4.3 Neutrino scattering via neutral currents

We first examine the various possible neutrino-scattering cross-sections in which the *charged* current intervenes (see figures 5.3(c) and (d)): at low energies the purely charged-current contributions are ($y := E'_e/E_\nu$ in the laboratory frame, initial electron at rest, $E'_e =$ final electron energy):

$$\frac{d\sigma_{CC}(\nu_e e \to \nu_e e)}{dy} = \frac{G_F^2 s}{\pi} \qquad (LL \to LL, \Rightarrow J = 0) \qquad (5.2.36a)$$

and

$$\frac{d\sigma_{CC}(\bar\nu_e e \to \bar\nu_e e)}{dy} = \frac{G_F^2 s}{\pi}(1-y)^2 \quad (RL \to RL, \Rightarrow J = 1). \qquad (5.2.36b)$$

Exercise 5.2.1. *Neglecting the electron mass (where permissible), show that $y = \frac{1}{2}(1 + \cos\theta_e)$, where θ_e is the electron centre-of-mass scattering angle with respect to the (anti-)neutrino beam direction. Thus, show that the $J = 1$ cross-section given above behaves as $(1 - \cos\theta_e)^2$.*

Recall that for an intermediate vector particle, we normally expect an angular dependence of the form $1 + \cos^2 \theta_e$. The difference here is that only one helicity state of the intermediate vector boson is available, corresponding to the amplitude $1 - \cos \theta_e$. The other helicity *would* have provided an amplitude $1 + \cos \theta_e$ and then the combination *would* have given

$$\tfrac{1}{2}(1 - \cos \theta_e)^2 + \tfrac{1}{2}(1 + \cos \theta_e)^2 = 1 + \cos^2 \theta_e, \tag{5.2.37}$$

as expected for a normal vector boson.

These are to be compared with the corresponding *neutral*-current cross-sections (see figure 5.14). For $W^\pm$ we have $g_L = 1$ and $g_R = 0$ while for Z^0 exchange they are as listed just above. Thus, for purely neutral-current contribution, we have

$$\frac{d\sigma_{NC}(\nu_e e \rightarrow \nu_e e)}{dy} = \frac{G_F^2 s}{\pi} \left[g_L^2 + g_R^2 (1 - y)^2 \right], \tag{5.2.38a}$$

$$\frac{d\sigma_{NC}(\bar{\nu}_e e \rightarrow \bar{\nu}_e e)}{dy} = \frac{G_F^2 s}{\pi} \left[g_R^2 + g_L^2 (1 - y)^2 \right]. \tag{5.2.38b}$$

Note that whereas the electron may be either L or R here, the ν_e is only L and the $\bar{\nu}_e$ only R.

Now, for the scattering of muon neutrinos off electrons, only the neutral-current contributions survive and so

$$g_L = -\tfrac{1}{2} + \sin^2 \theta_W \qquad \text{and} \qquad g_R = \sin^2 \theta_W. \tag{5.2.39}$$

Adding the charged-current diagrams for the electron-neutrino case, we have

$$\nu_e: \begin{cases} g_L = -\tfrac{1}{2} + \sin^2 \theta_W + 1 = \tfrac{1}{2} + \sin^2 \theta_W, \\ g_R = \sin^2 \theta_W + 0 = \sin^2 \theta_W, \end{cases} \tag{5.2.40a}$$

$$\bar{\nu}_e: \text{ as above with } g_L \leftrightarrow g_R. \tag{5.2.40b}$$

The value deduced from the experimental comparison of these cross-sections in the case of muon neutrinos is $\sin^2 \theta_W = 0.2324 \pm 0.0083$ (Vilain *et al* 1994), in good agreement with the measured M_W / M_Z ratio. Note, as always, that due account must be made for important quantum corrections.

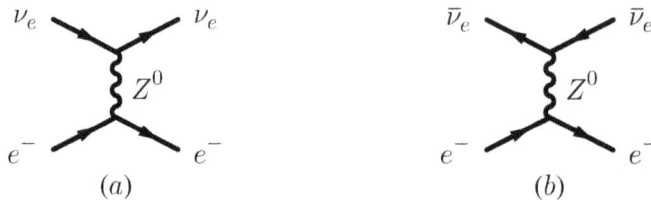

Figure 5.14. The neutral-current interactions between electrons and (a) electron neutrinos and (b) electron antineutrinos.

5.2.4.4 Polarisation asymmetries

A further independent method for extracting $\sin \theta_W$ is provided by polarisation asymmetries measured inelastic electron–nucleon scattering. The first such experiments were performed at SLAC in 1978 using a polarised electron beam of 19–22 GeV, provided by the linear accelerator (Prescott *et al* 1978). The process studied was electron–deuteron DIS, with an unpolarised deuteron target:

$$e^-_{L,\,R} + d \rightarrow e^- + X. \tag{5.2.41}$$

In addition to the dominant QED photon-exchange diagram, there is also a weak neutral-current contribution coming from Z^0 exchange. Interference between the two allows a measurable parity-violating asymmetry

$$\mathcal{A} := \frac{\sigma_R - \sigma_L}{\sigma_R + \sigma_L}. \tag{5.2.42}$$

The two contributing amplitudes are

$$\mathcal{M}_{\mathrm{EM}} \sim \frac{e^2}{q^2} \quad \text{and} \quad \mathcal{M}_{\mathrm{W}} \sim G_{\mathrm{F}}. \tag{5.2.43}$$

The electromagnetic contribution dominates the denominator of the asymmetry, whereas in the numerator it cancels between σ_R and σ_L, leaving the interference term to dominate. The resulting asymmetry may thus be estimated as

$$\mathcal{A} \sim \frac{2\,G_{\mathrm{F}}\,q^2}{e^2} \sim \frac{2\left(10^{-5}/m_p^2\right)q^2}{4\pi/137} \tag{5.2.44}$$
$$\sim 10^{-4}\,q^2 \qquad (\text{for } q^2 \text{ in GeV}^2).$$

More precisely, as a function of $y = E'_e/E_e$ (laboratory energies), we find

$$\mathcal{A} = -\frac{9G_{\mathrm{F}}q^2}{20\sqrt{2}\,\pi\alpha}\left[c_1 + c_2\,\frac{1 - (1 - y)^2}{1 + (1 - y)^2}\right], \tag{5.2.45}$$

where

$$c_1 = 1 - \frac{20}{9}\sin^2\theta_W \quad \text{and} \quad c_1 = 1 - 4\sin^2\theta_W. \tag{5.2.46}$$

The value obtained was $\sin^2\theta_W = 0.22 \pm 0.02$, again, in good agreement with other determinations.

5.2.5 The Higgs boson

'It shouldn't be a Higgs field. If it's anybody's, it should be Goldstone field, I think. When Nambu wrote his short paper in 1960, Jeffrey Goldstone of Cambridge University, who was visiting CERN, heard about it. He then

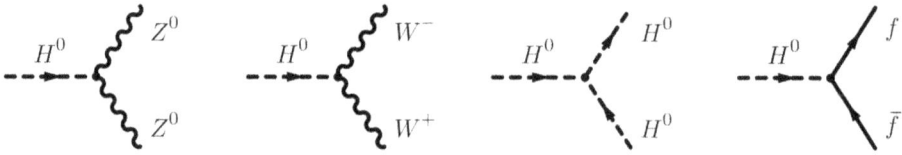

Figure 5.15. Examples of the Higgs trilinear couplings to gauge and matter fields in the standard electroweak theory (f stands for any massive fermion).

> *wrote a paper which was conceptually similar to what Nambu had done,*
> *but a simpler model.'*
>
> Peter Higgs—BBC interview, 17/04/2013

The final *prediction* of the Glashow–Salam–Weinberg model that we shall examine is the existence of the Higgs boson. The Higgs mechanism, as applied to the construction of the Glashow–Salam–Weinberg electroweak model, unavoidably leads to a neutral, massive, scalar boson H^0. The model also determines very precisely the form of the Higgs-particle interactions with the other fields in the theory, generating a number of trilinear and quadrilinear couplings, such as those in figure 5.15. It therefore couples to *all* other fields in the theory and to itself; it will even couple (albeit very weakly) to any massive neutrino.

5.2.6 Prior to discovery

This was the last particle in the SM to be experimentally detected and the demonstration of its existence represented a major quest. The problem was approached in two mutually independent ways:

- its direct production and detection through the final decay-product system configuration for specific decay channels,
- its indirect contribution in quantum corrections to the various accurately measured electroweak processes and parameters via combined fits.

First of all, the direct *non*-detection of the process $e^+e^- \to Z^0H^0$ at LEP already placed a lower limit on the mass (Beringer *et al* 2012):

$$m_H > 114.4 \text{ GeV} \qquad (95\%\text{CL}). \qquad (5.2.47)$$

On the other hand, indirect evidence from consideration of electroweak quantum corrections actually placed both upper and lower limits:

$$54 \text{ GeV} < m_H < 219 \text{ GeV} \qquad (95\%\text{CL}), \qquad (5.2.48)$$

with a central value of around 100 GeV, which was thus already effectively excluded by direct searches.

Direct searches at LEP were performed by checking the following two possible production processes:

$$e^+e^- \to Z^0 \to \begin{cases} H^0 + \ell^+ + \ell^-, \\ H^0 + \nu_\ell + \bar{\nu}_\ell, \end{cases} \qquad (5.2.49)$$

where $\ell = e$, μ or τ, as shown in figure 5.16. The branching ratio for such a channel in Z^0 decay may be calculated:

$$10^{-4} \gtrsim \frac{\Gamma(Z^0 \to H^0\ell^+\ell^-)}{\Gamma_{\text{tot}}^{Z^0}} \gtrsim 3 \times 10^{-6} \text{ for } 10\,\text{GeV} \lesssim m_{\text{H}} \lesssim 50\,\text{GeV} \quad (5.2.50)$$

and $B_\nu \simeq 2B_\ell$. The LEP I data thus led to a lower limit of about 60 GeV. LEP II raised the centre-of-mass energy to 200 GeV and was therefore able to search for the direct channel $e^+e^- \to Z^0 \to Z^0H^0$. In this way a lower limit of about 114 GeV was finally obtained.

LHC is, however, a proton–proton machine and although the colliding quarks evidently do not have all of the laboratory 14 TeV available, given the mass window already suggested, it would have been possible to produce the Higgs particle directly via, for example, the process depicted in figure 5.17 ($pp \to H^0 + X$). This is a good channel for LHC, in which both beams are protons, as both the u and d are then valence quarks. Naturally, the Higgs boson must be detected through its decay products. If, as appeared to be evermore likely, $m_{\text{H}} \geqslant 2m_{Z^0}$ then the two following decay channels would have become possible:

$$H^0 \to Z^0 + Z^0 \text{ and } W^+ + W^-. \quad (5.2.51)$$

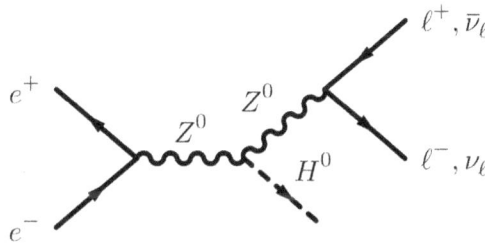

Figure 5.16. One of the simplest and most sensitive processes for Higgs-particle production and detection at LEP ($\ell = e$, μ or τ).

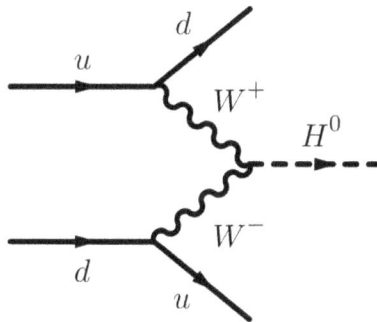

Figure 5.17. Production of the Higgs boson in proton–proton collisions via the so-called W^+W^- fusion process.

The two weak bosons must be detected via their decay products; in the case of W^+W^- and subsequent leptonic decays, two of the final-state particles would be neutrinos and escape detection. The simplest and cleanest signal for Higgs-boson production (in this mass range) thus involves a final state of four charged leptons:

$$p + p \rightarrow H^0 + X$$
$$ \hookrightarrow Z^0 + Z^0 \rightarrow \ell^+ + \ell^- + \ell^+ + \ell^-. \tag{5.2.52}$$

The final state may be one of $\mu^+\mu^-\mu^+\mu^-$, $\mu^+\mu^-e^+e^-$ or $\mu^+\mu^-\mu^+\mu^-$. The so-called *gold-plated* channel is the four-muon final state. Such a process allows for Higgs-particle searches in the range $200\,\text{GeV} \lesssim m_H \lesssim 500\,\text{GeV}$. Of course, the mass may be measured by reconstructing the invariant mass of the four final-state muons, for which we expect to see a classic Breit–Wigner distribution. The branching ratio here is

$$\frac{\Gamma(H^0 \rightarrow 4\ell)}{\Gamma(H^0 \rightarrow 2Z^0)} \simeq 4\%. \tag{5.2.53}$$

Note that for m_H very large, the width becomes comparable to the mass and the Breit–Wigner resonance shape is lost with the signal merging invisibly into the standard continuum background.

Should the Higgs boson have been much lighter than the threshold for double weak-boson production, the search would have become, perhaps surprisingly, rather more difficult. This is because the signal would have had be extracted from processes with hadronic final states, for example $H^0 \rightarrow b\bar{b}$. In this case the b quarks give rise to a pair of jets and such signals risk being swamped by a dominating standard QCD jet-production background. A possible alternative in this case is the rare but very distinctive $H^0 \rightarrow \gamma\gamma$ channel depicted in figure 5.18. In the SM the branching ratio for this decay is about 10^{-3} and it is dominated by the two processes shown in figure 5.18. The two diagrams represent virtual-particle loops, in which the only requirement is a charge for the circulating field. Thus, both quarks and leptons may contribute, together with the charged weak bosons $W^\pm$.

5.2.7 Discovery

Early in 2012 both ATLAS (Aad *et al* 2012) and CMS (Chatrchyan *et al* 2012) published papers on the search for the Higgs boson at LHC. The results indicated an excess of events (with respect to background) in the region of $m_H \simeq 125\,\text{GeV}$, at a combined level of 4.3σ. At around the same time the two Tevatron experiments

Figure 5.18. The probably dominant contributions to Higgs-boson decay into a photon pair; f may be any charged fermion (since it must couple to the photon).

CDF and D0 (Aaltonen *et al* 2012) also announced results in agreement with the CERN data although statistically less significant.

The channels in which evidence for Higgs-boson production was eventually found are

$$Z^0 + Z^{0*} \to \ell^+ + \ell^- + \ell^+ + \ell^-,$$
$$W^\pm + W^{\mp*} \to e + \nu + \mu + \nu,$$
$$\gamma + \gamma, \tag{5.2.54}$$
$$b + \bar{b},$$
$$\tau^+ + \tau^-.$$

The channels involving $W^\pm$ and Z^0 bosons both require that at least one of the two intermediate particles be off-shell (indicated * above) as the Higgs boson turns out to be too light to decay into a pair of on-shell bosons. The world average experimental values for the Higgs mass shortly after its discovery was (Olive *et al* 2014):

$$m_{\text{H}} = 125.7 \pm 0.4\,\text{GeV}. \tag{5.2.55}$$

Note that for a 125 GeV mass, the SM prediction for the width is (Dittmaier *et al* 2011)

$$\Gamma_{\text{H}}^{\text{SM}} = 4.07 \pm 0.16\,\text{MeV}, \tag{5.2.56}$$

which is nearly three orders of magnitude smaller than the LHC experimental mass resolution, far too small to be measured directly by the LHC experiments. For a review of Higgs-particle searches, see e.g. Bernardi and Herndon (2014).

5.2.8 Precision LHC measurements

Since its discovery in 2012 and up to the present date (mid-2024) the two major LHC experiments with a focus on studying the Higgs-particle properties (ATLAS and CMS) have produced, respectively, about 9 and 8 million Higgs particles. These correspond to integrated luminosities per experiment of approximately 5 fb^{-1} at 7 TeV, 20 fb^{-1} at 8 TeV, and up to 139 fb^{-1} at 13 TeV. However, only a very small fraction (much less than a percent) of the bosons produced were experimentally identifiable, but that still led to many thousands of recognisable Higgs events in their various forms.

The two primary channels for Higgs-boson study are $H \to \gamma\gamma$ (see figure 5.18) and $H \to Z^0 Z^0 \to 4\ell$ (see equation (5.2.52)). That two so very different production mechanisms should provide perfectly coinciding measurements is a clear indication of the validity of the theory. The large data sample provides many tests of the theory and also measurements of *SM* parameters; we shall now examine a selection. A useful compendium of theoretical tools may be found in Dittmaier *et al* (2011).

5.2.8.1 The Higgs boson mass
ATLAS and CMS have published combined results on their Higgs-mass measurements (Aad *et al* 2015b):

$$m_H = 125.09 \pm 0.21 \text{ (stat)} \pm 0.11 \text{ (syst) GeV.} \tag{5.2.57}$$

This result is an average of the two experiments and of the two channels that provide the best resolution: $H \to \gamma\gamma$ and $H \to Z^0 Z^0 \to 4\ell$, where the 4ℓ may be any of the possible combinations of $\mu^+\mu^-$ and e^+e^- pairs. The combined results for m_H in the two separate channels are

$$m_H^{\gamma\gamma} = 125.07 \pm 0.25 \text{ (stat)} \pm 0.14 \text{ (syst) GeV}$$

and $\tag{5.2.58}$

$$m_H^{4\ell} = 125.15 \pm 0.37 \text{ (stat)} \pm 0.15 \text{ (syst) GeV.}$$

As already noted, the excellent agreement between these two unrelated channels (the first proceeding via a charged-particle loop and the second via intermediate neutral weak bosons) is strong evidence that this is indeed the Higgs field of the SM.

There is a small *caveat* though to these measurements: the mass is always taken as the Breit–Wigner peak value of the centre-of-mass energy. As noted in the discussions of the Breit–Wigner formalism (see appendix C.6.3), care must be taken with regard to interference effects. Although the Higgs peak itself is very narrow, interference with any non-resonant background may cause a shift in the peak position (Dixon and Siu 2003). The effect in the $\gamma\gamma$ channel is predicted to be only a few tens of MeV for a Higgs boson with a width around the SM in the 4ℓ channel it is expected to be much smaller and so, for the time being it may be neglected. In any case, the most up-to-date value for the mass is (Navas *et al* 2024)

$$m_H = 125.20 \pm 0.11 \text{ GeV.} \tag{5.2.59}$$

Note that the error here includes a scale factor of 1.4 as per the PDG standard practice in the case of discrepant data sets.

5.2.8.2 The Higgs boson width

As just mentioned, the predicted width is much less that the current LHC energy resolution. And so other means must be found to measure it (Sirunyan *et al* 2019). The technique adopted exploits the 4ℓ channel and compares so-called '*off-shell*' and '*on-shell*' production, where for on-shell the final-state invariant-mass is restricted to the region $105 \text{ GeV} < m_{4\ell} < 140 \text{ GeV}$, the off-shell region being $220 \text{ GeV} < m_{4\ell}$. Since the on-shell cross-section should be essentially independent of the width, but far off-shell it will be proportional to Γ, with some assumptions as to the general behaviour of the cross-sections and interference between the various mechanisms, an extraction of the width may be accomplished (Caola and Melnikov 2013). The combined LHC analysis leads to (Sirunyan *et al* 2019)

$$\Gamma_H = 2.8 \, ^{+2.2}_{-1.7} \text{ MeV,} \tag{5.2.60}$$

which is compatible with the SM prediction.

As far as individual branching fractions are concerned, it is not yet possible to measure absolute values and only upper limits are deduced. However, ratios of production strengths with respect to to SM predictions are derived.

Table 5.3. The ratios of observed rate to predicted SM event rate for different of Higgs-boson final states; from Navas *et al* (2024).

Final state	Ratio to SM
Average	1.13 ±0.05
WW^*	1.19 ±0.12
ZZ^*	1.01 ±0.07
$\gamma\gamma$	1.10 ±0.07
$b\bar{b}$	0.98 ±0.12
$\mu^+\mu^-$	1.19 ±0.34

The Higgs signal strength in any given final state X is given by the cross-section multiplied by the branching ratio for the channel normalised to the SM value: $\sigma \cdot \mathcal{B}(H \to X)/\sigma \cdot \mathcal{B}(H \to X)^{\text{SM}}$, calculated for the measured Higgs mass. For completeness, in table 5.3 we list just a few of the better known values. All are in good agreement with the predictions. In particular, the agreement with couplings of the W and Z shows that the Higgs boson does indeed couple to the mass. In the SM the interaction Lagrangian for the W and Z and Higgs takes the form

$$\mathcal{L}_{\text{int}} = \frac{2m_W^2}{v}HW_\mu^+W^{-\mu} + \frac{m_Z^2}{v}HZ_\mu Z^\mu, \qquad (5.2.61)$$

where $v = 246$ GeV is the Higgs vacuum expectation value.

5.2.8.3 The Higgs boson spin and parity

The theorem (Landau 1948, Yang 1950) states that a massive on-shell spin-1 particle cannot decay into a pair of identical massless spin-1 particles. Therefore, assuming that the $\gamma\gamma$ signal truly represents a pair of photons and that neither is e.g. a collinear pair, the observation of this final state rules out the spin-1 possibility.

Study of the angular distributions of leptons in the ZZ and WW channels then leads to a positive parity assignment and the exclusion of various models, including the possibilities of $J^P = 1^\pm$, which are excluded at confidence levels >97.8%. The specific $J^P = 2^+$ case is excluded at a confidence level >99.9% (Aad *et al* 2013), as is the pseudoscalar $J^P = 0^-$ hypothesis (Chatrchyan *et al* 2013); i.e. by far the most probable assignment is $J^P = 0^+$ (Aad *et al* 2015a, Khachatryan *et al* 2015).

5.2.8.4 Other Higgs properties

The electroweak theory outlined leads to the various couplings displayed in figure 5.15; in particular it predicts the trilinear self-interaction show in the third diagram there, a quadrilinear self-interaction is also predicted. The relative coupling constants are also precisely predicted; the relevant part of the interaction Lagrangian is

$$\mathcal{L}_{\text{int}} = \frac{3m_H^2}{v}H^3 + \frac{3m_H^2}{v^2}H^4. \qquad (5.2.62)$$

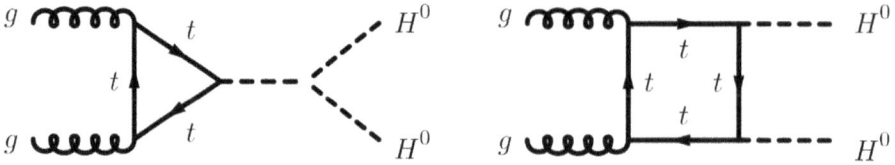

Figure 5.19. Two example contributions to double-Higgs production.

An important test of the theory will be the measurement of double-Higgs (or diHiggs) production. Unfortunately, there are various sources of double-Higgs production, many of which do not involve the trilinear self-coupling. In figure 5.19 we display two example contributions to double-Higgs production. Both involve gluon fusion and proceed via a top-quark loop (the strongest coupling to the Higgs boson); only the first though involves the trilinear Higgs self-coupling. So far only experimental limits have been published (Aad *et al* 2023), but these are compatible with the SM.

5.3 The CKM matrix

Perhaps one of the richest areas to emerge in hadronic physics in the 1990s is that of the CKM matrix. Recall that this matrix describes the unitary transformation between the asymptotic quark (or mass) eigenstates and those of the weak interaction. Since we do not as yet have a more complete theory, the CKM matrix parametrises our ignorance of any possible link between the electroweak theory and the physics of the strong interaction (or QCD). As such, it is presently seen by many as an important window onto possible physics beyond the SM. First of all, it is the best candidate we have for the origin of CP violation, which should then be described by just a single imaginary phase appearing in various transition matrix elements. Secondly, and perhaps more importantly at this point, the property (or requirement) of unitarity can be tested experimentally.

We should note that the existence of a non-trivial (i.e. non-diagonal) mixing matrix also requires that the quarks all have different masses (at least separately within the two classes of up and down types). While this is already experimentally verified, it also indicates that any measured CP-violating effects will be proportional to the mass(-squared) differences.

Before proceeding let us briefly examine the present status of the experimental determination of the CKM matrix. We should first define its elements:

$$V_{\text{CKM}} = \begin{pmatrix} V_{ud} & V_{us} & V_{ub} \\ V_{cd} & V_{cs} & V_{cb} \\ V_{td} & V_{ts} & V_{tb} \end{pmatrix}. \tag{5.3.1}$$

As we have shown, in its most general form, this matrix may be parametrised by three Euler angles and just one imaginary phase. A standard choice of representation is the following (Chau and Keung 1984)[9]:

---

[9] The original parametrisation of Kobayashi and Maskawa was arranged somewhat differently.

$$V_{CKM} = \begin{pmatrix} 1 & 0 & 0 \\ 0 & c_{23} & s_{23} \\ 0 & -s_{23} & c_{23} \end{pmatrix} \begin{pmatrix} c_{13} & 0 & s_{13}e^{-i\delta_{CP}} \\ 0 & 1 & 0 \\ -s_{13}e^{i\delta_{CP}} & 0 & c_{13} \end{pmatrix} \begin{pmatrix} c_{12} & s_{12} & 0 \\ -s_{12} & c_{12} & 0 \\ 0 & 0 & 1 \end{pmatrix}$$

$$= \begin{pmatrix} c_{12}c_{13} & s_{12}c_{13} & s_{13}e^{-i\delta_{CP}} \\ -s_{12}c_{23} - c_{12}s_{23}s_{13}e^{i\delta_{CP}} & c_{12}c_{23} - s_{12}s_{23}s_{13}e^{i\delta_{CP}} & s_{23}c_{13} \\ s_{12}s_{23} - c_{12}c_{23}s_{13}e^{i\delta_{CP}} & -c_{12}s_{23} - s_{12}c_{23}s_{13}e^{i\delta_{CP}} & c_{23}c_{13} \end{pmatrix}, \tag{5.3.2}$$

where $s_{ij} = \sin\theta_{ij}$ and $c_{ij} = \cos\theta_{ij}$, while δ_{CP} is the single allowed phase, which may then be responsible for CP violation. Note also that, by suitable global rotation of both quark bases, all the angles may be taken to lie in the range $[0, \pi/2]$ so that all sines and cosines are non-negative. For antiquarks $\delta_{CP} \to -\delta_{CP}$.

If we now restrict consideration to the first two families (i.e. the first two rows and columns), then we just have the Cabibbo matrix, with θ_{12} being the Cabibbo mixing angle (we should set $\theta_{13} = 0 = \theta_{23}$ and thus the phase terms disappear). Experimentally, we have already noted that this angle is small; that is, the diagonal elements are near to unity and are much larger than those off-diagonal. Extending the discussion to the 3×3 case, this hierarchy continues and we find that the elements furthest from the diagonal are smallest; in other words,

$$\sin\theta_{12} \gg \sin\theta_{23} \gg \sin\theta_{13}. \tag{5.3.3}$$

Indeed, if we measure the scale of smallness of the near off-diagonal terms via a parameter λ, then those further off-diagonal are order λ^3. This observation leads to an alternative parametrisation due to Wolfenstein (1983):

$$V_{CKM} = \begin{pmatrix} 1 - \lambda^2/2 & \lambda & A\lambda^3(\rho - i\eta) \\ -\lambda & 1 - \lambda^2/2 & A\lambda^2 \\ A\lambda^3(1 - \rho - i\eta) & -A\lambda^2 & 1 \end{pmatrix} + O(\lambda^4). \tag{5.3.4}$$

The parameter λ is then essentially $\sin\theta_C$ and A is of order unity; see Navas *et al* (2024) for the precise definitions and relations. The current best fits to the world experimental data for the moduli of the single elements separately give (Navas *et al* 2024)

$$V_{CKM} = \begin{pmatrix} 0.973\,67 \pm 0.000\,32 & 0.224\,31 \pm 0.000\,85 & 0.003\,82 \pm 0.000\,20 \\ 0.221 \pm 0.004 & 0.975 \pm 0.006 & 0.0422 \pm 0.0005 \\ 0.0086 \pm 0.0002 & 0.0415 \pm 0.0009 & 1.010 \pm 0.027 \end{pmatrix}. \tag{5.3.5}$$

There is clearly too little information as yet to provide a precise value directly for the last element (V_{tb}). However, imposing unitarity leads to an overall more precise determination of the entire matrix (CKMfitter group):

$$V_{CKM} = \begin{pmatrix} 0.974\,358^{+0.000\,049}_{-0.000\,054} & 0.224\,98^{+0.000\,23}_{-0.000\,22} & 0.003\,730^{+0.000\,044}_{-0.000\,048} \\ 0.224\,84^{+0.000\,23}_{-0.000\,21} & 0.973\,509^{+0.000\,054}_{-0.000\,059} & 0.041\,60^{+0.000\,20}_{-0.000\,58} \\ 0.008\,573^{+0.000\,046}_{-0.000\,158} & 0.040\,88^{+0.000\,20}_{-0.000\,66} & 0.999\,1248^{+0.000\,0268}_{-0.000\,0074} \end{pmatrix}. \tag{5.3.6}$$

Note, in particular, that $|V_{tb}|$ is thus constrained to be extremely close to unity.

5.3.1.1 The unitarity triangles

Let us begin with the question of unitarity; the relevant equation is simply:

$$V_{CKM}^{\dagger} V_{CKM} = 1. \tag{5.3.7}$$

Since in the SM this is a 3×3 matrix equation, it represents nine equations or constraints. The three '*diagonal*' equations each have a left-hand side involving the square of one large component, which dominates the sum and the comparison with the large right-hand side. They are therefore rather difficult to test experimentally. Indeed, although the present experimental data indicate a minor discrepancy, nothing serious can yet be inferred: from Navas *et al* (2024), we have (Chakraborty *et al* 2024)

$$
\begin{aligned}
|V_{ud}|^2 + |V_{us}|^2 + |V_{ub}|^2 &= 0.9984(7), \\
|V_{cd}|^2 + |V_{cs}|^2 + |V_{cb}|^2 &= 1.001(12), \\
|V_{ud}|^2 + |V_{cd}|^2 + |V_{td}|^2 &= 0.9971(20), \\
|V_{us}|^2 + |V_{cs}|^2 + |V_{ts}|^2 &= 1.003(12).
\end{aligned}
\tag{5.3.8}
$$

In contrast, the six '*off-diagonal*' equations dilute the dominance of the large diagonal components and are therefore less critical. It can thus even be hoped, for example, that possible physics *beyond* the SM might be made manifest via the *non-vanishing* of these sums of products. As the off-diagonal equations each contain three terms, they may be expressed as triangles in the complex plane, each term representing one side. Let us recall first the Wolfenstein parametrisation (5.3.4), which, to highlight the relative order of magnitude of the elements, may be schematically simplified as

$$V_{CKM} \approx O \begin{pmatrix} 1 & \lambda & \lambda^3 \\ \lambda & 1 & \lambda^2 \\ \lambda^3 & \lambda^2 & 1 \end{pmatrix}, \tag{5.3.9}$$

where the scale parameter $\lambda \approx O(0.2)$. We immediately see that, should the hierarchy apparent in (5.3.9) continue, any eventual fourth generation would have only a very small (if not negligible) impact on unitarity tests.

Let us first examine the 'up–up-type' products. The 'u–c' product is thus seen to contain two terms $O(\lambda)$ and one $O(\lambda^5)$, the 'c–t' product contains two terms $O(\lambda^2)$ and one $O(\lambda^4)$, while the 'u–t' product has all three terms $O(\lambda^3)$. The first does not depend on the top quark, but contains widely mixed orders of λ and will thus have poor sensitivity to possible unitarity violation. The second both depends on the top quark and contains mixed orders of λ and is thus again of little use. In the last, however, although the three terms are comparable, they *all* involve a t-quark element and are thus difficult to measure with precision.

The 'down–down-type' products are thus seen to be the best suited for unitarity tests. The most commonly used triangle is that generated by the 'd–b' product:

$$V_{ud} V_{ub}^* + V_{cd} V_{cb}^* + V_{td} V_{tb}^* = 0, \tag{5.3.10}$$

where again all three terms are $O(\lambda^3)$, but only the last involves t-quarks elements. This product then turns out to be the most sensitive to CP violation. Since the order of magnitude of the terms is similar, the best determined, $V_{cb}\,V_{cd}^*$, may be taken as a reference length and used to rescale all three: thus,

$$\frac{V_{ud}\,V_{ub}^*}{V_{cd}\,V_{cb}^*} + 1 + \frac{V_{td}\,V_{tb}^*}{V_{cd}\,V_{cb}^*} = 0. \tag{5.3.11}$$

The middle term above (now unity) may then be naturally placed along the positive real axis, running from the origin to the point $(1, 0)$. Considering the other two as complex numbers, the sum above then represents a triangle in the complex plane. This leads to the geometric representation shown in figure 5.20. The new variables $\bar{\rho}$ and $\bar{\eta}$ are equivalent to ρ and η up to corrections of order λ^2, which are thus order λ^4 corrections to the matrix elements themselves. The angles of the triangle are just the phases of the various ratios of elements:

$$\phi_1 = \beta = \arg\left(\frac{V_{cd}\,V_{cb}^*}{V_{td}\,V_{tb}^*}\right), \tag{5.3.12a}$$

$$\phi_2 = \alpha = \arg\left(\frac{V_{td}\,V_{tb}^*}{V_{ud}\,V_{ub}^*}\right), \tag{5.3.12b}$$

$$\phi_3 = \gamma = \arg\left(\frac{V_{ud}\,V_{ub}^*}{V_{cd}\,V_{cb}^*}\right). \tag{5.3.12c}$$

First and foremost, it should be experimentally verified whether the three terms do indeed form a non-trivial triangle (i.e. they do not collapse trivially to a single line). This would then demonstrate that the CP-violating phase is indeed non-zero. Secondly, we should note that the three angles are given by different combinations of matrix elements and are therefore experimentally independent. Unitarity of the matrix evidently requires that the sum of the angles be exactly $180°$ and that the two upper sides end at the same point; if either condition were found not to hold, then we

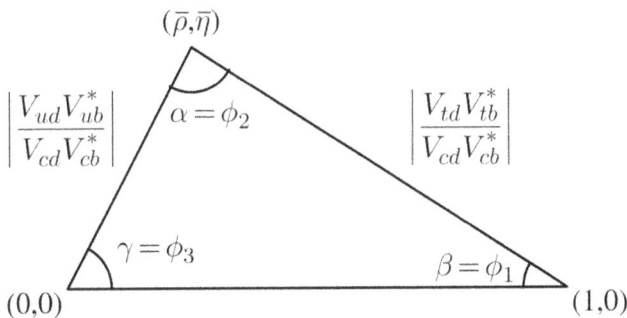

Figure 5.20. A schematic representation of the 'd–b' CKM-matrix unitarity triangle.

should have a signal for physics beyond the SM. The most obvious case would be a fourth family of fermions.

A further observation is that the six possible triangles so formed all have the same area. An obvious necessary condition for the CKM matrix to generate CP violation is that the area should be non-zero. Now, it is a fairly simple exercise in geometry to show that it is given by half the invariant J, derived by Jarlskog (1985):

$$J = \Im m \left(V_{ij} \, V_{kl} \, V_{il}^* \, V_{kj}^* \right) / \sum_{mn} \varepsilon_{ikm} \varepsilon_{jln}. \tag{5.3.13}$$

A more specific example expression is

$$J = \Im m \left(V_{us} \, V_{cs}^* \, V_{cb} \, V_{ub}^* \right). \tag{5.3.14}$$

Note that the definition is entirely phase-convention independent and J is thus indeed an invariant. In terms of the general parametrisations of the CKM matrix given earlier, we have

$$J = c_{12} \, c_{23} \, c_{13}^2 \, s_{12} \, s_{23} \, s_{13} \sin \delta_{\text{CP}} \simeq A^2 \, \lambda^6 \, \eta. \tag{5.3.15}$$

The present value (from global fits) for this parameter is (CKMfitter group)

$$J = 3.115^{+0.047}_{-0.059} \times 10^{-5}. \tag{5.3.16}$$

The fact that the Jarlskog invariant involves elements from *all three* families is intimately related to the observation that for less than three families there can be no CP violation at the level of the CKM matrix. In other words, if an experimental (physical) quantity depends on elements involving less than three families then, by means of a suitable, unitary transformation of the matrix itself, any CP-violating phase can be rotated away. Conversely, this also implies that for an experimentally measurable quantity to be sensitive to CP violation all three families must contribute to the process. For example, in the case of the $K^0 - \bar{K}^0$ system a three-family GIM-like mechanism would actually cause all such effects to cancel and it is the large mass of the t quark that partially deactivates the cancellation.

5.3.1.2 Measurement of the CKM-matrix elements

We shall now briefly outline how the magnitudes of at least some of the CKM-matrix elements are determined experimentally. As usual, see Navas *et al* (2024) for up-to-date experimental values and technical discussions and for the most recent global extractions, see the (CKMfitter group).

$|V_{ud}|$: The most precise determinations of V_{ud} are provided by the so-called super-allowed $0^+ \rightarrow 0^+$ nuclear beta decays (for a more complete review, see Hardy and Towner 2009). Recall that, whereas (vector) Fermi transitions involve an $e\nu_e$ pair in a spin-zero state, (axial-vector) Gamow–Teller transitions produce a spin-one pair and cannot therefore contribute to the superallowed decays considered, which are thus purely vector transitions. Combining the fifteen most precise determinations leads to (Navas *et al* 2024)

$$|V_{ud}| = 0.973\ 67 \pm 0.000\ 32. \tag{5.3.17}$$

The error is dominated by theoretical uncertainties in the nuclear corrections.

The neutron lifetime measurement also affords a precise determination of $|V_{ud}|$. The theoretical uncertainties are very small here too but the extraction of the CKM element requires precise knowledge of the ratio between the axial-vector and vector couplings (g_A/g_V), which is measured to comparable precision via the decay angular distributions. Finally, the theoretically very clean charged-pion decay $\pi^+ \rightarrow \pi^0 e^+ \nu$ may also be used; however, present experimental precision is not yet competitive.

$|V_{us}|$: The magnitude of V_{us} is typically extracted from either semileptonic kaon decays or strangeness-changing semileptonic hyperon decays. Considerable experimental effort has been made in recent years with regard to the former. High-statistics measurement of $B(K^+ \rightarrow \pi^0 e^+ \nu)$ and a number of measurements of neutral-kaon branching ratios, form factors, and lifetime have been performed. Form-factor input is also required: the presently rather precise theoretical value $f^+(0) = 0.9698 \pm 0.0017$ is now available. The kaon semileptonic decay rates then lead to

$$|V_{us}| = 0.2233 \pm 0.0005. \tag{5.3.18}$$

The lattice-QCD calculation of the ratio between the kaon and pion decay constants allows extraction of $|V_{us}/V_{ud}|$ from $K \rightarrow \mu\nu$ and $\pi \rightarrow \mu\nu$ (we recall the dominance of the μ over e mode). The KLOE (Frascati) measurement of the $K^\pm \rightarrow \mu^\pm \nu$ branching ratio (Ambrosino *et al* 2006), combined with the theoretical

$$f_K/f_\pi = 1.1932 \pm 0.0021, \tag{5.3.19}$$

leads to

$$|V_{us}| = 0.2250 \pm 0.0004, \tag{5.3.20}$$

where accuracy is limited by poor knowledge of the decay-constant ratio. The PDG 2024 quotes an average of the two above determinations:

$$|V_{us}| = 0.224\ 31 \pm 0.000\ 85. \tag{5.3.21}$$

The determination from hyperon decays has long lacked comparable theoretical understanding although in recent years it has received new input from both experiment and theory. In analogy with the strangeness-conserving decays, the vector form factor is protected against first-order SU(3)-breaking effects by the (Ademollo and Gatto, 1964). Therefore, the ratio between the axial-vector and vector form factors (often denoted in this context as g_1/f_1) may again be taken as experimental input, thereby circumventing the problem of accounting for SU(3)-breaking effects in the axial-vector contribution. The best present extraction is

$$|V_{us}| = 0.2250 \pm 0.0027, \tag{5.3.22}$$

which does not though include estimates of the theoretical uncertainty due to second-order SU(3) breaking.

Hadronic τ decays and leptonic kaon decays provide further determinations of $|V_{us}|$. The τ-decay branching ratio into strange particles gives $|V_{us}| = 0.2207 \pm 0.0014$ (Amhis *et al* 2023).

$|V_{cd}|$: Various measurements have now been performed around the world at high-statistics so-called '*charm factories*'; these are e^+e^- machines operating at an energy corresponding to the mass of the $\psi(3770)$, which decays predominantly into $D\bar{D}$ ('*beauty-factory*' experiments operating at the $\Upsilon(4S)$ mass also provide such data). The modern precise determinations thus come from semileptonic charm-meson β-like decays ($c \to d\ell\nu$), combined with lattice QCD calculations of the relevant form factors. The average of the measurements of $D \to \pi\ell\nu$ decays by BES III, CLEO-c, BaBar and Belle gives $|V_{cd}| = 0.2140 \pm 0.0029 \pm 0.0133$, where the first uncertainty is experimental, and the second theoretical.

The determination of $|V_{cd}|$ is also possible from the purely leptonic decays $D^\pm \to \mu^\pm\nu$. The quark subprocesses are $c\bar{d} \to W^+ \to \mu^+\nu$ and its charge conjugate. Averaging recent BES III precision measurements with an earlier CLEO measurement leads to $|V_{cd}| = 0.2173 \pm 0.0051 \pm 0.0007$.

Older determinations were based on neutrino and antineutrino DIS interactions with nucleons. The difference in the ratio of double- to single-muon production by neutrino and antineutrino beams is proportional to the charm production cross-section off valence d quarks, and therefore to $|V_{cd}|^2$ times the average semileptonic branching ratio of charm mesons, $\mathcal{B}_\mu$ (see figure 5.21 for the quark-model interpretation)[10]. In the muon-neutrino case, the most probable process involves the transition $d \to u$ since it is both Cabibbo favoured (i.e. proportional to $|V_{ud}|$) and proportional to the relatively large valence d-quarks density; this leads to a single muon in the final state. However, there is also the Cabibbo-suppressed transition $d \to c$ (i.e. proportional to $|V_{cd}|$); the final c-quarks decay then sometimes leads to a second muon. Note that the theoretical uncertainties in the d-quarks density cancel in the ratio. There is, of course, some background due to the presence of s in the proton, but these are sea quarks and therefore much less probable.

This method was used by various neutrino DIS experiments: in the 1980s CDHS followed in the 1990s by CCFR and CHARM II. The PDG 2024 quotes an average of these results as

Figure 5.21. The neutrino and antineutrino interactions with a nucleon leading to the production of single-muon (u and $\bar{u}$ final states) and double-muon (c and $\bar{c}$) events.

---

[10] Once again we see that helicity-flip amplitudes favour production of muons over electrons in D-meson decays.

$$\mathcal{B}_\mu |V_{cd}|^2 = (0.463 \pm 0.034) \times 10^{-2}. \tag{5.3.23}$$

Moreover, data from the more recent CHORUS experiment at CERN are now sufficiently precise to also extract $\mathcal{B}_\mu$ directly and combining their results with those of the other experiments leads to $\mathcal{B}_\mu = 0.087 \pm 0.005$ and $|V_{cd}| = 0.230 \pm 0.011$, finally giving an overall average

$$|V_{cd}| = 0.221 \pm 0.004. \tag{5.3.24}$$

$|V_{cs}|$: An analogous determination of $|V_{cs}|$ from neutrino and antineutrino scattering is much less precise than in the previous case, since it requires knowledge of the s-quarks density, which is relatively small and not so well measured. Other approaches must therefore be adopted.

The direct determination of $|V_{cs}|$ is possible from semileptonic D or leptonic D_s decays, but again substantial theoretical input from hadronic matrix-element calculations is required. The use of $D_s^+ \rightarrow \ell^+\nu$ requires a lattice-QCD calculation of the decay constant fD_s; while for semileptonic D decays, form factors are required, which depend on the lepton-pair invariant mass. Lattice-QCD calculations can predict the both normalisation and shape of the form factors in $D \rightarrow K\ell\nu$ and $D \rightarrow \pi\ell\nu$. These theoretical results and the isospin-averaged semileptonic widths or rates provide (Navas *et al* 2024)

$$|V_{cs}| = 0.975 \pm 0.006. \tag{5.3.25}$$

Real $W^\pm$ decays are also sensitive to $|V_{cs}|$ and such measurements were made at LEP-II, where, recall, real $W^\pm$ pairs were produced (a total of approximately 10 000 such pairs were recorded). In fact, the $W^\pm$ branching ratios depend on all six CKM matrix elements involving quarks lighter than M_W. For each lepton flavour we have

$$\mathcal{B}(W \rightarrow \ell\bar{\nu}_\ell) = \frac{1}{3}\left[1 + \left(1 + \frac{\alpha_s(m_W)}{\pi}\right)\sum_{u,c,d,s,b}|V_{ij}|^2\right]^{-1}, \tag{5.3.26}$$

where we have also included the first-order QCD correction. The factor $1/3$ here is for both the three lepton channels and the three quark colours. Assuming lepton universality, the theoretical result

$$\mathcal{B}(W \rightarrow \ell\bar{\nu}_\ell) = (10.83 \pm 0.07 \pm 0.07)\% \tag{5.3.27}$$

implies

$$\sum_{u,c,d;s,b}|V_{ij}|^2 = 2.002 \pm 0.027, \tag{5.3.28}$$

which is a precise test of unitarity. However, $|V_{cs}|$ can only be extracted directly from flavour-tagged measurements; i.e. by examining the strange-particle content of the reconstructed jets. The LEP experiment DELPHI measured tagged $W^+ \rightarrow c\bar{s}$ decays and obtained (Abreu *et al* 1998)

$$|V_{cs}| = 0.94^{+0.32}_{-0.26} \pm 0.13. \tag{5.3.29}$$

$|V_{cb}|$: Naturally, this element can be extracted from semileptonic B-meson decays to charm states. Inclusive determinations use the semileptonic $B \to D\ell\bar{\nu}$ decay-rate measurement combined with the leptonic-energy and the hadronic invariant-mass spectra. The basis of the calculation is the so-called heavy-quark expansion, via which the total rate and moments of differential energy and invariant-mass spectra are expressed as expansions in inverse powers of the heavy-quark masses. Since the dependence on m_b, m_c and other parameters occurring at subleading order is different for different moments of the lepton energy, the large number of measured moments overconstrains the parameters and tests the consistency of the approach. Inclusive measurements have been performed using B mesons from Z^0 decays at LEP and in other e^+e^- machines (the so-called B-factories) operating at the $\Upsilon(4S)$-mass energy, $\mathcal{B}(\Upsilon(4S) \to B\bar{B}) > 96\%$. At LEP the large boost of B mesons from Z^0 decays allows determination of the moments throughout phase space, which is not otherwise possible, but the high statistics provided by the B-factories leads to more precise determinations. Averaging over all the measurements leads to $|V_{cb}| = 0.0422 \pm 0.0005$.

Semileptonic B decays into D and D^* also permit exclusive determinations. In the limit $m_{b,c} \gg \Lambda_{\text{QCD}}$, using again the heavy-quark expansion, all form factors are furnished by a single so-called Isgur–Wise function (Isgur and Wise 1989), which is a function of the scalar product of the four-velocities, $w = v \cdot v'$, of the initial- and final-state hadrons. Heavy-quark symmetry determines the normalisation of the rate at $w = 1$, the maximum momentum transfer to the leptons, and $|V_{cb}|$ is obtained from an extrapolation to $w = 1$.

Combining the two above determinations and scaling the error by a factor $\sqrt{\chi^2} = 3$ to take into account the poor agreement between the two approaches[11], the PDG 2024 quotes

$$|V_{cb}| = 0.0411 \pm 0.0012. \tag{5.3.30}$$

$|V_{ub}|$: The natural determination of $|V_{ub}|$ from the very CKM-suppressed inclusive $B \to X_u\ell\bar{\nu}$ decays, where X_u represents any meson containing a u quarks, suffers (less CKM-suppressed) large $B \to X_c\ell\bar{\nu}$ backgrounds. In general, in phase-space regions where the charm background may be excluded for kinematic reasons there are unknown non-perturbative contributions: the so-called shape functions. In contrast, the non-perturbative physics for $|V_{cb}|$ is encoded in a few parameters. At leading order in Λ_{QCD}/m_b there is only one shape function, which may be extracted from the photon-energy spectrum in $B \to X_s\gamma$ and applied to spectra in $B \to X_u\ell\bar{\nu}$.

Alternatively, the measurements may be extended to the $B \to X_c\ell\bar{\nu}$ region to render them more inclusive and thus reduce the theoretical uncertainties. Analyses of the electron-energy endpoint from CLEO, BaBar and Belle quote $B \to X_u e\bar{\nu}$ partial rates for $|p_e| \geqslant 2.0$ GeV and 1.9 GeV, which are well below the charm endpoint. The large and pure B–$\bar{B}$ samples produced at B-factories permit the selection of $B \to X_u\ell\bar{\nu}$ decays in events where the recoiling $\bar{B}$ is fully reconstructed. Using such a full-reconstruction tag method, the four-momenta of both the leptonic and hadronic systems can be extracted.

---

[11] This is the standard PDG procedure for dealing with discrepant data.

Exclusive channels may also be used, but then form factors are needed. The better experimental signal-to-background ratios are offset by smaller yields. The $B \to \pi \ell \bar{\nu}$ branching ratio is now known to 5% precision and lattice-QCD calculations of the $B \to \pi \ell \bar{\nu}$ form factor for $q^2 > 16 \text{GeV}^2$ have been performed. So-called light-cone QCD sum rules are applicable for $q^2 < 14 \text{GeV}^2$ and yield somewhat smaller values for $|V_{ub}|$. The theoretical uncertainties in the extractions of $|V_{ub}|$ from inclusive and exclusive decays are somewhat different.

The two procedures (inclusive and exclusive) lead to (see Navas *et al* (2024) for an explanation of the errors)

$$|V_{ub}| = (4.13 \pm 0.12 \, {}^{+0.13}_{-0.14} \pm 0.18) \times 10^{-3} \qquad \text{(inclusive)}, \qquad (5.3.31a)$$

$$|V_{ub}| = (3.70 \pm 0.10 \pm 0.12) \times 10^{-3} \qquad \text{(exclusive)}. \qquad (5.3.31b)$$

The PDG 2024 thus quotes the following overall average value:

$$|V_{ub}| = (3.82 \pm 0.20) \times 10^{-3}, \qquad (5.3.32)$$

which is dominated by the exclusive measurement and includes a scaled error, multiplied by $\sqrt{\chi^2} = 1.4$.

$|V_{td}|, |V_{ts}|$ **and** $|V_{tb}|$**:** The CKM matrix elements involving the t quark are rather more difficult to access, first and foremost owing to the limited number of t quarks so-far produced (and detected) in laboratory experiments; indeed, it is unlikely that processes involving top quarks will lead to very precise measurements in the near future. However, the top quark does play an important rôle in the intermediate states in B–$\bar{B}$ oscillation phenomena and also in higher-order corrections (coming from penguin diagrams, see e.g. figure 5.9). Present experimental precision does now allow more-or-less meaningful measurements of these very small elements to be performed:

$$|V_{td}| = (8.6 \pm 0.2) \times 10^{-3} \quad \text{and} \quad |V_{ts}| = (41.5 \pm 0.9) \times 10^{-3}. \qquad (5.3.33)$$

On the other hand, the Tevatron experiments CDF and D0 have had access to the element $|V_{tb}|$ through top-quark pair-production and subsequent weak decay. The measurement of the branching ratio $\mathcal{B}(t \to Wb)$ places some limits on its value, though not stringent, especially in view of the expectation that it should in any case be very close to unity. They have also found evidence for single top-quark production. However, more recently, ATLAS and CMS have been able to measure single top-quark production cross-sections and have thus extracted $|V_{tb}|$ directly. The basic partonic subprocess studied is the annihilation of a quark and antiquark of different flavours via s-channel production of a virtual intermediate $W^{\pm}$, which subsequently decays into $\bar{b}t$ or $b\bar{t}$. Special techniques are necessary to extract the single top-quark signal from the large background. The combined Tevatron and LHC cross-section measurements can be used to provide a direct extraction of the CKM matrix element $|V_{tb}|$:

$$|V_{tb}| = 1.010 \pm 0.027 \qquad (5.3.34)$$

5.3.1.3 The CP-violating phase and unitarity-triangle angles

The angles of the unitarity triangle are evidently non-trivial, i.e. the triangle is not flat, *if and only if* CP violation has its origins in the CKM matrix itself. It is therefore evident that their measurement requires the study of CP-violating effects. Different processes provide more-or-less direct access to different angles and thus it is, in principle, possible to verify that the sum of the three angles is indeed 180°. The field is in continual evolution and here we shall limit ourselves to a presentation of the current picture. In figure 5.22 the combined world constraints on the unitarity triangle are displayed as 95% CL allowed regions in $\bar{\rho}$ and $\bar{\eta}$ plane. The global agreement is excellent. Indeed, we note that a slight discrepancy owing to earlier values of $|V_{ud}|$ has now disappeared.

The final global-fit values for the unitarity-constrained CKM matrix elements were already shown in equation (5.3.6), while, a constrained fit for the standard parametrisation using three Euler angles and the CP phase δ, gives (Navas *et al* 2024):

$$\sin\theta_{12} = 0.225\,01 \pm 0.000\,68, \qquad \sin\theta_{23} = 0.041\,83\,^{+0.000\,79}_{-0.000\,69},$$
$$\sin\theta_{13} = 0.003\,732\,^{+0.000\,090}_{-0.000\,085} \qquad \delta = 1.147 \pm 0.026, \tag{5.3.35}$$

where the angle hierarchy is manifest and the CP phase (in radians here) is seen to be large. In terms of the Wolfenstein parameters defined earlier, we have:

$$\lambda = 0.224\,98\,^{+0.000\,23}_{-0.000\,21}, \quad A = 0.8215\,^{+0.0047}_{-0.0082},$$
$$\bar{\rho} = 0.1562\,^{+0.0112}_{-0.0040}, \qquad \bar{\eta} = 0.3551\,^{+0.0051}_{-0.0057}. \tag{5.3.36}$$

Note, of course, that λ is none other than $\sin\theta_{12}$.

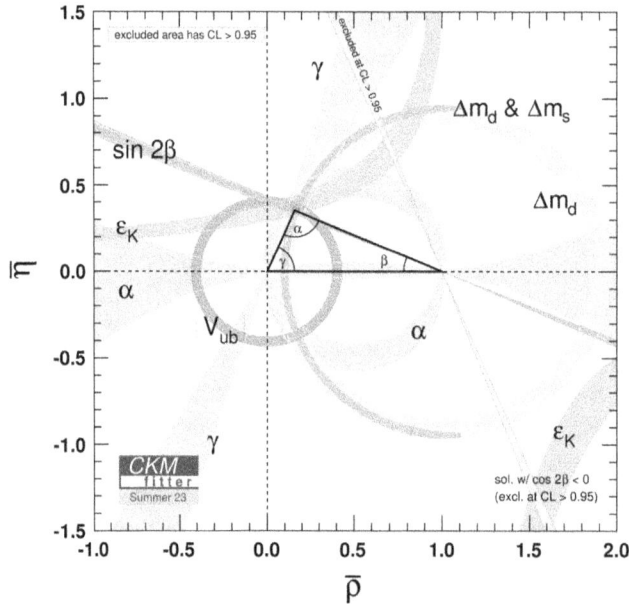

Figure 5.22. World constraints on the unitarity triangle represented in the $\bar{\rho}$–$\bar{\eta}$ plane. The shaded areas are 95%CL intervals for the various parameters and the small red ellipse around the upper vertex indicates the overall constraint on its position. Figure courtesy of CKM fitter group, CC By 4.0.

What should be stressed, however, is that while all measurable CP-violating effects are exceedingly small (recall e.g. that the $K_L^0 \to \pi^+\pi^-$ branching fraction is just 2%), the CP-violating phase parameter itself is found to be of order unity. The point is simply that measurable effects are either indirect (as in K_L^0 decay or in any case proportional to products of the sines of small mixing angles.

5.3.1.4 Measurement of CP violation

We have already examined CP violation in some detail and here we shall just give one more recent example: namely, that observed in the decays of neutral meson B mesons into charged pion pairs. Just as with the $K^0 - \bar{K}^0$ system the B^0 and $\bar{B}^0$ are distinct (particle and antiparticle) states, but which share common decay channels; in particular, the $\pi^+\pi^-$ channel. This implies the possibility of oscillation, which has already been observed by various experiments.

It also allows the study of CP violation by comparing the decay probabilities, which would be equal were CP respected. The so-called B-factories are e^+e^- machines running at a centre-of-mass energy corresponding to the mass of the $\Upsilon(4S)$, which decays nearly 50% into $B^0\bar{B}^0$. The two beauty mesons are produced back-to-back in the laboratory in a coherent state. If one decays into $\pi^+\pi^-$ ($\mathcal{B} \simeq 5 \times 10^{-6}$), then the other may be used as a tag by examining its decay products. Vertex reconstruction allows measurement of the time at which each of the two meson B mesons decayed and thus the time-dependent decay rate:

$$P(\Delta t, \eta) = \frac{e^{-\Delta t/\tau_B}}{4\tau_B} \left\{ 1 + q \left[\mathcal{A}_{CP} \cos \Delta m \Delta t + \mathcal{S}_{CP} \sin \Delta m \Delta t \right] \right\}, \qquad (5.3.37)$$

where Δt is the proper-time interval between the two decays, τ_B is the B^0 lifetime, Δm the mass difference between the two mass eigenstates (cf. K_1 and K_2 in the $K^0 - \bar{K}^0$ system) and the signature q is ± 1 for $B^0/\bar{B}^0$. As always, there is the possibility of both direct (determined by $\mathcal{A}_{CP}$) and indirect or mixing-induced (determined by $\mathcal{S}_{CP}$) CP violation.

The decay under consideration proceeds at the quark level via the decay $b \to u\bar{u}d$ and charge conjugate. It is thus sensitive to the CKM matrix angle $\alpha = \phi_2$ defined in equation (5.3.12). Recall that a non-zero values implies CP violation. The relevant tree-level and penguin-type diagrams are displayed in figure 5.23.

If we neglect the penguin contribution, then there can only be mixing-induced CP violation and $\mathcal{S}_{CP} = \sin 2\phi_2$. The penguin contribution will however shift the measurement by, say, $\Delta\phi_2$ and so now

$$\mathcal{S}_{CP} = \sqrt{1 - \mathcal{A}_{CP}^2} \sin(2\phi_2 + 2\Delta\phi_2). \qquad (5.3.38)$$

In order to disentangle the penguin and tree contributions, a simple isospin analysis is sufficient (Gronau and London 1990).

Consider the full set of $B \to \pi\pi$ amplitudes: $B^0 \to \pi^+\pi^-$, $\pi^0\pi^0$ and $B^+ \to \pi^+\pi^0$ (i.e. those for $\bar{b}$ quarks with either a u or d quark spectator). First of all, owing to Bose–Einstein statistics (section B.2) the above two-pion final states can only have

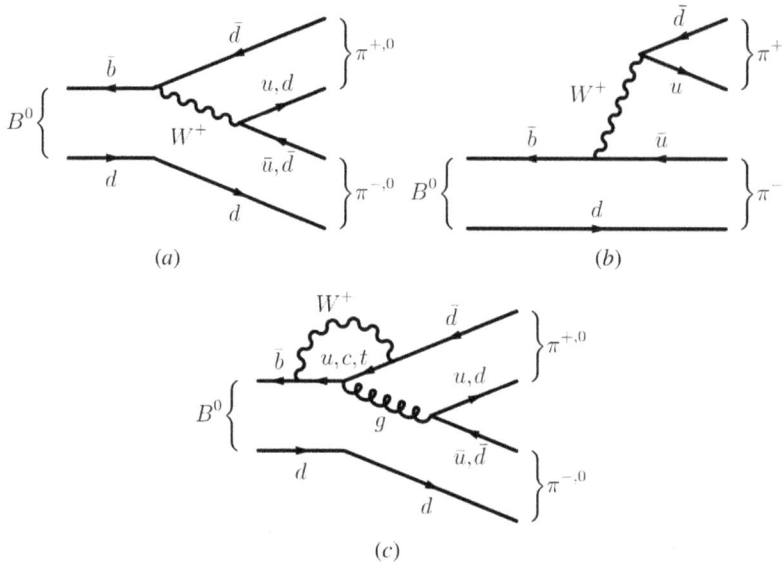

Figure 5.23. The tree-level (a) and (b), and penguin-type (c) diagrams for $B^0 \to \pi^+\pi^-$.

$I = 0$ or 2[12]. Now, whereas both $I = 0$ and 2 are accessible in the tree-level diagrams of figures 5.23(a) and (b), $I = 2$ is excluded in the gluon-mediated penguin diagram of figure 5.23(c). Finally, the $\pi^+\pi^0$ final state can have only $I = 2$. Therefore, the decay $B^+ \to \pi^+\pi^0$ can only proceed via the tree-level diagram.

Let us represent the three relevant amplitudes as A^{00}, A^{+-} and A^{+0}. These are to be expressed then as superpositions of the two possible isospin amplitudes A_0 and A_2. Consider that the pion states must always be symmetrised: thus,

$$\pi^+\pi^- = \frac{1}{\sqrt{2}}\left[\pi_1^+\pi_2^- + \pi_1^-\pi_2^+\right] \tag{5.3.39}$$

and likewise for the $\pi^+\pi^0$ state. Therefore, the relevant Clebsch–Gordan coefficients give

$$\frac{1}{\sqrt{2}}A^{+-} = A_2 - A_0, \qquad A^{00} = 2A_2 + A_0 \quad \text{and} \quad A^{+0} = 3A_2. \tag{5.3.40}$$

The system is overconstrained and we obtain the following complex triangle relation:

$$0 = \frac{1}{\sqrt{2}}A^{+-} + A^{00} - A^{+0}. \tag{5.3.41a}$$

Likewise, there is a similar relation for the charge-conjugate processes:

$$0 = \frac{1}{\sqrt{2}}\bar{A}^{+-} + \bar{A}^{00} - \bar{A}^{-0}. \tag{5.3.41b}$$

---

[12] See e.g. the discussion on $\omega^0 \to \pi\pi$ in section 4.2.2.

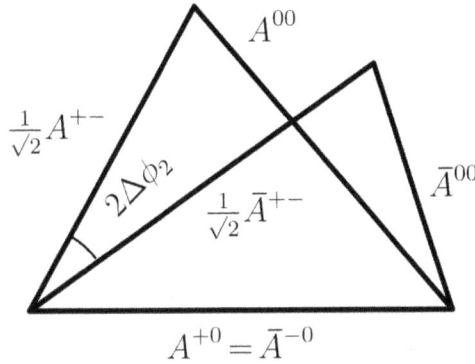

Figure 5.24. The complex triangle relations expressed in equation (5.3.41).

Since $B^+ \to \pi^+\pi^0$ has only the tree-level contribution $A^{+0} = \bar{A}^{-0}$ and they may be taken as a common base for the triangles, as shown in figure 5.24. The triangles and ϕ_2 are completely determined by the branching ratios $\mathcal{B}(B^0 \to \pi^+\pi^-)$, $\mathcal{B}(B^0 \to \pi^0\pi^0)$ and $\mathcal{B}(B^+ \to \pi^+\pi^0)$, and the three non-zero CP-violation parameters $\mathcal{A}_{CP}(B^0 \to \pi^+\pi^-)$, $\mathcal{S}_{CP}(B^0 \to \pi^+\pi^-)$ and $\mathcal{A}_{CP}(B^0 \to \pi^0\pi^0)$. Therefore, the shift in angle ϕ_2 may be determined and thus too the angle itself.

The Belle collaboration at the KEKB asymmetric-energy (3.5 on 8 GeV) e^+e^- collider has collected a total 772×10^6 $B\bar{B}$ pairs at the $\Upsilon(4S)$ resonance and obtains the following results for the CP-violation parameters:

$$\mathcal{A}_{CP}(B^0 \to \pi^+\pi^-) = +0.33 \pm 0.06 \text{ (stat)} \pm 0.03 \text{ (syst)}$$
$$\text{and} \tag{5.3.42}$$
$$\mathcal{S}_{CP}(B^0 \to \pi^+\pi^-) = -0.64 \pm 0.08 \text{ (stat)} \pm 0.03 \text{ (syst)},$$

confirming their own previous results and those of other similar experiments. They also exclude the range $23.8° < \phi_2 < 66.8°$ at the $1\,\sigma$ level.

5.4 Neutrino masses, mixing and oscillation

For a very recent, comprehensive and up-to-date review on the general topic of neutrino physics; see e.g. Sajjad Athar *et al* (2022). The current phenomenological situation regarding neutrino masses, mixing, and oscillation, is also presented in e.g. de Salas *et al* (2018) and, as always, the PDG 2024 review.

In the SM, the masses of all three neutrino species are rigorously zero. However, strong experimental evidence to the contrary has now been gathered. The important point is that, if at least one neutrino is indeed massive, the possibility of a mixing matrix *à la* CKM naturally arises and thus too of leptonic CP violation. Moreover, quantum oscillation phenomena may be expected to occur (Pontecorvo 1957).

In the section we shall examine the evidence for non-zero neutrino masses together with the theoretical and experimental consequences. As this is a rapidly evolving subject with continuing experimental and theoretical updates, the best current source of further information is probably the dedicated article found in the PDG review (Navas *et al* 2024).

As already explained, in the SM, neutrinos are massless. This conventional assignment is closely related to the observation that only (left-) right-handed (anti-) neutrinos appear to be found in Nature; or rather, only left-handed neutrinos are involved in the observed interactions. A standard Dirac mass term in the Lagrangian describing the propagation of a fermion explicitly connects left- and right-handed states and might, in principle, therefore violate such observations. Although it should be remembered that it is the nature of the interaction that selects the left-handed neutrinos and so sterile right-handed neutrinos could still exist. To understand this intuitively, note that a massive neutrino would necessarily travel at less than the speed of light and thus there would exist frames moving at higher velocities in which the neutrino would appear to be travelling in the *opposite* direction. The spin vector would still point in the same direction and therefore its helicity would appear inverted. We should note that it is possible to introduce a so-called Majorana mass term for the neutrino, which then becomes its own antiparticle, without altering the so-far observed phenomenology.

However, recent data provide strong evidence for oscillations between the different neutrino states. Such a phenomenon is possible if and only if at least one of the three neutrino species has a non-zero mass. In fact, as we saw in the $K^0 - \bar{K}^0$ system, the oscillation process requires a mass *difference* and so we are only able to conclude that there is a difference between the masses of at least two neutrino states. Over the next few years more precise and detailed measurements should clarify the issue.

5.4.1 Neutrino masses and limits

Oscillation phenomena aside, other experimental data do place upper limits on neutrino masses. By far the most stringent are those on the electron-neutrino mass and come from nuclear β-decay studies. In particular, the end-point (in the Kurie plot) of the β-decay spectrum is lowered if the emitted neutrino is massive (by just the energy equivalent to its rest mass). Recall that the experimental data on tritium decay provide an upper limit of around 1 eV (Aseev *et al* 2011). As we shall now see, the phenomenon of oscillation is, in principle, rather more sensitive to mass and thus provides a significant and interesting window onto this elusive property of neutrinos.

5.4.2 Neutrino oscillation—theory

Let us consider the simple case of two neutrino states with different masses (not necessarily both non-zero). Call the two *mass* eigenstates states $\nu_{1,2}$. Now, the two states we call, for example, $\nu_{e,\mu}$ are *weak-interaction* eigenstates, which may then be expressed as superpositions of the mass eigenstates, thus:

$$\nu_e = \cos\theta\, \nu_1 + \sin\theta\, \nu_2 \tag{5.4.1a}$$

and

$$\nu_\mu = -\sin\theta\,\nu_1 + \cos\theta\,\nu_2. \tag{5.4.1b}$$

This may be compared to the case of the $K^0 - \bar{K}^0$ system. An important difference here is that typically $E_\nu \gg m_\nu$ and therefore *momentum* eigenstates should be employed.

Consider an electron-neutrino produced in a weak interaction at instant $t = 0$:

$$|\nu_e, \boldsymbol{p}\rangle = \cos\theta\,|\nu_1, \boldsymbol{p}\rangle + \sin\theta\,|\nu_2, \boldsymbol{p}\rangle. \tag{5.4.2}$$

We now suppress the momentum variable and write the corresponding state evolved to the instant t as

$$|\nu_e, t\rangle = a_1(t)\cos\theta\,|\nu_1\rangle + a_2(t)\sin\theta\,|\nu_2\rangle, \tag{5.4.3}$$

where the time-dependent coefficients are

$$a_i(t) = e^{iE_i t} \text{ with } E_i = \sqrt{p^2 + m_i^2} \text{ for } i = 1, 2. \tag{5.4.4}$$

Since $m_1 \neq m_2$, for a given well-defined momentum $\boldsymbol{p}$, $E_1 \neq E_2$.

Exercise 5.4.1. *Following similar steps as in the discussion of $K^0 - \bar{K}^0$ oscillation, but using momentum eigenstates, show that the probability for a state initially produced at $t = 0$ as, say, an electron-neutrino to become a muon neutrino at time t is*

$$P(\nu_e \to \nu_\mu; t) = \sin^2 2\theta \, \sin^2\left(\tfrac{1}{2}(E_2 - E_1)t\right).$$

The energy difference may be rewritten as follows (using $E_i^2 = p^2 + m_i^2$):

$$E_2 - E_1 = \frac{E_2^2 - E_1^2}{E_2 + E_1} = \frac{\Delta m^2}{2E_\nu}, \tag{5.4.5}$$

where $\Delta m^2 = m_2^2 - m_1^2$ and E_ν is the average of the two energies. We see that the oscillation depends on two parameters: the mixing angle $\sin\theta$ (which determines the amplitude) and the mass-squared difference Δm^2 (which determines the frequency), both must be non-zero. Indeed, in the limit $E_\nu \gg m$ (taking $v_\nu \simeq c$), the time-dependent transition probability may be rewritten as

$$P(\nu_e \to \nu_\mu; t) \simeq \sin^2 2\theta \, \sin^2\left(\frac{1.27\,\Delta m^2\,L(t)}{E_\nu}\right), \tag{5.4.6}$$

where Δm^2 is in eV^2, $L(t)$ is the distance travelled in metres and E_ν is the neutrino energy in MeV.

Already in 1957, Bruno Pontecorvo had hypothesised the phenomenon of oscillation for neutrinos and since it naturally leads to a loss of electron-neutrino flux, he suggested it (Pontecorvo, 1967) as a possible explanation of the so-called solar-neutrino problem, to which we shall now turn.

5.4.3 Neutrino oscillation—experiment

The first hints of oscillation phenomena in neutrinos are to be found in the measurements of the solar-neutrino flux in the experimental work of Davis (1964)[13] at the Homestake Gold Mine in Lead, South Dakota (the largest and deepest, 2438 m, gold mine in the USA); see also the related theoretical calculations of Bahcall (1964) and a review by Davis (1994).

Using two 500-gallon (1900-litre) tanks of tetrachloroethylene (C_2Cl_4, a fluid commonly used for dry-cleaning) as detectors, via a chemical-extraction process, Davis (1964) measured the production rate of ^{37}Ar by the neutrino-capture reaction

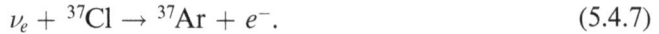

$$\nu_e + {}^{37}\text{Cl} \rightarrow {}^{37}\text{Ar} + e^-. \tag{5.4.7}$$

The ^{37}Ar so created subsequently decays back to ^{37}Cl via K-capture; the activity is a measure of the quantity of argon produced and thus the number of neutrino interactions, from which the flux may be deduced. The findings were

$$\Phi_\nu^{\text{expt}} \simeq 0.4\ \Phi_\nu^{\text{th}}. \tag{5.4.8}$$

It should be stressed, however, that the neutrinos detected by Davis (1964) were of relatively high energy ($\gtrsim 0.814$ MeV, the reaction threshold); i.e. those generated by the PP-II and PP-III branches and the carbonnitrogenoxygen cycle, but *not* the main PP-I branch. The temperature dependence of the predicted flux is particularly strong for these chains (see e.g. Bahcall and Ulmer 1996):

$$\Phi_\nu^{\text{th}} \propto T_{\text{core}}^{10}\ (^7\text{Be}), \quad T_{\text{core}}^{24}\ (^8\text{B}), \quad T_{\text{core}}^{20}\ (\text{CNO}). \tag{5.4.9}$$

Such sensitivity to model-dependent details rendered the results very suspect. It was thus necessary to carry out experiments on the lower-energy neutrinos; i.e. those produced in the PP-I chain. One such measurement was performed by the GALLium EXperiment (GALLEX) collaboration, which ran between 1991 and 1997 at the Gran Sasso National Laboratory. The detector consisted of 30 t of gallium, which is sensitive to very low-energy neutrinos; the process exploited was

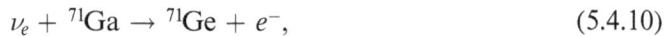

$$\nu_e + {}^{71}\text{Ga} \rightarrow {}^{71}\text{Ge} + e^-, \tag{5.4.10}$$

for which the threshold is 0.23 MeV. It was therefore sensitive to neutrinos produced in the PP-I chain, for which the upper energy limit is 0.42 MeV.

Through the above weak interaction, the PP-I solar neutrinos transformed less than one gallium nucleus per day into germanium, which was again extracted chemically and detected via observation of its radioactive decay products. The

---

[13] The 2002 Nobel Prize in Physics was awarded one half jointly to Raymond Davis Jr and Masatoshi Koshiba 'for pioneering contributions to astrophysics, in particular for the detection of cosmic neutrinos', and the other half to Riccardo Giacconi 'for pioneering contributions to astrophysics, which have led to the discovery of cosmic X-ray sources'.

Standard Solar Model theory in this case is much less susceptible to uncertainty. Most of the energy irradiated by the Sun is produced by the PP-I chain (with only weak temperature dependence), in which each cycle produces about 25MeV and exactly two electron neutrinos of up to 0.42 MeV. Since the power output of the Sun is 2.40×10^{45} eV s^{-1}, the fusion process must have a rate of about 10^{38} cycles s and so we may expect a solar-neutrino production rate $R_\nu^\odot \simeq 2 \times 10^{38} \nu_e$ s^{-1}. The expected solar-neutrino flux at Earth is then

$$\Phi_\nu^{\text{Earth}} \simeq \frac{R_\nu^\odot}{4\pi r_{S-E}^2} \simeq 6 \times 10^{10} \nu_e / \text{cm}^2 / \text{s}. \tag{5.4.11}$$

The GALLEX collaboration recorded just around two thirds of this figure (see e.g. Hampel *et al* 1999), a discrepancy of approximately 7σ; see too the results of the similar Soviet–American Gallium Experiment (SAGE) (Abdurashitov *et al* 1999). Davis (1964) and Bahcall's (1964) results were thus substantially confirmed.

However, the full significance of Davis' (1964) findings went long before being truly understood, owing to various theoretical problems surrounding their interpretation. First of all, the question of the Standard Solar Model is central (especially for CNO neutrinos): the comparison is always with a theoretically estimated flux based on our description of what happens inside the Sun. Clearly though, the GALLEX and SAGE results essentially swept away such doubts, as the sensitivity of the PP-I dominance to the temperature is not so delicate.

There is, however, a further source of uncertainty: namely, our limited knowledge of the nuclear neutrino-capture cross-section for these processes and energies. Owing to the very small interaction probabilities, neutrino measurements are always fraught with difficulties and so the cross-sections used to estimate the neutrino fluxes (for all experiments) are again essentially based on theoretical estimates, which may or may not be correct, depending as they do on our limited knowledge of nuclear wave-functions; recall e.g. Equation (2.1.13). For many years there was thus a tendency to dismiss the results as being due to inaccuracies in the theoretical description of either the Sun or the low-energy nuclear neutrino cross-sections.

If, however, the results are taken at face value, we are forced to seek a particle-physics explanation. The most natural exploits the phenomenon of quantum oscillation[14] (already observed in other systems, e.g. $K^0 - \bar{K}^0$). That is, some of the electron neutrinos produced by the Sun might effectively transform in-flight into the other two types and thus avoid detection.

Some time after the GALLEX and SAGE results were published, Sudbury Neutrino Observatory (SNO, in Canada) produced convincing evidence for such oscillation effects. The SNO detector consisted of 1000 t of heavy water (D$_2$O), surrounded by a shield of ordinary water, approximately 9600 photomultiplier tubes (PMTs), all held in a 6-metre radius spherical container situated more than 6 km underground in the Creighton Mine in Sudbury, Ontario.

---

[14] The 2015 Nobel Prize in Physics was awarded jointly to Takaaki Kajita and Arthur B McDonald 'for the discovery of neutrino oscillations, which shows that neutrinos have mass'.

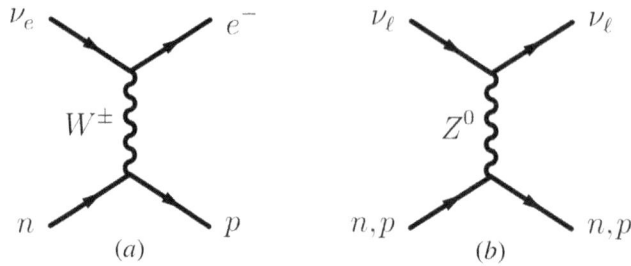

Figure 5.25. Feynman diagrams for the two types of processes contributing to the SNO experiment: (a) charged-current $\nu_e n \to e p$ and (b) neutral-current $\nu_\ell N \to \nu_\ell N$, where ℓ may represent any one of the three charged leptons.

The deuterium in the target can interact with neutrinos via both charged- ($W^\pm$) and neutral-current (Z^0) exchange (see figure 5.25); and is therefore sensitive to all neutrino types. This experiment was thus able to confirm the theoretical overall solar-neutrino flux, while at the same time measuring the individual electron-neutrino flux.

The two types of interactions involved are the following.

Charged-current: A neutrino may convert a neutron into a proton and electron via $W^\pm$ exchange (a standard β-process), see figure 5.25(a); it is the electron that is then detected. Solar neutrinos do not have sufficient energy to produce either muons or tau leptons (with their large masses) and thus only electron neutrinos can interact in this way.

Neutral-current: A neutrino may break up a deuteron into a neutron and proton via Z^0 exchange (just as an electron may break up a nucleus via γ-exchange), see figure 5.25(b). The neutrino is not absorbed or transformed and therefore all three neutrino flavours can contribute with equal probability. The neutrons so produced may be captured by a deuterium nucleus (in the heavy water) or a proton (in the ordinary water), with subsequent γ-ray emission. Note that for the rather low energies involved, a sufficiently good description would be given by the four-point contact interaction of the original Fermi theory, augmented with a similar *ad hoc* neutral current–current interaction describing elastic processes such as $\nu_\ell N \to \nu_\ell N$, where $N = p$ or n (see figure 5.25(b)).

Neutrino-oscillation phenomena are also possible in fluxes produced by nuclear reactors and accelerators provided that a suitable baseline is chosen (as determined by the energy, see equation (5.4.6)) and moreover the fluxes of neutrinos created in cosmic-ray collisions in the atmosphere may be compared between those produced immediately above the apparatus and also on the far side of the Earth. Indeed, oscillation phenomena have now been confirmed by, among others, the Super-Kamiokande experiment in Japan (see Fukuda *et al* 1998, Ashie *et al* 2004, 2005, for example). The data on atmospheric neutrinos are consistent with two-flavour $\nu_\mu \leftrightarrow \nu_\tau$ oscillations, $\sin^2 2\theta > 0.92$ and

$$1.5 \times 10^{-3} < \Delta m^2 < 3.4 \times 10^{-3}\,\text{eV}^2 \text{ at the 90\% CL.} \qquad (5.4.12)$$

Table 5.4. The mass-difference sensitivity of the different types of neutrino-oscillation experiments.

Source	Type	$\langle E \rangle$(MeV)	L(km)	Δm^2 limit (eV^2)
Reactor	$\bar{\nu}_e$	~ 1	100	10^{-5}
Accelerator	ν_μ, $\bar{\nu}_\mu$	$\sim 10^3$	10^3	10^{-3}
Atmosphere	$\nu_{e,\mu}$, $\bar{\nu}_{e,\mu}$	$\sim 10^3$	10^4	10^{-4}
Sun	ν_e	~ 1	$\sim 10^8$	10^{-11}

However, the available data do not yet indicate which neutrino superposition is the most massive and different mass hierarchies are presently allowed.

Using equation (5.4.6), we can immediately evaluate the relevant parameters and thus the sensitivity of the different types of experimental situations. These are shown in table 5.4. The baseline L is determined by the distance between the point of production and the detector.

5.4.4 The MSW matter effect

An important phenomenon that is at first sight perhaps rather surprising is the so-called Mikheyev–Smirnov–Wolfenstein (MSW) effect (Mikheyev and Smirnov 1985, Wolfenstein, 1978); that is, the effect of matter on neutrino oscillation (see also Parke 1986, Rosen and Gelb 1986). The central point is the observation that ordinary matter contains large numbers of electrons but essentially no muons or taus. Therefore, whereas muon and tau neutrinos (of not too high energy) may only interact with matter via neutral-current exchange, electron neutrinos may also interact elastically via the charged-current. This leads to a coherent forward scattering and an effect analogous to the optical phenomenon of refraction experienced by photons propagating in a medium, whereby they acquire an effective mass and thus slow down.

Indeed, similarly to the case in optics, the coherent elastic forward scattering induces a phase difference, a refractive index, or equivalently, a neutrino effective mass. Since only electron neutrinos and antineutrinos are so affected, such an effective mass can dramatically alter the oscillation properties, which depend on mass-squared differences. In particular, even in the complete absence of neutrino masses, oscillation becomes the possible. Certainly, oscillations will be enhanced in environments rich in electrons, such as the Sun, but even passage through the Earth can have non-negligible consequences.

It turns out that the effect is important for higher-energy neutrinos; the threshold in the Sun is at around 2 MeV. This means that the solar-neutrinos of various origins suffer differently. The Borexino experiment (Basilico *et al* 2023) has observed neutrinos produced by the *pp* chain (<0.42 MeV), ^{7}Be (0.86 MeV), *pep* (1.44 MeV), and ^{8}B (<15 MeV) separately and has thus been able to verify the MSW phenomenon. GALLEX and SAGE have also obtained compatible results.

5.4.5 Neutrino mixing

We shall now briefly examine the full three-flavour case and the consequent 3×3 neutrino mixing matrix also in the light of recent experimental findings. We have already seen in the quark sector that three-flavour mixing has a much richer structure than a simple two-component model and that it also allows for CP violation. Once non-zero masses are admitted, then we should contemplate a mixing matrix *à la* CKM, which is known as the PMNS matrix (Maki *et al* 1962).

The oscillation probability depends on the neutrino energy, the source–detector distance, the relevant mass-squared difference $\Delta m_{ij}^2 := m_i^2 - m_j^2$ and mixing angle, see equation (5.4.6). Note that there are only two independent mass-squared differences (usually taken as Δm_{12}^2 and Δm_{13}^2) and, indeed, one neutrino may still be massless. Conventionally, Δm_{12}^2 refers to the smaller difference, which is that responsible for the solar-neutrino deficiency and the reactor $\bar{\nu}_e$ oscillations observed by the KamLAND experiment in Japan. Note that there are two distinct possibilities for the mass orderings:

$$\text{natural hierarch (NH)} - m_1 < m_2 < m_3,$$

$$\text{inverted hierarch (IH)} - m_3 < m_1 < m_2.$$

For completeness, we shall now give the full three-flavour oscillation formula; that is, the probability of neutrino oscillation between two flavour eigenstates ν_α and ν_β, $P(\nu_\alpha \to \nu_\beta)$, which is the probability of detecting a neutrino of flavour β at a distance $L \simeq ct$ from a source producing neutrinos of flavour α and energy E:

$$P(\nu_\alpha \to \nu_\beta) = \delta_{\alpha\beta} - 4 \sum_{i<j} \Re e \left[U_{\alpha i} U_{\beta i}^* U_{\alpha j}^* U_{\beta j} \right] \sin^2 \left(\frac{\Delta m_{ji}^2 L}{4E} \right)$$

$$+ 2 \sum_{i<j} \Im m \left[U_{\alpha i} U_{\beta i}^* U_{\alpha j}^* U_{\beta j} \right] \sin \left(\frac{\Delta m_{ji}^2 L}{2E} \right), \tag{5.4.13}$$

where $\Delta m_{ji}^2 := m_j^2 - m_i^2$. In the case of antineutrinos, U and U^* are exchanged, which has the effect of changing the sign of the last term. We thus see that, in principle, oscillation experiments can extract all of the PMNS-matrix angles, the CP-violating phase and the mass-squared differences. Note in particular, the difference between $P(\nu_\alpha \to \nu_\beta)$ and $P(\bar{\nu}_\alpha \to \bar{\nu}_\beta)$ for $\alpha \neq \beta$ is sensitive to CP violation.

The PMNS matrix may be parametrised as in equation (5.3.2), where we recall that $\delta_{\text{CP}} \to -\delta_{\text{CP}}$ for antineutrinos. There are thus four new parameters to be determined experimentally: three Euler angles and one imaginary phase (which, if non-zero, would naturally lead to CP violation in the lepton sector). The CP-violating phase now starts be slightly constrained and the now numerous experiments performed on neutrino oscillations lead to the list of values shown in table 5.5. The present combined world data, covering as they do the various possibilities, thus clearly indicate the existence of full three-neutrino mixing. To have a feeling for the

Table 5.5. Best-fit parameters from a global three-neutrino oscillation analysis (Esteban *et al* 2020); in some cases slightly different results are obtained for the assumptions of natural (NH) and inverted (IH) hierarchies, otherwise the single common value is given. The mass-squared difference $\Delta m_{3\ell}^2 = \Delta m_{31}^2 > 0$ for NH and $\Delta m_{32}^2 < 0$ for IH. Table reproduced from NuFIT.

	NH		IH
$\sin^2 \theta_{12}$		$0.303^{+0.012}_{-0.011}$	
$\sin^2 \theta_{23}$	$0.572^{+0.018}_{-0.023}$		$0.578^{+0.016}_{-0.021}$
$\sin^2 \theta_{13}$	$0.022\,03^{+0.000\,56}_{-0.000\,59}$		$0.022\,19^{+0.000\,60}_{-0.000\,57}$
δ_{CP}	$197°^{+42}_{-25}$		$286°^{+27}_{-32}$
Δm_{21}^2		$7.41^{+0.21}_{-0.20} \times 10^{-5}\mathrm{eV}^2$	
$\Delta m_{3\ell}^2$	$2.511^{+0.028}_{-0.027} \times 10^{-3}\mathrm{eV}^2$		$-2.498^{+0.032}_{-0.025} \times 10^{-3}\mathrm{eV}^2$

overall appearance of the PMNS mixing matrix, we show the results of the NuFIT global analysis (Esteban *et al* 2020) in terms of the 3σ CL (99.7% C.L.) ranges for the magnitudes of the elements:

$$
\begin{pmatrix}
0.803-0.845 & 0.514-0.578 & 0.142-0.155 \\
0.233-0.505 & 0.460-0.693 & 0.630-0.779 \\
0.262-0.525 & 0.473-0.702 & 0.610-0.762
\end{pmatrix}. \tag{5.4.14}
$$

The natural hierarchy is now slightly favoured over the inverted hierarchy, but nothing conclusive may be said yet. What does stand out clearly is that the mixing is much greater than in the quark sector, where recall the diagonal elements are all close to unity and those far off are very small.

References

Aad G *et al* (ATLAS Collab) 2012 *Phys. Lett.* **B710** 49

Aad G *et al* (ATLAS Collab) 2013 *Phys. Lett.* **B726** 120

Aad G *et al* (ATLAS Collab) 2015a *Eur. Phys. J.* **C75** 476 [Erratum: Eur.Phys.J.C 76, 152 (2016)]

Aad G *et al* (ATLAS, CMS Collabs) 2015b *Phys. Rev. Lett.* **114** 191803

Aad G *et al* (ATLAS Collab) 2023 *Phys. Lett.* **B843** 137745

Aaltonen T *et al* (CDF and D0 Collabs) 2012 *Phys. Rev. Lett.* **109** 071804

Abdurashitov J N *et al* (SAGE Collab) 1999 *Phys. Rev. Lett.* **83** 4686

Abreu P *et al* (DELPHI Collab) 1998 *Phys. Lett.* **B439** 209

Acosta D *et al* (CDF Collab) 2005 *Phys. Rev. Lett.* **95** 031801

Ademollo M and Gatto R 1964 *Phys. Rev. Lett.* **13** 264

Ambrosino F *et al* (KLOE Collab) 2006 *Phys. Lett.* **B632** 76

Amhis Y S *et al* (HFLAV Collab) 2023 *Phys. Rev.* **D107** 052008

Anderson P W 1963 *Phys. Rev.* **130** 439

Aseev V N *et al* (Troitsk Collab) 2011 *Phys. Rev.* **D84** 112003

Ashie Y *et al* (Super-Kamiokande Collab) 2004 *Phys. Rev. Lett.* **93** 101801

Ashie Y *et al* (Super-Kamiokande Collab) 2005 *Phys. Rev.* **D71** 112005

Bahcall J N 1964 *Phys. Rev. Lett.* **12** 300

Bahcall J N and Ulmer A 1996 *Phys. Rev.* **D53** 4202

Basilico D *et al* (BOREXINO Collab) 2023 *Phys. Rev.* **D108** 102005

Beringer J *et al* (Particle Data Group) 2012 *Phys. Rev.* **D86** 010001

Bernardi G and Herndon M 2014 *Rev. Mod. Phys.* **86** 479

Bouchiat M-A and Bouchiat C 1997 *Rep. Prog. Phys.* **60** 1351

Caola F and Melnikov K 2013 *Phys. Rev.* **D88** 054024

Chakraborty B, Gilman A, Hoferichter M and Koval M 2024 *Presented at the 12th. Int. Workshop on the CKM Unitarity Triangle (Santiago de Compostela, September 2023)*

Charles J *et al* (CKM fitter Group) 2004 hep-ph/0406184; updated results and plots available at http://ckmfitter.in2p3.fr

Chatrchyan S *et al* (CMS Collab) 2012 *Phys. Lett.* **B710** 26

Chatrchyan S *et al* (CMS Collab) 2013 *Phys. Rev. Lett.* **110** 081803

Chau L-L and Keung W-Y 1984 *Phys. Rev. Lett.* **53** 1802

Coleman S R and Weinberg E 1973 *Phys. Rev.* **D7** 1888

Davis R Jr 1964 *Phys. Rev. Lett.* **12** 303

Davis R Jr 1994 *Prog. Part. Nucl. Phys.* **32** 13

de Salas P F, Forero D V, Ternes C A, Tórtola M and Valle J W F 2018 *Phys. Lett.* **B782** 633

Dittmaier S *et al* (LHC Higgs Cross section Working Group) 2011 *CERN Yellow Rep.* 2011–002; 2012–002; 2013-004; 2017-002

Dixon L J and Siu M S 2003 *Phys. Rev. Lett.* **90** 252001

Ellis J R, Gaillard M K, Nanopoulos D V and Rudaz S 1977 *Nucl. Phys.* **B131** 285; *erratum* ibid **B132**, 541

Elsener K *et al* 1984 *Phys. Rev. Lett.* **52** 1476

Englert F and Brout R 1964 *Phys. Rev. Lett.* **13** 321

Esteban I, Gonzalez-Garcia M C, Maltoni M, Schwetz T and Zhou A 2020 *JHEP* **09** 178

Feynman R P and Gell-Mann M 1958 *Phys. Rev.* **109** 193

Fukuda Y *et al* (Super-Kamiokande Collab) 1998 *Phys. Rev. Lett.* **81** 1562

Georgi H and Glashow S L 1972 *Phys. Rev. Lett.* **28** 1494

Glashow S L 1961 *Nucl. Phys.* **22** 579

Goldstone J 1961 *Nuovo Cim.* **19** 154

Gronau M and London D 1990 *Phys. Rev. Lett.* **65** 3381

Guralnik G S, Hagen C R and Kibble T W B 1964 *Phys. Rev. Lett.* **13** 585

Haeberli W and Holstein B R 1995 *Symmetries and Fundamental Interaction in Nuclei* ed E M Henley and W C Haxton (Singapore: World Scientific) p 17

Hampel W *et al* (GALLEX Collab) 1999 *Phys. Lett.* **B447** 127

Hardy J C and Towner I S 2009 *Phys. Rev.* **C79** 055502

Hasert F J *et al* (Gargamelle Neutrino Collab) 1973a *Phys. Lett.* **B46** 121

Hasert F J *et al* (Gargamelle Neutrino Collab) 1973b *Phys. Lett.* **B46** 138 *Nucl. Phys.*, 1

Higgs P W 1964 *Phys. Lett.* **12** 132; *Phys. Rev. Lett.* **13** 508

Isgur N and Wise M B 1989 *Phys. Lett.* **B232** 113; **B237**, 527

Jarlskog C 1985 *Phys. Rev. Lett.* **55** 1039; *Z. Phys.*, **C29**, 491

Khachatryan V *et al* (CMS Collab) 2015 *Phys. Rev.* **D92** 012004

Kibble T W B 2009 *Scholarpedia* **4** 8741 *reprinted in Int. J. Mod. Phys.*, 6001

Landau L D 1948 *Dokl. Akad. Nauk. SSSR* **60** 207

Maki Z, Nakagawa M and Sakata S 1962 *Prog. Theor. Phys.* **28** 870

Maxwell J C 1865 *Phil. Trans. R. Soc.* **155** 459

Mikheyev S P and Smirnov A Y 1985 *Sov. J. Nucl. Phys.* **42** 913

Nambu Y and Jona-Lasinio G 1961 *Phys. Rev.* **122** 345; **124**, 246

Navas S *et al* (Particle Data Group) 2024 *Phys. Rev.* **D110** 030001

Neubeck K, Schober H and Wäffler H 1974 *Phys. Rev.* **C10** 320

Olive K A *et al* (Particle Data Group) 2014 *Chin. Phys.* **C38** 090001

Parke S J 1986 *Phys. Rev. Lett.* **57** 1275

Pontecorvo B 1957 *Zh. Eksp. Teor. Fiz.* **33** 549 *transl. Sov. Phys. JETP* **6** 429

Pontecorvo B 1967 *Zh. Eksp. Teor. Fiz.* **53** 1717

Pontecorvo B 1968 *Sov. Phys. JETP* **26** 984

Prescott C Y *et al* 1978 *Phys. Lett.* **B77** 347

Quigg C 2007 *Rep. Prog. Phys.* **70** 1019

Rosen S P and Gelb J M 1986 *Phys. Rev.* **D34** 969

Sajjad Athar M *et al* 2022 *Prog. Part. Nucl. Phys.* **124** 103947

Sakurai J J 1958 *Nuovo Cim.* **7** 649

Salam A 1968 *Proc. of Elementary Particle Theory: Relativistic Groups and Analyticity (Aspenäsgården, May 1968)* N Svartholm (Almqvist and Wiksell) no. 8 *The Nobel Symposium Series* p 367

Sirunyan A M *et al* (CMS Collab) 2019 *Phys. Rev.* **D99** 112003

Sudarshan E C G and Marshak R E 1958 *Phys. Rev.* **109** 1860

't Hooft G 1971 *Nucl. Phys.* **B35** 167

Vainshtein A I, Zakharov V I and Shifman M A 1975 *JETP Lett.* **22** 55 *Pis'ma Zh. Eksp. Teor. Fiz.* **22** 123; Shifman, M.A., Vainshtein, A.I. and Zakharov, V.I., *Nucl. Phys.*, **B120** (1977) 316

Vanderleeden J C and Boehm F 1970 *Phys. Rev.* **C2** 748

Vilain P *et al* (CHARM-II Collab) 1994 *Phys. Lett.* **B335** 246

Weinberg S 1967 *Phys. Rev. Lett.* **19** 1264

Wolfenstein L 1978 *Phys. Rev.* **D17** 2369

Wolfenstein L 1983 *Phys. Rev. Lett.* **51** 1945

Yang C-N 1950 *Phys. Rev.* **77** 242

Chapter 6

Beyond the Standard Model
(where we might be going)

'We are not to tell nature what she's gotta be.

...

She's always got better imagination than we have.'

Richard P Feynman

It is natural to inquire as to what might lie beyond the structures described here. First of all, as shown in the previous chapter, the Standard Model (SM) describes all known phenomenon very well, within experimental and theoretical errors. It does, however, fail in certain aspects. It does not explain the precise values of the various parameters of Nature: particle masses, coupling constants and mixing angles; it does not explain the precise values of the CP-violating phase or why indeed P is violated at the 100% level; moreover, it does not include gravity. The complete theory should do all of this (and maybe more), the problem is though to make a first step beyond the present framework. Thus, on the one hand the search is now underway for experimental indications of missing pieces while on the other the theoretical effort is to imagine other possible scenarios.

In this final chapter then we shall deal with some of the possible extensions of the SM. That is, some of the ideas we have as to a more complete theory and the experimental endeavour aimed at uncovering new physics. We should first perhaps stress the amazing success and precision of the SM: the many theoretical predictions for experimentally measurable quantities and the overall agreement might, at first sight, suggest that there is little more to be discovered or explained. However, while the successes are undeniable, we also have many pointers to the fact that the theory

doi:10.1088/978-0-7503-5759-3ch6

cannot yet be considered complete. Just as we have seen that the weak-interaction theory created by Fermi (and its more immediate extension by Gamow and Teller), despite describing perfectly well many low-energy phenomenon, already held in itself the seeds of inconsistency, we shall now demonstrate that, for similar reasons, the present picture cannot be the full story.

The first and most obvious question is that of unification. The history of physics contains many examples of combining existing apparently unrelated phenomena into unified pictures. The most notable and best known example probably being that of Maxwell's unification of the previously apparently unrelated theories of electricity and magnetism into the single theory of classical electromagnetism. At present, we have three apparently unrelated theories (strong and weak nuclear and electro-magnetic) all with different coupling constants (i.e. strengths), not to mention gravity. We should stress here that the Glashow–Salam–Weinberg theory of the electroweak interaction is not a true unification as there remain two apparently unrelated couplings. And so, although there is no compelling necessity, it would seem natural to ask if they cannot be accommodated in a single global theory (perhaps even including gravity).

There are also a number of essentially theoretical difficulties that, while evidently not affecting the wonderful agreement between theory and presently available experimental data, would seem to suggest that there is something more to be discovered. The question of renormalisation raises the problem of so-called fine tuning. The quantum corrections inherent to all quantum field theories imply that the values of any physical parameters initially inserted into the Lagrangian (or Hamiltonian) a given system will not correspond to what is measured experimen-tally. In particular, if the theory contains scalar fields (such as the Higgs boson), then e.g. particle masses will generally take on physical values close to the high-energy cut-off of the theory (e.g. the grand-unification or Planck scale). Thus, in order to maintain the small known values of the quark and lepton masses, very fine but unjustified cancellations must be arranged between unphysical parameters. Various solutions to this problem exist and all imply the existence of presently unknown particles.

In this chapter then we shall examine some of the more experimentally accessible, though necessarily theoretically driven, approaches to extending the SM and their phenomenological implications, together with the experimental searches proposed or underway. We shall start though with a discussion of possible new phenomena involving neutrinos: namely, the possibility of CP-violation in the leptonic sector and neutrinoless double β-decay. In particular, the interest in attempts to uncover the possible Majorana nature of the neutrino.

6.1 New neutrino physics

The neutrino phenomena presented in the previous chapter, though already beyond the strict SM, no longer constitute *new* physics and are essentially also now encompassed by an only marginally modified version. The remaining open questions regard more intimately the specific nature of the neutrino and its interactions; in particular, its possible Majorana *versus* Dirac nature, but also its rôle in CP violation.

Since the question of neutrino masses is technically beyond the SM, though not exactly new, it seems appropriate to mention the so-called 'seesaw' mechanism, which can explain the particularly small neutrino masses as compared to the other fermions of the SM. We shall thus now briefly outline the idea.

6.1.1 The seesaw neutrino mass

We have already mentioned the so-called hierarchy or fine-tuning problem regarding the renormalisation effects on the general particle masses. However, there is a more immediate and purely phenomenological observation in relation to neutron masses: why are they so small with respect to even the lightest charged lepton or quark (the electron or up quark)? Recall that $m_{\nu_e} < 0.8$ eV (cf. $m_e \sim 0.5$ MeV or $m_u \sim 2$ MeV).

A simple and elegant construction that naturally generates such large differences is the so-called *seesaw* mechanism (Yanagida 1979). There are various versions, each requiring an extension of the SM. We shall just examine the simplest version here, the type 1 seesaw. This model posits the existence of at least two right-handed neutrinos, which however, do not interact with the other fields; i.e. they are *inert* or *sterile*. It also requires the existence of some very high-energy scale, which is usually associated with the grand-unification scale. Exploiting this high scale as a lever, the final physical states of the model are one very light left-handed neutrino and one very heavy right-handed state, which is thus unobservable, as phenomenologically required.

The starting point is given by considering the properties of certain simple matrices:

$$\mathsf{M} = \begin{pmatrix} 0 & A \\ A & B \end{pmatrix}, \qquad (6.1.1)$$

the eigenvalues of which are

$$m_{\pm} = \frac{B \pm \sqrt{B^2 + 4A^2}}{2}. \qquad (6.1.2)$$

Since the product $m_- \, m_+ = -A^2$ is independent of B, we see that by a judicious choice of A and B, one may be made very small, forcing the other to be very large; hence the term 'seesaw'. Therefore, if we assume $B \gg A$, we have

$$m_- \simeq -A^2/B \quad \text{and} \quad m_+ \simeq B. \qquad (6.1.3)$$

Consider now the two possible neutrino two-component spinors, η_L and η_R. There are three possible spinor bilinears or mass-like terms (together with their Hermitian conjugates):

$$\tfrac{1}{2} B' \eta_L^\dagger \eta_L, \quad \tfrac{1}{2} B \eta_R^\dagger \eta_R \quad \text{and} \quad \tfrac{1}{2} A \eta_L^\dagger \eta_R. \qquad (6.1.4)$$

Written in a more compact and suggestive form, these become

$$\left(\eta_L^\dagger, \ \eta_R^\dagger \right) \begin{pmatrix} B' & A \\ A & B \end{pmatrix} \begin{pmatrix} \eta_L \\ \eta_R \end{pmatrix}. \tag{6.1.5}$$

Now, we must ask the origins of the various elements of the above mass matrix. In general, the gauge symmetry of the SM would effectively force all of them to be zero. However, the Higgs mechanism may certainly create a non-zero value for A (via a standard Yukawa-type coupling), which may then be expected to be of order the Higgs-field vacuum expectation value, i.e. ≈ 246 GeV (see equation (5.2.32)). Whereas $B' = 0$ is protected by the various symmetries, B (a Majorana mass term) may be generated by the large possible radiative corrections inherent in such theories containing scalars and therefore could be expected to be of the order of the typical SM grand-unification theory (GUT) scale 10^{15} GeV (see section 6.2.1). Taking then $A \approx 100$ GeV, we obtain

$$m_- \simeq -A^2/B \approx 0.001 \text{ eV}. \tag{6.1.6}$$

Note that the corresponding eigenvector is

$$\nu_- \simeq \eta_L + \frac{A}{B} \, \eta_R; \tag{6.1.7}$$

i.e. it is almost purely the left-handed neutrino we desire ($A/B \approx 10^{-13}$). The other eigenstate ν_+ is an almost purely right-handed neutrino with a mass of the order of the GUT scale. Note finally that lepton-number conservation is violated by the Majorana term, which corresponds though to a very large mass and the effect is thus highly suppressed.

6.1.2 Neutrinoless double β-decay

An interesting prospect arising in the eventuality of non-zero neutrino masses is that of double β-decay in which *no neutrinos* are emitted. This becomes possible *if and only if* neutrinos are Majorana type fermions[1]. The idea (Majorana, 1937) was that *neutral* spin-½ particles may be described by a *real* version of the Dirac equation, the Majorana equation. In this way, since particle and antiparticle wave functions are related by complex conjugation, the antineutrino would be identical to the neutrino. That is, the right-handed antineutrino and left-handed neutrino would be equivalent; and precisely a Majorana mass term allows for conversion of one into the other. Such a possibility is not as exotic as it may seem and in certain attempts at formulating grand unified theories (see later) it is a natural result. The important phenomenological implication is that the neutrino is then its own antiparticle.

Consider first the process of double β-decay, proposed by Goeppert-Mayer (1935) and which, although very rare, has now been observed for a number of nuclei. In such a nuclear process, two essentially independent standard single β-decays occur more-or-less simultaneously. The decay rate may be calculated in the usual manner

---

[1] See appendix B.1.3 for a brief theoretical discussion of the Majorana equation and its consequences.

via Fermi's golden rule (and is independent of the Dirac or Majorana nature of the neutrinos); the differential rate is

$$\frac{d^3\Gamma^{2\nu\beta\beta}(E_1, E_2, \cos\theta)}{dE_1 dE_2 \cos\theta} \propto G_F^4 \cos^4\theta_C \, F(Z', E_1) \, F(Z', E_2) \, E_1 \, p_1 \, E_2 \, p_2$$

$$\times (E_0 - E_1 - E_2)^5 \, (1 - \beta_1\beta_2 \cos\theta),$$

(6.1.8)

where $F(Z', E)$ is the Fermi function (with Z' the charge of the final-state nucleus), 1 and 2 indicate the two emitted electrons, $E_{1,2}$ their total energies, $p_{1,2}$ momenta, $\beta_{1,2}$ velocities, θ is the angle between the two electrons and E_0 is the initial–final nuclear-mass difference; recall that E_0 is not quite the Q-value:

$$E_0 := (M_i - M_f) = Q + 2m_e.$$

(6.1.9)

There are some 35 naturally occurring nuclides for which double β-decay is energetically allowed, of which 14 have been observed experimentally. Note though that, clearly, double β-decay can be detected experimentally *if and only if* the single β-decay channel is forbidden or is highly suppressed, as is the case for ^{48}Ca and ^{96}Zr. There are also a very few cases in which double K-capture has been observed: ^{130}Ba, ^{78}Kr and ^{124}Xe.

The first experimental confirmation of such processes (though not their direct observation) dates back to 1950, when Inghram and Reynolds investigated the decay ^{130}Te $\rightarrow$ ^{130}Xe $+ 2\beta^- + 2\bar{\nu}$ indirectly via a *geochemical* approach. Approximately 34% of naturally occurring tellurium is ^{130}Te. They measured the excess of ^{130}Xe (a stable nuclide) found in geologically old tellurium ores (crystals of Bi_2Te_3 recovered from deep underground) and deduced an approximate half-life for this double β-decay mode of $\sim 1.4 \times 10^{21}$ yr (the modern value is about $\sim 7.9 \times 10^{20}$ yr). It was not until 1987 that the first direct laboratory observation (at U.C. Irvine) of double β-decay was finally made (Elliott *et al* 1987). Using a time projection chamber, the experiment studied the decay ^{82}Se $\rightarrow$ ^{82}Kr $+ 2\beta^- + 2\bar{\nu}$. They quoted the measured half-life as $1.1^{+0.8}_{-0.3} \times 10^{20}$ yr. At that time it was the slowest naturally occurring decay process to have ever been observed directly in the laboratory. Since then, a number of experiments have observed standard double β-decay in other nuclides. All have very long half-lives, greater than 10^{18} yr. The measured half-lives range from $\sim 7 \times 10^{18}$ yr for ^{100}Mo to $\sim 2 \times 10^{21}$ yr for ^{136}Xe.

Already in 1939 Furry had suggested that, if neutrinos were Majorana fermions, then double β-decay could occur without neutrino emission, via the process now called *neutrinoless double β-decay* ($0\nu2\beta$-decay). Such a decay would proceed via the mutual annihilation of the two neutrinos produced, made possible since a Majorana neutrino is its own antiparticle. Indeed, the early theoretical predictions even gave this form as the more probable of the two. As we now know, however, this is not true; indeed none of the experiments aimed at detecting the neutrinoless process has so far produced positive results, leading to a lower bound on the half-life of approximately 10^{25} yr.

Such a decay would proceed via the mutual annihilation of the two neutrinos produced, made possible since a Majorana neutrino is its own antiparticle. If indeed

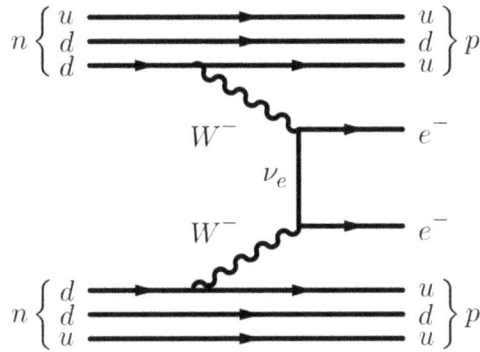

Figure 6.1. Neutrinoless double β-decay: two neutrons in the same nucleus simultaneously undergo single β-decay; however, instead of the usual emission of two neutrinos, they share so-to-speak the same Majorana-type neutrino as a virtual exchange particle.

the neutrino were its own antiparticle, then the two neutrinos normally emitted could in effect behave as a single virtual exchange field via the diagram shown in figure 6.1; for a more complete theoretical discussion see, e.g. Vergados *et al* (2012) or Bilenky and Giunti (2015). In standard double β-decay, the two neutrinos emitted carry off a large fraction of the energy released and so the total measured energy of the electron (or positron) pair has a broad spectrum; by contrast in the neutrinoless version all the energy released is carried away by the electron pair. Note that in these decays lepton number is, of course, *not* conserved.

The theoretical single differential decay rate is given by

$$\frac{d\Gamma^{0\nu\beta\beta}}{dE_1} = \frac{|m_{\beta\beta}|^2}{m_e^2} |\mathcal{M}^{0\nu}|^2. \tag{6.1.10}$$

Note that it is now a three-body decay, with one heavy and two very light final-state particles. The effective Majorana mass $m_{\beta\beta}$ here is given by

$$m_{\beta\beta} = \sum_i U_{ei}^2 m_i, = c_{12}^2 c_{13}^2 e^{2i\alpha_1} m_1 + c_{13}^2 s_{12}^2 e^{2i\alpha_2} m_2 + s_{13}^2 m_3. \tag{6.1.11}$$

where the m_i are the neutrino mass eigenvalues, U is the PMNS neutrino mixing matrix introduced in section 5.4.5, with s_{ij} and c_{ij} the related Euler-angle sines and cosines, and $\alpha_{1,2}$ the now two independent phases possible in the Majorana case[2]. The nuclear matrix element $\mathcal{M}^{0\nu}$ is proportional to a similar phase-space factor as before:

$$|\mathcal{M}^{0\nu}|^2 \propto G_F^4 \cos^4 \theta_C \ F(Z', E_1) \ F(Z', E_2) \ E_1 \ p_1 \ E_2 \ p_2, \tag{6.1.12}$$

---

[2] Since the Majorana equation does not possess the usual U(1) symmetry, it contains one less phase degree of freedom (for eliminating phases in the matrix).

where now only one of the electron energies is independent, since $E_0 = E_1 + E_2$, with E_0 as above. A then quite standard calculation leads to the following half-life (see e.g. Dolinski *et al* 2019):

$$t_{1/2}^{0\nu\beta\beta} \simeq \left[\frac{0.01 \text{ eV}}{m_{\beta\beta}} \right]^2 10^{27-28} \text{ yr.} \qquad (6.1.13)$$

Now, as seen from the global-analysis results presented earlier, the neutrino-mass region of interest is for values just around ~ 0.01 eV. Thus, if we take a benchmark ten-yr running time and a minimal requirement of ten events for credible observation, we see that a quantity of material containing of the order of $\sim 10^{27-28}$ neutrinoless double β-decaying nuclei would be necessary. For a nominal atomic mass of 100, this is equivalent to a detector mass of the order of a tonne. This is thus the mass scale of most present and planned experiments on neutrinoless double β-decay (see again e.g. Dolinski *et al* 2019).

Exercise 6.1.1. *Using the above quantities as a basis, make a more precise estimate of the mass of material required for an experimental determination of the rate for neutrinoless double β-decay.*

Many experiments are underway or have been proposed to detect such rare events but so far a positive signal has yet to be observed. A comprehensive and historical discussion of the experimental search for neutrinoless double β-decay may be found in Barabash (2011).

As a final aside, we should also mention the natural partner process known as double-electron (or -K) capture; e.g.

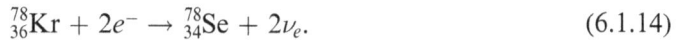

$$_{36}^{78}\text{Kr} + 2e^- \rightarrow _{34}^{78}\text{Se} + 2\nu_e. \qquad (6.1.14)$$

Once again, if the neutrino is of the Majorana type, then the following corresponding neutrinoless process becomes a possibility; e.g.

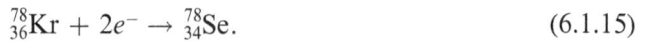

$$_{36}^{78}\text{Kr} + 2e^- \rightarrow _{34}^{78}\text{Se}. \qquad (6.1.15)$$

And again, this has yet to be seen experimentally.

6.1.3 Neutrino CP violation

As mentioned in section 2.7.8, Sakharov (1967) identified CP violation as a necessary condition to explain the baryon asymmetry of the Universe. However, the level of CP violation found in the quark sector is generally considered insufficient. Unfortunately, this is the only form of CP violation so far experimentally observed. The PMNS lepton mixing matrix (Maki *et al* 1962) allows though for a CP-violating phase and it has been shown that the consequent leptonic CP violation could generate a matter–antimatter imbalance through the process known as '*leptogenesis*' (Fukugita and Yanagida 1986).

Such CP violation is accessible via the study of oscillations between muon and electron neutrinos (and antineutrinos), as performed by the Tokai-to-Kamioka (T2K, Abe *et al* 2014) and NuMI Off-axis ν_e Appearance (NOvA, Acero *et al* 2019) experiments, exploiting accelerator-produced neutrino and antineutrino beams.

The T2K experiment compares beams of muon neutrinos (ν_μ) and antineutrinos ($\bar{\nu}_\mu$), with energies around 0.6 GeV, produced alternately by 30 GeV proton collisions on graphite. The beam then travels 295 km through the Earth between the J-PARC accelerator in Tokai and the T2K 50-kt water detector Super-Kamiokande. The similar NOvA experiment uses the main injector at Fermilab to produce a bema of neutrinos aimed at a 14-kt detector 800 km away in Ash River, Minnesota; a smaller 300-t detector is also placed close to the production point for comparison.

An indication of CP violation has been reported by the T2K Collaboration (Abe *et al* 2020): a significantly higher fraction of electron neutrinos (ν_e) were detected from the ν_μ beams, than were electron antineutrinos ($\bar{\nu}_e$) from the $\bar{\nu}_\mu$ beams. While this is clear evidence of CP violation, the results are not yet precise enough to determine the magnitude. Moreover, the NOvA experiment has yet to see evidence of CP violation in neutrino oscillations (Acero *et al* 2022) and there appears to be a slight tension between NOvA and T2K.

6.2 Grand unified theories

In the earlier chapters we have seen how the attempt to correctly describe the weak interaction and, in particular, the massive intermediate bosons it requires led to the development of a (quasi) *unified* theory of the electromagnetic and weak interactions. The concept of a spontaneously broken symmetry allows the bosons to acquire a mass *without* violating the local gauge invariance of the theory and leads to the final $U(1)_{EM}$ and $SU(2)_W$ symmetries. Note though that since there are two distinct coupling constants, g and g' in the previous sections, we cannot speak of a true unification.

Moreover, this does not yet include quantum chromodynamics (QCD), the theory of the strong interaction. The successes of the electroweak theory immediately sparked attempts to include the SU(3) gauge symmetry of QCD. The idea is a little more challenging than a mere extension of $U(1)_Y \otimes SU(2)_W$ to the larger $U(1)_Y \otimes SU(2)_W \otimes SU(3)_{QCD}$. It would, in reality, be preferable to have a *single* gauge group with thus a *single* coupling constant, which is what is known as a *grand unified theory* (GUT). Simple examples of groups containing the above product are SU(5) or SO(10).

6.2.1 The running coupling constants

The problem of the single coupling is almost automatically solved when we recall that the three known constants actually *vary* with energy scale: the largest α_{QCD} decreases, as does that associated with the $SU(2)_W$ (but more slowly), whereas the smallest α_Y (more-or-less α_{QED}) increases, see figure 6.2. Indeed, examination of the theoretical variation suggests that they should all have a roughly similar value, α_{GUT},

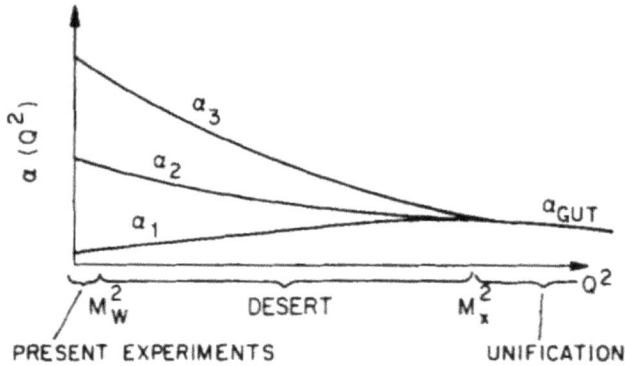

Figure 6.2. An early sketch of the possible running of the three SM coupling constants to a common value, α_{GUT}, at high energy. From bottom to top, the curves are, respectively, for the QED (α_1) weak (α_2) and strong (α_3) coupling constants; figure reproduced from Langacker (1984), with the permission of AIP Publishing.

for $E \sim 10^{15}$ GeV. We might thus imagine that the larger symmetry exists at around this high scale, but that, for lower energies, it is somehow broken spontaneously down to the product group we now perceive: $U(1)_Y \otimes SU(2)_W \otimes SU(3)_{QCD}$.

6.2.2 Proton decay

In the SM, the proton is the lightest (spin-half) baryon and must therefore be absolutely stable[3]. Now, any enlargement of the gauge group to e.g. SU(5) Georgi and Glashow (1974), SO(10) Georgi *et al* (1975), E_6 Gürsey *et al* (1976), or similar, always necessarily implies the introduction of additional gauge degrees of freedom: new (heavy) gauge bosons (usually indicated X). As we have seen, the SM contains 12 bosons (the photon, the three W/Z bosons and eight gluons). Since e.g. SU(N) has $N^2 - 1$ generators, we immediately deduce that even a minimal SU(5) extension would require 24 generators and therefore twelve new gauge fields, which should all have masses at least of order of the GUT scale M_X. That is, they would be responsible for interactions that would appear super-weak at the typical energy scales even of present-day high-energy accelerator experiments.

Now, some of these new fields would also inevitably couple quarks with leptons (in much the same way that the $W^{\pm}$ couple charged leptons and neutrinos or up-type and down-type quarks) and thus allow for quark- or baryon- and lepton-number-changing transitions. In particular, the proton would no longer be stable since the quarks inside it could be converted into leptons and antiquarks, thus leading typically to decay modes such as $p \to \pi^0 e^+$ or $\pi^0 \mu^+$ (see figure 6.3). In these models, while the baryon and lepton numbers (B and L) are separately violated, the difference $B - L$ is conserved and so the lepton produced here is either a positron or a μ^+. Using e.g. a water-Čerenkov detector, the experimental signal would thus be

---

[3] Note here that we are referring to free protons and thus β^+-decay ($p \to n e^+ \nu_e$), which can only take place inside a nucleus, is not to be considered true proton decay in this sense.

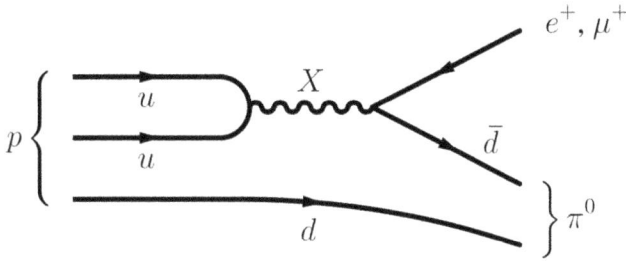

Figure 6.3. The Feynman diagram representing the proton-decay processes $p \to \pi^0 e^+$ and $\pi^0 \mu^+$ via an intermediate X-boson.

three Čerenkov rings: one due to the positron emitted and two from the two photons produced by the π^0 decay.

Based on the picture depicted in figure 6.3, a rough dimensional-based estimate of the proton lifetime in such minimal GUT models may be obtained by combining Sargent's rule (see the footnote following equation (2.1.20)), i.e. $\Gamma \propto Q^5 \approx m_p^5$, with an intermediate X-boson propagator of mass m_X (which has a typical GUT value of $m_X \sim 10^{15}$ GeV), implying an amplitude factor m_X^{-2}. The interaction then implies that the amplitude acquires a further factor of α_{GUT}, which has a typical value of approximately 0.02. Put all together, this gives $\Gamma_p \approx \alpha_{\text{GUT}}^2 m_p^5 / m_X^4$, thus leading to $\tau_p \approx O(10^{30-31} \text{ yr})$.

Indeed, in all such minimal grand-unified models the proton lifetime is estimated to be no greater than about 10^{31} yr, whereas present 90%-CL lower limits coming from *non*-observation of proton decay in the currently most sensitive dedicated experiment, the 27.2 kt Super-Kamiokande, are (Takenaka *et al* 2020) 2.4×10^{34} yr for the $\pi^0 e^+$ mode and 1.6×10^{34} yr for $\pi^0 \mu^+$. However, there are versions of such theories in which the proton lifetime could be as long as 10^{36} years, but again this should be within the reach of future experiments. A planned future experiment, Hyper-Kamiokande, is expected to increase these limits by a factor of 5–10. A comprehensive review covering the question of proton decay in the various types of GUTs may be found in Nath and Fileviez Pérez (2007), while a more a recent brief review is given in Ohlsson (2023).

6.2.3 Renormalisation problems

There are further problems with the *naïve* extensions to obtain a GUT. First of all, even in the simpler electroweak theory, problems to do with renormalisation still remain. The presence of a scalar field (necessary for the Higgs mechanism) upsets the usual renormalisation programme: the contributions coming from virtual scalar loops tend to shift the masses of the particle spectrum up to the upper momentum cut-off, which here should be of order $E \sim 10^{15}$ GeV. In order to arrive at the masses of the known quarks and leptons, we therefore have to invoke very delicate cancellations between different contributions. That is, the unphysical (bare) masses need to be fine-tuned to many decimal places in order that the difference of two large numbers be the small number required; this is known as the hierarchy or fine-tuning

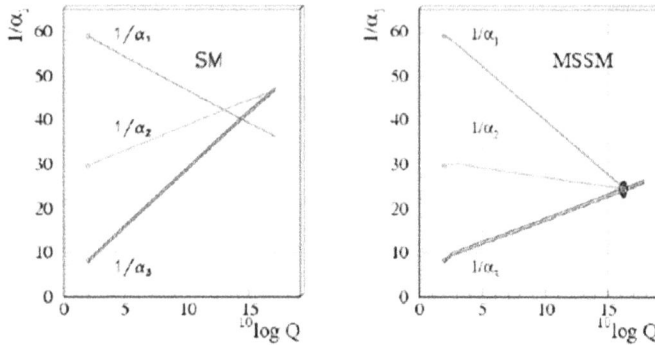

Figure 6.4. The running of the three coupling constants for the SM on the left and the so-called *minimal supersymmetric* SM on the right. From top to bottom, the curves are, respectively, for the QED, weak and strong coupling constants; figure reproduced from Kazakov (2001) CC BY 4.0.

problem. Moreover, close examination of the running of the three coupling constants reveals that they do *not* all meet at a single energy and thus such a *naïve* grand unification is not strictly speaking possible (see figure 6.4).

6.3 Supersymmetry

The difficulties encountered in constructing GUTs, as discussed above, suggest some missing ingredient. There are many possibilities, such as models with composite quarks and/or leptons (see e.g. D'Souza and Kalman 1992). Here we shall just mention the presently most favoured: namely, *supersymmetry*, originally proposed by Wess and Zumino (1974). In a nutshell, the idea is that the fermions and bosons of the theory should be related by a supersymmetry transformation. While consideration of pure quantum eletrodynamics (QED) with only the electron might allow such a possibility (i.e. relating the photon to the electron), the full SM does not possess such symmetrical numbers of fermion and boson degrees of freedom.

The idea then in supersymmetric extensions of the SM is to enlarge the spectrum of particles and symmetries by including a symmetry operation that transforms fermions into bosons and vice versa. It is then necessary to introduce an entire family of supersymmetric partners for the existing particles. Thus, to every quark there corresponds a pair of scalar *squarks* (matching the two quark helicity degrees of freedom), while for every gauge boson there is a fermionic *gaugino*. Since no such particle states have ever been observed, they must evidently be very heavy.

With a suitable spectrum (typical masses should be of the order of a TeV but not much more), both the fine-tuning and GUT-point problems may be solved. The fine-tuning is now automatic: the 'opposite' spin nature means that the contributions in quantum corrections enter with the opposite sign and thus cancellations are guaranteed. This has to do with Fermi–Dirac statistics: loop diagrams with a fermion circulating acquire an extra minus sign with respect to boson loops. Such cancellations will not be exact however, owing precisely to the necessarily large masses of the hypothesised supersymmetric particles. Indeed, this observation helps to place an upper limit on the likely masses, in order that the cancellations should not be lost altogether.

Secondly, the extra contributions alter the running of the coupling constants, which can now be arranged to all coincide at one particular energy. The new unification point is typically of order 10^{16} GeV (see figure 6.4). Both solutions are achieved if and only if the *sparticle* masses are around 1 TeV (Amaldi *et al* 1991). Such a limit suggests that they may well be within reach of LHC.

6.4 Superstrings and beyond the four dimensions

In the early eighties, working on string models of hadronic interactions, Green and Schwarz (1982) realised that supersymmetric versions of such theories automatically contained a spin-2 gauge field in their particle spectrum, a perfect candidate for the graviton. This immediately led to the hope for a truly GUT, including gravity and thus resolving the long-standing problem of quantising Einstein's theory of general relativity.

As already mentioned, the idea of a string-like interaction appears natural in the description of hadrons and phenomenological models enjoyed discrete success (Rebbi 1974; Veneziano 1974); a string-tension of 1 GeV fm^{-1} and therefore a length of order 1 fm was quite reasonable. However, on the purely theoretical side they suffer severe inconsistencies (known as '*anomalies*') unless embedded in a 26-dimensional world, leaving them somewhat unappealing. Green and Schwarz (1984)[4] realised that the new supersymmetric versions, so-called superstring theories, required '*only*' ten dimensions to avoid the anomalies. At the same time the choice of gauge group appeared restricted to SO(10) or $E_8 \otimes E_8$; both are ideal candidates for a GUT. Moreover, it was also realised that the excess six dimensions could be conveniently '*compactified*' (or rolled up) into closed forms and since the natural length (mass) scale for such a string is determined by the Planck scale 10^{-20} fm (10^{19} GeV), they would effectively be unresolvable at the scales experimentally accessible. For a general theoretical review see, e.g., Green *et al* (1987).

What then might be the phenomenological implications of such theories? Unfortunately, the present understanding of superstring theory does not permit any phenomenological predictions and, with the natural scales mentioned above, it is hard to imagine any immediately accessible physical consequences. This does, however, bring us to our final consideration on physics beyond the SM: the question of extra dimensions. In a somewhat different context, but with a similar idea of extra compact dimensions, the hypothesis of unification, including gravity, at the weak scale (Arkani-Hamed *et al* 1998) leads to a standard weak-interaction phenomenology. The SM fields fields only propagate in four dimensions, whereas the graviton also propagates in the extra dimensions, which are here characterised by a weak scale of order 1 TeV. There should then be a marked transitional behaviour: $1/r^2 \rightarrow 1/r^{2+n}$ for n extra dimensions. It is thus argued that already at LHC energies

---

[4] In 1990, Edward Witten became the first physicist to be awarded a Fields Medal by the International Mathematical Union, for his mathematical insights in physics.

such deviations from standard behaviour might be observable via missing-energy effects (for recent results see, e.g., Aad *et al* 2013; Chatrchyan *et al* 2012).

In conclusion then, the possibilities for physics beyond the SM are numerous and there a good reasons to hope that some signal may appear in the experiments already being performed at LHC. That said, there is at present no indication of the correct or even a more likely direction to follow and we shall thus have to leave no stone unturned and patiently analyse all data with an eye to possible deviation from SM predictions.

6.5 Dark matter and energy

A question that has grown considerably in importance, both theoretically and experimentally, over recent years is that of the general and large-scale composition of the Universe. While the problem arises in the context of astrophysics and the observations of stellar motion, a far cry from the world of quarks and leptons that we have been studying so far, as we shall now see, it naturally leads to the quest for a particle-physics answer.

6.5.1 Dark matter

Two questions naturally arise: first and foremost, why should we even consider such a possibility? And secondly, even if it should exist in important quantities, how might its presence be verified? A very complete and recent review of the observational, experimental and theoretical status of dark matter can be found in Cirelli *et al* (2024).

6.5.1.1 The discovery of dark matter
The concept of dark matter has a very long history. Many of the early considerations were somewhat speculative and based on misconceptions or incorrect estimates (of e.g. the age of the Sun). Among those who had discussed the idea were Lord Kelvin and Henri Poincaré. Their conclusions were, however, very imprecise by modern standards. Kelvin posed the problem first by questioning whether many stars might actually be dark and therefore invisible. This is, of course, quite a reasonable hypothesis. The real conundrum then is how the presence of such matter could be confirmed despite its being, by definition, invisible and at that time undetectable.

The first notable paper on the subject is probably that of Zwicky (1933), who provided an early answer. By applying the virial theorem (see appendix A.1) to the Coma cluster (containing over 1000 identified galaxies), Zwicky obtained evidence of what he called '*dunkle Materie*' (dark matter). From the brightness and number of galaxies observed, he estimated that they should have a velocity dispersion of around 80 km s^{-1}. In contrast, the actual measured velocities gave a value near to 1000 km s^{-1}. In other words, they were moving so fast that they should have escaped the gravitational field of the cluster, which was evidently not the case. His calculations led him to deduce a total mass of around 400 times that of the visible objects. It turns out that, owing in part to an old and inaccurate value of the Hubble constant, his estimates were incorrect by an order of magnitude. The basic conclusion was, however, valid: the cluster contained far more invisible than visible matter.

The acclaimed definitive proof of the existence of dark matter come from the work of Rubin and Ford (1970) and Freeman (1970) in the 1960s and 70s. Studying galaxy rotation curves with a new spectrograph technique to accurately measure the velocity curve of edge-on spiral galaxies, they found that galaxies must typically contain a quantity of dark matter around six times the visible mass (Rubin and Ford 1970; Freeman 1970).

There are now many other independent observations fully corroborating the early deductions: weak gravitational lensing, large-scale structure formation, CMB anisotropy and acoustic peaks etc (see e.g. Cirelli *et al* 2024). The conclusions are embodied in the standard Lambda Cold Dark Matter Model (ΛCDM) of cosmology, where the mass–energy content of the Universe is given as 5% visible matter, 26.8% dark matter and 68.2% dark energy.

6.5.1.2 The search for dark matter

Since it is safe to assume that the Earth is situated inside a region containing dark matter, we may also assume that there is a continuous flux through any and all terrestrial detectors. The main conceptual difficulty lies in our ignorance as to the precise nature of dark matter particles: any assumption we make might be wrong and the detectors considered therefore ineffective. A real practical difficulty lies in the very low expected scattering rates, which may be less than one event per ton of target material per year. Moreover, the expected energy deposition is likely to be limited to the range eV to keV. The implicit background problem then requires underground experiments employing ultra-pure materials.

As with standard detection techniques, kinematics plays an important rôle in determining the probable energy deposition. The two basic possible scatterers are atomic electrons and atomic nuclei. From our earlier discussions of scattering kinematics, we know that a light projectile impinging on a much heavier object will deposit only a negligible amount of its kinetic energy. Moreover, we know that the typical velocity to be expected is $\beta \approx O(10^{-3})$. Non-relativistic kinematics is thus appropriate. It is then a simple exercise to show that the typical energy transfer ν in an elastic collision between a projectile of mass M and a stationary target of mass m is given by

$$\nu = \frac{\beta^2\, mM}{2(M + m)}. \tag{6.5.1}$$

Thus, the best case is when the projectile has a similar mass to the target, which means $O(10 - 100\,\text{GeV})$ for nuclei and $O(1\,\text{MeV})$ for electrons, with energy deposition, respectively, a few keV and a few eV. A *naïve*, back-of-the-envelope calculation, using typical neutrino-interaction cross-section as an example, gives possible event rates of the order of a few per kg of material per year.

We may also estimate the cross-sections to be expected in the case of different types of interaction. A typical strong-interaction cross-section may be set as

$\Lambda_{QCD}^{-2} \approx 10^{-26}$ cm^2. Note that for such a hadronic cross-section, the particle would actually be stopped in the atmosphere and would never reach an underground detector, but in any case such dark matter has essentially been excluded. A neutrino-like weakly interacting massive particle (WIMP) would have $\sigma \approx 10^{-38}$ cm^2, while Higgs-boson exchange would give $\sigma \approx 10^{-43}$ cm^2. The present experimental limit on the nucleon cross-section is $\sigma \lesssim 10^{-46}$ cm^2. All of these are therefore essentially excluded, unless the coupling constants involved are very small for some reason.

6.5.1.3 Axion and axion-like dark matter

A more specific dark-matter candidate is the axion, together with other so-called axion-like particles (ALPs); for a general review, see e.g. Choi *et al* (2021). Since the axion is generated as Goldstone boson of a global symmetry that is broken at some high scale Λ_a (typical a GUT scale, say 10^{15-16} GeV), its couplings to the SM fields are well determined (see e.g. Graham and Rajendran 2013), as in the following interaction Lagrangian terms, which should each be multiplied by a coupling constant of $O(1)$:

$$\frac{1}{\Lambda_a} a\, F^{\mu\nu}\tilde{F}_{\mu\nu}, \quad \frac{1}{\Lambda_a} a\, G^{a\mu\nu}\tilde{G}_{\mu\nu}^a, \quad \frac{1}{\Lambda_a} a\, \bar{\psi}_f \sigma_{\mu\nu}\gamma_5\psi_f F^{\mu\nu}, \quad \frac{1}{\Lambda_a}\partial_\mu a\, \bar{\psi}_f \gamma^\mu\gamma_5\psi_f, \quad (6.5.2)$$

where a is the axion field, $F_{\mu\nu}$ and $G_{\mu\nu}^a$ are, respectively, the QED and QCD field-strength tensors and ψ_f is any SM fermion. That is the first and second represent normal couplings to photons and gluons, the third is a coupling to an electric dipole momentum (as there might be in the case of strong CP violation), the last is an axial coupling that would cause something similar to the photoelectric effect, but coupling to the spin of the fermion. In the case of quarks, this last requires knowledge of the spin-dependent structure functions, see equation (3.3.6), which have now been measured rather well (see e.g. Anselmino *et al* 1995). Moreover, the various astrophysical bounds on the interactions in equation (6.5.2) exclude values of $f_a \lesssim 10^9$ GeV. On theoretical grounds we expect the value to be set by the scale of new physics; e.g. the GUT scale (10^{15-16} GeV) or the Planck scale (10^{19} GeV).

Now, if they were true Goldstone bosons, they would be massless and could not contribute to dark matter. However, they could acquire a mass if the associated global symmetry were anomalous (as is the case of pions for example). If this were, say, a QCD anomaly then a mass could be generated by non-perturbative QCD dynamics: $m_a \approx \Lambda_{QCD}^2/\Lambda_a$, which we rewrite as

$$m_a \approx 6 \times (10^{16}\text{ GeV}/\Lambda_a) \times 10^{-10}\text{ eV}. \quad (6.5.3)$$

Most experiments aimed at directly detecting axions exploit the first of the couplings in equation (6.5.2) and thus use electromagnetic fields via e.g. microwave cavities. The difficulty here is that, in order to enhance the signal, the cavity must be in resonance with the axion mass. That is, the cavity should be the approximate size of the axion Compton wavelength $1/m_a$. For the typical GUT scale suggested above, $1/m_a \approx 300$ m!

6.5.2 Dark energy

The true nature of dark energy is entirely unknown, with only very speculative suggestions. It is a purely hypothetical form of energy necessary to explain the accelerating rate of the expanding Universe, for which it is required to constitute around 68%–70% of the total energy–matter density of the Universe. Dark energy is presumably homogeneous throughout the Universe; it is certainly not at all dense and is thought to interact only via gravity. Therefore, since it does not represent any form of massive matter and only has a density equivalent of approximately $10^{-27}\,\mathrm{kg\,m^{-3}}$, it should be almost undetectable in any foreseeable laboratory experiments. Although some hint of dark energy has been reported, later experiments have essentially ruled this out (Aprile *et al* 2022). We are thus forced to leave discussion of this topic to the more specific texts available.

6.6 The effective-field-theory approach

It seems appropriate to close the theoretical discussions with what is now generally considered the more pragmatic approach to the search for physics beyond the SM. Let us first be clear on the nature of the problem. As we have stressed on many occasions here, the SM as presently formulated (i.e. including the possibility of non-zero neutrino masses) appears to describe all known phenomena rather well[5]. We are thus left searching for unpredicted phenomena or deviations from the predicted values of SM parameters. The question is how to guide such searches.

The traditional approaches may be divided into two broad categories: '*top-down*' and '*bottom-up*'. The first, preferred for many years, relies on the existence of some sort of '*Theory of Everything*' or nearly everything (*e.g. SuSy, SuGra*, superstrings etc), which would hopefully be fully specified by requirements of internal consistency and which should produce the SM and its results as a low-energy limit. Such an approach has now been largely abandoned as fruitless: almost infinitely many such theories are plausible and are generally associated with very high energy scales, rendering extraction of phenomenological consequences for laboratory based experiments arduous if not impossible.

The second approach is based on what are known generically as '*effective field theories*'. Unwittingly perhaps, the reader has already been exposed to an example of just such an approach earlier in this volume: namely, the Fermi theory of the weak interaction (see section 2.1.2). Fermi's strategy was to construct an (effective) interaction (operator) by exploiting the then known (particle) fields; i.e. *n, p, e* and ν. His four-field interaction operator was the simplest and lowest-dimension interaction possible. The extension by Gamow and Teller (1936) was little more than the inclusion of all possible operators of the same dimension (by coupling via the Pauli spin matrices) and determining experimentally which were actually present. Reading the passage depicted in figure 5.1 in the opposite direction to that indicated there, we see that the higher-scale, more complete theory, containing higher-mass

---

[5] While at any given point in time over the past decades, there may have arisen various small discrepancies that could have been signals of failure, none has yet stood the test of time.

fields, is reduced to a low-energy effective theory in which the high-mass fields have been integrated out[6].

The approach is thus to include all possible higher-order effective interaction terms to the SM Lagrangian in the form of a systematic expansion in powers of mass. The idea is that, by dimensional analysis, the higher-mass operators should have coefficients containing suitable inverse powers of the mass scale (usually denoted Λ) of the true full global theory. Since this scale is presumably very high (probably at least 10^{15} GeV, see section 6.2.1), this implies an ever-increasing suppression of higher-order operators. That is, we have a very useful expansion.

Now, any standard Lagrangian (or Lagrangian density) has dimension four[7], which means that the new composite terms begin at dimension five. That is, we may imagine the following extension of the SM Lagrangian:

$$\mathcal{L}_{\text{SMEFT}} = \mathcal{L}_{\text{SM}} + \sum_{\forall i, n \geqslant 5} \frac{C_i \mathcal{O}_i^{(n)}}{\Lambda^{n-4}}, \tag{6.6.1}$$

where the dimensionless Wilson coefficients C_i are, in principle, $O(1)$. There is only one possible type of dimension-five operator, which induces lepton-number violation and also neutrino Majorana masses at a level compatible with present limits and oscillation measurements if $\Lambda \approx O(10^{15} \text{ GeV})$.

The standard general analysis thus begins at dimension six. The choice of operators is usually limited by the requirements of Lorentz and gauge invariance, as we have no reason to assume their violation. This unfortunately still leaves a very large number of possible terms. The advantage though of such an approach is that it make no reference to any specific model and therefore does not introduce any theoretical bias. A rather complete and pedagogical review, together with some example applications, may be found in Falkowski (2023); see also Brivio and Trott (2019).

6.7 Future colliders

It seems appropriate in this final chapter on physics beyond the SM to conclude with a discussion of the current experimental outlook and, in particle, the future of collider physics. Normally, this might not have found a place in such a volume as it would risk becoming out-of-date too rapidly. However, after decades of continuous and regular increases in accelerator energies, we are now set for a much longer wait for the next significant developments (see figure 6.5), which will then presumably have rather long lifetimes.

There can be no doubt that progress in experimental high-energy particle physics over the decades has been determined to a large extent by continued improvement in

---

[6] The expression 'integrated out' makes reference to the Feynman path-integral representation of a field theory and the idea that it may be reduced to a low-energy effective theory by expanding and performing the integral over the high-mass fields.

[7] Note that all spin-1/2 fermion spinor fields have dimension 3/2, while all gauge fields have dimension one.

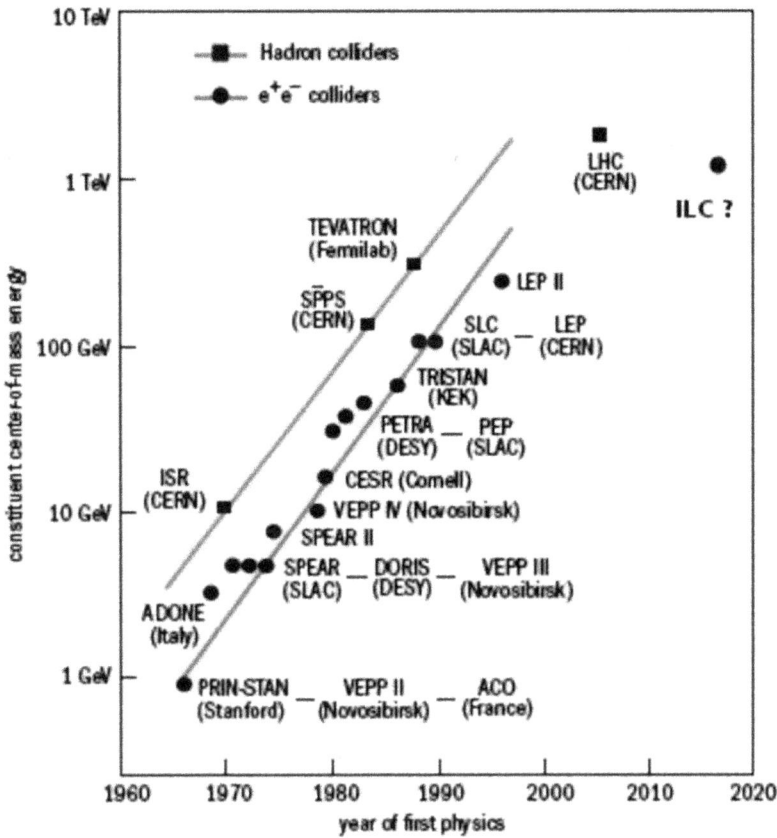

Figure 6.5. A so-called *Livingston plot* of the centre-of-mass energy of particle accelerators (lepton and hadron colliders) versus time. In the case of hadron machines, the energies have been adjusted to account for quark and gluon constituents carrying only a fraction of the proton energy ($E_q \approx \frac{1}{6}E_p$); figure from Spentzouris *et al* (2008). Copyright IOP Publishing Ltd. All rights reserved.

accelerator capabilities. Not only have the beam energies increased by orders of magnitude, but also their intensities.

The late-sixties SLAC studies on the structure of the proton might be considered the first true high-energy particle-physics laboratory experiments. The energy available then was about 2 GeV in an electron beam incident on a hydrogen (proton) target; the equivalent total centre-of-mass energy was thus also about 2 GeV. The present-day CERN facility, with its Large Hadron Collider (LHC), which produced the first collisions in 2010, now reaches a total centre-of-mass energy close to 14 TeV[8]; that is, an increase of nearly four orders of magnitude in just over four decades. However, as noted, this historical trend appears to have recently slowed down. That said, there are a number of projects worldwide considering possible future machines of various types. We shall now briefly list a few of these.

---

[8] N.B. This is the *pp* centre-of-mass energy, not that of the interacting *qq* or *qq̄*.

6.7.1 The International Muon Collider Collaboration (IMCC)

We begin by mentioning a current project investigating the feasibility of producing and exploiting very high-energy colliding muon beams (MuCol). Why muons? For a given energy, synchrotron radiation (one of the principal energy limitations in circular electron–positron accelerators) is proportional to the inverse fourth power of the particle mass and therefore, since $m_e/m_\mu \simeq 1/207$, there is a relative suppression factor of the losses in a muon machine with respect to an equivalent electron ring of about 0.5×10^{-9}. Although the muon is unstable, its lifetime of $1.6 \; \mu s$, combined with relativistic time dilation, would allow it to complete approximately $150 \, B$ revolutions in a ring of which about half contains bending magnets with an average field of B tesla[9]. Of natural concern though are the high background levels due to the inevitable presence of the muon-beam decay products.

A muon collider running at a centre-of-mass energy of around $\sqrt{s} = 10$ TeV and with a luminosity of order 10^{35} cm^{-2} s^{-1} would have a sufficiently high Higgs-boson production rate to directly measure its trilinear and quadrilinear self-couplings, thus enabling precision studies of the Higgs self-interactions. And, for example, it has been demonstrated that the measurement of the process $\mu^+\mu^- \to H\nu\bar{\nu} \to b\bar{b}\nu\bar{\nu}$ is feasible with a precision on $\sigma(H\nu\bar{\nu}) \cdot \mathcal{B}(Hb\bar{b})$ at the 1% level for $\sqrt{s} = 3$ TeV, which is competitive with other proposed machines. The present proposal aims at a two-stage development over periods of $15 + 15$ years: stage 1 (2) reaching 3 (10) TeV centre-of-mass energies, with an estimated luminosity of around $2 \; (20) \times 10^{34}$ cm^{-2} s^{-1} and a beam lifetime of 1039 (1558) revolutions. The size of the ring would be 4.5 (10) km in circumference.

Further information may be found at the IMCC website.

6.7.2 The Future Circular Collider (FCC)

In December 2018, CERN (Abada *et al* 2019) unveiled the blueprint for a 90- km circular collider, the FFC, reaching a *pp* centre-of-mass energy of 100 TeV, intended to search for physics beyond the SM and study the Higgs boson in detail. The design report for the FCC presents the project for an underground particle accelerator that would be connected to the existing LHC and operate as both a hadron and lepton collider.

It envisages e^+e^-, *pp*, *ep* and heavy-ion programmes, with the aim of studying electroweak, Higgs and strong interactions, the top quark and heavy-flavour physics, as well as phenomena beyond the SM, such as various dark-matter scenarios. The e^+e^- collider (FCC-ee) would operate at various centre-of-mass energies, producing around 5×10^{12} Z^0 bosons (91 GeV), 10^8 W^+W^- pairs (160 GeV), 10^6 Higgs bosons (240 GeV) and 10^6 $t\bar{t}$ pairs (350–365 GeV). In its *pp*-collider phase (FCC-hh), it is expected tobproduce more than 10^{10} Higgs bosons; it could

---

[9] Muon storage rings do already exist: *e.g.* the Muon $g - 2$ experiment at FNAL uses a 14-m diameter electromagnet to maintain a secondary muon beam in an orbit of the same size. The purpose is precision measurement (via precession) of the muon magnetic moment, a parameter of fundamental importance to our understanding of elementary particle theory. Note, they are not employed to produce collisions.

also operate with heavy ions (e.g. Pb–PB at 39 TeV). Optionally, the FCC-eh, with 50 TeV protons colliding with 60 GeV electrons would provide 3.5 TeV ep collisions.

Further information may be found at the FCC website.

6.7.3 The Circular Electron–Positron Collider (CEPC)

In September 2018, the CEPC Study Group (2023) of Chinese physicists presented the conceptual design for a CEPC, a 100 km ring e^+e^- collider, capable of reaching a centre-of-mass energy of 240 GeV. With such an energy the CEPC would serve as a Higgs factory, generating more than two million Higgs particles. The aim is thus to study the Higgs-boson properties as precisely as possible, including mass, spin, CP nature, couplings, etc. The design also allows for operation at 91 GeV to provide a Z^0-factory and at 160 GeV as a W-factory, producing around 10^{12} Z^0 bosons and about 15×10^6 W^+W^- pairs.

Moreover, it has the potential to be upgraded to a proton–proton and heavy-ion collider, the Super Proton–Proton Collider, and reach unprecedented high energies in an attempt to discover new physics beyond the SM. The SPPC and CEPC could actually operate simultaneously and thus permit both ep and electron–ion programmes. The initial centre-of-mass energy of the SPPC would be 75 TeV with a nominal luminosity of 5×10^{34} cm^{-2} s^{-1}; higher luminosities are possible. A final upgrade phase for the SPPC is envisaged reaching centre-of-mass energies of 125–150-TeV. The timeline currently envisages first running in 2035 for the CEPC and 2045 for the SPPC.

Further information may be found at the CEPC website.

6.7.4 The International Linear Collider (ILC)

Plans for an International Linear Collider to be hosted in the mountainous regions of northern Japan as a global project (Aihara *et al* 2019) are currently being considering by the Japanese government.

A linear collider has the advantages of avoiding the synchrotron radiation losses that plague e^+e^- ring colliders and of permitting straightforward and continuous energy upgrades, which would allow the ILC to remain competitive for a very long period. Moreover, the necessary technology is already available.

The aim of the ILC would be to produce electron–positron collisions with highly polarised beams (80%) at an initial centre-of-mass energy of 250 GeV (with possible extensions up to 1 TeV). The project foresees two linear accelerators approximately 20 km long. The ILC aims to collide electrons and positrons, with collisions being produced at a rate of nearly 7000 per second. The initial focus would be on high-precision and model-independent measurements of the Higgs-boson couplings, with an expected precision of the order of a very few percent. The construction phase could begin very soon and is envisaged to require about 9 years.

Further information may be found at the ILC website.

6.7.5 The Compact Linear Collider (CLIC)

Finally, there is a proposal for an electron–positron, so-called Compact Linear Collider, Aicheler *et al* (2019), which would aim to reach 3 TeV, requiring an acceleration gradient of 100 MeV m^{-1}, in order to provide high sensitivity to physics beyond the SM. Again, the linear design avoids the synchrotron radiation problem and permits progressive energy upgrades. The accelerator is foreseen in three stages, at centre-of-mass energies of 380 GeV, 1.5 TeV and 3 TeV, for a total length ranging from 11 km to 50 km (more than ten times the final length of SLAC).

Precise measurements of e.g. the top quark will be possible at all three stages and would provide sensitivity to potential physics beyond the SM. The proposed starting date for construction of the first CLIC energy stage is 2026. This first beams should thus be available by 2035, initiating a physics programme that should continue for the following 25 to 30 years.

Further information may be found at the CLIC website.

References

Aad G *et al* (ATLAS Collab) 2013 *JHEP* **1304** 075

Abada A *et al* (FCC Collab) 2019 *Eur. Phys. J.* **C79** 474 *Eur. Phys. J. ST* **228** 261; **228** 755; **228** 1109

Abdallah W *et al* (CEPC Study Group) 2023 arXiv:2312.14363

Abe K *et al* (T2K Collab) 2014 *Phys. Rev. Lett.* **112** 061802

Abe K *et al* (T2K Collab) 2020 *Nature* **580** 339 *erratum ibid* **583** E16

Acero M A *et al* (NOvA Collab) 2019 *Phys. Rev. Lett.* **123** 151803

Acero M A *et al* (NOvA Collab) 2022 *Phys. Rev.* **D106** 032004

Aicheler M *et al* 2019 arXiv:1903.08655

Aihara H *et al* ILC Collab 2019 arXiv:1901.09829

Amaldi U, de Boer W and Fürstenau H 1991 *Phys. Lett.* **B260** 447

Anselmino M, Efremov A and Leader E 1995 *Phys. Rep.* **261** 1 *erratum ibid* **281** 399

Aprile E *et al* (XENON Collab) 2022 *Phys. Rev. Lett.* **129** 161805

Arkani-Hamed N, Dimopoulos S and Dvali G R 1998 *Phys. Lett.* **B429** 263

Barabash A S 2011 *Phys. Atom. Nucl.* **74** 603

Bilenky S M and Giunti C 2015 *Int. J. Mod. Phys.* **A30** 1530001

Brivio I and Trott M 2019 *Phys. Rep.* **793** 1

Chatrchyan S *et al* (CMS Collab) 2012 *JHEP* **1209** 094

Choi K, Im S H and Sub Shin C 2021 *Annu. Rev. Nucl. Part. Sci.* **71** 225

Cirelli M, Strumia A and Zupan J 2024 arXiv:2406.01705

Dolinski M J, Poon A W P and Rodejohann W 2019 *Annu. Rev. Nucl. Part. Sci.* **69** 219

D'Souza I A and Kalman C S 1992 *Preons: Models of Leptons, Quarks and Gauge Bosons as Composite Objects* (Singapore: World Scientific)

Elliott S R, Hahn A A and Moe M K 1987 *Phys. Rev. Lett.* **59** 2020

Falkowski A 2023 *Eur. Phys. J.* **C83** 656

Freeman K C 1970 *Astrophys. J.* **160** 811

Fukugita M and Yanagida T 1986 *Phys. Lett.* **B174** 45

Furry W H 1939 *Phys. Rev.* **56** 1184

Gamow G and Teller E 1936 *Phys. Rev.* **49** 895

Georgi H, Wolfe H C and Carlson C E 1975 *Proc. of the Williamsburg Meeting of APS/DPF—Particles and Fields 1974 AIP Conf. Proc.(Williamsburg, Sept. 1974)* **vol 23**; H C Wolfe and C E Carlson *AIP Conf. Proc.* p 575

Georgi H and Glashow S L 1974 *Phys. Rev. Lett.* **32** 438

Goeppert-Mayer M 1935 *Phys. Rev.* **48** 512

Graham P W and Rajendran S 2013 *Phys. Rev.* **D88** 035023

Green M B and Schwarz J H 1982 *Phys. Lett.* **B109** 444

Green M B and Schwarz J H 1984 *Phys. Lett.* **B149** 117

Green M B, Schwarz J H and Witten E 1987 *Superstring Theory. Vol. 1: Introduction Vol. 2: Loop Amplitudes, Anomalies and Phenomenology* (Cambridge: Cambridge University Press)

Gürsey F, Ramond P and Sikivie P 1976 *Phys. Lett.* **60B** 177

Inghram M G and Reynolds J H 1950 *Phys. Rev.* **78** 822

Kazakov D I 2001 *Proc. of the 2000 European School of High-Energy Physics—ESHEP 2000 CERN Yellow Rep. School Proc.(Caramulo, Aug. Set. 2000)* **vol 2001-003**; N Ellis and J D March-Russell *CERN Yellow Rep. School Proc.* p 125

Langacker P 1984 *Proc. of the Annual Meeting of the Division of Particles and Fields of the APS (Blacksburg, Sept. 1983) AIP Conf. Proc.* **vol 112** p 251

Majorana E 1937 *Nuovo Cim.* **14** 171 *transl. by L. Maiani in Ettore Majorana Scientific Papers*, ed G F Bassani (Springer-Verlag, 2006) p 201

Maki Z, Nakagawa M and Sakata S 1962 *Prog. Theor. Phys.* **28** 870

Nath P and Fileviez Pérez P 2007 *Phys. Rep.* **441** 191

Ohlsson T 2023 *Proc. of the 30th. Int. Conf. on Neutrino Physics and Astrophysics–Neutrino 2022 (Seoul, 2022); Nucl. Phys.* **B993** 116268

Rebbi C 1974 *Phys. Rep.* **12** 1

Rubin V C and Ford W K Jr. 1970 *Astrophys. J.* **159** 379

Sakharov A D 1967 *Pis'ma Zh. Eksp. Teor. Fiz.* **5** 32 *reprinted in Sov. Phys. Usp.* **34** 417

Sargent B W 1933 *Proc. R. Soc. Lond.* **A139** 659

Spentzouris P *et al* 2008 *J. Phys. Conf. Ser.* **125** 012005

Takenaka A *et al* (Super-Kamiokande Collab) 2020 *Phys. Rev.* **D102** 112011

Veneziano G 1974 *Phys. Rep.* **9** 199

Vergados J D, Ejiri H and Šimkovic F 2012 *Rep. Prog. Phys.* **75** 106301

Wess J and Zumino B 1974 *Nucl. Phys.* **B70** 39

Witten E 1984 *Phys. Lett.* **B149** 351

Yanagida T 1979 *Proc. of the Workshop on the Unified Theories and the Baryon Number in the Universe (Tsukuba, Feb. 1979) Conf. Proc.* **vol C7902131**; O Sawada and A Sugamoto p 95

Zwicky F 1933 *Helv. Phys. Acta.* **6** 110

IOP Publishing

An Introduction to Elementary Particle Phenomenology
(Second Edition)

Philip G Ratcliffe

Appendix A

Background notes

*'Notes aren't a record of my thinking process. They **are** my thinking process.'*

Richard Feynman

A.1 The virial theorem

Here we briefly recall the virial theorem, which may be used to relate the potential and kinetic energies of physical systems. The basic requirement is a stable interacting system of particles, which might be electrons in an atom, nuclei in a star, stars in a galaxy etc. Mathematically, for a general system of point masses with positions r_i subject to applied forces F_i, the theorem is expressed in its most general form as

$$\langle K \rangle = -\tfrac{1}{2}\sum_i \langle F_i \cdot r_i \rangle, \tag{A.1.1}$$

where the left-hand side is the time-averaged total kinetic energy of the system and the right-hand side is called the '*virial*' of the system and is related to the total potential energy generated by the interactions between the constituents of the system. To see this, let us restrict our attention to the simple but common case of particles or objects subject to conservative central forces $F_i = -\nabla U(r_i)$, where $U(r_i)$ is the potential energy of the ith particle. Equation (A.1.1) therefore becomes

$$\langle K \rangle = \tfrac{1}{2}\sum_i \langle r_i \cdot \nabla U(r_i) \rangle. \tag{A.1.2}$$

Note that this equation may be exploited in either direction. That is, if we know the form and the value of the potential involved, then we can estimate the mean kinetic energy, even of a complex system. On the other hand, if we can measure the

doi:10.1088/978-0-7503-5759-3ch7

kinetic energies of the objects in consideration and we know the form of the potential binding them, then we can estimate the potential or binding energy, which will allow an estimate of the strength of the potential.

Now, in many situations the potential U follows some simple power law; thus, $U(r_i) \propto r_i^n$. This leads to

$$2\langle K \rangle = n\langle U \rangle. \tag{A.1.3}$$

For example, a Coulomb-like potential has $n = -1$. The virial theorem then finally gives a very simple relation:

$$2\langle K \rangle = -\langle U \rangle. \tag{A.1.4}$$

Recall that the potential energy in such a case is negative; it is in fact minus the binding energy of the system.

A.2 A list of physical constants and units

We provide here short lists of useful physical constants and units; a more extensive compilation may be found in the PDG 2024 (Navas *et al* 2024). There follows a brief note on the numerical value of α_{QED}.

A.2.1 Fundamental physical constants

Constant	Symbol	Approximate value
Speed of light in vacuum	c	2.998×10^8 m s^{-1}
Planck constant	h	6.625×10^{-34} J $\cdot$ s
	$\hbar$	1.055×10^{-34} J $\cdot$ s
		6.582×10^{-22} MeV $\cdot$ s
	$\hbar c$	197.3 MeV $\cdot$ fm
		197.3 eV $\cdot$ nm
		1.973 keV $\cdot$ Å
	$(\hbar c)^2$	0.389 GeV$^2 \cdot$ mb
Boltzmann constant	k	1.381×10^{-23} J K^{-1}
		8.617×10^{-5} eV K^{-1}
Avogadro number	N_A	6.022×10^{23} mol^{-1}
Elementary charge	e	1.602×10^{-19} C
Permeability of vacuum	μ_0	1.257×10^{-6} H m^{-1}
Permittivity of vacuum	ε_0	8.854×10^{-12} fm^{-1}
Fine-structure constant	$\alpha = \frac{e^2}{4\pi\varepsilon_0 \hbar c}$	$1/137.036$
Rydberg constant	$R_\infty = \frac{m_e c \alpha^2}{2h}$	1.097×10^7 m^{-1}
Rydberg energy	$\frac{1}{2}m_e c^2 \alpha^2$	13.61 eV
Fermi coupling constant	$G_F/(\hbar c)^3$	1.166×10^{-5} GeV^{-2}
Gravitational constant	G	6.674×10^{-11} m^3 kg^{-1} s^{-2}

Magnetic flux quantum	$\phi_0 = \frac{h}{2e}$	2.068×10^{-15} Wb
Electron mass	m_e	0.911×10^{-30} kg
		510.999 keV/c^2
Proton mass	m_p	1.673×10^{-27} kg
		938.272 MeV/c^2
Neutron mass	m_n	1.675×10^{-27} kg
		939.565 MeV/c^2
Proton–electron mass ratio	m_p/m_e	1836
Energy–mass equivalence	$E/M = c^2$	8.988×10^{16} J kg^{-1}
		5.610×10^{35} eV kg^{-1}

A.2.2 Common physical units

Unit	Symbol	Value
Electronvolt	eV	1.602×10^{-19} J
ångstrom	Å	1×10^{-10} m
Femtometre/fermi	fm	1×10^{-15} m
Atomic mass unit	u (amu)	1.661×10^{-27} kg
		931.494 MeV c^2
Barn	b	10^{-28} m^2
		100 fm^2
Becquerel	Bq	1 decay/s
Curie	Ci	3.7×10^{10} decays s^{-1}
Bohr magneton	$\mu B = \frac{e\hbar}{2m_e}$	9.274×10^{-24} J T^{-1}
		5.788×10^{-5} eV T^{-1}
Nuclear magneton	$\mu N = \frac{e\hbar}{2m_p}$	5.051×10^{-27} J T^{-1}
		3.152×10^{-8} eV T^{-1}
Parsec	pc	3.084×10^{16} m

Reference

Navas S *et al* (Particle Data Group) 2024 *Phys. Rev.* **D110** 030001

An Introduction to Elementary Particle Phenomenology
(Second Edition)

Philip G Ratcliffe

Appendix B

Quantum mechanics

'Nature isn't classical, dammit, and if you want to make a simulation of nature, you'd better make it quantum mechanical, and by golly it's a wonderful problem, because it doesn't look so easy.'

Richard P Feynman

B.1 Relativistic quantum mechanics and the Dirac equation

B.1.1 The Klein–Gordon equation

Historically, the first attempts at a relativistic formulation of quantum mechanics (QM) are due independently to Klein (1927) and Gordon (1926) although earlier both Fock and Schrödinger had considered such a possibility. Starting from the Einstein energy–momentum relation

$$E^2 = \boldsymbol{p}^2 c^2 + \mathrm{m}^2 c^4, \tag{B.1.1}$$

the canonical approach of simply transforming the variables E and $\boldsymbol{p}$ into operators maybe applied as usual:

$$E \rightarrow i\hbar \frac{\partial}{\partial t} \text{ and } \boldsymbol{p} \rightarrow -i\hbar \boldsymbol{\nabla}. \tag{B.1.2}$$

This is represented more conveniently and compactly in four-vector notation as follows:

$$p^2 = p^\mu p_\mu = \mathrm{m}^2 c^2, \tag{B.1.3}$$

where now

$$p^\mu := (E/c, \boldsymbol{p}) \text{ and e.g. } x^\mu := (ct, \boldsymbol{x}). \tag{B.1.4}$$

doi:10.1088/978-0-7503-5759-3ch8

The operator substitution then becomes

$$p^\mu := (E/c, \boldsymbol{p}) \to i\hbar\, \partial^\mu := i\hbar\, \frac{\partial}{\partial x_\mu} \equiv i\hbar\left(\frac{1}{c}\frac{\partial}{\partial t}, -\boldsymbol{\nabla}\right). \tag{B.1.5}$$

Note the negative sign in front of the spatial components here.

This leads to the following Lorentz-covariant wave equation

$$\left(\partial_\mu\partial^\mu + \frac{m^2 c^2}{\hbar^2}\right)\phi = 0. \tag{B.1.6}$$

For an even more compact notation, it is customary to define the d'Alembertian or wave operator:

$$\Box := \partial_\mu\partial^\mu \tag{B.1.7}$$

and thus write (setting now $\hbar = 1 = c$)

$$(\Box + m^2)\phi = 0. \tag{B.1.8}$$

This is the Klein–Gordon equation. Historically, the well-known problems with negative particle probability densities, closely associated with the possibility of negative-energy solutions (energy enters *squared* in the Einstein equation), led to its initially being abandoned and to Dirac's famous alternative (Dirac 1928)[1].

B.1.2 The Dirac equation

In a nutshell, Dirac's idea was an attempt to avoid negative energies by effectively taking the square-root of the Klein–Gordon equation (in order to linearise in p^μ) and writing

$$\gamma^\mu p_\mu\, \psi(x) = m\, \psi(x), \tag{B.1.9}$$

where γ^μ is some new and unknown vector object (to be determined), necessary to render the left-hand side a scalar quantity, as is the right-hand side. Then, in order that the operator version $\gamma^\mu p_\mu \triangleq m$ should agree with Einstein's relation $p^2 = m^2$ for any p^μ, we need a Clifford (or Dirac) algebra: $\{\gamma^\mu, \gamma^\nu\} = g^{\mu\nu}$. Indeed, it is easy to see that with such an algebra we have

$$\gamma^\mu p_\mu \gamma^\nu p_\nu = \tfrac{1}{2}\{\gamma^\mu, \gamma^\nu\}p_\mu p_\nu = g^{\mu\nu}p_\mu p_\nu = p^2. \tag{B.1.10}$$

The simplest way to represent such *anticommuting* objects γ^μ is via matrices; the minimal representation has rank four and may be constructed block-wise with the aid of the Pauli σ-matrices. An explicit form (originally due to Dirac (1928), now the standard representation) is

---

[1] The 1933 Nobel Prize in Physics was awarded equally to Erwin Schrödinger and Paul Adrien Maurice Dirac 'for the discovery of new productive forms of atomic theory'.

$$\gamma^0 = \begin{pmatrix} 1 & 0 \\ 0 & -1 \end{pmatrix} \text{ and } \gamma = \begin{pmatrix} 0 & \sigma \\ -\sigma & 0 \end{pmatrix}, \tag{B.1.11}$$

where the sub-matrices are 2×2. The first immediate consequence is that the wave functions must be represented by *four*-component spinors (not to be confused with Lorentz four-vectors). The indices on such a spinor, as too those (implicit) on the matrices γ^μ, are often referred to as Dirac indices and the space over which they run, Dirac space. Note that the spinor may not be thought of as any sort of vector since, for example, we find that a spatial rotation through 2π reverses its sign.

We now simply list some of the basic properties of the γ-matrices as defined in the Dirac representation:

$$\gamma^{0\dagger} = \gamma^0, \quad \gamma^\dagger = -\gamma, \quad \text{or } \gamma_\mu = \gamma^{\mu\dagger} = \gamma^0 \gamma^\mu \gamma^0. \tag{B.1.12}$$

Finally then, the relativistic wave equation or Dirac (1928) equation is ($\hbar = 1 = c$)

$$[i\gamma^\mu \partial_\mu - m\mathbb{1}]\psi(x) = 0, \tag{B.1.13}$$

where $\mathbb{1}$ is just the rank-four unit matrix and ψ must now be a four-*component spinor*. Introducing the so-called '*slashed*' shorthand, this may be rewritten in a more compact form (also eliminating the matrix $\mathbb{1}$):

$$[i\slashed{\partial} - m]\psi(x) = 0. \tag{B.1.14}$$

As we might imagine from the presence of the Pauli matrices, the different spinor components have to do with the spin states of the electron.

If we consider the coupling to a classical electromagnetic field (via $i\partial^\mu \to i\partial^\mu - eA^\mu$, with $A^\mu = (\Phi, \boldsymbol{A})$, equivalent to $E \to E - e\Phi$ and $\boldsymbol{p} \to \boldsymbol{p} - e\boldsymbol{A}$) and then take the non-relativistic limit, we find that the two-by-two block form of the Dirac (1928) equation reduces to the Schrödinger equation augmented with the Pauli construction to describe the coupling of the electron *intrinsic* magnetic moment. Note first that the Φ term couples through γ^0, which has a positive (negative) upper (lower) block. The lower spinor components thus behave as an oppositely charged field, the positron.

This is most conveniently achieved by decomposing the four-component spinor into upper and lower two-component parts, thus

$$\psi \equiv \begin{pmatrix} \tilde{\phi} \\ \tilde{\chi} \end{pmatrix}, \tag{B.1.15}$$

we have reinstated the 'c' factor to highlight the non-relativistic limit we seek. The Dirac equation (1928) coupled to an electromagnetic field then becomes ($\boldsymbol{\pi} := \boldsymbol{p} - e\boldsymbol{A}$)

$$i\hbar \frac{\partial}{\partial t} \begin{pmatrix} \tilde{\phi} \\ \tilde{\chi} \end{pmatrix} = \boldsymbol{\sigma} \cdot \boldsymbol{\pi} c \begin{pmatrix} \tilde{\chi} \\ \tilde{\phi} \end{pmatrix} + e\Phi \begin{pmatrix} \tilde{\phi} \\ \tilde{\chi} \end{pmatrix} + mc^2 \begin{pmatrix} \tilde{\phi} \\ \tilde{\chi} \end{pmatrix}. \tag{B.1.16}$$

In the non-relativistic limit the rest-mass energy dominates the temporal dependence of the wave functions and we thus write

$$\begin{pmatrix} \tilde{\phi} \\ \tilde{\chi} \end{pmatrix} \equiv e^{-\frac{i}{\hbar} mc^2 t} \begin{pmatrix} \phi \\ \chi \end{pmatrix}, \tag{B.1.17}$$

where now the two-component spinors ϕ and χ will be relatively slowly varying functions of time. The Dirac (1928) equation then becomes

$$i\hbar \frac{\partial}{\partial t} \begin{pmatrix} \phi \\ \chi \end{pmatrix} = \boldsymbol{\sigma} \cdot \boldsymbol{\pi} c \begin{pmatrix} \chi \\ \phi \end{pmatrix} + e\Phi \begin{pmatrix} \phi \\ \chi \end{pmatrix} - 2mc^2 \begin{pmatrix} 0 \\ \chi \end{pmatrix}. \tag{B.1.18}$$

Neglecting all terms that are small compared to mc^2 (i.e. all kinetic and interaction energies), the lower equation here may be approximated as

$$\chi \simeq \frac{\boldsymbol{\sigma} \cdot \boldsymbol{\pi}}{2mc} \phi \tag{B.1.19}$$

We thus see that the lower components χ in this approximation are to be considered *small* as compared to the upper components ϕ. Roughly speaking, we have $\chi/\phi \sim v/c$. Note, moreover, that the upper and lower components are therefore not, in fact, independent.

Inserting the expression for χ into the upper equation leads to

$$i\hbar \frac{\partial}{\partial t} \phi = \left[\frac{(\boldsymbol{\sigma} \cdot \boldsymbol{\pi})^2}{2m} + e\Phi \right] \phi. \tag{B.1.20}$$

Now, a simple identity satisfied by the Pauli matrices is

$$\boldsymbol{\sigma} \cdot \boldsymbol{a} \, \boldsymbol{\sigma} \cdot \boldsymbol{b} = \boldsymbol{a} \cdot \boldsymbol{b} + i\boldsymbol{\sigma} \cdot (\boldsymbol{a} \wedge \boldsymbol{b}). \tag{B.1.21}$$

Thus,

$$\begin{aligned} \boldsymbol{\sigma} \cdot \boldsymbol{\pi} \, \boldsymbol{\sigma} \cdot \boldsymbol{\pi} &= \pi^2 + i\boldsymbol{\sigma} \cdot (\boldsymbol{\pi} \wedge \boldsymbol{\pi}) \\ &= \pi^2 - \frac{e\hbar}{c} \boldsymbol{\sigma} \cdot \boldsymbol{B}. \end{aligned} \tag{B.1.22}$$

Finally then, we obtain the following *non*-relativistic two-component equation for ϕ:

$$i\hbar \frac{\partial \phi}{\partial t} = \left[\frac{\left(\boldsymbol{p} - \frac{e}{c} \boldsymbol{A} \right)^2}{2m} - \frac{e\hbar}{2mc} \boldsymbol{\sigma} \cdot \boldsymbol{B} + e\Phi \right] \phi, \tag{B.1.23}$$

which is immediately recognised as Pauli's extension of the Schrödinger equation to include the spin degree-of-freedom of the electron and, in particular, its magnetic moment. The two components of ϕ clearly correspond to the two spin projections (*up* and *down*, say) of the spin-1/2 electron.

Indeed, simplifying the equation further and keeping only the leading terms of the interaction with the magnetic field (which is normally weak owing to the factor $1/c$), we have

$$i\hbar \frac{\partial \phi}{\partial t} = \left[\frac{\boldsymbol{p}^2}{2m} - \frac{e}{2mc} (\boldsymbol{L} + 2\boldsymbol{S}) \cdot \boldsymbol{B} + e\Phi \right] \phi, \qquad (B.1.24)$$

where, as usual, $\boldsymbol{L} = \boldsymbol{r} \wedge \boldsymbol{p}$ is just the orbital angular momentum of the electron and $\boldsymbol{S} = \frac{1}{2}\hbar\boldsymbol{\sigma}$ is the electron spin operator with eigenvalues $\pm\frac{1}{2}\hbar$. And thus the g-factor for the electron g_e is *predetermined* to be precisely two, as found experimentally[2].

> '*When I realised that the equation contained the spin of the electron, and also the magnetic moment—everything needed for the properties of the electron—it was a surprise. A complete surprise.*'
>
> P A M Dirac

At this point, we should mention that, in the light of the interpretation of the Dirac equation as containing antimatter, the Klein–Gordon equation was soon resurrected. It is found to describe correctly the QM of a spin-zero (i.e. scalar or pseudoscalar) field, such as the Higgs boson, pion or kaon.

We further find that the various transformations of spatial inversion (or parity), time reversal and charge conjugation are obtained via multiplication of the spinor ψ by suitable combinations of the γ^μ (together with any other necessary transformations, e.g. $\boldsymbol{x} \to -\boldsymbol{x}$ or $t \to -t$). In particular, the parity operation, besides sending $\boldsymbol{x} \to -\boldsymbol{x}$, requires the spinors to be multiplied by γ^0. The sign difference between the upper and lower blocks of this matrix leads to the opposite parity assignment for fermions and antifermions.

Wishing to have a natural (positive definite) expression for the probability density, such as $\psi^\dagger\psi$, we are led to define the conserved current corresponding to the Dirac (1928) equation as

$$j^\mu = (\rho, \boldsymbol{j}) := \bar{\psi} \, \gamma^\mu \, \psi, \qquad (B.1.25)$$

where the *conjugate* spinor is $\bar{\psi} := \psi^\dagger\gamma^0$ and therefore the natural probability density is $\rho = j^0 = \psi^\dagger\psi$, which is now positive definite, as it should be. In electromagnetism the coupling between the electron and the electromagnetic field is then perfectly well described by an interaction of the form $j \cdot A$. Taking this as a template for particle interactions, we immediately realise it is not unique: the most general form for a 'current' is $\bar{\psi} \, \Gamma \, \psi$, where Γ may be any one of a number of matrices spanning the Dirac (1928)-spinor space.

Briefly, as the reader may easily verify, the free-particle plane-wave solutions to the Dirac (1928) equation take the following form:

$$\psi(x) = w(p, s)\mathrm{e}^{-\mathrm{i}\varepsilon \, p \cdot x}, \qquad (B.1.26)$$

where the sign $\varepsilon = \pm$ will be explained shortly, $w(p, s)$ is a space–time constant, four-component spinor (coefficient), containing both energy–momentum (p) and spin (s) information and which satisfies

---

[2] To be precise, owing to quantum-mechanical corrections, the value is not exactly two. Nevertheless, the measured and calculated values coincide perfectly to a very high precision.

$$\varepsilon \, \gamma^\mu p_\mu \, w(p, s) = m \, w(p, s), \tag{B.1.27}$$

in which p^μ is no longer an operator. The sign of the exponent determines two types of solutions. In fact, the Dirac (1928) equation does not eliminate the negative-energy solutions, but gives them a meaning: antimatter. Thus, according as to whether $\varepsilon = \pm$ the spinor $w(p, s)$ takes on different forms.

Exercise B.1.1. *Verify that the free-particle plane-wave solutions to the Dirac equation do indeed take the form shown in equation (B.1.26), with the spinor $w(p, s)$ satisfying equation (B.1.27).*

The general form of a free-particle state is then given by a spinor $w(p, s)$ of the form

$$w(p, s) = \begin{cases} u(p, s) = \begin{pmatrix} \sqrt{E + m}\,\mathbb{1} \\ \sqrt{E - m}\,\sigma \cdot \hat{p} \end{pmatrix} \otimes \chi_s \text{ (positive-energy)}, \\[3mm] v(p, s) = \begin{pmatrix} \sqrt{E - m}\,\sigma \cdot \hat{p} \\ \sqrt{E + m}\,\mathbb{1} \end{pmatrix} \otimes \chi_s \text{ (negative-energy)}, \end{cases} \tag{B.1.28}$$

with a standard two-component (spin up or down) spinor χ_s simply

$$\begin{pmatrix} 1 \\ 0 \end{pmatrix} \quad \text{and} \quad \begin{pmatrix} 0 \\ 1 \end{pmatrix}. \tag{B.1.29}$$

B.1.3 The Majorana equation

The Dirac (1928) equation, equation (B.1.13), naturally has a mass term of the so-called Dirac type. That is, as already noted, it permits the phenomenon of helicity flip or, more technically, mixes left- and right-handed states. To see this, let us examine the mass-term contribution to the relative Lagrangian:

$$\mathcal{L}_m = -m \, \bar{\psi} \, \psi. \tag{B.1.30}$$

Taking into account that $\bar{\psi} := \psi^\dagger \gamma^0$ and $\psi_\pm = \frac{1}{2}(1 \pm \gamma_5)\psi$ with $\psi = \psi_+ + \psi_-$, it may be re-expressed as[3]

$$\mathcal{L}_m = -m \, (\bar{\psi}_+ \psi_- + \bar{\psi}_- \psi_+). \tag{B.1.31}$$

---

[3] This is just the well-known helicity-flip possibility in a vector gauge theory that is, however, relatively suppressed for light fermions.

There is, though, an alternative form of relativistic equation due to Majorana (1937), which has some interesting properties and implications for phenomenology. Note first that the Dirac equation written in the Dirac basis is complex and therefore so too must be the wave-function. This allows it to describe a possibly charged fermion field. In the Majorana basis, given by

$$\tilde{\gamma}^0 = \begin{bmatrix} 0 & \sigma_2 \\ \sigma_2 & 0 \end{bmatrix}, \quad \tilde{\gamma}^1 = \begin{bmatrix} i\sigma_1 & 0 \\ 0 & i\sigma_1 \end{bmatrix}, \quad \tilde{\gamma}^2 = \begin{bmatrix} 0 & \sigma_2 \\ -\sigma_2 & 0 \end{bmatrix}, \quad \text{and} \quad \tilde{\gamma}^3 = \begin{bmatrix} i\sigma_3 & 0 \\ 0 & i\sigma_3 \end{bmatrix}, \quad \text{(B.1.32)}$$

the matrices are all real, since σ^1 and σ^3 are both real whereas σ^2 is imaginary. Therefore, the equation is real and so too may be the wave-function. A real wave-function normally represents a neutral state; it then becomes possible that the particle in question be its own antiparticle (just as e.g. the photon or the neutral pion, but not, of course, e.g. the neutron).

This then allows for a reformulation as the Majorana equation (1937):

$$i\tilde{\gamma}^\mu \partial_\mu \psi(x) - m\, \mathbb{1}\psi^{\mathcal{C}}(x) = 0, \quad \text{(B.1.33)}$$

where note now that the $\tilde{\gamma}$-matrices are necessarily in the Majorana representation and $\psi^{\mathcal{C}} := \mathcal{C}\psi$ is just the charge conjugate of ψ, which in the Majorana basis is defined as

$$\psi^{\mathcal{C}} := i\psi^*. \quad \text{(B.1.34)}$$

Since both ψ and $\psi^{\mathcal{C}}$ appear in the Majorana equation, the field must be *neutral*; otherwise, it would violate charge conservation.

With this observation, the only immediate candidate Majorana fermions are the neutrinos, all other elementary fermions being charged. Although this may seem to be in marked contrast with the Dirac scheme that the other fermions must follow, there are forms of grand unified theories in which the neutrinos emerge naturally as Majorana particles. Note that in the same paper Majorana equation (1937) also demonstrated that such a formulation does not alter, for example, the standard β-decay rates. Moreover, this construction permits the resolution of the β-decay problems with a minimum of new states (defined as the number of degrees of freedom in the relative equation of motion): instead of the four Dirac-fermion states (electron/positron, poitive/negative helicity) a Majorana neutrino only has two (a left-handed neutrino and right-handed antineutrino). However, if neutrinos were indeed of the Majorana type, then the door would be opened onto a number of new phenomena: e.g. neutrinoless double β-decay (as discussed in the main text) and various other lepton-number violating meson and charged-lepton decays.

B.1.4 Relativistic currents

So far there are two obvious possibilities for constructing generalised currents: namely, $\bar{\psi}\gamma^\mu\psi$ and $\bar{\psi}\mathbb{1}\psi$ (the latter would couple to a scalar or pseudoscalar 'force field').

However, this 16-dimensional Dirac (1928) space offers three other different possibilities. First of all, let us construct the following special matrix[4]:

$$\gamma_5 \equiv \gamma^5 := i\gamma^0\gamma^1\gamma^2\gamma^3 = \begin{bmatrix} 0 & 1 \\ 1 & 0 \end{bmatrix}. \tag{B.1.35}$$

This immediately provides five more independent matrices: γ_5 and $\gamma_5\gamma^\mu$. Finally, it is conventional to include the antisymmetric product (with six independent components) $\sigma^{\mu\nu} := \frac{i}{2}[\gamma^\mu, \gamma^\nu]$ [5].

There are then five distinct possible types of generalised currents:

$$\bar{\psi}\,\mathbf{1}\,\psi, \quad \bar{\psi}\,\gamma_5\,\psi, \quad \bar{\psi}\,\gamma^\mu\,\psi, \quad \bar{\psi}\,\gamma_5\gamma^\mu\,\psi, \quad \bar{\psi}\,\sigma^{\mu\nu}\,\psi, \tag{B.1.36}$$

which take the names, respectively, of **s**calar, **p**seudoscalar, (polar) **v**ector, **a**xial (or pseudo-)vector and **t**ensor. These names reflect their properties under Lorentz transformations and spatial inversion. Each current also has specific behaviour under temporal inversion and charge conjugation.

From the preceding discussion, we see that the vector current is related to the four-momentum p^μ of a particle, whereas examination of the rôle of γ_5 and $\gamma_5\gamma^\mu$ reveals that the axial-vector current is related to intrinsic spin. Given the above form of the spinor solutions, we find that $\bar{\psi}\gamma_5\gamma^\mu\psi$ measures precisely the spin s^μ of a particle. The other currents, however, have no simple physical (classical) interpretation, while the precise form of vector and axial-vector currents suggests how to proceed in order to construct parity-violating matrix elements.

Exercise B.1.2. *Verify the anticommutation relation* $\{\gamma_5, \gamma^\mu\} = 0$.

Exercise B.1.3. *Verify that* $\bar{\psi}\,\gamma^\mu\psi$ *and* $\bar{\psi}\,\gamma^\mu\gamma_5\psi$ *do indeed give the momentum* p^μ *and spin* s^μ *of the particle, respectively, and find the constants of proportionality.*

B.2 Spin statistics

The rôle of spin and statistics in physics is central to many phenomena, from low-energy condensed-matter systems to high-energy particle interactions. Here we shall briefly outline the question of the properties and behaviour of composite systems of particles with regard to the question of Bose (1924) and Einstein (1924) *versus* Fermi (1926) and Dirac (1926) statistics.

B.2.1 Spin statistics in quantum mechanics

In QM composite systems are often described by superpositions of wave functions and we are thus led to consider the question of symmetry under interchange of the

---

[4] Note that, as defined, in four dimensions we have $\{\gamma_5, \gamma^\mu\} = 0$.
[5] From the definition of γ_5, it is easy to show that $\gamma_5\sigma^{\mu\nu}$ is not a further independent set.

sub-states when these represent indistiguishable particles. That is, since e.g. $|1, 2\rangle$ and $|2, 1\rangle$ represent the same physical state when particles 1 and 2 are identical, it is natural to consider the two possible constructions:

$$\frac{1}{\sqrt{2}}\left[|1, 2\rangle + |2, 1\rangle\right] \text{ and } \frac{1}{\sqrt{2}}\left[|1, 2\rangle - |2, 1\rangle\right]. \tag{B.2.1}$$

In the first case there is clearly no issue if the two particles occupy the same quantum state, as the resulting interference will be constructive. However, in the second case the interference is destructive and therefore such a state will not physically exist; i.e. it is forbidden for two such particles to occupy the same quantum state.

Based on empirical deductions (but also theoretical arguments (Pauli 1940), the following associations may be established: the symmetric solution is ascribed to the case of identical bosons and gives rise to Bose−Einstein statistics; whereas the antisymmetric superposition applies to identical fermions, leading to Fermi−Dirac statistics and to the Pauli exclusion principle (1925). For completeness, we note that in statistical physics these deductions lead to the different distribution laws obeyed by the two particle types. First of all, in the distinguishable-particle (or non-quantum) case, we have

$$\text{Boltzmann:} \quad n_j \propto \frac{1}{e^{\varepsilon_j/kT}}, \tag{B.2.2a}$$

whereas for indistinguishable particles,

$$\text{Bose−Einstein:} \quad \frac{n_i}{g_i} = \frac{1}{e^{(\varepsilon_i - \mu)/kT} - 1}, \tag{B.2.2b}$$

$$\text{Fermi−Dirac:} \quad \frac{n_i}{g_i} = \frac{1}{e^{(\varepsilon_i - \mu)/kT} + 1}, \tag{B.2.2c}$$

where n_i is the number of particles in state i, g_i the degeneracy of state i, e_i the energy of the ith state, μ the chemical potential, k the Boltzmann constant and T the absolute temperature. Note that in the above, whereas j labels individual particle states, i labels energy levels, not individual states.

B.2.2 Spin statistics in particle physics

The peculiar forms of Bose−Einstein (Bose 1924; Einstein 1924) and Fermi−Dirac (Fermi 1926; Dirac 1926) statistics have various important and detectable repercussions in elementary-particle interactions.

B.2.2.1 $\pi^- d \rightarrow nn$

As discussed in section 2.2.3, the process $\pi^- d \rightarrow nn$ may be used to determine the parity of the pion. The theoretical analysis exploits the antisymmetry of the final state containing two identical fermions, which constrains the overall parity of the pair to be negative.

B.2.2.2 $\rho^0 \to \pi\pi$

Consider next ρ^0 decay into two pions; there are two possible final states, $\pi^+\pi^-$ and $\pi^0\pi^0$. At first sight, both appear to be equally probable (up to possibly differing isospin Clebsch–Gordan coefficients); however, the symmetry requirement for the latter final state is a severe restriction. Strong-interaction processes conserve parity and this imposes important constraints. We see in exercise 4.2.2 how the symmetry of the $\pi^0\pi^0$ final state contrasts the requirement of $L = 1$, necessary to conserve the negative parity of the ρ^0. Thus, whereas ρ^0 may decay strongly into $\pi^+\pi^-$, the $\pi^0\pi^0$ channel is highly suppressed.

B.2.2.3 Loop diagrams

The higher-order quantum corrections to various physical quantities (e.g. coupling constants, masses and cross-sections) invariably involve loop contributions (in Feynman diagrams). A consequence of the antisymmetry requirement for fermion fields is that each internal fermion loop implies an overall extra factor -1 for the given amplitude. Thus, for example, the gluon- and fermion-loop contributions to the quantum chromodynamics (QCD) β-function have opposite signs, which explains the differing scaling behaviour in QCD and quantum electrodynamics, see section 3.4.2.

A further consequence is highlighted by the possible rôle of supersymmetry in the Standard Model. Especially when there are scalar fields (e.g. the Higgs) present in a theory, the problem of quantum corrections or renormalisation renders phenomenology unmanageable; e.g. particle masses tend to all be shifted to the highest scales in the theory. If, however, the theory possesses a supersymmetry, which implies the presence of a boson for every fermion degree of freedom and *vice versa*, then general renormalisation effects cancel out (up to the possible fermion–boson mass differences).

B.2.2.4 $D \to K\pi$

The symmetry requirements are also important for Feynman diagrams when considering external (i.e. initial or final) states containing identical particles. The known matter particles are all fermions and we shall primarily consider the case of final-state effects. Now, in the case of final states containing identical fermions a relative minus sign must be included between diagrams for which there is an exchange of any two such identical fermions The immediate and somewhat obvious consequence is destructive interference or so-called Pauli blocking.

A concrete example is given by considering decays of certain mesons containing c: $D^0 \to K^-\pi^+$ and $D^+ \to K^0\pi^+$. While pions and kaons are clearly different particles, at least some of the quarks of which they are composed may be identical. To see this, in figure B.1 we display the underlying Feynman diagrams for these two decays. Close examination of the second reveals the presence of two identical fermions, the two $\bar{d}$ quarks. There is therefore a further diagram, figure B.1, in which the two are exchanged. We thus see that the two amplitudes of figures B.1(*b*) and (*c*) differ only by interchange of the two $\bar{d}$ quarks. Note, however, that the colour flow in the two diagrams is then different. In amplitude (*b*) the two separate finale-state quark–antiquark pairs are

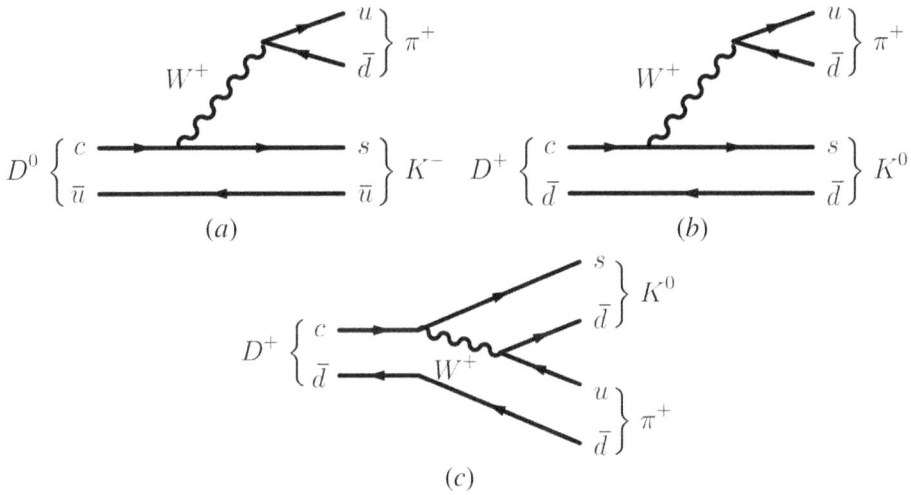

Figure B.1. The Feynman diagrams for the two decays (a) $D^0 \to K^- \pi^+$ and (b) $D^+ \to K^0 \pi^+$.

automatically colour neutral; however, in amplitude (c) the crossing over of the $\bar{d}$ quarks will introduce a suppression factor $\frac{1}{3}$ (analogous to that mentioned in section 3.4.1 for the DY process), since only one third of the $u\bar{d}$ pair produced will have the correct colour charges to neutralise the other two and form the final colourless mesons. The effect is thus reduced to a partial (Pauli) cancellation.

The measured lifetimes of these two charmed mesons are (Navas *et al* 2024):

$$\tau_{D^0} = 0.410(1) \times 10^{-12} \text{ s} \tag{B.2.3a}$$

and

$$\tau_{D^+} = 1.033(5) \times 10^{-12} \text{ s.} \tag{B.2.3b}$$

There are thus clearly other effects that should also be taken into account and there are, of course, many other channels to be included. In any case, Pauli suppression of this type is very common.

B.3 The discrete symmetries C, P and T

As stressed at various points during the lectures, the concept of symmetry plays a central rôle in the development of physics in general and particularly in elementary particle physics. While many of the symmetries encountered are continuous (e.g. spatial and temporal translations, rotations etc), there are three fundamental discrete symmetries, which we shall discuss here within the context of a quantum-mechanical description of particle interactions. These symmetry transformations are the operations of:

$\mathcal{C}$: transforming particle into antiparticle and vice versa;
$\mathcal{P}$: spatial inversion, i.e. $\mathbf{x} \to -\mathbf{x}$;
$\mathcal{T}$: time reversal, i.e. $t \to -t$.

Recall that the first two are *linear* transformations, whereas the last is *antilinear*; that is, together with the obvious coordinate transformation, we must apply complex conjugation to all $\mathbb{C}$-number parameters (e.g. masses, coupling constants etc) involved.

Since we shall need to deal with current–current interactions, it will be useful to know in advance the transformation properties of the possible currents under the action of the above operations. Generalised currents may be constructed from bilinears of spin-half fields. We shall thus now examine each separately within the context of the Dirac equation.

B.3.1 Charge-conjugation invariance (C)

The obvious way in which the presence of an antiparticle may be detected theoretically (and indeed experimentally) is through the relative sign of its coupling to an external classical electromagnetic field $A^{\mu} = (\Phi, A)$. Indeed, the Dirac equation for an electron in the presence of a real (for simplicity) electromagnetic field is

$$\left[i\gamma_{\mu}\partial^{\mu} + e\gamma_{\mu}A^{\mu} - m \right]\psi(x) = 0. \tag{B.3.1}$$

The sign adopted for the electromagnetic coupling is, of course, purely conventional. It does determine, though, the corresponding (opposite) sign for the coupling of a positron. Indeed, consider the Coulomb part of the potential $A^{0}(x)$: it couples via γ^{0}, which, as we have seen has positive entries for the two upper diagonal elements and negative below. The two lower spinor components thus couple to the electric field with opposite sign and hence we have the anti-electron or positron.

If the physics of the positron is to be the same as that of the electron, we must seek a transformation ($\mathcal{C}$) that results in the following equation:

$$\left[i\gamma_{\mu}\partial^{\mu} - e\gamma_{\mu}A^{\mu} - m \right]\psi^{C}(x) = 0. \tag{B.3.2}$$

Now, the action of complex conjugation evidently only affects the sign of the first term (for a real electromagnetic field). Recall too the following relation (which is a direct result of the Clifford algebra and does *not* depend on the particular representation adopted):

$$\gamma^{0}\gamma_{\mu}\gamma^{0} = \gamma_{\mu}^{\dagger}. \tag{B.3.3}$$

The following relation is, however, representation-dependent and holds only in the standard Dirac representation:

$$\gamma^{2}\gamma_{\mu}\gamma^{2} = -\gamma_{\mu}^{\mathsf{T}}, \tag{B.3.4}$$

where T indicates matrix transposition. Thus, by applying complex conjugation to the Dirac equation (B.3.1) and then multiplying from the left by $\gamma^{0}\gamma^{2}$, we can easily verify that the precise form shown in equation (B.3.2) is obtained with the identification

$$\mathcal{C}: \psi(x) \rightarrow \psi^{C}(x) := i\gamma^0\gamma^2 \, \bar{\psi}^{\, T}(x), \tag{B.3.5}$$

where the conjugate spinor is defined as $\bar{\psi} := \psi^\dagger\gamma^0$. Note that the presence of the factor 'i' corresponds to an arbitrary, but conventional, phase choice.

B.3.2 Spatial-inversion invariance (P)

Following a similar procedure to that of the previous section we shall now derive the form of the operator generating spatial inversion. The starting point will again be the Dirac equation (B.3.1), although the coupling to an electromagnetic field is now superfluous. The transformation of the equation under x to $-x$ may be represented by simply replacing ∂^μ with ∂_μ (since the lowering of the index implies a sign change in the spatial components). Now, the effect of γ^0, already noted above, may be equally expressed as

$$\gamma^0\gamma_\mu\gamma^0 = \gamma^\mu. \tag{B.3.6}$$

Thus, we recover the original equation via the identification

$$\mathcal{P}: \psi(x) \rightarrow \psi^{P}(x') := \gamma^0 \, \psi(x'), \quad \text{with } x' = (t, \, -x), \tag{B.3.7}$$

where once again the implicit phase choice is conventional.

An immediate consequence of the above form for the parity transformation is that, since the matrix γ^0 is block-diagonal ± 1, the parities of the upper and lower components of ψ are opposite. That is, fermion and antifermion have opposite parities. By convention, the parity of fermions is chosen positive and antifermions negative. This particular choice has, of course, no physical consequence as fermions are always produced in fermion–antifermion pairs (for which the overall intrinsic parity is predetermined to be -1).

B.3.3 Time-reversal invariance (T)

Finally, we turn to the case of symmetry under time reversal. Note first that the transformation $t \rightarrow -t$ also implies exchange of initial and final states. Since, as remarked above, this transformation has the peculiar property of being antilinear, let us start with the simpler case of the Schrödinger equation for a free particle:

$$i\frac{\partial}{\partial t} \, \psi(t, \, x) = -\frac{1}{2m} \, \nabla^2\psi(t, \, x). \tag{B.3.8}$$

The eigen-solutions are plane-waves and may be written as

$$\psi(t, \, x) = u(p)e^{-i(Et-p\cdot x)}, \tag{B.3.9}$$

where, of course, the energy and momentum satisfy $E = p^2/2m$. It should be immediately obvious that the first choice of simply changing t to $-t$ in the above does not satisfy the original equation, nor indeed does it even correspond to a particle with momentum $-p$, as it should for time reversal. However, the choice

$$\psi^*(-t, \, x) = \psi_0^* e^{-i(Et+p\cdot x)} \tag{B.3.10}$$

respects all requirements. We thus see the necessity for an antilinear (complex conjugation) operator. That is, the transformation $t \rightarrow -t$ is accompanied by complex conjugation applied to all $\mathbb{C}$-number quantities.

Now, complex conjugation applied to the Dirac equation (B.3.1) leads to

$$\left[-i\gamma_\mu^{\mathsf{T}} \partial^\mu - m \right] \bar{\psi}^{\mathsf{T}}(t, \boldsymbol{x}) = 0, \tag{B.3.11}$$

where, as always, T simply stands for matrix transposition and once again the anticommutation properties of the γ matrices with γ^0 have been exploited. Multiplication from the left by γ^0 together with the transformation $x^\mu \rightarrow x'^\mu := (-t, \boldsymbol{x})$ then leads to

$$\left[i\gamma_\mu^{\mathsf{T}} \partial'^\mu - m \right] \gamma^0 \, \bar{\psi}^{\mathsf{T}}(-t', \boldsymbol{x}') = 0. \tag{B.3.12}$$

All that remains is to find the unitary transformation (which must exist) between the Dirac bases for γ_μ and γ_μ^{T}: it is simply $i\gamma^1\gamma^3$, where the factor 'i' is again conventional. Thus, the time-reversal operation is given by

$$\mathcal{T}: \psi(x) \rightarrow \psi^{\mathcal{T}}(x') := i\gamma^1\gamma^3 \, \bar{\psi}^{\mathsf{T}}(x'), \quad \text{with } x' = (-t, \boldsymbol{x}). \tag{B.3.13}$$

B.3.3.1 T and complex potentials

As a closing remark to this section, let us illustrate the rôle of a complex contribution to the potential describing particle interactions. While there is evidently no counterpart in classical mechanics, in QM all quantities are potentially complex. Typically, when particles (or radiation) may be emitted or absorbed (created or destroyed), an imaginary contribution to the scattering matrix elements is found.

Consider schematically the temporal evolution of a state of definite energy E,

$$\phi(t, \boldsymbol{x}) = a(\boldsymbol{x}) e^{-iEt}. \tag{B.3.14}$$

The probability density is just $\rho(t, \boldsymbol{x}) := \phi^*\phi = |a(\boldsymbol{x})|^2$, which in this case is time independent. If, however, we introduce an imaginary contribution, $-\frac{1}{2}i\Gamma$ say, to the energy, that is $E \rightarrow E - \frac{1}{2}i\Gamma$, then something interesting occurs:

$$\rho(t, \boldsymbol{x}) = |\phi(t, \boldsymbol{x})|^2 = |a(\boldsymbol{x})|^2 e^{-\Gamma t}. \tag{B.3.15}$$

The probability density thus follows the usual decay law, with rate Γ.

Note that the resulting time dependence evidently violates time-reversal invariance. That is, an imaginary phase in the interaction Hamiltonian automatically provokes a violation of T. This may be immediately understood as a direct consequence of the *anti*linearity property of the temporal inversion operator.

B.3.4 Fermi–Dirac statistics and Pauli exclusion

As already mentioned, the spinor solutions to the Dirac equation display peculiar properties under spatial rotations, which reflect the Pauli exclusion principle (Pauli 1925).

Let us first demonstrate that a spatial rotation through 2π reverses the sign. It can be shown that a spatial rotation through an angle θ about, e.g., the z-axis for spinors in Dirac theory is achieved via the following unitary operator:

$$U_R(\theta) = \exp\left[\frac{\mathrm{i}}{2}\theta\sigma^{12}\right], \qquad (B.3.16)$$

where $\sigma^{\mu\nu} := \frac{\mathrm{i}}{2}[\gamma^\mu, \gamma^\nu]$. The factor ½ in the above expression immediately leads to the famous result that a spinor representing a spin-½ object changes sign on rotation through an angle 2π.

This fact is closely related to the Pauli exclusion principle and the Fermi–Dirac statistics obeyed by fermions. An important consequence of the Pauli exclusion principle is found processes where the final states contain identical fermions. In particular, there are cases where although the physically detected particles (say mesons) are different the underlying quark substructure is such that there, are in fact, two (or more) identical quarks in the final state. On second or field quantisation we find that the Feynman rules include a relative minus sign between diagrams that differ only by interchange of identical fermions lines. This then naturally leads to destructive interference in those regions of phase space in which the two fermions have similar energies and momenta. Another important consequence is a relative minus sign between diagrams in which a closed boson loop is replaced by a fermion.

B.3.5 Dirac-spinor bilinears and $\mathcal{CPT}$

Armed with the previously derived transformation operators, it is now easy to determine the transformation property of any spinor-field bilinear and thus any current we may wish to employ in the description of particle interactions. The natural electric four-current associated with the Dirac equation, is shown by standard methods to be $j^\mu := \bar{\psi}\gamma^\mu\psi$, the temporal component of which is $\rho = \psi^\dagger\gamma^0\gamma^0\psi = \psi^\dagger\psi$, a natural and positive-definite probability density.

However, we also wish to describe all possible interactions (including the weak and strong nuclear forces) and thus we must consider all possible currents. The complete basis of generalised currents is $S = \bar{\psi}\psi$, $P = \bar{\psi}\gamma_5\psi$, $V = \bar{\psi}\gamma^\mu\psi$, $A = \bar{\psi}\gamma^\mu\gamma_5\psi$ and $T = \bar{\psi}\sigma^{\mu\nu}\psi$. Their transformation properties under the discrete transformations $\mathcal{C}$, $\mathcal{P}$ and $\mathcal{T}$ are summarised in table B.1.

Table B.1. The properties of the five spinor-bilinear currents (S, P, A, V and T) under the discrete transformations $\mathcal{C}$, $\mathcal{P}$ and $\mathcal{T}$.

	S	P	V	A	T
$\mathcal{C}$	+	+	−	+	−
$\mathcal{P}$	+	−	+	−	+
$\mathcal{T}$	+	−	+	+	−
$\mathcal{CPT}$	+	+	−	−	+

In order to construct a current–current interaction, it is necessary to combine two (or possibly more) currents by *completely* contracting or saturating the indices. For the currents listed here, it is immediately noticeable that the CPT signature is $(-1)^{n_i}$, where n_i is the number of indices. Since complete saturation implies n_i even, it follows that it is *impossible* to obtain a product that is overall CPT odd. In other words, within our present knowledge and method of constructing (current–current) interactions in field theory, CPT cannot be violated. That is not to say, that it is absolutely impossible, but simply that we do not know how. This is what is meant by the CPT theorem.

For completeness, let us mention that a *possible* consequence of CPT violation might be a difference between particle and antiparticle masses. At present, the most stringent limits come from the study of the $K^0 - \bar{K}^0$ system (which is examined in detail in section 2.7.3):

$$\left| \frac{m_{K^0} - m_{\bar{K}^0}}{m_{K^0}} \right| \leqslant 4 \times 10^{-19} \quad (95\%\text{CL}). \tag{B.3.17}$$

Note, in contrast, that any of C, P or T may be violated individually (or in pair products) by a suitable choice of interfering currents; e.g. the product $V \cdot A$ violates both C and P but not T. What is not included above is the possibility of a complex coupling (as in the elements of the V_{CKM} matrix, see section 2.7.2). Such a contribution would naturally induce a violation of time-reversal invariance, which is rather difficult (though not impossible) to detect experimentally. For this reason and since the conservation of CPT requires a simultaneous compensating violation of the product CP, we normally talk of CP violation and not T violation, although the two are entirely equivalent in this context.

B.3.6 *C and P of simple composite systems*

Many simple composite systems, such as positronium (an e^+e^- bound state) and $q\bar{q}$ pairs, but also two- or multi-pion final states, may possess well-defined symmetry properties under the operations of $\mathcal{C}$ and $\mathcal{P}$. In theories in which these symmetries are respected such properties naturally lead to the idea of associated conserved quantum numbers C and P, even for composite objects (the same is true too for T). Such discrete quantum numbers are *multiplicative* in nature and thus in the case of a composite system all the relevant quantum numbers of the parts must simply be multiplied together. We shall now present a few instructive examples.

From the foregoing discussion on CPT, we see that, by complementarity, once the properties with respect to C and P are understood an explicit discussion of T is superfluous. That said, for example, in condensed matter physics a consequence of T-invariance for systems of many fermions is the so-called Kramers degeneracy, whereby even certain highly disordered systems must have degenerate energy eigenstates.

B.3.6.1 Charge conjugation in composite systems
First of all, note that the quantum number C can clearly only be associated with neutral systems, such as neutral $q\bar{q}$ states.

$C_{\pi^+\pi^-}$

Let us start by considering the two-pion state $\pi^+\pi^-$. The action of C is to interchange the two pions and this will introduce a factor $(-1)^L$, where L is the orbital angular momentum quantum number, owing to the parity of the spatial part of the wave-function. Since there is no other effect, we have $C_{\pi^+\pi^-} = (-1)^L$.

$C_{e^+e^-}$

Next we examine the case of positronium. This is a little more complicated owing to the spin effects. For the spatial exchange of the electron and positron there is the same factor $(-1)^L$ above. However, the spin part of the wave-function must also be considered: for the spin-0 singlet state this is antisymmetric and for the spin-1 triplet, symmetric. This leads to another factor conveniently expressed as $-(-1)^S$. Finally, the full relativistic theory of electrons generates a further -1 for every interchange of two identical fermion or antifermion states. Putting all this together, we obtain a charge-conjugation quantum number $C_{e^+e^-} = (-1)^{L+S}$.

B.3.6.2 Parity in composite systems

Here we should recall that all particles (elementary or not) may be ascribed an intrinsic parity: for fermions this is not determined absolutely, but Dirac theory predicts it to be opposite for fermion and antifermion. A fermion–antifermion pair thus have overall negative intrinsic parity, which must then be multiplied by the parity of the relative spatial wave-function. In the case of bosons a scalar particle has, by definition $P = +1$, whereas a pseudoscalar (such as a pion) has $P = -1$.

$P_{\pi^+\pi^-}$

Given an even number of pions, it is not necessary to know the intrinsic parity of the pion since in this case we have $P^2 = +1$, whatever the value of $P = \pm 1$. We therefore need only consider the parity of the spatial wave-function, which leads to $P_{\pi^+\pi^-} = (-1)^L$.

$P_{e^+e^-}$

As noted above the product of intrinsic parities for the e^+e^- pair is -1, which again must be multiplied by the spatial contribution. We thus have $P_{e^+e^-} = -(-1)^L$.

B.3.6.3 The J^{PC} classification of mesons

We can now study the classification of mesons ($q\bar{q}$ states) in terms of the three quantum numbers J, P and C. Let us examine the angular momentum first. The purely spin part may be either spin-0 (antisymmetric, singlet) or spin-1 (symmetric, triplet). The orbital angular momentum part is naturally any integer from zero up. The J^{PC} assignments may then be deduced easily from the previous analysis of the positronium case.

The lowest-lying neutral mesons (π^0, K^0, η and η') are $L = 0$, $S = 0$ states, which must then have $J^{PC} = 0^{-+}$. While C is clearly not defined for a charged meson, the J^P assignments given still hold. The slightly heavier s-wave $S = 1$ states (ρ, K^*, ϕ and ω) have $J^{PC} = 1^{--}$. For increasing mass the J^{PC} quantum numbers then follow a natural sequence. A complete list of all the known mesonic states and their J^{PC} assignments may be found in the PDG 2024 (Navas *et al* 2024).

Exercise B.3.1. *From the foregoing classification, show that for a standard $q\bar{q}$ state, the assignments $J^{PC} = 0^{-+}$, 0^{++}, 1^{+-}, 1^{++}, 1^{--}, 2^{-+}, 2^{++}, 2^{--} etc are all allowed, whereas 0^{--}, 0^{+-}, 1^{-+}, 2^{+-}, 3^{-+} etc are not admissible.*

Finally, for completeness, we might add that the photon has $J^{PC} = 1^{--}$, the gluons (being colour charged, with therefore indefinite C) have $J^P = 1^-$, whereas the spin-1 $W^{\pm}$ and Z^0 weak bosons have neither P nor C well-defined. The neutral Higgs boson in the Standard Model has $J^{PC} = 0^{++}$.

B.4 Isospin and $SU(2)$

B.4.1 Isospin in nuclear physics

In nuclear and particle physics a number of symmetries are apparent. One of the simplest is the existence of a large number of so-called *mirror nuclei*: that is, pairs of nuclei that differ only by interchange of the number of protons and neutrons. An example is

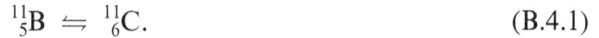

$$^{11}_{5}\text{B} \;\rightleftharpoons\; ^{11}_{6}\text{C}. \tag{B.4.1}$$

While the chemical properties of ^{11}B and ^{11}C atoms are obviously rather different, the nuclei are very similar indeed. When we take into account the variation due to the effects of Coulomb repulsion, we might even say they are identical, as far as the strong interaction is concerned, that is.

Evidently, such a symmetry must have to do with a corresponding symmetry at the nucleon level. That is, we assume it to be just the manifestation of a deeper proton–neutron symmetry. In fact, at the hadronic level in general we see much the same sort of mirror behaviour in various particles:

- The masses of the neutron and the proton are very similar; indeed, although the proton has a positive charge and the neutron is neutral, they are almost identical in all other respects. In fact, inasmuch as electromagnetic effects may be ignored with respect to the strong interaction and taking into account that, as we now know, there is a small up–down quark mass difference, they might be considered as two different states of the same fundamental field.
- The strong interaction between any pair of nucleons is identical, independently of whether they are protons or neutrons. That is, the proton–proton, proton–neutron and neutron–neutron forces are the same. Again, to see this phenomenologically, the electromagnetic effects must first be subtracted.
- In a similar fashion, the three known pion states π^+, π^0 and π^- are also very similar. Indeed, the two charged pions have exactly the same mass, while the neutral pion is just slightly lighter. Moreover, apart from very systematic differences, which are in fact *explained* by the symmetry picture we shall now discuss, their strong interactions with matter (protons and neutrons) are also the same.

Heisenberg (1932) thus introduced the notion of *isotopic spin*[6](or *isobaric spin*) to explain these observations. The standard contraction of the name is now *isospin*.

Now, we know from QM that when the Hamiltonian of a system possesses a discrete symmetry, e.g. with respect to spatial inversion, this manifests itself through a degeneracy of the energy states of the system. Consider, for example, the various energy levels of the hydrogen atom. In particle physics mass is equivalent to energy (since $E = mc^2$) and so the near mass degeneracy of the neutron and proton indicates a symmetry of the Hamiltonian describing the strong interactions. The neutron does have a slightly higher mass and so the degeneracy is not exact. However, here (as the case would be in general for QM) the appearance of a symmetry may be imperfect as it can be perturbed by other forces, giving rise to slight differences between otherwise degenerate states. Indeed, the proton is charged, whereas the neutron is not and therefore electromagnetism must play a different rôle.

Heisenberg noted that the mathematical description of the observed symmetry was very similar to the symmetry structure of orbital angular momentum or *spin*. In mathematical terms, the isospin symmetry is due to an invariance of the strong-interaction Hamiltonian under the action of the (Lie) group SU(2). The neutron and the proton are placed in a doublet (a spin-½ or fundamental representation) of SU(2).

B.4.2 Isospin in particle physics

The mathematical structure (or algebra) is then quite simply that of the usual angular momentum. Isospin is described by two quantum numbers: the total isospin I and the spin projection along the quantisation axis I_3. The concept may now be extended to other particles. As noted, the proton and neutron both have $I = \frac{1}{2}$; the proton has $I_3 = +\frac{1}{2}$ or 'isospin up', whereas the neutron has $I_3 = -\frac{1}{2}$ or 'isospin down'. The pions naturally belong to the $I = 1$ triplet and are assigned an adjoint (rank-3) representation of SU(2), with π^+, π^0 and π^- having $I_3 = +1$, 0 and -1, respectively.

In Dirac notation, for the nucleon pair, we write

$$|p\rangle = |\tfrac{1}{2}, +\tfrac{1}{2}\rangle \text{ and } |n\rangle = |\tfrac{1}{2}, -\tfrac{1}{2}\rangle, \tag{B.4.2}$$

while the pion triplet becomes

$$|\pi^+\rangle = |1, +1\rangle, \quad |\pi^0\rangle = |1, 0\rangle \text{ and } |\pi^-\rangle = |1, -1\rangle. \tag{B.4.3}$$

The pairs of quantum numbers above then have the same *mathematical* significance as the j, m pairs for angular-momentum states.

---

[6] The term *isotopic spin* was coined later though, by Wigner (1937), in recognition of the mathematical similarity to angular-momentum spin: there are two equivalent states (precisely the proton and neutron), just as there are two spin states of the electron. Moreover, the algebra describing the symmetry turned out to be SU(2), again, just as that of spin-1/2 states.

An important consequence of isospin symmetry and its mathematical structure is the possibility to apply Clebsch–Gordan coefficients to constructing combinations (or the composition) of particles. A simple example is the strong decay of the so-called Δ resonances (spin-$\frac{3}{2}$, isospin-$\frac{3}{2}$), which may be generically described as $\Delta \to N\pi$ (N being a nucleon, p or n). For concreteness, let us consider the state Δ^+, whose *isospin* designation is indicated as $|\frac{3}{2}, \frac{1}{2}\rangle$. There are two distinct possible final states: $p\pi^0$ and $n\pi^+$ or $|\frac{1}{2}, +\frac{1}{2}\rangle|1, 0\rangle$ and $|\frac{1}{2}, -\frac{1}{2}\rangle|1, 1\rangle$, respectively. Now, a glance at a table of Clebsch–Gordan coefficients tells us that a spin-$\frac{3}{2}$ state may be decomposed into the following combination of spin-one and spin-$\frac{1}{2}$ objects:

$$|\tfrac{3}{2}, \tfrac{1}{2}\rangle = \sqrt{\tfrac{2}{3}}\, |\tfrac{1}{2}, +\tfrac{1}{2}\rangle|1, 0\rangle + \sqrt{\tfrac{1}{3}}\, |\tfrac{1}{2}, -\tfrac{1}{2}\rangle|1, 1\rangle. \qquad (B.4.4)$$

The squares of the coefficients provide the branching fractions: namely, $\frac{2}{3}$ into $p\pi^0$ and $\frac{1}{3}$ into $n\pi^+$. These fractions are experimentally well verified.

B.4.3 Isospin in the quark model

At the quark level it is natural to treat the up and down quarks as the proton and neutron above, which then become composite states of isospin (three isospin one-half objects may be combined to form a total isospin of either $\frac{1}{2}$ or $\frac{3}{2}$). The pions are just the isospin-one composition of a quark and antiquark. There is, though, a subtlety here, which we shall now briefly explain.

Translating this into quarks we see that we must consider two doublets: the u- and d-quark pair and their antiquark pair. These are conventionally taken as

$$\begin{pmatrix} u \\ d \end{pmatrix} \quad \text{and} \quad \begin{pmatrix} -\bar{d} \\ \bar{u} \end{pmatrix}. \qquad (B.4.5)$$

The ordering is chosen so that the $I_3 = \frac{1}{2}$ state is uppermost and the minus sign has the effect of rendering the two representations identical. This is only possible for SU(2); that is, the anti-triplet in SU(3), say, necessarily has a different representation with respect to the triplet.

Let us now briefly show that this does indeed work as suggested. Isospin symmetry requires that the physics of quarks should not change under a generalised rotation in isospin space:

$$\begin{pmatrix} u' \\ d' \end{pmatrix} = \mathcal{U} \begin{pmatrix} u \\ d \end{pmatrix}, \qquad (B.4.6)$$

where the (possibly complex) unitary matrix $\mathcal{U}$ may be represented generically as

$$\mathcal{U} = \begin{pmatrix} a & b \\ -b^* & a^* \end{pmatrix}, \quad \text{with} \quad (|a|^2 + |b|^2 = 1). \qquad (B.4.7)$$

The corresponding transformation for antiquarks is then

$$\begin{pmatrix} \bar{u}' \\ \bar{d}' \end{pmatrix} = \mathcal{U}^* \begin{pmatrix} \bar{u} \\ \bar{d} \end{pmatrix}. \tag{B.4.8}$$

Now, it turns out for this case that there exists a unitary transformation that takes $\mathcal{U}$ into $\mathcal{U}^*$: namely,

$$\mathcal{U}^* = V \mathcal{U} V^\dagger \quad \text{with} \quad V = \begin{pmatrix} 0 & -1 \\ 1 & 0 \end{pmatrix}. \tag{B.4.9}$$

We then have

$$\begin{pmatrix} -\bar{d}' \\ \bar{u}' \end{pmatrix} = \mathcal{U} \begin{pmatrix} -\bar{d} \\ \bar{u} \end{pmatrix}; \tag{B.4.10}$$

i.e. the quark and antiquark doublets given in equation (B.4.5) transform in the same way and this is therefore the conventional choice of representation.

B.4.4 G-parity

We may now introduce the concept of G-parity, useful in determining the symmetry properties of pions. We start from the observation that, whereas the neutral pion is its own antiparticle and so may be assigned a signature under charge conjugation, the same is clearly not possible for the $\pi^\pm$; however, it is possible to exploit the isospin symmetry just discussed to define a new operator $\mathcal{G}$ (Lee and Yang 1956) for which they too are eigenstates. We may simply employ the isospin operators to undo, as it were, the $\mathcal{C}$ operation by a 180° rotation (conventionally chosen to be) about the second axis in isospin space (via $\mathcal{I}_2$):

$$\mathcal{G} := \mathcal{C} e^{i\pi \mathcal{I}_2}. \tag{B.4.11}$$

G-parity is therefore another multiplicatively conserved quantum number (with eigenvalues $G = \pm 1$), a sort of generalisation of C-parity to certain meson isospin multiplets. Note though that whereas $\mathcal{C}$ exchanges quark and antiquark, $e^{i\pi \mathcal{I}_2}$ exchanges u and d quarks.

It is easy to see that since the strong interaction conserves charge and isospin then it must also conserve G whereas the opposite holds for the weak and electromagnetic interactions; it thus permits the derivation of useful selection rules. G-parity is thus of particular use in describing strong interactions involving pions and nucleons, although it is restricted to systems with zero baryon number (e.g. nucleon–antinucleon, but not single baryons). More precisely, the requirement for applicability to any given multiplet is that overall it must have zero electric charge, baryonic number, strangeness and charm etc.

Let us now briefly examine the G-parity of the pions and nucleons. The standard phase definitions, brought over from normal spin and related to the minus sign appearing in equation (B.4.5), lead to

$$e^{i\pi \mathcal{I}_2}|I, I_3\rangle = (-1)^{I-I_3}|I, -I_3\rangle, \tag{B.4.12}$$

where I_3 here is the eigenvalue. Applying this to the pion triplet then gives

$$e^{i\pi \mathcal{I}_2}|\pi^\pm\rangle = +|\pi^\pm\rangle \quad \text{and} \quad e^{i\pi \mathcal{I}_2}|\pi^0\rangle = -|\pi^0\rangle. \tag{B.4.13}$$

The π^0 decays into two photons implying $\mathcal{C}|\pi^0\rangle = +|\pi^0\rangle$ and so $\mathcal{G}|\pi^0\rangle = -|\pi^0\rangle$. It is then conventional to choose the C phases for the charged pion transformations to maintain the same sign over the entire multiplet; thus, $G_\pi = -1$ and therefore for an n-pion state $G_{n\pi} = (-1)^n$, which must be multiplied by the spatial wave-function signature. Thus, e.g. for generic two-pion states ($\pi^0\pi^0$ or $\pi^+\pi^-$) we have

$$\mathcal{G}|\pi, \pi\rangle = (-1)^{L+I}|\pi, \pi\rangle. \tag{B.4.14}$$

We can do the same for a nucleon–antinucleon system: by analogy with spatial rotations, the above isospin rotation for an $I_3 = 0$ system should behave as

$$e^{i\pi \mathcal{I}_2}|I, 0\rangle = (-1)^I|I, 0\rangle \tag{B.4.15}$$

and, as in the case of positronium, we should have

$$\mathcal{C}|L, S\rangle = (-1)^{L+S}|L, S\rangle. \tag{B.4.16}$$

We therefore find

$$\mathcal{G}|N, \bar{N}\rangle = (-1)^{L+S+I}|N, \bar{N}\rangle. \tag{B.4.17}$$

Exercise B.4.1. *Derive the preceding general expressions for G in the case of fermion–antifermion and boson–antiboson systems.*

Again, by choosing the conventional phases accordingly, this may be extended to systems with $I_3 \neq 0$ to write more generally $G_{N, \bar{N}'} = (-1)^{L+S+I}$. Finally, other mesons such as ρ, ω, ϕ etc may be considered. These spin-one vector mesons all couple to the photon and therefore have $C = -1$. For the isoscalars ω, ϕ and also J/ψ, Υ etc, the rotation in isospin space has no effect and so for these $G = -1$, whereas the isospin triplet ρ will behave as the pion leading to $G_\rho = +1$.

Armed with these assignments, we can immediately examine the consequences for various decay and scattering processes. Consider first the strong decays of the above mesons:

$$\rho \to 2\pi \qquad\qquad \rho \not\to 3\pi \tag{B.4.18}$$

$$\omega, \phi \to 3\pi \qquad\qquad \omega, \phi \not\to 2\pi \tag{B.4.19}$$

Table B.2. The selections rules for $\bar{p}n \to 2\pi$ and 3π. Transitions forbidden by P and angular momentum conservation are indicated ×and those forbidden by G conservation are indicated −; all others are allowed; table reproduced from Lee and Yang (1956).

State	$^1S_0$	$^3S_1$	$^1P_1$	$^3P_0$	$^3P_1$	$^3P_2$
$I^G\ J^P$	$1^-\ 0^-$	$1^+\ 1^-$	$1^+\ 1^+$	$1^-\ 0^+$	$1^-\ 1^+$	$1^-\ 2^+$
2π	×		×	−	×	−
3π		−	−	×		

$$J/\psi,\ \Upsilon \to \begin{cases} n\pi & (n\ \text{odd}) \\ \rho n\pi & (n\ \text{even}) \\ \omega n\pi & (n\ \text{odd}) \end{cases} \qquad J/\psi,\ \Upsilon \nrightarrow \begin{cases} n\pi & (n\ \text{even}) \\ \rho n\pi & (n\ \text{odd}) \\ \omega n\pi & (n\ \text{even}) \end{cases} \qquad (\text{B.4.20})$$

We can also examine the case of the isoscalar η^0 meson. Its (electromagnetic) decay into two photons indicates that $C = +1$ and therefore $G = +1$ too since the isospin rotation has no effect. Thus, $\eta^0 \to 3\pi$ is not an allowed strong decay. However, $J^P_{\eta^0} = 0^-$ and therefore the two-pion state is also forbidden, thus leaving no allowed strong-decay channel and explaining the exceedingly narrow width $\Gamma_{\eta^0} = 1.31 \pm 0.05$ keV. As a final example, let us consider annihilation of anti-protons and neutrons into multi-pion final states (the subject of the original paper by Lee and Yang (1956)). Since a $\bar{p}n$ pair has $I_3 = -1$, the isospin assignment of the initial state must be $|1, -1\rangle$ and therefore $G = (-1)^{L+S+1}$. We should also consider spatial parity: $P_{N\bar{N}} = (-1)^{L+1}$ (recall that fermions and antifermions have opposite parity). The combined requirements of conservation of G, P and angular momentum then lead to the selection rules shown in table B.2

Exercise B.4.2. *Derive the preceding expressions for the intrinsic G-parity of the pion isospin triplet and the η^0.*

Of course, G-parity considerations do not lead to any more information than those of C and isospin put together, but G-parity does allow a more general, immediate and direct application where possible.

B.5 The Yukawa potential

We shall now provide a simple derivation of the Yukawa potential. For comparison, the starting point will, however, be Maxwell's equations (1865) for a classical electromagnetic field, which may be expressed as

$$\nabla \cdot E = \rho, \qquad \nabla_{\times}E + \frac{\partial B}{\partial t} = 0,$$
$$\nabla \cdot B = 0, \qquad \nabla_{\times}B - \frac{1}{c^2}\frac{\partial B}{\partial t} = j. \qquad (\text{B.5.1})$$

If we now define the scalar and vector potentials V and A by

$$\boldsymbol{B} =: \boldsymbol{\nabla} \wedge \boldsymbol{A} \qquad (B.5.2)$$

and

$$\boldsymbol{E} =: -\boldsymbol{\nabla} V - \frac{\partial \boldsymbol{A}}{\partial t},$$

then the four equations reduce to the following two:

$$-\boldsymbol{\nabla}^2 V - \frac{\partial}{\partial t}(\boldsymbol{\nabla} \cdot \boldsymbol{A}) = \rho \qquad (B.5.3)$$

and

$$\left(-\boldsymbol{\nabla}^2 + \frac{1}{c^2}\frac{\partial^2}{\partial t^2}\right)\boldsymbol{A} + \boldsymbol{\nabla}\left(\boldsymbol{\nabla} \cdot \boldsymbol{A} + \frac{1}{c^2}\frac{\partial V}{\partial t}\right) = \boldsymbol{j}.$$

With the Lorentz gauge condition,

$$\boldsymbol{\nabla} \cdot \boldsymbol{A} = -\frac{1}{c^2}\frac{\partial V}{\partial t}, \qquad (B.5.4)$$

these simplify to

$$\left(-\boldsymbol{\nabla}^2 + \frac{1}{c^2}\frac{\partial^2}{\partial t^2}\right)V = \rho \qquad (B.5.5a)$$

and

$$\left(-\boldsymbol{\nabla}^2 + \frac{1}{c^2}\frac{\partial^2}{\partial t^2}\right)\boldsymbol{A} = \boldsymbol{j}, \qquad (B.5.5b)$$

which, introducing $A^\mu := (cV, \boldsymbol{A})$ and $j^\mu := (c\rho, \boldsymbol{j})$, may be conveniently rewritten in four-vector notation as a manifestly relativistic wave equation (the Lorentz condition is then just $\partial^\mu A_\mu = 0$):

$$\partial^2 A^\mu = j^\mu.$$

We now look for solutions in the static, spherically symmetric configuration, which is the case of a point charge placed at the origin. Moving over to spherical polar coordinates, for the electrostatic potential away from the origin (i.e. in free space, where both $\rho = 0$ and $\boldsymbol{j} = 0$), we have, from equation (B.5.5a),

$$\frac{1}{r^2}\frac{\partial}{\partial r}\left(r^2 \frac{\partial V(r)}{\partial r}\right) = 0. \qquad (B.5.6)$$

It is then not difficult to show that the most general spherically symmetric solution takes the classic form of the Coulomb potential:

$$V(r) = \frac{V_0}{r}, \qquad (B.5.7)$$

where, of course, the precise value of V_0 is determined by the magnitude of the charge at the origin. The full set of Maxwell's equations (1865) is, of course, not necessary to obtain the Coulomb potential, but we shall now show how this derivation may be paralleled by that of a massive scalar field in a relativistic approach.

Let us therefore compare the above with a relativistic version of QM. The first and simplest approach is due to (Klein 1927 and Gordon 1926). The natural starting point is the standard relativistic energy–momentum relation for a free particle:

$$E^2 = \boldsymbol{p}^2 c^2 + m^2 c^4, \tag{B.5.8}$$

where m is the rest mass of the particle we wish to describe. In the standard approach to quantisation, kinematical variables, such as E and $\boldsymbol{p}$, are replaced with the corresponding operators:

$$E \rightarrow i\hbar \frac{\partial}{\partial t} \quad \text{and} \quad \boldsymbol{p} \rightarrow -i\hbar \boldsymbol{\nabla}. \tag{B.5.9}$$

These substitutions into (B.5.8) lead to the following relativistic wave equation:

$$\left(\hbar^2 \frac{\partial^2}{\partial t^2} - \hbar^2 c^2 \boldsymbol{\nabla}^2 + m^2 c^4 \right) \phi(t, \boldsymbol{x}) = 0, \tag{B.5.10a}$$

with four-vector version

$$\left(\partial^2 + \frac{m^2 c^2}{\hbar^2} \right) \phi(x) = 0, \tag{B.5.10b}$$

known as the (Klein 1927 and Gordon 1926) equation; it replaces the Schrödinger equation in the relativistic regime and describes a massive scalar field.

We note immediately the similarity with Maxwell's (1865) equation for the electric scalar potential, which is seen to be just the Klein–Gordon equation with $m = 0$. We are thus led to consider the full Klein–Gordon equation as a possible description of a theory similar to QED, but with a *massive* exchange field. Seeking then the static, spherically symmetric solutions to the Klein–Gordon equation, i.e.

$$\frac{1}{r^2} \frac{\partial}{\partial r} \left(r^2 \frac{\partial V(r)}{\partial r} \right) - \frac{m^2 c^2}{\hbar^2} V(r) = 0, \tag{B.5.11}$$

it is again not difficult to show that the most general solutions are

$$V(r) = \frac{V_0}{r} e^{\pm mcr/\hbar}. \tag{B.5.12}$$

Only the exponentially decaying form (with negative exponent) can be physically acceptable and rewriting this as

$$V(r) = \frac{V_0}{r} e^{-r/a}, \tag{B.5.13}$$

which is just the Yukawa potential. We see that the effective range of such a potential is

$$a = \hbar/mc. \tag{B.5.14}$$

Reconsidering then Maxwell's (1865) equations, we see that the solution for the electric scalar potential is equivalent to a massless field ($m = 0$), which implies $a = \infty$. That is, the Coulomb potential may be viewed as the massless limit of Yukawa theory or, vice versa, Yukawa theory is a massive version of electrodynamics[7].

Note finally that, while the $1/r$ behaviour, common to both theories, depends principally on the space–time dimensionality (i.e. that we live in precisely three spatial dimensions), the exponential behaviour is seen to stem from the mass of the exchange particle in question. The physical significance of this may be understood in terms of the Heisenberg uncertainty principle; a *virtual* exchange field of mass $m \neq 0$ may exist without truly violating energy conservation for a time interval not in excess of $\hbar/mc^2$. In this time lapse it can travel, at most, a distance $=c \times \hbar/mc^2$, which is then just the range a indicated in the above exponential form.

B.6 The double well and quantum oscillation

The double potential well in QM nicely demonstrates one of the more surprising phenomena associated with quantisation: namely, *quantum oscillation*. Consider the situation in which there are two identical square wells (in one dimension for simplicity) sufficiently separated so that the form of the solution to the Schrödinger equation locally in the neighbourhood of either well is not appreciably affected by the presence of the other. In other words, the solution within the regions of the wells is very similar to that of a single isolated well. The form of the well and the x dependence of the corresponding two lowest-energy eigenstates are represented in figure B.2.

The solutions can be represented generically as

$$\psi_{1,2}(x, t) = u_{1,2}(x)e^{-\frac{i}{\hbar}E_{1,2}t}, \tag{B.6.1}$$

where the precise form of $u_{1,2}(x)$ is entirely irrelevant for the present purposes.

Now, these represent the 'unperturbed' eigenstates of the system and are those with independent temporal evolution. However, if now some 'interaction' with the system introduces a particle into *one* of the wells (the left, say), the state induced will not correspond to any single pure eigenstate. In other words, a different basis is necessary to describe this external interaction:

$$u_{\mathrm{L,R}}(x) = \frac{1}{\sqrt{2}}\left[u_1(x) \pm u_2(x) \right]. \tag{B.6.2}$$

---

[7] We are, of course, closing an eye here to the very important question of local gauge invariance, which is intimately related to the massless nature of the photon and other gauge fields.

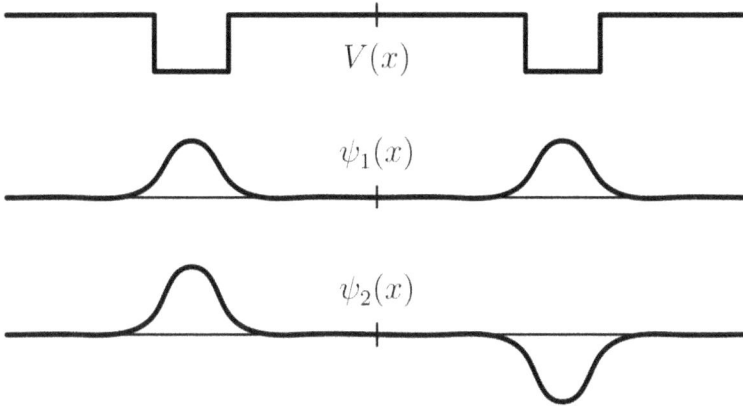

Figure B.2. The double potential well and the spatial dependence of the corresponding two lowest-energy eigenstates, $E_1 > E_0$.

Such a state describes the system at the instant $t = 0$, when the particle is introduced into the left-hand well. Taking now the initially left-hand case for definiteness, the evolution at a later time is given by

$$\psi_L(x, t) = \frac{1}{\sqrt{2}}\left[u_1(x)e^{-\frac{i}{\hbar}E_1 t} + u_2(x)e^{-\frac{i}{\hbar}E_2 t} \right]. \tag{B.6.3}$$

However, if we wish to know the probability of finding the particle in one or other well, we should decompose over the L/R basis:

$$\psi_L(x, t) = \frac{1}{2}\left[\left(u_L(x) + u_R(x)\right)e^{-\frac{i}{\hbar}E_1 t} + \left(u_L(x) - u_R(x)\right)e^{-\frac{i}{\hbar}E_2 t} \right]$$
$$= \frac{1}{2}\left[u_L(x)\left(e^{-\frac{i}{\hbar}E_1 t} + e^{-\frac{i}{\hbar}E_2 t}\right) + u_R(x)\left(e^{-\frac{i}{\hbar}E_1 t} - e^{-\frac{i}{\hbar}E_2 t}\right) \right]. \tag{B.6.4}$$

Introducing now the average energy $\langle E \rangle := \frac{1}{2}(E_1 + E_2)$ and the energy difference $\Delta := E_2 - E_1$, we finally obtain

$$\psi_L(x, t) = \left[u_L(x)\cos\left(\frac{\Delta}{2\hbar} t\right) - iu_R(x)\sin\left(\frac{\Delta}{2\hbar} t\right) \right]e^{-\frac{i}{\hbar}\langle E \rangle t}. \tag{B.6.5}$$

We see that the coefficient of, for example, $u_L(x)$, which determines the probability of finding the particle in the left-hand well, oscillates in time:

$$\mathcal{P}_L(t) = \cos^2\left(\frac{\Delta}{2\hbar} t\right) = \frac{1}{2}\left[1 - \cos\left(\frac{\Delta}{\hbar} t\right) \right]. \tag{B.6.6}$$

Note that the frequency is thus $\Delta/\hbar$ (and not half that). In other words, the particle effectively oscillates between the two potential wells with a frequency determined by the difference of the natural frequencies associated with the two states involved. Note too the close parallel with the case of coupled oscillators in classical mechanics.

To conclude this section, let us underline a common aspect of such phenomena: the original physical system possesses a symmetry (in this case under parity or spatial inversion), which the state created externally does *not* respect (i.e. it is not an eigenstate of) the basic Lagrangian. This is a sufficient (and in fact necessary) condition for the induced mixing, which lies at the heart of the quantum oscillation phenomenon.

References

Boltzmann L 1868 *Wiener Berichte* **58** 517

Bose S N 1924 *Z. Phys.* **26** 178

Dirac P A M 1926 *Proc. R. Soc. Lond.* **A112** 661

Dirac P A M 1928 *Proc. R. Soc. Lond.* **A117** 610

Einstein A 1924 *Sitzung. d. Preuß. Ak.* **22** 261 **A117** 610

Fermi E 1926 *Rend. Acc. Naz. Lincei* **3** 145

Gordon W 1926 *Z. Phys.* **40** 117

Heisenberg W 1932 *Z. Phys.* **78** 156 587

Klein O 1927 *Z. Phys.* **41** 407

Lee T-D and Yang C-N 1956 *Lett. Nuovo Cim.* **10** 749

Majorana E 1937 *Nuovo Cim.* **14** 171 *transl. by L Maiani* in *Ettore Majorana Scientific Papers*, ed G F Bassani (Springer-Verlag, 2006) p 201

Maxwell J C 1865 *Phil. Trans. R. Soc. (London)* **155** 459

Navas S *et al* (Particle Data Group) 2024 *Phys. Rev.* **D110** 030001

Pauli W 1925 *Z. Phys.* **31** 765

Pauli W 1940 *Phys. Rev.* **58** 716

Wigner E 1937 *Phys. Rev.* **51** 106

IOP Publishing

An Introduction to Elementary Particle Phenomenology
(Second Edition)

Philip G Ratcliffe

Appendix C

Scattering theory

'He who obtains has little.
He who scatters has much.'

Lao Tzu

C.1 Scattering kinematics

In this section we shall briefly recall various useful notions and relations for describing the kinematics of generic scattering processes. Since we shall mostly be limited to two-body processes, we shall only present formulæ for two-body kinematics. More general expressions may be found int the review article on kinematics in PDG 2024 (Navas *et al* 2024, p 743).

C.1.1 Mandelstam variables

For a simple generic two-body process in which particles 1 and 2 interact and re-emerge as particles 3 and 4, energy–momentum conservation is simply expressed (using standard relativistic four-vector notation) as

$$p_1^{\mu} + p_2^{\mu} = p_3^{\mu} + p_4^{\mu}. \tag{C.1.1}$$

In the centre-of-mass system, where particles 1 and 2 have equal and opposite three-momenta (as do particles 3 and 4), we have that the initial total four-momentum squared

$$(p_1 + p_2)^2 = (E_1 + E_2)^2 \tag{C.1.2}$$

is just the centre-of-mass energy squared (i.e. $E_1 + E_2 = E_{CM}$). As this is a commonly used quantity, following Mandelstam (1958), it is convenient to define it as

doi:10.1088/978-0-7503-5759-3ch9

$$s := (p_1 + p_2)^2 = (p_3 + p_4)^2, \tag{C.1.3a}$$

$$t := (p_1 - p_3)^2 = (p_2 - p_4)^2, \tag{C.1.3b}$$

$$u := (p_1 - p_4)^2 = (p_2 - p_3)^2, \tag{C.1.3c}$$

where we have also included the other two standard Mandelstam variables. Note that, whereas s must be positive, both t and u are normally negative (this is easy to check). Moreover, the three variables are not independent, being related by

$$s + t + u = \sum_{i=1}^{4} m_i^2, \tag{C.1.4}$$

where m_i is the mass of particle i (again, this is not difficult to demonstrate).

In the centre-of-mass system, where the initial pair of three-momenta have the same magnitude (as do the final pair), we have

$$|p_1| = |p_2| = \frac{1}{2\sqrt{s}} \sqrt{\lambda(s, m_1^2, m_2^2)}, \tag{C.1.5a}$$

$$|p_3| = |p_4| = \frac{1}{2\sqrt{s}} \sqrt{\lambda(s, m_3^2, m_4^2)}, \tag{C.1.5b}$$

where $\lambda(x, y, z)$ is the Källén function (1964) defined as

$$\lambda(x, y, z) := x^2 + y^2 + x^2 - 2xy - 2yz - 2zx. \tag{C.1.6}$$

The energies are given by

$$E_1 = \frac{s + m_1^2 - m_2^2}{2\sqrt{s}} \quad \text{and} \quad E_2 = \frac{s + m_2^2 - m_1^2}{2\sqrt{s}}, \tag{C.1.7}$$

the final-state energies being obtained by substituting $1 \rightarrow 3$ and $2 \rightarrow 4$ in the above. The variable t (or u) may be used in place of the centre-of-mass scattering angle θ (between particles 1 and 3). For example, in the particularly simple case of the elastic scattering of two equal-mass (m) particles we have

$$\cos\theta = 1 + \frac{2t}{s - 4m^2}. \tag{C.1.8}$$

It is often necessary to transform between the laboratory and the centre-of-mass frames. This is a comparatively simple Lorentz boost and the following two formulæ relating the energies and momenta are often handy:

$$E_{\text{CM}} = \sqrt{s} = \sqrt{m_1^2 + m_2^2 + 2E_{1\text{lab}}m_2}, \tag{C.1.9a}$$

$$p_{\text{CM}} = p_{\text{lab}} \frac{m_2}{\sqrt{s}}, \tag{C.1.9b}$$

where we have taken particle 1 to be the projectile and particle 2 the target at rest in the laboratory frame.

C.2 Electron scattering

Through the study of α-particle scattering, Rutherford (1911) arrived at a new understanding of the atom and its internal structure. However, the α-particle is not a point-like object and can also interact via the strong force. Thus, the details of a small nuclear target obtained via such scattering will be clouded by the internal structure of the probe used. On the other hand, the electron only has electromagnetic interactions (the weak interaction may usually be neglected in comparison) and so far it has exhibited a purely point-like behaviour (at least up to the highest energies presently available).

For low-energy scattering ($E \ll m_W$, where the weak force is particularly suppressed) the interaction of the electron with a nucleon or nucleus is governed purely by the theory of quantum electrodynamics and is therefore completely known. This makes the electron an ideal probe to study the internal structure of the nucleus and, going deeper, of the nucleon.

C.2.1 Non-relativistic point-like elastic scattering

The simplest example of scattering with electrons is the elastic case. Here energy and momentum are transferred from an electron to a nucleus (or nucleon) exchange via (single) photon exchange and the final nucleus remains intact (see figure C.1). The four-momenta in the problem are then the initial (final) electron momentum k^μ (k'^μ) and the initial (final) nucleus momentum p^μ (p'^μ). Conservation of energy and momentum requires

$$k^\mu + p^\mu = k'^\mu + p'^\mu. \tag{C.2.1}$$

Since the momentum of the final nucleus is not usually measured, it is convenient to rewrite this as

$$p'^2 = M^2 = (k^\mu - k'^\mu + p^\mu)^2, \tag{C.2.2}$$

where M is the nuclear mass. It is then straightforward to derive the relation

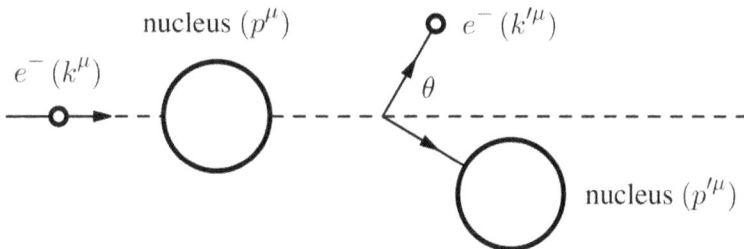

Figure C.1. Electron–nucleus elastic scattering.

$$EM = E'E(1 - \cos\theta) + E'M, \qquad\qquad (C.2.3)$$

where E and E' are the laboratory-frame initial and final electron energies, respectively, θ the laboratory-frame electron scattering angle and we have neglected the electron mass, note that in the laboratory frame $p^\mu = (M, 0)$. This may then be rearranged to express E' as a function of E and θ:

$$E' = \frac{E}{1 + (1 - \cos\theta)E/M}. \qquad\qquad (C.2.4)$$

which provides the well-known result that, for *elastic* scattering, the final energy is determined by the angle (and *vice versa*).

Let us now simplify to the non-relativistic limit (we shall treat the relativistic case later). The scattering cross-section for electrons is then given by the Rutherford (1911) formula, with the obvious substitution $z = -1$ for the electron:

$$\frac{d\sigma^{e^-}}{d\Omega} = \frac{1}{16}\left(\frac{Z\alpha}{E_\infty}\right)^2 \operatorname{cosec}^4\frac{\theta}{2}. \qquad\qquad (C.2.5)$$

However, this is the cross-section for a point-like target, whereas we may wish to study the charge *distribution* inside the nucleus or nucleon; we shall discuss later how a distributed charge modifies the formula. Let us first consider though the modifications due to relativistic effects.

C.2.2 Relativistic elastic scattering—the Mott formula

To resolve the internal structure of a nucleus (i.e. to be sensitive to energy dependence in the form factors, see later), we require the wavelength of the exchange photon to be small compared to the nuclear size. Since $\hbar c \sim 200$ MeV, we deduce that the energy required is of the order of 100s of MeV. This implies that the electrons will certainly be relativistic, in which case we should really perform a calculation based on the Dirac equation. The relativistic calculation for a point-like object, first performed by Mott (1929), leads to the following cross-section:

$$\frac{d\tilde\sigma}{d\Omega}^{\text{Mott}} = \left(1 - \beta\sin^2\frac{\theta}{2}\right)\frac{d\sigma}{d\Omega}^{\text{Ruther}}, \qquad\qquad (C.2.6)$$

where the tilde indicates that this is not yet the full (high-energy) Mott formula since we are still neglecting the nuclear recoil; we shall call this the reduced Mott formula. The new factor is due to the conservation of angular momentum and the rôle played by the spin of the electron. Note that in the ultra-relativistic limit, where $\beta \to 1$, the spin factor becomes simply $\cos^2\frac{\theta}{2}$. The large-angle Mott cross-section falls off more rapidly than that of Rutherford; indeed, for $\theta = 180°$ it vanishes. Note though that the effect conveniently factorises.

Let us just take a moment to try and understand this behaviour in physical terms. For Dirac theory in the relativistic limit, $\beta \to 1$ (which is evidently equivalent to the limit $m \to 0$), it turns out that the helicity or projection of the particle spin onto the direction of motion $h := \hat{s} \cdot \hat{p}$ is a conserved quantum number if the interactions are of a purely vector or axial-vector type (e.g. via photon or weak-boson exchange). Indeed, starting from the Dirac equation, it can be shown that helicity-flip amplitudes are proportional to m/E, where E is some characteristic energy scale of the interaction (e.g. the centre-of-mass energy). We thus speak of right- and left-handed fermions as having $h = \pm 1$, respectively, and in the massless limit they cannot flip (i.e. the two helicities do not communicate). Consider now the extreme cases of forward and backward scattering, in which the incoming electron collides and either continues unaltered or returns in the direction from where it came ($\theta = 0$ or π in figure C.2). Now, assuming a spin-zero nucleus, since any orbital angular momentum between the electron–nucleus pair $\boldsymbol{L} = \boldsymbol{r} \wedge \boldsymbol{p}$ must lie in the plane orthogonal to $\boldsymbol{p}$, the spin of the electron must be conserved on its own. For forward scattering this is trivially the case and for backward scattering it is evidently impossible. For intermediate cases, we need to understand how spin-projection eigenstates are constructed for arbitrary directions. For a state of positive helicity travelling in a direction θ with respect to the chosen quantisation axis ($\hat{z}$ say), we find

$$|+, \theta\rangle = \cos \tfrac{\theta}{2} |+, z\rangle + \sin \tfrac{\theta}{2} |-, z\rangle. \qquad (C.2.7)$$

Thus, the amplitude $\langle +, \theta|+, z\rangle = \cos \tfrac{\theta}{2}$ and, squaring, we have the Mott result. Of course, if the nucleus also possesses an intrinsic angular momentum (due to internal motion or spins of the constituent nucleons), then the situation is a little more complex and we also need to understand the mechanism by which the nucleus, as a whole, may change or flip its spin projection.

Finally we should take into account the recoil of the struck nucleus and the consequent modification of the final-state phase space. The final, full Mott (1929) formula is then (in the target rest frame)

$$\frac{d\sigma}{d\Omega}^{\text{Mott}} = \frac{E'}{E} \frac{d\tilde{\sigma}}{d\Omega}^{\text{Mott}} = \frac{E'}{E}\Big(1 - \beta \sin^2 \tfrac{\theta}{2}\Big)\frac{d\sigma}{d\Omega}^{\text{Ruther}}. \qquad (C.2.8)$$

Again, note the fortunate factorisation of all the new effects.

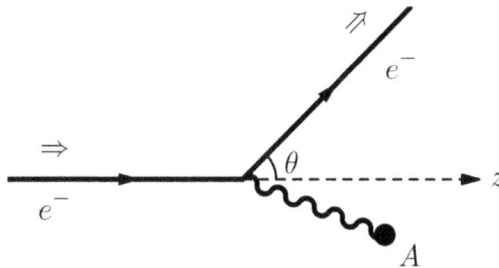

Figure C.2. Helicity conservation in high-energy scattering processes; θ is the electron scattering angle (the blob A represents the recoil target nucleus).

C.3 Form factors

C.3.1 Elastic scattering off a distributed charge

In describing the case of scattering off a distributed charge, it will be helpful to make two simplifying approximations:

(i) $E_e \ll m_A$—in order to neglect the nuclear recoil,

(ii) $Z\alpha \ll 1$—to permit the Born approximation (i.e. single-photon exchange).

We shall also consider the quantum-mechanical treatment, as the concepts of interference and coherence will become relevant here. Our starting point will therefore be Fermi's golden rule (see also (C.6.47)):

$$W = \frac{\sigma v_a}{V} = \frac{2\pi}{\hbar} \left| \langle \psi_f | \mathcal{H}_{\text{int}} | \psi_i \rangle \right|^2 \frac{\mathrm{d}n}{\mathrm{d}E_f}, \tag{C.3.1}$$

where the third factor in the last line is the density of final states and E_f is the total final energy $K + m_A$ (K is the final kinetic energy of the electron) and, since m_A is constant, $\mathrm{d}E_f = \mathrm{d}E' = \mathrm{d}E$.

We first need the description of the initial and final electron states, which we shall naturally take to be plane-waves:

$$\psi_i = \frac{\mathrm{e}^{\frac{\mathrm{i}}{\hbar} \boldsymbol{p} \cdot \boldsymbol{x}}}{\sqrt{V}} \quad \text{and} \quad \psi_f = \frac{\mathrm{e}^{\frac{\mathrm{i}}{\hbar} \boldsymbol{p}' \cdot \boldsymbol{x}}}{\sqrt{V}}, \tag{C.3.2}$$

where the normalisation is one particle in a volume V (of course, V will not appear in the final answer). With this choice, the density of final states is

$$\mathrm{d}n(p) = V \frac{p^2 \mathrm{d}p \, \mathrm{d}\Omega}{(2\pi\hbar)^3}, \tag{C.3.3}$$

where $p := |\boldsymbol{p}|$ (recall that we are using a non-relativistic approximation). We therefore have

$$\frac{\mathrm{d}\sigma}{\mathrm{d}\Omega} = \frac{V^2 E'^2}{(2\pi)^2} \left| \mathcal{M}_{fi} \right|^2 \frac{1}{\hbar^4 v}. \tag{C.3.4}$$

We thus need to find the transition matrix element $\mathcal{M}_{fi} = \langle \psi_f | \mathcal{H}_{\text{int}} | \psi_i \rangle$.

For a non-relativistic electron (with charge $-e$) the interaction is given in terms of the electromagnetic scalar potential $\phi(\boldsymbol{x})$: $\mathcal{H}_{\text{int}}(\boldsymbol{x}) = e\phi(\boldsymbol{x})$. We therefore have (suppressing $\hbar$ for clarity)

$$\langle \psi_f | \mathcal{H}_{\text{int}} | \psi_i \rangle = \frac{e}{V} \int \mathrm{d}^3 \boldsymbol{x} \, \mathrm{e}^{\mathrm{i}\boldsymbol{q} \cdot \boldsymbol{x}} \phi(\boldsymbol{x}), \tag{C.3.5}$$

where we have defined $\boldsymbol{q} := \boldsymbol{p} - \boldsymbol{p}'$, the *momentum transfer*. It is immediately obvious that this is none other than the Fourier transform of the potential $\phi(\boldsymbol{x})$, which is in turn determined by the charge-density distribution $\rho(\boldsymbol{x})$ that generates it:

$$\nabla^2 \phi(\boldsymbol{x}) = -\rho(\boldsymbol{x}). \tag{C.3.6}$$

The plane-wave form allows us to rewrite the expression for the matrix element directly in terms of $\rho(\boldsymbol{x})$ by noting that

$$\nabla^2 e^{i\boldsymbol{q}\cdot\boldsymbol{x}} = -q^2 e^{i\boldsymbol{q}\cdot\boldsymbol{x}}. \tag{C.3.7}$$

If we now apply Green's theorem or integration by parts, we obtain

$$\langle \psi_f | \mathcal{H}_{\text{int}} | \psi_i \rangle = \frac{e}{V q^2} \int d^3\boldsymbol{x}\, e^{i\boldsymbol{q}\cdot\boldsymbol{x}} \rho(\boldsymbol{x}), \tag{C.3.8}$$

where $q := |\boldsymbol{q}|$.

It is convenient to define a normalised density $f(\boldsymbol{x})$,

$$\rho(\boldsymbol{x}) =: Ze\, f(\boldsymbol{x}), \tag{C.3.9}$$

such that

$$\int d^3\boldsymbol{x}\, f(\boldsymbol{x}) = 1. \tag{C.3.10}$$

We thus obtain

$$\langle \psi_f | \mathcal{H}_{\text{int}} | \psi_i \rangle = \frac{Z\alpha}{V q^2} \int d^3\boldsymbol{x}\, e^{i\boldsymbol{q}\cdot\boldsymbol{x}} f(\boldsymbol{x}). \tag{C.3.11}$$

The integral on the right-hand side, the Fourier transform of the charge density, is known as the *form factor*:

$$F(\boldsymbol{q}) := \int d^3\boldsymbol{x}\, e^{i\boldsymbol{q}\cdot\boldsymbol{x}} f(\boldsymbol{x}). \tag{C.3.12}$$

Putting everything together, we have the differential cross-section:

$$\frac{d\sigma}{d\Omega} = \frac{Z^2 \alpha^2 E'^2}{q^4} |F(\boldsymbol{q})|^2. \tag{C.3.13}$$

It is easy to show (neglecting the electron mass) that

$$q^2 = (\boldsymbol{p} - \boldsymbol{p}')^2 = 4EE' \sin^2 \frac{\theta}{2} \tag{C.3.14}$$

and thus we may finally write

$$\frac{d\sigma}{d\Omega} = \frac{Z^2 \alpha^2}{16E^2 \sin^4 \frac{\theta}{2}} |F(\boldsymbol{q})|^2, \tag{C.3.15a}$$

or

$$= |F(\boldsymbol{q})|^2 \frac{d\sigma}{d\Omega}^{\text{Ruther}}. \tag{C.3.15b}$$

In other words, the substructure of the nucleon has the effect of introducing a multiplicative form factor $F(\boldsymbol{q})$, which simply modulates the cross-section. From the definition, we see that the standard Rutherford cross-section is recovered for a point-like distribution, which is just a δ-function (for which the Fourier transform is just unity). Note that this is also the limiting case for low-energy scattering; for $q \ll \hbar/r_{\text{nucl}}$ or $\lambda \gg r_{\text{nucl}}$, $f(x)$ does not vary appreciable over the nuclear volume. For q large, however, the q-dependence of $F(q)$ makes itself felt and thus changes the energy dependence with respect to that of the point-like formula.

C.3.2 The phenomenology of form factors

We have just seen that the effect of an extended charge distribution is factorisable into a form factor, which only depends on the momentum transfer and which, being a Fourier transform, contains (at least in principle) all necessary information on the charge distribution. That is, if we were able to measure $F(q)$, by comparing data with the point-like Mott expression,

$$\frac{\mathrm{d}\sigma^{\,\text{expt}}}{\mathrm{d}\Omega} = |F(q)|^2 \frac{\mathrm{d}\sigma^{\,\text{Mott}}}{\mathrm{d}\Omega} \,. \qquad \text{(C.3.16)}$$

over the *entire* range of q from zero to infinity, we could then perform the inverse Fourier transform to obtain $f(x)$. Needless to say, this is impossible; the momentum transfer is always limited by the beam (or centre-of-mass) energy. However, the lack of higher frequencies (or shorter wavelengths) simply translates into a lack of resolution. Example experimentally measured cross-sections are displayed in figure C.3. Note the pattern of maxima and minima, reminiscent of diffraction in classical optics. Note also that the rapid fall-off with angle severely limits the maximum q effectively available. The first measurements of this type were made in the fifties at SLAC, with a beam energy of around 500 MeV (for which the effective absolute resolution is $\lambda_{\text{min}} \sim 0.4$ fm).

In comparison with the point-like Mott cross-section, scattering off a continuous extended charge distribution falls off very rapidly with growing $|q|$. At very small angles, for which the momentum transfer is kinematically constrained to be low, the exchange photon has a very long wavelength and thus does not resolve the internal nuclear structure. In this region there is no difference between a point-like and extended charge distribution. However, if the beam energy is sufficient, with growing angle the photon wavelength decreases and thus the coherence volume it 'sees', thus reducing the effective charge with which it interacts. The wavelength may eventually become short enough to resolve the internal structure of point-like charges. In this case the cross-section is much less suppressed, even out to large angles and does not fall off as rapidly as in the unresolved, extended case. This comparison will be important in the discussion of very high-energy electron scattering off a single proton and the question of the proton substructure.

C.3.3 Fitting form factors to trial functions

By appealing to model parametrisations for plausible charge-density distributions, the related parametrisations of the form factors may be obtained. Such parametrisations are then compared to the data to extract the parameters. Note that the

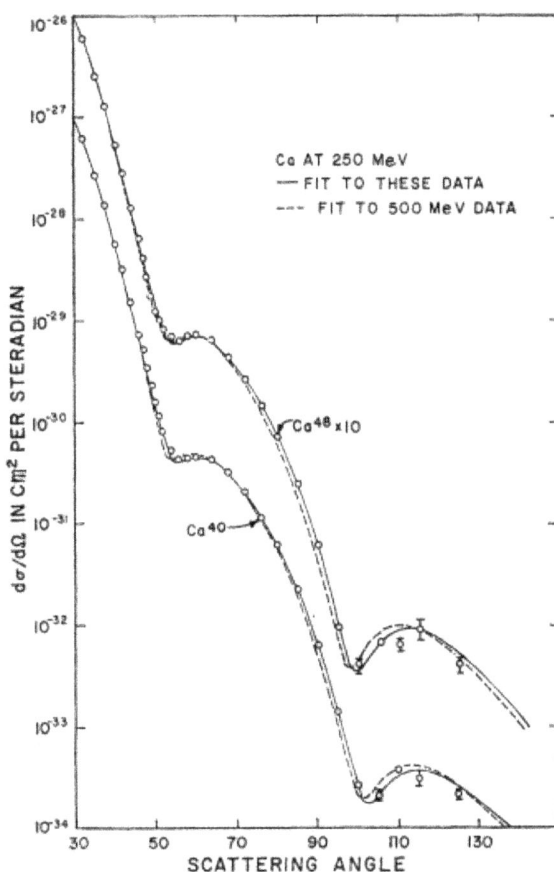

Figure C.3. Experimental data on electron–nucleus elastic scattering. The two curves with data represent the differential cross-section for 250 MeV elastic electron scattering off ^{40}Ca and ^{48}Ca. The diffractive dips are clearly visible; figure reproduced from Frosch *et al* (1968), copyright (1968) by the American Physical Society.

large-angle dependence provides information on the internal structure while as $\theta \to 0$ (for which eventually $\lambda > R_{\mathrm{nucl}}$) we should see a return to the typical q^{-4} behaviour. By exploiting spherical symmetry, the angular dependence may be integrated out to simplify the expression for $F(\boldsymbol{q})$. This thus leaves

$$F(q^2) = \frac{4\pi}{q} \int_0^\infty \mathrm{d}r \, r \sin qr \, f(r), \qquad (\mathrm{C}.3.17)$$

where $q = |\boldsymbol{q}|$. Note that the density $f(r)$ is then normalised as

$$4\pi \int_0^\infty r^2 \, \mathrm{d}r \, f(r) = 1. \qquad (\mathrm{C}.3.18)$$

In table C.1 we provide a list of typical functional forms used. As an example, consider the case of a uniform sphere; the first minimum lies at $qR \simeq 4.5$. Thus, referring to the graph in figure C.3, we find $R \simeq 2.5$ fm for ^{12}C. Moreover, the fact

Table C.1. A collection of possible forms of charge distribution inside the nucleus together with the corresponding form factors. In all cases R represents a measure of the nuclear radius.

Form	$f(r)$	$F(q^2)$	Behaviour
Point	$\delta(r)/4\pi$	1	Constant
Exponential	$\frac{1}{8\pi R^3}\,e^{-r/R}$	$\frac{1}{(1+q^2R^2)^2}$	'Dipole'
Gaussian	$\frac{1}{(2\pi)^{3/2}R^3}\,e^{-\frac{1}{2}r^2/R^2}$	$e^{-\frac{1}{2}q^2R^2}$	Gaussian
Sphere	$\frac{3}{4\pi R^3}\quad(r<R)$	$\frac{(\sin\rho-\rho\cos\rho)}{\rho^3}\quad(\rho:=qR)$	Oscillatory

that the minima are not as sharp as is predicted for a uniform sphere indicates the existence of a 'soft' outer 'skin' of finite depth.

C.3.4 Physical interpretation

Let us conclude this section by providing a physical interpretation of the form-factor effects. For wavelengths much greater than the size of an extended target object, the latter is not resolved and acts as an effective point-like charge. However, for a given wavelength there is a limited region over which scattering may be coherent; that is, over which the subregions all interfere constructively. A simple comparison of trajectories reveals that the size of this region is of the order of one wavelength. Thus, for objects much larger than the wavelength used, only a small fraction of their total charge effectively contributes to the scattering. This leads to a rapid decrease of the cross-section with decreasing wavelength, or correspondingly increasing energy–momentum transfer.

Note that these ideas all clearly carry over to consideration of high-energy electron scattering of single-nucleon targets, such as hydrogen. In this case for elastic *ep* scattering, we find that the so-called 'dipole' form factor gives a good description of the effective charge distribution. This means that, for Q^2 much larger than the inverse size of the nucleon, the cross-section should fall off very rapidly; i.e. it will have an extra factor $\sim Q^{-4}$ with respect to the natural point-like behaviour.

C.4 Quasi-elastic scattering

Let us now consider a case intermediate between pure elastic scattering and the process known as deeply inelastic scattering (DIS, in which the proton is completely broken up into many pieces). The process we wish to consider is thus called *quasi-elastic scattering*. As we have already seen, elastic-scattering kinematics imposes a one-to-one relation between the scattering angle and the energy of the outgoing electron. We should thus expect a single spectral line. Let us examine what is observed in practice for scattering off a nucleus at high energy. The example we shall exploit is that of scattering off a helium nucleus (see figure C.4). Two prominent features are to be found in the spectrum. The first is a pronounced spike at about 375 MeV due to elastic electron–helium scattering; this may be readily checked from the

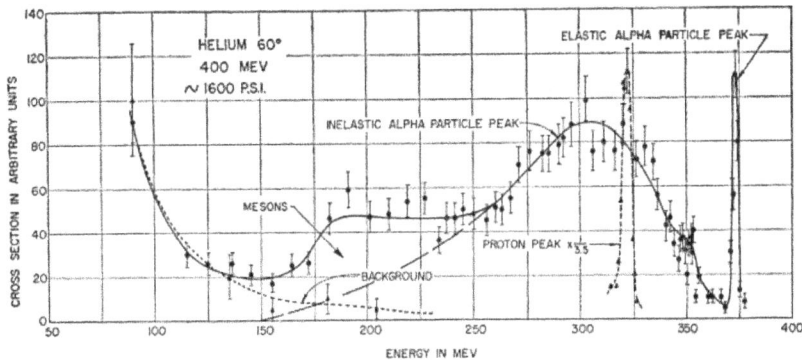

Figure C.4. The cross-section for quasi-elastic scattering of electrons off helium as a function of the outgoing-electron energy for a beam energy of 400 MeV at a scattering angle of 60°; figure reproduced from Hofstadter (1956), copyright (1956) by the American Physical Society.

position of the spike with respect to the initial energy and fixed scattering angle using equation (C.2.4). Then there is rather a broad peak centred a little above 300 MeV. In addition, a larger (but scaled-down) spike corresponding to the proton elastic peak at a little under 325 MeV has been superimposed for comparison (we can again check the correct energy–angle relation). The energy difference between the two spikes precisely reflects the mass difference between a helium nucleus and a proton. The question then remains as to the nature of the broader structure mentioned, which nevertheless has the clear form of a peak. We shall now show that this may be attributed to scattering off a proton bound inside the nucleus; the proton is then ejected from the nucleus and this is what is known as *quasi-elastic* scattering.

Exercise C.4.1. *Invert relation (C.2.3) derived earlier to provide a formula for the target mass as a function of the electron initial and final energies and scattering angle. Thus, calculate the masses corresponding to the two sharp peaks and deduce that their origins really are point-like elastic scattering off a free proton and a helium nucleus.*

We know that, for example, the Fermi-gas model makes rather precise predictions: the nucleons lie in a potential well of approximate depth 40 MeV with a Fermi level corresponding to a momentum of the order of 250 MeV. The depth of the well represents an energy that must be supplied above and beyond the kinematical needs while the Fermi motion will induce smearing of the total centre-of-mass energy, leading to a smearing of the final-state spectrum. Let us now examine in detail how this works. The process we wish to study has a *three*-body final state: the electron, the proton and the recoiling nuclear remnant (see figure C.5). We begin by introducing the necessary kinematic variables in the laboratory frame:

$$\boldsymbol{p} = \text{initial electron momentum,} \tag{C.4.1a}$$

$$\boldsymbol{p}' = \text{final electron momentum,} \tag{C.4.1b}$$

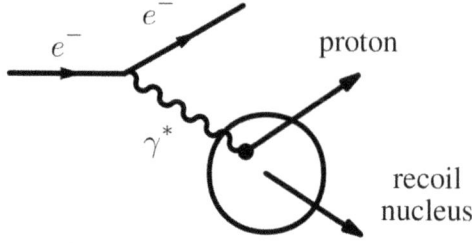

Figure C.5. A schematic view of quasi-elastic electron–nucleus scattering: via exchange of a virtual photon a proton is ejected from the nucleus.

$$P = \text{initial struck proton momentum,} \tag{C.4.1c}$$

$$P' = \text{final struck proton momentum,} \tag{C.4.1d}$$

$$\begin{aligned} q &= p - p' \\ &= \text{momentum transfer.} \end{aligned} \tag{C.4.1e}$$

Finally, a useful variable is $\nu := E_e - E_e'$, the energy transfer in the laboratory frame. It is then a simple exercise to show that we have

$$\nu = \frac{q^2}{2M} + V_0 + \frac{2qP\cos\alpha}{2M}, \tag{C.4.2}$$

where $q = |q|$, $P = |P|$ and α is the angle between the vectors q and P.

Exercise C.4.2. *Derive the above expression for the energy transfer ν. Hint: relativistic kinematics is not necessary.*

Now, assuming the initial momentum P to be distributed uniformly inside the Fermi sphere ($|P| \leqslant p_F$, the Fermi momentum), both $\langle \nu \rangle$ and its mean deviation σ_ν can be easily evaluated:

$$\begin{aligned} \langle \nu \rangle &= 2\pi \int_0^{p_F} p^2\,\mathrm{d}p \int_{-1}^{1} \mathrm{d}\cos\alpha \left[\frac{q^2}{2M} + V_0 + \frac{2qP\cos\alpha}{2M} \right] \Big/ \left[\frac{4\pi p_F^3}{3} \right] \\ &= \frac{q^2}{2M} + V_0 \end{aligned} \tag{C.4.3}$$

and

$$\begin{aligned} \sigma_\nu &:= \sqrt{\langle \nu^2 - \langle \nu \rangle^2 \rangle} = \frac{q}{M}\sqrt{\langle P^2 \rangle \cdot \langle \cos^2 \alpha \rangle} = \frac{q}{M}\sqrt{\tfrac{1}{3}\langle P^2 \rangle} \\ &= \frac{1}{\sqrt{5}}\frac{qp_F}{M}, \end{aligned} \tag{C.4.4}$$

where we have used the standard result that, for a uniform spherically symmetric momentum distribution, $\langle P^2 \rangle = \frac{3}{5}p_F^2$.

Exercise C.4.3. *From the quasi-elastic peak position and width displayed in figure C.4, estimate the corresponding values of V_0, p_F and E_F (the Fermi momentum and energy).*

We should now comment on the implicit approximation made here: the fact that the cross-section may be calculated as though the nucleons (or protons in this case) were *free* inside the nuclear volume is not entirely trivial. If the struck particle sits in a potential well with a strong spatial dependence, then it is presumably permanently subject to forces comparable to the scattering potential itself and that should therefore be added to those operating during the scattering process. However, the data lead to a picture in which the nucleons apparently lie in a potential well with an effectively *flat* bottom. That is, deep inside the nucleus they are not subject to any forces except for the brief moments in which they make contact with the boundary or surface of the nucleus and thus we are justified in making what is known as the *impulse approximation*.

It is perhaps also worth mentioning that the same approximation is used in describing the scattering of electrons off single quarks inside the nucleon (a process known as *deeply inelastic scattering*). In this case the struck quark does not actually materialise as a free particle in the laboratory owing to the absolute confining effect of the strong interaction. Nevertheless, calculations (first performed by Feynman) in the impulse approximation describe the data surprisingly well. In this case the explanation is not the triviality of the potential, but a phenomenon known as *asymptotic freedom*. According to this property of quantum chromodynamics, the strength of the interaction decreases with increasing energy scale or, equivalently, at short distances. Thus, provided the energy of the probe is sufficiently large, then the time and distance scales become such that the struck quark can interact only weakly with the parent nucleon.

A more quantitative and rigorous way of defining the impulse approximation is in terms of interaction times. The struck objects, bound inside a more complex structure (be they nucleons inside the nucleus or quarks inside a nucleon) move freely for a time that may be roughly estimated as the diameter of the surrounding structure divided by their average velocity. The interaction has a time scale which may be estimated as the inverse of the energy transfer. The impulse approximation is then justified if the interaction time is less than that of the mean free motion.

Exercise C.4.4. *Estimate the mean free time (i.e. the time between 'collisions' with the outer nuclear 'wall') for a nucleon bound inside an iron nucleus and thus estimate the energy transfer necessary to guarantee applicability of the impulse approximation.*

C.5 The partial-wave expansion

C.5.1 Scattering in perturbation theory

In quantum mechanics scattering by weak finite-range potentials may be treated in a perturbative manner. That is, we assume to have solved the free-field equations and that the scattering potential maybe treated as a small perturbation. In this section we shall present a simplified derivation of the cross-section for the case of a spherically symmetric potential. The basis is the knowledge we have of solutions to the Schrödinger equation in such a case; the angular part is provided by just the spherical harmonic functions already encountered in the solution of the hydrogen atom. Moreover, for a finite-range potential the asymptotic form of the radial wave-function is also predetermined.

We consider then the simple case of a particle of momentum k incident on a spherically symmetric potential $V(r)$, centred at the origin. The initial state, in the absence of $V(x)$, may be represented as a plane-wave

$$u_0(x) = e^{ik \cdot x}. \tag{C.5.1}$$

We suppress the normalisation since our aim is to calculate a cross-section, which is a ratio of fluxes. In the presence of the scattering potential, but *outside* its range, the full solution to the Schrödinger equation will be of the form

$$u(x) \xrightarrow{r \to \infty} e^{ik \cdot x} + f(\theta, \phi) \frac{1}{r} e^{ikr} \tag{C.5.2}$$

where θ and ϕ are the polar scattering angles in the laboratory frame with respect to k, which we then take to be along the z-axis, while $r = |x|$. That is, we have the incident plane-wave plus an outgoing spherical wave centred on the origin.

In perturbation theory the form of $f(\theta, \phi)$ can always be calculated from first principles if the potential is known. We shall just quote the result here: in the Born (or leading-order) approximation we have

$$
\begin{aligned}
f(\theta, \phi) &= -\frac{2m}{\hbar^2} \frac{(2\pi)^3}{4\pi} \langle k' | V | k \rangle \\
&= -\frac{2m}{\hbar^2} \frac{(2\pi)^3}{4\pi} \int d^3x' \frac{e^{-ik' \cdot x'}}{(2\pi)^{3/2}} V(x') \frac{e^{+ik \cdot x'}}{(2\pi)^{3/2}},
\end{aligned} \tag{C.5.3}
$$

where the final-state momentum k' is direct along (θ, ϕ) and energy-momentum conservation requires that $k = k'$. Gathering together exponentials and constant factors, this becomes

$$f(\theta, \phi) = -\frac{2m}{4\pi\hbar^2} \int d^3x' e^{i(k - k') \cdot x'} V(x'). \tag{C.5.4}$$

That is, as remarked elsewhere in these notes, the Born approximation to the scattering amplitude is just a Fourier transform of the potential, in the variable $q = k - k'$, the three-momentum transfer. In practice then, we may model a potential and calculate the corresponding phase shifts; alternatively, at least in principle, measured phase shifts may be used to reconstruct (or fit) a potential model.

Exercise C.5.1. *Using the Yukawa potential,*

$$V_{\text{Yuk}}(r) \equiv \frac{V_0 e^{-\mu r}}{r},$$

and taking the limit $\mu \to 0$, with $V_0 \to Zz\alpha$, show that the quantum result for Rutherford scattering takes on the same form as the classical calculation.

Often, however, we do not possess *a priori* a realistic model of the scattering potential, e.g. for pion scattering off a nucleon. In such cases we require a suitably parametrised description of the cross-section. In what follows we shall exploit our knowledge of the general solution to the Schrödinger equation in the case of a spherically symmetric potential to construct a faithful and simple parametrisation, which may then be fit to experimental data.

C.5.2 The partial-wave formula

The cross-section is defined in terms of the ratio of the scattered flux divided by the incident flux. Recall that in quantum mechanics the flux corresponding to a wave-function solution $u(x)$ to the Schrödinger equation has the form

$$j(x) = \frac{\hbar}{2mi} \left[u^*(x) \nabla u(x) - u(x) \nabla u^*(x) \right]. \tag{C.5.5}$$

In equation (C.5.2) for $u(x)$, we should consider the two parts as spatially separated: the incident wave will, in practice, be collimated along a narrow region around the z-axis, where the outgoing spherical wave will have little weight. The incident flux (j_{inc}) therefore simply corresponds to the plane-wave piece while the scattered flux (j_{scatt}) corresponds to the other, with no interference terms (which are negligible as far as the present discussion is concerned, but see later). The partial cross-section for scattering into a given solid angle $d\Omega(\theta, \phi)$ is defined by

$$
\begin{aligned}
d\sigma &:= \frac{\text{scattered flux in } d\Omega}{\text{incident flux}} \\
&= \frac{j_{\text{scatt}} \, r^2 d\Omega}{j_{\text{inc}}} = |f(\theta, \phi)|^2 d\Omega.
\end{aligned}
\tag{C.5.6}
$$

The differential cross-section is therefore quite simply

$$\frac{d\sigma}{d\Omega} = |f(\theta, \phi)|^2. \tag{C.5.7}$$

We thus see that the object containing all the necessary information is just the angular modulation $f(\theta, \phi)$.

The spherical symmetry of the problem suggests transformation to a basis of solutions in terms of spherical harmonic functions. In the case of a plane-wave, we have the following decomposition

$$e^{i\mathbf{k}\cdot\mathbf{x}} = e^{ikr\cos\theta} = \sum_{\ell=0}(2\ell + 1)i^\ell\, j_\ell(kr)\, P_\ell(\cos\theta), \qquad (C.5.8)$$

where $P_\ell(\cos\theta)$ are Legendre polynomials and $j_\ell(kr)$ are spherical Bessel functions, the solutions to the reduced radial equation. Note that the imaginary factor may also be rewritten as

$$i^\ell \equiv e^{\frac{1}{2}i\ell\pi} \qquad (C.5.9)$$

and asymptotically the Bessel functions take the form

$$j_\ell(kr) \xrightarrow{r\to\infty} \frac{\sin\left(kr - \frac{1}{2}\ell\pi\right)}{kr} = \frac{1}{2ikr}\left[e^{+i\left(kr-\frac{1}{2}\ell\pi\right)} - e^{-i\left(kr-\frac{1}{2}\ell\pi\right)}\right]. \qquad (C.5.10)$$

Now, considering the full wave-function in the presence of the scattering potential, we may make a similar expansion and write for r large

$$u(\mathbf{x}) = \sum_{\ell=0}^{\infty}(2\ell + 1)i^\ell\frac{1}{2ikr}\left[e^{+i\left(kr-\frac{1}{2}\ell\pi+2\delta_\ell\right)} - e^{-i\left(kr-\frac{1}{2}\ell\pi\right)}\right]P_\ell(\cos\theta). \qquad (C.5.11)$$

In writing this expression we have taken into account certain simple general properties of the solutions to the Schrödinger equation. First of all, each term in the sum corresponds to a component of well-defined orbital angular momentum $\ell\hbar$ and, since angular momentum is conserved, term-by-term the partial amplitudes cannot change in magnitude but only in phase. Moreover, the two terms in square brackets on the right-hand side, represent outgoing and incoming waves, respectively. Only the outgoing wave has had the opportunity to interact with the potential $V(\mathbf{x})$ and thus only these components may have a phase shift, $2\delta_\ell(E_k)$, the factor 2 is for later convenience. Note that the phase shift can (and therefore will) only be a function of the incident beam energy.

Comparing now (C.5.2) and (C.5.11), taking into account (C.5.8), we find

$$f(\theta, \phi) = \sum_{\ell=0}^{\infty}(2\ell + 1)\,i^\ell\frac{1}{2ik}\left[e^{i\left(-\frac{1}{2}\ell\pi+2\delta_\ell\right)} - e^{i\left(-\frac{1}{2}\ell\pi\right)}\right]P_\ell(\cos\theta) \qquad (C.5.12)$$

and using the expression for i^ℓ given earlier, this becomes

$$= \sum_{\ell=0}^{\infty}(2\ell + 1)\,\frac{1}{2ik}\left[e^{2i\delta_\ell} - 1\right]P_\ell(\cos\theta) \qquad (C.5.13)$$

$$= \frac{1}{k}\sum_{\ell=0}^{\infty}(2\ell + 1)e^{i\delta_\ell}\sin\delta_\ell\, P_\ell(\cos\theta). \qquad (C.5.14)$$

Let us make a few comments: firstly, we see that there is no ϕ dependence, as would be expected for a spherically symmetric potential; since $\boldsymbol{k}$ is directed along the z-axis, there can be no z-component of orbital angular momentum. Secondly, we see that all the scattering information is contained in the phase-shifts $\delta_\ell(E_k)$; they determine not only the phase of each partial amplitude (through the factor $e^{i\delta_\ell}$), but also the magnitude (through the factor $\sin \delta_\ell$). Finally, let us remark on the validity of the above formula: it may be used for two-body scattering, provided we employ the centre-of-mass system and reduced mass etc, and even carries over to the relativistic case.

The importance of the partial-wave expansion is that, as just stated, it provides a useful (and, in particular, model-independent) parametrisation of the scattering amplitude, even when nothing is known about the potential. This permits unbiased analysis of experimental scattering data, which may then be compared with theoretical predictions. As it stands though, the expansion contains an infinite number of parameters (the phase shifts δ_ℓ for $\ell = 0, 1, \ldots, \infty$), which would render any experimental fit totally impracticable. However, for a finite-range potential (say $r < a$) and finite incident momentum, the maximum orbital angular momentum that can be generated is

$$\ell_{\max} \hbar \approx a \, |\boldsymbol{k}|. \tag{C.5.15}$$

Therefore, recalling that $\hbar c \sim 200$ MeV, we see that, for example, a typical nuclear-potential of range $O(1 \text{ fm})$ and a beam of momentum 200 MeV could generate up to $\ell = O(1)$. In general then, low-energy scattering off a finite-range potential involves only a very limited number of partial waves. Incidentally, this also explains why the expansion only has limited application in high-energy hadronic physics: a beam of momentum 20 GeV, say, would involve up to $\ell = O(100)$ terms in the expansion. Note that with a small number of partial waves contributing (each with its own angular dependence), the phase shifts may be extracted directly from the measured angular distributions.

C.5.3 The optical theorem

We can now provide a simplified proof of a very important theorem in scattering. Let us first calculate the total cross-section:

$$\sigma = \int d\Omega \frac{d\sigma}{d\Omega} = \int d\Omega \, |f(\theta, \phi)|^2$$

$$= \int d\Omega \left| \frac{1}{k} \sum_{\ell=0}^{\infty} (2\ell + 1) e^{i\delta_\ell} \sin \delta_\ell \, P_\ell(\cos \theta) \right|^2 ,$$

which, using the orthogonality of the Legendre polynomials, reduces to

$$= \frac{4\pi}{k^2} \sum_{\ell=0}^{\infty} (2\ell + 1) \sin^2 \delta_\ell. \tag{C.5.16}$$

Consider now the forward amplitude, i.e. for $\theta = 0$ (or $\cos \theta = 1$):

$$f(0) = \frac{1}{k}\sum_{\ell=0}^{\infty}(2\ell + 1)e^{i\delta_\ell}\sin \delta_\ell, \qquad (C.5.17)$$

where we have used the fact that $P_\ell(1) = 1$ for all ℓ. The imaginary part of this last expression is just the total cross-section, up to a factor $4\pi/k$:

$$\sigma = \frac{4\pi}{k}\mathcal{J}m\, f(0). \qquad (C.5.18)$$

This is precisely the optical theorem: the *total cross-section* is proportional to the *forward scattering amplitude*. Its validity actually extends beyond the simple proof provided here.

The apparent contradiction in a left-hand side that is, by definition, proportional to an amplitude *squared* and a right-hand side, *linear* in the amplitude may be reconciled by carefully considering the origin of the scattering cross-section. In equation (C.5.2) we see that the general form of the wave-function is a sum of two terms:

$$\psi(\boldsymbol{x}) \propto e^{i\boldsymbol{k}\cdot\boldsymbol{x}} + \frac{e^{ikr}}{r}f(\theta, \phi). \qquad (C.5.19)$$

Now, the flux *loss* in the forward direction (which is clearly proportional to the total cross-section) must be due to the *interference* between these two terms (to see this, consider the difference in flux along the z-axis between points before and after the scattering centre) and thus is indeed linear in $f(\theta, \phi)$. In fact, if we perform the calculation in this way, it becomes clear that the theorem is very general and holds even in the presence of *inelastic* scattering (or absorption). As might be imagined, the expression *optical theorem* is borrowed from classical optics, where the phenomenon is well known: a bright central spot that appears at a certain distance behind a black disc diffracting a light source of suitable wavelength.

The effect may be thought of as the remnant of the *shadow* of the scatterer. In optics it was discovered independently by von Sellmeier (1871) and Strutt (1871). In other words, an object that causes any scattering at all must have a non-zero forward scattering amplitude, hence the bright central spot.

C.6 Resonances and the Breit–Wigner form

C.6.1 Resonances in classical mechanics

In classical mechanics the equation of motion for a forced oscillator subject to friction is

$$m\ddot{x} + \gamma\dot{x} + kx = F\cos \omega t, \qquad (C.6.1)$$

Defining the natural frequency of the oscillator as $\omega_0 = \sqrt{k/m}$, the solution is

$$x = x_{\max}\cos(\omega t + \phi), \qquad (C.6.2)$$

where the phase difference ϕ is given by

$$\tan \phi = \frac{-\gamma \omega}{m(\omega^2 - \omega_0^2)} \qquad (C.6.3)$$

and the oscillation *amplitude* is

$$x_{\max} = \frac{F}{[m^2(\omega^2 - \omega_0^2)^2 + \gamma^2 \omega^2]^{1/2}}. \qquad (C.6.4)$$

The total energy ($E_{\mathrm{kin}} + E_{\mathrm{pot}}$) of the oscillator is thus

$$E = \frac{kF^2}{[m^2(\omega^2 - \omega_0^2)^2 + \gamma^2 \omega^2]}. \qquad (C.6.5)$$

Notice that the presence of dissipation (in the form of friction, $\gamma \neq 0$) tames the otherwise divergent behaviour for $\omega = \omega_0$.

C.6.2 Breit–Wigner resonances in quantum mechanics

A similar behaviour occurs in quantum mechanics for the production of intermediate so-called resonant (virtual) states when the natural energy of the virtual state is near to that of the real energy of the system. A formal description may be provided by considering the variation of phase shifts in the partial-wave decomposition of scattering amplitudes (see appendix C.5). We shall start from the standard form of the elastic-scattering amplitude in quantum mechanics, taken for some particular partial wave ℓ (recall ℓ is the total orbital angular momentum quantum number):

$$f_\ell = \frac{\hbar}{2ip}(2\ell + 1)(a_\ell e^{2i\delta_\ell} - 1)P_\ell(\cos \theta), \qquad (C.6.6)$$

where p is the centre-of-mass initial-state momentum, a_ℓ is the amplitude of the of the ℓ-th partial scattered wave ($0 \leqslant a_\ell \leqslant 1$, $a_\ell < 1$ implies absorption) and δ_ℓ is the so-called phase-shift, which contains all relevant information on the scattering potential.

In the purely elastic case, i.e. with *zero* absorption (i.e. $a_\ell = 1$), the corresponding partial cross-section is then

$$\sigma_\ell^{\mathrm{el}} = \frac{\pi \hbar^2}{p^2}(2\ell + 1)|e^{2i\delta_\ell} - 1|^2. \qquad (C.6.7)$$

This expression has a maximum whenever $\delta_\ell = \left(n + \frac{1}{2}\right)\pi$ with n integer. The maximum value is

$$\sigma_\ell^{\max} = (2\ell + 1)\frac{4\pi \hbar^2}{p^2}. \qquad (C.6.8)$$

Note that for the case of *total* absorption (*i.e.* $a_\ell = 0$), the cross-section is just a quarter of this.

Experimentally, cross-sections are often observed with a well-pronounced peak at some particular centre-of-mass energy. The question then is how we might describe such effects in the absence of a complete theory of the interaction involved. From the above formula, we immediately deduce that if the cross-section attains a maximum at some energy for some partial wave, then the corresponding phase-shift evidently passes through a value $(n + \frac{1}{2})\pi$, so that $\cot \delta_\ell$ passes through a zero. Now, the only independent variable for each individual partial wave is the centre-of-mass energy (the angular dependence is already coded into the spherical harmonics) so the phase-shift δ_ℓ is only a function of E. Thus, those *energies* for which $\delta_\ell = (n + \frac{1}{2})\pi$ and for which the cross-section is therefore maximal, correspond to *resonances*. We shall assume that such a resonance is sufficiently well separated from any others so that it dominates the cross-section for $E \sim E_0$, the resonant energy. It is instructive to study the behaviour of the amplitude in the neighbourhood of such a point.

The identity

$$\frac{e^{2i\delta} - 1}{2i} \equiv \frac{1}{\cot \delta - i} \qquad (C.6.9)$$

may be used to rewrite the elastic partial-wave amplitude as

$$f_\ell = \frac{\hbar/p}{(\cot \delta_\ell - i)}. \qquad (C.6.10)$$

At the resonance $\cot \delta_\ell = 0$; therefore, performing a Taylor expansion about this point in the energy E and retaining only the leading term, linear in $E - E_0$, we may write

$$\cot \delta_\ell(E) \simeq -\frac{2}{\Gamma}(E - E_0). \qquad (C.6.11)$$

The sign choice is conventional but physically motivated: indeed, suppose the phase-shift *grows* with energy in the neighbourhood of the resonance. Therefore, $\cot \delta_\ell(E)$ *decreases* and the parameter Γ is *positive*. Inserting this into equation (C.6.10) leads to the standard Breit–Wigner form for the amplitude:

$$f_\ell(E) = -\frac{\hbar}{p} \frac{\Gamma/2}{(E - E_0) - i\Gamma/2}(2\ell + 1)P_\ell(\cos \theta). \qquad (C.6.12)$$

That is, the elastic cross-section is described by the form

$$\sigma_\ell \simeq \frac{4\pi\hbar^2}{p^2}(2\ell + 1)\frac{(\Gamma/2)^2}{(E - E_0)^2 + (\Gamma/2)^2}. \qquad (C.6.13)$$

More simply, this may be rewritten as

$$\sigma_\ell \simeq \sigma_\ell^{\max} \frac{(\Gamma/2)^2}{(E - E_0)^2 + (\Gamma/2)^2}. \qquad (C.6.14)$$

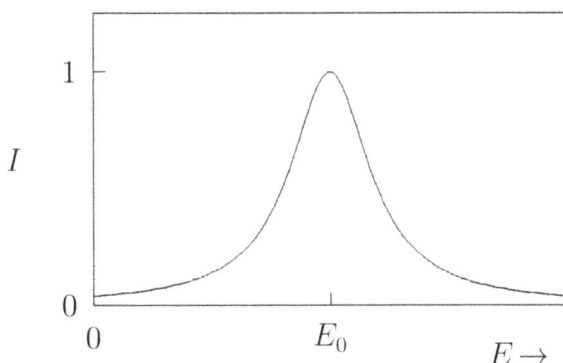

Figure C.6. The standard Breit–Wigner resonance shape; the underlying behaviour of σ_ℓ^{max} has been factored out.

For any production process that passes through a (well-isolated) resonant channel, we thus find that the spectrum or cross-section assumes the Breit–Wigner form shown in figure C.6 (Breit and Wigner 1936). Note that, in practice, the peak behaviour is superimposed over the energy dependence of σ_ℓ^{max}, which from equation (C.6.8), for example is typically a rapidly falling function of energy.

Finally, including the multiplicity factors associated with spin states in the case of initial particles of arbitrary spin, the Breit–Wigner approximation to the total cross-section for particles 1 and 2 scattering via a resonance R may be expressed as

$$\sigma_R \simeq \frac{4\pi\hbar^2}{p^2} \frac{(2J+1)}{(2s_1+1)(2s_2+1)} \frac{(\Gamma_{12}/2)(\Gamma/2)}{(E-E_R)^2 + ((\Gamma/2)^2}, \qquad \text{(C.6.15)}$$

where the denominators $(2s_{1,2}+1)$ provide the usual *average* over the initial-state spins $s_{1,2}$ and J is the spin of the resonance R. The numerator factor Γ_{12} represents the *partial* width for resonance two-body decay into particles 1 and 2 (this may be better understood considering the Feynman diagrams to be introduced in the following subsection). The possible multiplicity factors for any final-state spins are subsumed (indeed, summed not averaged) in the numerator Γ.

C.6.3 Breit–Wigner resonances in quantum field theory

We shall now try to motivate such a form in particle physics without recourse to detailed calculation. First, recall that for a metastable state the decay rate is the inverse of the mean life (up to factors of $\hbar$ and c): $\Gamma = \tau^{-1}$. Note that here we are not necessarily limited to discussing purely elastic processes, the main aspect is the resonant nature of the intermediate state. The probability density for the decaying state then takes the following form:

$$\mathcal{P}(t) \propto e^{-\Gamma t}. \qquad \text{(C.6.16)}$$

In quantum mechanics this should simply be the squared modulus of the wave-function describing the state. Thus, we are led to the following form for the time dependence of the wave-function:

$$\psi(t) \propto e^{-iEt}e^{-\frac{1}{2}\Gamma t} = e^{-i\left(E - i\frac{1}{2}\Gamma\right)t}. \tag{C.6.17}$$

Considering a particle of physical mass m_0 in its rest frame, the total energy E may be replaced by m_0. The wave-function is then seen to represent a state of *complex* mass $m \equiv m_0 - i\Gamma/2$.

If we now make the *plausible* step of using this mass in the propagators appearing in any Feynman diagram where such an unstable particle might propagate internally, we are led to the following substitution (assuming $\Gamma \ll m_0$):

$$\frac{1}{p^2 - m^2} \longrightarrow \frac{1}{p^2 - m_0^2 + im_0\Gamma}. \tag{C.6.18}$$

It can be shown, by explicit calculation, that the effect of the self-interaction induced by the decay channels (i.e. the possibility of temporary spontaneous fluctuations into the decay final states) is precisely this. However, the full armoury of quantum field theory is needed to attack such a problem.

As an example, let us finally examine the effect of such a substitution on the interaction probability of the process $e^+e^- \to Z^0 \to \mu^+\mu^-$ (see figure C.7). At lowest order in perturbation theory the propagator associated with the intermediate Z^0 state is normally:

$$\frac{1}{q^2 - m_Z^2}, \tag{C.6.19}$$

where q^μ is the Z^0 four-momentum and m_Z its mass. According to the above discussion, we should thus adopt the following form:

$$\frac{1}{s - m_Z^2 + im_Z\Gamma_Z}, \tag{C.6.20}$$

where use has been made of the Mandelstam variable

$$s := (p_1 + p_2)^2 \equiv q^2 \equiv E_{\mathrm{CM}}^2. \tag{C.6.21}$$

The interaction probability is proportional to the modulus squared of the amplitude and thus we should really examine

$$\left| \frac{1}{s - m_Z^2 + im_Z\Gamma_Z} \right|^2 = \frac{1}{(s - m_Z^2)^2 + m_Z^2\Gamma_Z^2}. \tag{C.6.22}$$

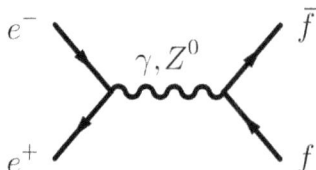

Figure C.7. The Feynman diagram for an example particle-physics interaction proceeding via an s-channel intermediate state.

Note once again that the presence of dispersion (in this case particle decay) tames a potential divergence for $E_{CM} = m_Z$. This also demonstrates that higher-order corrections are not merely a theoretical luxury to be easily foregone. The form shown in equation (C.6.22) is relevant for relativistic field-theory calculations; noting that for energies near the *pole* mass ($E_{CM} \sim m_Z$)

$$
\begin{aligned}
s - m_Z^2 &= E_{CM}^2 - m_Z^2 \\
&= (E_{CM} + m_Z)(E_{CM} - m_Z) \simeq 2m_Z(E_{CM} - m_Z),
\end{aligned}
\tag{C.6.23}
$$

we readily obtain the standard Breit–Wigner form:

$$
\propto \frac{1}{(E_{CM} - m_Z)^2 + \frac{1}{4}\Gamma_Z^2}.
\tag{C.6.24}
$$

The final complete form for $e^+e^- \to Z \to ab$ is then

$$
\sigma_{e^+e^- \to Z \to ab} \simeq \frac{4\pi\hbar^2}{p^2} \frac{(2s_Z + 1)}{(2s_1 + 1)(2s_2 + 1)} \frac{(\Gamma_{12}/2)(\Gamma_{ab}/2)}{(E - m_Z)^2 + (\Gamma_{tot}/2)^2},
\tag{C.6.25}
$$

where s_Z and $s_{1,2}$ are the spins of the Z^0, electron and positron, respectively. The partial widths in the numerator, Γ_{12} and Γ_{ab} are those for Z^0 decay into e^+e^- and ab final states while for the denominator the total width Γ_{tot} must be used in *all* cases.

One further refinement is necessary for greater precision in those cases where the momenta of the outgoing particles varies appreciably over the width Γ of the resonance. Since Γ may also be correctly interpreted as the decay rate, it will depend on the final-state momentum $|\boldsymbol{p}|$ (evaluated in the decay rest-frame). It can be shown by explicit calculation (see Flatté (1976)) that, for a resonance of mass M, the generally correct form is

$$
\Gamma(s) = \left| \frac{\boldsymbol{p}}{\boldsymbol{p}_0} \right|^{2\ell+1} \Gamma(M^2),
\tag{C.6.26}
$$

where $\Gamma(M^2)$ is the on-shell decay rate, i.e. for $s = M^2$; $\boldsymbol{p}$ ($\boldsymbol{p}_0$) is the off-shell (on-shell) final-state momentum and ℓ is the intrinsic spin of the resonance. The effect of this is seen in a certain skew of the otherwise symmetric Breit–Wigner form. If not taken into account, it would induce a shift in the extracted mass of the resonance and also lead to an incorrect value of the nominal width.

A few final observations are in order before concluding. Here we have only considered the simplified case of a single resonance contributing to a given channel. Now, while it is true that only one intermediate state is likely to be *resonant* at any one energy, if two or more resonances are near in mass (i.e. with respect to their widths), then interference effects can become important. In such cases care must be taken to sum over all possible contributing *amplitudes*, after which the cross-section (or decay rate) may be calculated from the square of the total amplitude so obtained.

Note that it is quite likely that more than one process contributes to the overall width or rate Γ for the decay of any given resonance while we may only be interested experimentally in a particular channel. In such a case the procedure is quite simple: the width Γ appearing in the denominator, being effectively the imaginary part of the physical mass, must be taken as the *total* decay width. However, the width appearing in the numerator should be that corresponding to the particular channel under study.

Finally, there is evidently an implicit approximation in the derivation of the Breit–Wigner formula. Quite simply, the intermediate objects should not be too broad; we often speak of the 'narrow-resonance approximation', typically we require something like $\Gamma \ll M$. Now, while the above correction for the intrinsic energy dependence of the width goes some way to allowing even relatively broad states to be accurately included, this cannot completely take into account the non-elementary nature of many of the particles involved. The Breit–Wigner form has its foundation in a treatment of all processes as involving only elementary particles and although something can be done to include form-factor like effects there is no well-defined way in which to reliably accounted for the substructure of the resonance (nor indeed of the initial and/or final states).

C.6.4 Energy shift and decay width

In this section we shall examine the effect of a perturbing interaction on the energy levels to second order in perturbation theory. We shall find that the second-order energy shift may acquire an imaginary contribution, which will be shown to arise owing to the finite lifetime of the unstable state and indeed to be given precisely by the decay rate. Thus, we shall provide a rigorous justification of the Breit–Wigner form, in which the width of the resonance is given by the decay rate.

We shall consider the general case of a system governed by an unperturbed Hamiltonian $\mathcal{H}_0$ with eigenstates given by solutions of the Schrödinger equation,

$$i\hbar \frac{\partial}{\partial t} \psi_k(\boldsymbol{x},\, t) = \mathcal{H}_0 \psi(\boldsymbol{x},\, t), \tag{C.6.27}$$

taken to be $\psi_k(\boldsymbol{x},\, t)$ such that

$$\mathcal{H}_0 \psi_k(\boldsymbol{x},\, t) = E_k \psi_k(\boldsymbol{x},\, t). \tag{C.6.28}$$

For simplicity, we take them to be discrete, but the extension to continuous eigenvalues is straightforward. That is,

$$\psi_k(t) = u_k(\boldsymbol{x}) \, e^{-\frac{i}{\hbar} E_k t}, \tag{C.6.29}$$

where $u_k(\boldsymbol{x})$ is the wave-function embodying the spatial dependence and any implicit quantum numbers, which are all irrelevant to the present discussion. Note that time dependence is only present in the form of a phase and that the true state of the unperturbed system will be stationary.

We now add a time-dependent perturbation $\mathcal{V}(\boldsymbol{x},\, t)$. We shall thus be interested in the transition probability between eigenstates of the unperturbed system and also the perturbed energy eigenvalues. It will be convenient to take a perturbing potential of the form

$$\mathcal{V}(\boldsymbol{x}, t) = V(\boldsymbol{x})e^{\varepsilon t}, \tag{C.6.30}$$

where $V(\boldsymbol{x})$ is some small perturbing potential that is constant in time. Note that the perturbation so introduced is zero in the distant past ($t \to -\infty$) but grows slowly with time. Moreover, in the limit $\varepsilon \to 0$, the perturbation becomes constant. The effect of ε is then to allow an adiabatic introduction of the perturbing interaction. Our system is thus governed by the following Hamiltonian:

$$\mathcal{H} := \mathcal{H}_0 + \mathcal{V}(\boldsymbol{x}, t), \tag{C.6.31}$$

with $\mathcal{V}(\boldsymbol{x}, t)$ as above.

The question we now wish to address is the time evolution of the original eigenstates in the presence of the perturbation; i.e. the probability of transitions between the unperturbed states and, in particular, the decay rate of any given initially unperturbed state as induced by the perturbation. We therefore start with a generic wave-function $\psi(\boldsymbol{x}, t)$, a solution of the full perturbed Schrödinger equation:

$$i\hbar \frac{\partial}{\partial t} \psi(\boldsymbol{x}, t) = \mathcal{H}\psi(\boldsymbol{x}, t) \tag{C.6.32}$$

and express it as an expansion over the orthogonal set of unperturbed basis states:

$$\psi(\boldsymbol{x}, t) = \sum_k c_k(t)\psi_k(\boldsymbol{x}, t). \tag{C.6.33}$$

Inserting this into equation (C.6.32), we have

$$\sum_k i\hbar \dot{c}_k(t)u_k(\boldsymbol{x})e^{-\frac{i}{\hbar}E_k t} = \mathcal{V}(\boldsymbol{x}, t)\sum_k c_k(t)u_k(\boldsymbol{x})e^{-\frac{i}{\hbar}E_k t}, \tag{C.6.34}$$

where the dot indicates a time derivative. Projecting with $u_n^*(\boldsymbol{x})$ and integrating over $\boldsymbol{x}$ (summing and/or integrating over any possible implicit quantum numbers), we obtain

$$i\hbar \dot{c}_n(t)e^{-\frac{i}{\hbar}E_n t} = \int d^3 x\, u_n^*(\boldsymbol{x})\mathcal{V}(\boldsymbol{x}, t)\sum_k c_k(t)u_k(\boldsymbol{x})e^{-\frac{i}{\hbar}E_k t}. \tag{C.6.35}$$

Let us define

$$V_{nk} := \int d^3 x\, u_n^*(\boldsymbol{x})V(\boldsymbol{x})u_k(\boldsymbol{x}) \tag{C.6.36}$$

and thus write

$$i\hbar \dot{c}_n(t)e^{-\frac{i}{\hbar}E_n t} = e^{-\varepsilon t}\sum_k V_{nk} c_k(t)e^{-\frac{i}{\hbar}E_k t}. \tag{C.6.37}$$

Rearranging a little, we finally obtain a set of coupled differential equations for the coefficients $c_k(t)$:

$$\dot{c}_n(t) = \left(-\frac{i}{\hbar}\right) e^{-\varepsilon t} \sum_k V_{nk} c_k(t)\, e^{i\omega_{nk} t}, \tag{C.6.38}$$

where, for convenience, we have defined $\hbar\omega_{nk} := E_n - E_k$.

In general, such a system can only be solved perturbatively, i.e. as an expansion in powers of V. The easiest way to approach this is to begin by integrating the above with respect to time:

$$c_n(t) = c_n(-\infty) - \frac{i}{\hbar} \int_{-\infty}^{t} dt'\, e^{-\varepsilon t'} \sum_k V_{nk} c_k(t') e^{i\omega_{nk} t'}, \tag{C.6.39}$$

where $c_n(-\infty)$ are the initial values of the coefficients in the remote past (when the perturbation was absent).

In the remote past ($t \to -\infty$) the perturbation was not yet active and so we may now assume that the initial state was some pure eigenstate m, say; that is, $c_n(-\infty) = \delta_{nm}$. A formal solution may be obtained by iteratively substituting the entire right-hand side, for the coefficient $c_k(t')$ appearing there:

$$
\begin{aligned}
c_n(t) = \delta_{nm} &+ \left(-\frac{i}{\hbar}\right) \int_{-\infty}^{t} dt'\, e^{-\varepsilon t'} V_{nm} e^{i\omega_{nm} t'} \\
&+ \left(-\frac{i}{\hbar}\right)^2 \sum_k \int_{-\infty}^{t} dt'\, e^{-\varepsilon t'} \int_{-\infty}^{t'} dt''\, e^{-\varepsilon t''} V_{nk} e^{i\omega_{nk} t'} V_{km} e^{i\omega_{km} t''} + O(V^3).
\end{aligned}
\tag{C.6.40}
$$

Each term in the series expansion so obtained contains progressively higher powers of V. To first order in V and taking $n \neq m$, we have

$$c_n(t) \simeq -\frac{i}{\hbar} \int_{-\infty}^{t} dt'\, e^{-\varepsilon t'} V_{nm} e^{i\omega_{nm} t'}. \tag{C.6.41}$$

Performing the integral, we obtain

$$c_n(t) \simeq -\frac{i}{\hbar} V_{nm} \frac{1}{i\omega_{nm} - \varepsilon} e^{(i\omega_{nm} - \varepsilon)t}. \tag{C.6.42}$$

The transition rate from some given initial eigenstate m to state n is

$$w_{mn} := \frac{d}{dt} |c_n(t)|^2 \simeq \frac{|V_{nm}|^2}{\hbar^2} \left[\frac{2\varepsilon e^{2\varepsilon t}}{\omega_{nm}^2 + \varepsilon^2} \right]. \tag{C.6.43}$$

Now, a standard representation of the Dirac δ-function is

$$\lim_{\varepsilon \to 0} \frac{\varepsilon}{x^2 + \varepsilon^2} = \pi\, \delta(x). \tag{C.6.44}$$

We thus find

$$\lim_{\varepsilon \to 0} w_{mn} \simeq \frac{2\pi}{\hbar^2} |V_{nm}|^2 \delta(\omega_{nm}). \tag{C.6.45}$$

There may, of course, be (infinitely) many such states or a continuum and it is therefore natural to define a density of states $\rho(E)$, such that $\rho(E)\,dE$ is just the number of states in the energy interval $(E, E + dE)$. We may thus transform the sum over states into an integral and write

$$\sum_n \rightarrow \int E_n \rho(E_n). \tag{C.6.46}$$

Finally, using the δ-function to integrate, we obtain Fermi's golden rule:

$$w_{m \rightarrow \{\bar{n}\}} = \frac{2\pi}{\hbar}\, \overline{|V_{nm}|^2}\, \rho(E_n) \bigg|_{E_n = E_m}, \tag{C.6.47}$$

where $w_{m \rightarrow \{\bar{n}\}}$ indicates a sum over transitions from m to all energetically permitted states $n \neq m$, i.e. with $E_n = E_m$.

Thus far, we have only dealt with transitions from a state $|m\rangle$ to $|n\rangle$ with $n \neq m$ explicitly. We shall now examine the temporal evolution of the coefficient $c_m(t)$, i.e. for $n = m$. For our purposes, it will be necessary to evaluate the coefficient up to second order in the perturbation. For clarity, let us first evaluate each order separately; up to and including second order, we then have

$$c_m^{(0)}(t) = 1, \tag{C.6.48a}$$

$$c_m^{(1)}(t) = \left(-\frac{i}{\hbar}\right) V_{mm} \int_{-\infty}^{t} dt'\, e^{\varepsilon t'} = \left(-\frac{i}{\hbar}\right) V_{mm} \frac{e^{\varepsilon t}}{\varepsilon}, \tag{C.6.48b}$$

$$\begin{aligned}
c_m^{(2)}(t) &= \left(-\frac{i}{\hbar}\right)^2 \sum_k |V_{km}|^2 \int_{-\infty}^{t} dt' \int_{-\infty}^{t'} dt''\, e^{i\omega_{mk}t' + \varepsilon t'} e^{i\omega_{km}t'' + \varepsilon t''} \\
&= \left(-\frac{i}{\hbar}\right)^2 \sum_k |V_{km}|^2 \int_{-\infty}^{t} dt'\, e^{i\omega_{mk}t' + \varepsilon t'} \frac{1}{i\omega_{km} + \varepsilon} e^{i\omega_{km}t' + \varepsilon t'} \\
&= \left(-\frac{i}{\hbar}\right)^2 \sum_k |V_{km}|^2 \frac{1}{i\omega_{km} + \varepsilon} \int_{-\infty}^{t} dt'\, e^{2\varepsilon t'} \\
&= \left(-\frac{i}{\hbar}\right)^2 \sum_k |V_{km}|^2 \frac{1}{i\omega_{km} + \varepsilon} \frac{e^{2\varepsilon t}}{2\varepsilon} \\
&= \left(-\frac{i}{\hbar}\right)^2 |V_{mm}|^2 \frac{e^{2\varepsilon t}}{2\varepsilon^2} + \left(-\frac{i}{\hbar}\right)^2 \sum_{k \neq m} |V_{km}|^2 \frac{e^{2\varepsilon t}}{2\varepsilon(i\omega_{km} + \varepsilon)}.
\end{aligned} \tag{C.6.48c}$$

Putting this all together, we have

$$\begin{aligned}
c_m(t) \simeq 1 &+ \left(-\frac{i}{\hbar}\right) V_{mm} \frac{e^{\varepsilon t}}{\varepsilon} + \left(-\frac{i}{\hbar}\right)^2 |V_{mm}|^2 \frac{e^{2\varepsilon t}}{2\varepsilon^2} \\
&+ \left(-\frac{i}{\hbar}\right)^2 \sum_{k \neq m} |V_{km}|^2 \frac{e^{2\varepsilon t}}{2\varepsilon(i\omega_{km} + \varepsilon)}.
\end{aligned} \tag{C.6.49}$$

As always, we are interested in the transition rate (here this is the *loss* rate); we thus now take the time derivative. At this point we may set $\varepsilon = 0$ in the exponents (though not yet in the denominators) of the right-hand side,:

$$\dot{c}_m(t) \simeq \left(-\frac{i}{\hbar}\right)V_{mm} + \left(-\frac{i}{\hbar}\right)^2 |V_{mm}|^2 \frac{1}{\varepsilon} + \left(-\frac{i}{\hbar}\right)^2 \sum_{k \neq m} |V_{km}|^2 \frac{1}{i\omega_{km} + \varepsilon}. \quad (C.6.50)$$

The question of degeneracy now arises; i.e. the possibility that $E_k = E_m$ (or $\omega_{km} = 0$) for some $k \neq m$. This may be dealt with in a straightforward manner by following the standard procedure of diagonalising V_{km} in the degenerate subspace to eliminate the singular terms Here, for simplicity, we shall avoid this inessential complication by assuming non-degeneracy: i.e. for $k \neq m$, $E_k \neq E_m$.

Now, since we are neglecting all third- and higher-order terms, we may rewrite equation (C.6.50) as

$$\dot{c}_m(t) \simeq \left(-\frac{i}{\hbar}\right)\left\{V_{mm} + \left(-\frac{i}{\hbar}\right)\sum_{k \neq m}\frac{|V_{km}|^2}{i\omega_{km} + \varepsilon}\right\}c_m(t), \quad (C.6.51)$$

by noting that on the right-hand side, here it is sufficient to write $c_m(t)$ to *first* order only, since the multiplying expression is already first order in V. In other words (on the right-hand side, only), we may write approximately

$$c_m(t) \sim 1 + \left(-\frac{i}{\hbar}\right)\frac{1}{\varepsilon} V_{mm} + O(V^2). \quad (C.6.52)$$

We thus see that (*to this order in perturbation theory*) we have a simple first-order differential equation for $c_m(t)$, with a *constant* coefficient. To leading order in V, the coefficient is imaginary and so, taking into account the initial condition $c_m(-\infty) = 1$, we shall write

$$c_m(t) = e^{-\frac{i}{\hbar}W_m t}, \quad (C.6.53)$$

where the leading-order part of W_m will be real, although the higher-order contributions may be complex (and we shall indeed see that they are).

Expressing explicitly the expansion of W_m in powers of V,

$$W_m = W_m^{(1)} + W_m^{(2)} + \dots , \quad (C.6.54)$$

the first-order term is then simply

$$W_m^{(1)} = V_{mm}. \quad (C.6.55)$$

Perhaps not surprisingly, as obtained in time-independent perturbation theory, this is just the expectation value of the perturbing potential for the state under consideration. Let us now examine the second-order term:

$$W_m^{(2)} = -\frac{i}{\hbar}\sum_{k \neq m}\frac{|V_{km}|^2}{i\omega_{km} + \varepsilon} = -\frac{1}{\hbar}\sum_{k \neq m}\frac{|V_{km}|^2}{\omega_{km} - i\varepsilon}. \quad (C.6.56)$$

Yet another useful relation defining the Dirac δ-function is

$$\lim_{\varepsilon \to 0} \frac{1}{x \pm i\varepsilon} = \mathrm{PV}\frac{1}{x} \mp i\pi\delta(x), \tag{C.6.57}$$

where 'PV' indicates the principal value. This identity then provides the real and imaginary parts of $W_m^{(2)}$:

$$\mathcal{R}e\, W_m^{(2)} = -\mathrm{PV}\sum_{k \neq m}\frac{|V_{km}|^2}{E_k - E_m}, \tag{C.6.58a}$$

$$\mathcal{I}m\, W_m^{(2)} = -\pi\sum_{k \neq m}|V_{km}|^2\,\delta(E_k - E_m). \tag{C.6.58b}$$

The first of the two expressions above is just the standard second-order contribution in perturbation-theory to the shift in the energy eigenvalue for the original state, while the second is none other than Fermi's golden rule. Indeed, we may write

$$w_{m \to \{\bar{n}\}} = \frac{2\pi}{\hbar}\sum_{k \neq m}|V_{km}|^2\,\delta(E_k - E_m) = -\frac{2}{\hbar}\,\mathcal{I}m\, W_m^{(2)}. \tag{C.6.59}$$

This shift may be seen as the interaction-energy contribution due to the intermediate states. If we now define

$$\Gamma_m := -2\mathcal{I}m\, W_m, \tag{C.6.60}$$

then, from the solution of the differential equation for $c_m(t)$, we have

$$|c_m(t)|^2 = e^{-\frac{1}{\hbar}\Gamma_m t}. \tag{C.6.61}$$

And therefore the imaginary part of W_m (up to the factor -2) just corresponds to the decay rate or *width* (we shall explain the latter term shortly).

It is always important to verify the conservation of probability (to the relevant order in perturbation theory): for t small we have

$$|c_m(t)|^2 + \sum_{n \neq m}|c_n(t)|^2 \simeq \left(1 - \frac{1}{\hbar}\Gamma_m t\right) + w_{m \to \{\bar{n}\}}\, t = 1, \tag{C.6.62}$$

which is precisely the desired equality to second order in V.

Now, why do we refer to the quantity Γ, the imaginary part of W_m, as the *width*? To understand this, let us examine the Fourier transform of $c_m(t)$:

$$\tilde{c}_m(E) := \int dt\, e^{-\frac{i}{\hbar}Et} e^{-\frac{i}{\hbar}\left[E_m + \mathcal{R}e\, W_m - \frac{i}{2}\Gamma_m\right]t}$$

$$\propto \frac{1}{\left[E - (E_m + \mathcal{R}e\, W_m) + \frac{i}{2}\Gamma_m\right]}. \tag{C.6.63}$$

We thus find for the spectrum, the following distribution:

$$|\tilde{c}_m(E)|^2 \propto \frac{1}{\left[E - (E_m + \mathcal{Re}\ W_m)\right]^2 + \frac{1}{4}\Gamma_m^2}. \tag{C.6.64}$$

The above is now recognisable as the form obtained in appendix C.6 within the framework of the Breit–Wigner formalism. Clearly, Γ_m is precisely the *full width at half maximum* of the spectrum, which now has a peak centred on the *shifted* energy $(E_m + \mathcal{Re}W_m)$.

Furthermore, the decay law $|c_m(t)| \propto e^{-\Gamma_m t}$ implies a mean lifetime for the state of $\tau_m = 1/\Gamma_m$. Note that this is perfectly in accordance with (and indeed simply reflects) the uncertainty principle: $\Delta t \Delta E \geqslant \hbar$. That is, given a finite mean lifetime τ, the uncertainty principle places a limit on the precision with which the energy or equivalently mass of the state can be measured. The limiting precision is just $\hbar/\tau$, which is none other than the decay rate or width Γ. We thus expect the measured or rather perceived mass to vary (on average) over the interval $m \pm \frac{1}{2}\Gamma$.

C.7 Three-body final states and the Dalitz plot

An important by-product of the analysis performed by Dalitz (1953) on τ^+ decay into three pions is a parametrisation of three-body decays, now known as the Dalitz plot. A two-body decay is energetically trivial since the two final-state objects emerge back-to-back in the centre-of-mass frame, with fixed energies; i.e. every decay of a given type has exactly the same values for the kinematic variables. A three-body final state, however, allows for variation of all the energies and also relative directions (within certain kinematical constraints). It thus turns out that while still relatively simple to analyse, the information that may be extracted is very rich.

C.7.1 Three-body kinematics

In principle, a three-body final state has three four-vectors; i.e. 12 variables to define the configuration; not all are independent though. There are a number of constraints and redundancies that reduce this large number to just two. First of all, in the centre-of-mass system, since the three spatial momentum vectors ($k_{1,2,3}$, say) must sum to zero, the final-state is coplanar. A plane requires three Euler angles to define its orientation (which here is generally irrelevant unless there is, say, a known initial-state polarisation fixing a reference direction). The masses of the three final-state particles provide constraints between their momenta and energies. Overall four-momentum conservation provides a further four equations: $p^\mu = k_1^\mu + k_2^\mu + k_3^\mu$ (where the four-momenta are p^μ for the initial state and k_i^μ for the three final decay objects). We thus find a total of $3 + 3 + 4 = 10$ constraints and redundancies, leaving just two linearly independent variables.

An obvious possibility for these two variables would be two of the angles; however, a standard choice in non-relativistic kinematics is two of the final-state kinetic energies (T_i). In the non-relativistic case we have

$$T_1 + T_2 + T_3 = Q := M - m_1 - m_2 - m_3, \qquad (C.7.1)$$

for final-state masses m_i arising from the decay of a particle of mass M. In his original paper, Dalitz (1953) studied the very particular case of $K^\pm \to \pi^\pm \pi^+ \pi^-$, which has three equal-mass particles on the final state. In such a case, it is convenient to define the following two variables:

$$x := \sqrt{3}\,(T_1 - T_2)/Q \ \text{ and } \ y := (2T_3 - T_1 - T_2)/Q. \qquad (C.7.2)$$

It is straightforward to show that the (x, y) points defining the final states lie within the boundary of a circle of unit radius: $T_1/Q + T_2/Q + T_3/Q = 1$ (see figure C.8).

Exercise C.7.1. *Show that, for equal-mass final-state particles, the (x, y) points as defined in equation (C.7.2) do indeed lie within the boundary of a circle of unit radius:* $T_1/Q + T_2/Q + T_3/Q = 1.$

Moving over to a full relativistic formulation, it is natural to use the invariant masses of pairs of final-state particles. We have

$$m_{23}^2 := (k_2^\mu + k_3^\mu)^2 = (p^\mu - k_1^\mu)^2 = M^2 + m_1^2 - 2ME_1, \qquad (C.7.3a)$$

$$m_{31}^2 := (k_3^\mu + k_1^\mu)^2 = (p^\mu - k_2^\mu)^2 = M^2 + m_2^2 - 2ME_2, \qquad (C.7.3b)$$

$$m_{12}^2 := (k_1^\mu + k_2^\mu)^2 = (p^\mu - k_3^\mu)^2 = M^2 + m_3^2 - 2ME_3, \qquad (C.7.3c)$$

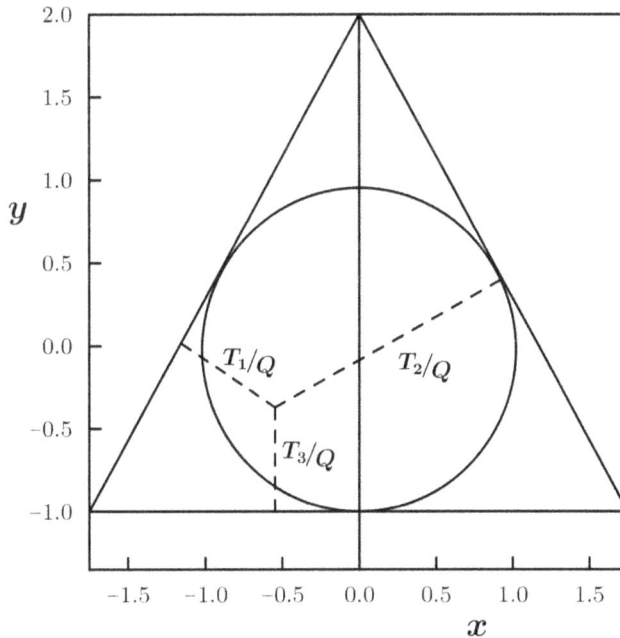

Figure C.8. The Dalitz plot expressed in the x–y plane, as defined in the main text.

with the mass-shell constraints $k_i^2 = E_i^2 - \boldsymbol{k}_i^2 = m_i^2$ and

$$m_{12}^2 + m_{23}^2 + m_{31}^2 = M^2 + m_1^2 + m_2^2 + m_3^2, \tag{C.7.4}$$

where the energies are defined in the centre-of-mass frame.

Exercise C.7.2. *Show that*

$$m_{12}^2 + m_{23}^2 + m_{31}^2 = M^2 + m_1^2 + m_2^2 + m_3^2,$$

which is therefore a constant.

The kinematical boundaries require a little work; for example, for fixed m_{23} and considering the limiting kinematical configurations in the 23 rest frame, we find the following limits for m_{13}:

$$\begin{aligned}
m_{13,\,\pm}^2 = m_1^2 + m_3^2 &+ \frac{1}{m_{23}^2}\Big[(M^2 - m_{23}^2 - m_1^2)(m_{23}^2 - m_2^2 - m_3^2) \\
&\pm \lambda^{1/2}(M^2, m_{23}^2, m_1^2)\,\lambda^{1/2}(m_{23}^2, m_2^2, m_3^2)\Big],
\end{aligned} \tag{C.7.5}$$

where λ was defined in equation (C.1.6).

C.7.2 Three-body Lorentz-invariant phase space

We may thus choose any two out of m_{12}, m_{23} and m_{31} as the independent variables to use, with the advantage in the relativistic case that they are invariants and may thus be measured in *any* inertial reference frame[1]. Starting from the full three-body Lorentz-invariant phase space,

$$\mathrm{dLIPS}_3 = \prod_{i=1,2,3} \frac{\mathrm{d}^3 \boldsymbol{k}_i}{2E_i}\, \delta^4(k_1^\mu + k_2^\mu + k_3^\mu - p^\mu), \tag{C.7.6}$$

integrating out the three redundant Euler angles defining the orientation of the decay plane and exploiting the three-momentum δ-functions, we have

$$\mathrm{dLIPS}_3 \propto \mathrm{d}E_1 \mathrm{d}E_2 \mathrm{d}E_3\, \delta(E_1 + E_2 + E_3 - M) = 2E_1 \mathrm{d}E_2, \tag{C.7.7}$$

where we have also used the relativistic $E\mathrm{d}E = k\mathrm{d}k$, with $k := |\boldsymbol{k}|$. Therefore, using also $\mathrm{d}E_1 \mathrm{d}E_2 = \mathrm{d}m_{23}^2 \mathrm{d}m_{13}^2 / 4M^2$, we finally obtain the full differential decay rate:

$$\mathrm{d}\Gamma = \frac{1}{(2\pi)^3} \frac{1}{32M^3} |\mathcal{M}|^2\, \mathrm{d}m_{23}^2\, \mathrm{d}m_{13}^2. \tag{C.7.8}$$

Note again that we may use any pair of m_{23}, m_{13} and m_{12}.

---

[1] This is important, for example, in the case of the *in-flight* decay of particles produced in high-energy collisions and which are therefore in motion in the laboratory frame.

Exercise C.7.3. *Starting from Fermi's golden rule for a three-body decay, derive the final form for the differential decay rate given in equation (C.7.8).*

Let us examine more closely the form of the decay rate given in equation (C.7.8). The phase-space part has no energy or angular dependence, any possible variation can only lie in the matrix element $\mathcal{M}$. This means that, for example, a non-resonant three-body contribution with no non-trivial dynamics will populate the Dalitz plot uniformly, see figure C.9. On the other hand, any energy or angular dependence of $\mathcal{M}$ will immediately stand out in the distribution in m_{23}^2 and m_{13}^2. In particular, if there is a resonance in, say, the 1–3 channel (i.e. an energetically accessible state corresponding to the combined quantum numbers of particles 1 and 3), then this will show up as a Breit–Wigner ridge in the event rate along the m_{23}^2 direction, for a fixed value of m_{13}^2 corresponding to the mass of the resonance.

C.7.3 An example of Dalitz-plot analysis

A typical example of how the Dalitz plot provides a revealing representation of the dynamics underlying a decay process is the case of a two-step decay of the form

$$X \to aY \to abc, \tag{C.7.9}$$

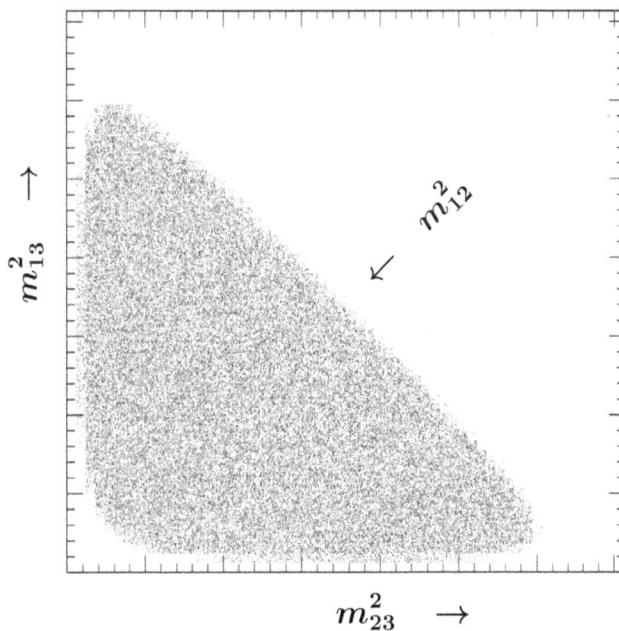

Figure C.9. The relativistic Dalitz plot expressed in the plane of two invariant final-state invariant masses squared (m_{23}^2 and m_{13}^2); the third (m_{12}^2) is equivalent to variations along the direction of the arrow.

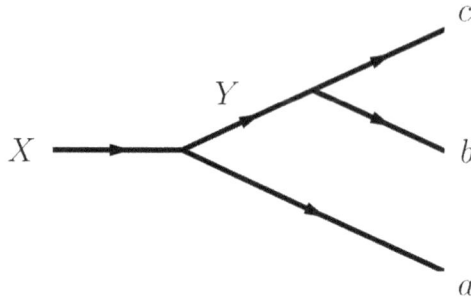

Figure C.10. The Feynman diagram representing a three-body decay ($X \to abc$) proceeding via an intermediate two-body resonance (Y) in the $b - c$ channel.

where Y is some intermediate resonance decaying very rapidly via the two-body process $Y \to bc$. That is, such a process may be interpreted in terms of a sum of possible contributions of the form shown in figure C.10. Note that for experimental analysis, a non-resonant three-body background is usually necessary too.

There will then be a consequent classic Breit–Wigner enhancement (see (C.6.12)) or band of excess events around the line $m_{bc}^2 = m_Y^2$, which will be clearly visible in the Dalitz plot. In figure C.11 we provide an experimental example: a process studied by the E687 experiment at Fermilab, the decay $D_s^+ \to K^+K^-\pi^+$ (D_s^+ is a $c\bar{s}$, $J = 0$, meson)[2], see Frabetti *et al* (1995). The D_s is identified via its mass, reconstructed from $(\sum k_i)^2$. The pronounced vertical band centred at $m_{KK}^2 \simeq 1.04$ GeV^2 is naturally interpreted as the Breit–Wigner resonance peak due to the $\phi^0(1020)$, an $s\bar{s}$, $J = 1$, ground state (*i.e.* $D_s^+ \to \phi^0\pi^+$ and $\phi^0 \to K^+K^-$), while the less clear broader horizontal band around $m_{K^-\pi^+}^2 \simeq 0.8$ GeV^2 corresponds to production of the $\bar{K}^{*0}(892)$, an excited $\bar{d}s$, $J = 1$, meson (i.e. $D_s^+ \to \bar{K}^{*0}K^+$ and $\bar{K}^{*0} \to K^-\pi^+$). Other resonances are required to fit the data, but lie outside the plot domain, one is the $\bar{K}^{*0}(1430)$, another excited $\bar{d}s$ state, but there is also need for the $f_0(980)$ and $f_0(1710)$; the latter is actually considered to be a serious candidate. In each of the visible bands we also notice the distribution in two separate blocks; we shall now see how this furnishes precise indications as to the spins of the resonances involved (also vital for their correct identification). Of course, the widths of the bands are also related more-or-less directly to the Breit–Wigner decay widths (Γ).

The precise form of a resonance band also provides direct information as to the spin of the intermediate resonance. We shall consider the simplest case; i.e. that in which the three final state particles a, b and c are all spin-zero mesons and the intermediate resonance Y in, say, the b–c channel has spin J. The decay amplitude will then have zeros corresponding to the Legendre polynomials $P_J(\cos\theta)$ associated with angular momentum J; a spin-J object will lead to precisely J zeroes in the Dalitz plot. Note, first of all, that the specific amplitude will now take the form

---

[2] Recall $K^+ = u\bar{s}$, $K^- = s\bar{u}$ and $\pi^+ = u\bar{d}$.

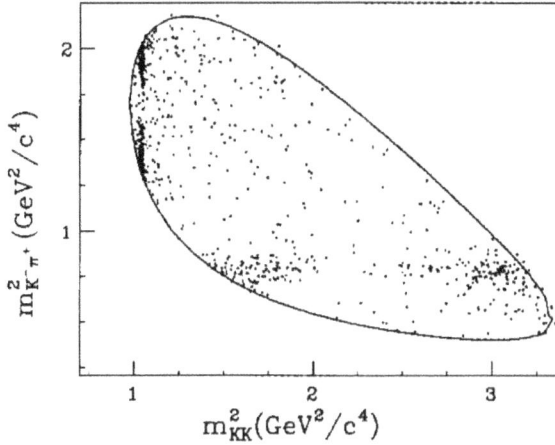

Figure C.11. A Dalitz plot for the decay $D_s^+ \rightarrow K^+ K^- \pi^+$ (the figure is taken from Frabetti *et al* (1995), copyright (1995) with permission from Elsevier).

$\mathcal{M} \sim \mathcal{M}_{\mathrm{BW}} \times P_J(\cos \theta)$, where the angle θ is defined in the Y rest frame, with respect to the direction of Y (or a) in the X rest frame. We recall the first three polynomials:

$$P_0(\cos \theta) = 1, \quad P_1(\cos \theta) = \cos \theta, \quad P_2(\cos \theta) = \tfrac{3}{2} \cos^2 \theta - \tfrac{1}{2}. \qquad \text{(C.7.10)}$$

And a simple calculation yields

$$\cos \theta = \frac{(m_{ac}^2)_{\max} + (m_{ac}^2)_{\min} - 2m_{ac}^2}{(m_{ac}^2)_{\max} - (m_{ac}^2)_{\min}} \qquad \text{(C.7.11)}$$

Exercise C.7.4. *Derive the formula given in equation (C.7.11).*

Therefore, whereas $P_0(\cos \theta)$ has no zeroes; $P_1(\cos \theta)$ has just one for $\cos \theta = 0$, i.e. for m_{bc}^2 at the midpoint between $(m_{bc}^2)_{\max}$ and $(m_{bc}^2)_{\min}$; the case $J = 2$ has two zeroes lying at $\cos \theta = \pm \frac{1}{\sqrt{3}}$ and so on. Going back to figure C.11, we see that both of the Breit–Wigner bands are distributed in two lobes, i.e. with just one zero, and therefore correspond to spin-one objects. The masses, widths and spins then lead unequivocally to the assignments already listed above.

A final interesting effect is that of constructive and destructive interference between overlapping resonances, which may be in the same channel (pairwise) or in a crossed channel. Recall the full complex amplitude for a Breit–Wigner resonance:

$$f_\ell(E) = -\frac{\hbar}{p} \frac{\Gamma/2}{(E - E_0) - i\Gamma/2} (2\ell + 1) P_\ell(\cos \theta). \qquad \text{(C.6.12)}$$

Note that between any two different resonant channels there is a further possible phase difference, $e^{i\phi}$ say, generated by non-perturbative bound-state dynamics. Thus, according to the value of ϕ, at the point of maximum overlap (if in the same channel) or intersection point (if in crossed channels) the amplitudes may interfere in a manner anywhere from in-phase to out-of-phase. For the Dalitz plot in, for example, the same-channel case this will lead to either an enhancement or a depletion in the region of maximum overlap.

In practice, the analysis is performed by modelling the distribution with basic two-body amplitudes given by magnitudes and phases (and possibly form-factors). These are then multiplied by the relevant Breit–Wigner amplitudes, which effectively take the place of propagators in the analogous Feynman diagrams. All possible contributing such compound amplitudes are then summed and squared to obtain the full Dalitz decay distribution. The input parameters are the (Breit–Wigner) masses and widths of the known possibly contributing resonances, while the analysis output provides amplitude magnitudes and relative phases for each of the initial two-body steps in the decays.

A final word of caution is in order before leaving this topic. It is natural to wish to create one-dimensional histograms of the data by projecting onto one of the two axes. Consider projecting onto, say, the $m^2_{K^+K^-}$ axis in figure C.11. While the Breit–Wigner peak of the ϕ^0 will clearly stand out well at 1.04 GeV^2, it is easy to see that the concentration in two lobes of the $\bar{K}^{*0}(892)$ will appear as two further separate resonance peaks at around 1.7 GeV^2 and 3.1 GeV^2. Likewise, projecting onto the $m^2_{K^-\pi^+}$ axis will highlight the $\bar{K}^{*0}(892)$ resonance at around 0.80 GeV^2, but will also create apparent peaks at around 1.5 GeV^2 and 1.9 GeV^2. These fake peaks are what are known in the jargon as '*kinematical reflections*' or '*shadows*'.

References

Breit G and Wigner E P 1936 *Phys. Rev.* **49** 519

Dalitz R H 1953 *Phil. Mag. Ser. 7* **44** 1068

Flatté S M 1976 *Phys. Lett.* **B63** 224

Frabetti P LE687 Collab *et al* 1995 *Phys. Lett.* **B351** 591

Frosch R F *et al* 1968 *Phys. Rev.* **174** 1380

Hofstadter R 1956 *Rev. Mod. Phys.* **28** 214

Källén G 1964 *Elementary Particle Physics* (Reading, MA: Addison-Wesley)

Mandelstam S 1958 *Phys. Rev.* **112** 1344

Mott N F 1929 *Proc. R. Soc. Lond.* **A124** 425

Navas S *et al* (Particle Data Group) 2024 *Phys. Rev.* **D110** 030001

Rutherford E 1911 *Phil. Mag.* **21** 669

Strutt J W and Hon 1871 *Phil. Mag. Ser. 4* **41** 107 274

von Sellmeier W 1871 *Annalen Phys.* **219** 272 *ibid* (1872) 386; *ibid* **221** (1872) 520; *ibid* **221** (1872) 399

IOP Publishing

An Introduction to Elementary Particle Phenomenology
(Second Edition)

Philip G Ratcliffe

Appendix D

Exercises

D.1 Exercises to chapter 1

Chapter 1 has no exercises.

D.2 Exercises to chapter 2

Exercise 2.1.1. *The decay* ^{210}Bi $\rightarrow$ ^{210}Po $+ e^- + \bar{\nu}_e$ *has a Q-value of 1.16* MeV; *calculate the maximum recoil kinetic energy of the daughter nucleus.*
   Do the same for $^{22}_{11}$Na $\rightarrow$ $^{22}_{10}$Ne $+ e^+ + \nu_e$, $Q = 4.38$ MeV *(take* $m_\nu = 0$). *What is the minimum recoil energy?*

Answer The maximum recoil is attained when the neutrino is produced with zero energy in the bismuth rest-frame; the energy released is then divided between the emitted electron and the recoiling product nucleus. To see this, share the three-momentum p of the recoiling nucleus between a parallel electron and neutrino pair, with, respectively, ξp and $(1 - \xi)p$ (by definition then, $0 < \xi < 1$). Assuming $m_\nu \approx 0$ and using non-relativistic kinematics for the nucleus, we have

$$\sqrt{\xi^2 p^2 + m_e^2} + (1 - \xi)p + \frac{p^2}{2M} = Q,$$

where M is the mass of the product nucleus. Taking the derivative with respect to ξ ($p' := \mathrm{d}p/\mathrm{d}\xi$) gives

$$\frac{\xi p(p + \xi p')}{\sqrt{\xi^2 p^2 + m_e^2}} + (1 - \xi)p' - p + \frac{pp'}{M} = 0.$$

doi:10.1088/978-0-7503-5759-3ch10

This leads to

$$p' := \frac{dp}{d\xi} = p\left[1 - \frac{\xi p}{\sqrt{\xi^2 p^2 + m_e^2}}\right]\left[1 - \xi + \frac{\xi^2 p}{\sqrt{\xi^2 p^2 + m_e^2}} + \frac{p}{M}\right]^{-1}.$$

For $m_e \neq 0$ and $0 < \xi < 1$, the right-hand side is always positive (by inspection) and so the maximum value of p is attained for $\xi = 1$; i.e. when the electron carries away all of the energy.

Denote p_{max} the absolute momenta of both the nucleus and electron in this configuration; energy conservation then gives

$$\sqrt{p_{max}^2 + m_e^2} + \frac{p_{max}^2}{2M} = Q.$$

Rearranging leads to

$$p_{max}^2 + m_e^2 = \left(Q - p_{max}^2 / 2M\right)^2$$
$$= Q^2 - Q p_{max}^2 / M + \left(p_{max}^2 / 2M\right)^2.$$

Since clearly $p_{max} = O(Q)$ and $Q \ll M$, we may neglect the last term on the right-hand side and so obtain

$$p_{max}^2 \simeq Q^2 \frac{1 - m_e^2/Q^2}{1 + Q/M}.$$

If instead we use fully relativistic formulæ, we have

$$(p_{Bi} - p_{Po})^2 = (p_e + p_\nu)^2.$$

For the limiting case, we have seen that $\boldsymbol{p}_\nu = 0$ and therefore

$$m_{Bi}^2 - 2m_{Bi}E_{Po} + m_{Po}^2 = m_e^2.$$

So we have

$$E_{Po} = \frac{m_{Bi}^2 + m_{Po}^2 - m_e^2}{2m_{Bi}}.$$

We may now substitute using $m_{Bi} = m_{Po} + m_e + Q$ to obtain the final answer.

On the other hand, the minimum recoil energy is clearly zero, when the electron and neutrino are emitted precisely back-to-back (with equal and opposite momenta absorbing all the energy).

Exercise 2.1.2. *We now know that the weak interaction is in fact mediated by the $W^\pm$ and Z^0 bosons; the former has a mass around ~80 GeV and the latter ~91 GeV. By appealing to the uncertainty principle, estimate the maximum distance that such particles may propagate as virtual intermediate states. Compare this with the wavelength associated with the typical momenta involved in β-decay.*

Answer The maximum propagation distance is determined by the time allowed by the uncertainty principle, which is just $\hbar/mc^2$; the distance is thus $\hbar c/mc^2$. With $mc^2 \simeq 10^5\,\text{MeV}$, we therefore find approximately $\sim 200/10^5\,\text{fm} = 2 \times 10^{-3}\,\text{fm}$. Typical exchange momenta and energies in β-decay are a few MeV and so the best resolution is of the order a few tens ($\simeq 200/\text{few}$) of fm; at such a scale even most nuclei appear almost point-like.

Exercise 2.1.3. *Derive expression* (2.1.26).

Answer The $(E_0 - E_e)$ behaviour in

$$K(E_e) := \sqrt{\frac{P(E_e)}{F(Z_f, E_e) \cdot E_e \cdot \sqrt{E_e^2 - m_e^2 c^4}}}, \tag{2.1.24}$$

comes from the factor $(E_0 - E_e)$ in the spectrum:

$$P(E_e) \propto F(Z_f, E_e)\, E_e\, \sqrt{E_e^2 - m_e^2 c^4}\, (E_0 - E_e)^2, \tag{2.1.23}$$

which actually arises from the term $E_\nu |\boldsymbol{p}_\nu|$, with $E_\nu = E_0 - E_e$. For a zero-mass neutrino, $|\boldsymbol{p}_\nu| = E_\nu$ and so this is just $E_\nu^2 = (E_0 - E_e)^2$, but for a massive neutrino this should be replaced by $E_\nu |\boldsymbol{p}_\nu| = E_\nu \sqrt{E_\nu^2 - m_\nu^2 c^4}$, leading directly to equation (2.1.26)

Exercise 2.2.1. *By considering the intrinsic parities of the proton and neutron (conventionally taken to be positive), together with their known orbital and spin alignments inside the deuteron, show that we do indeed expect $P_d = +1$.*

Answer If we conventionally take the parities of both the proton and neutron to be positive, the parity of the deuteron will be completely determined by the spatial wave-function of the system and thus, in particular, by the orbital angular momentum quantum number L. From the magnetic moments, we deduce that the spins are aligned, which may then account for the deuteron spin $J = 1$. Since the observed deuteron is naturally assumed to be the ground state of the proton–neutron system, we may also deduce that $L = 0$. The overall parity is thus $P_D = P_p P_n (-1)^L = +1$.

N.B. We may extend the discussion to include isospin: the deuteron has $I_3 = 0$, but may be either an $I = 0$ (singlet) or 1 (triplet). The spin and spatial wave-functions are both symmetric; the isospin part must therefore be antisymmetric and thus an $I = 0$ singlet.

Exercise 2.4.1. *Using the Dirac matrix algebra and the spinor structure given, show that the operators $P_{R/L} := \frac{1}{2}(1 \pm \gamma_5)$ project onto right- and left-handed helicity states, respectively; that is, $\frac{1}{2}(1 \pm \gamma_5)\,\psi = \psi_{R/L}$.*

Answer The helicity states should be those for which the polarisation is parallel or antiparallel to the particle motion. Without loss of generality, we choose the direction of motion as the z-direction; i.e. $\boldsymbol{p} = (0, 0, p)$ and the two spin parallel and antiparallel states ($\pm$) are given by $\chi = \begin{pmatrix} 1 \\ 0 \end{pmatrix}$ and $\begin{pmatrix} 0 \\ 1 \end{pmatrix}$. The full general spinor is

$$\begin{pmatrix} \sqrt{E+m}\ 1 \\ \sqrt{E-m}\ \sigma \cdot \hat{\boldsymbol{p}} \end{pmatrix} \otimes \chi_{\pm},$$

which, for our choice of $\boldsymbol{p}$, becomes

$$\begin{pmatrix} \sqrt{E+m}\ 1 \\ \sqrt{E-m}\ \sigma_3 \end{pmatrix} \otimes \chi_{\pm} = \begin{pmatrix} \sqrt{E+m}\ 1 \\ \pm\sqrt{E-m}\ 1 \end{pmatrix} \otimes \chi_{\pm}.$$

Let us examine the action of γ_5 on such states:

$$\gamma_5 \begin{pmatrix} \sqrt{E+m}\ 1 \\ \pm\sqrt{E-m}\ 1 \end{pmatrix} \otimes \chi_{\pm} = \begin{pmatrix} \pm\sqrt{E-m}\ 1 \\ \sqrt{E+m}\ 1 \end{pmatrix} \otimes \chi_{\pm} = \pm \begin{pmatrix} \sqrt{E-m}\ 1 \\ \pm\sqrt{E+m}\ 1 \end{pmatrix} \otimes \chi_{\pm}.$$

We thus see that specific spin-polarisation states correspond to helicity eigenstates only for $m = 0$. In that case we have, as desired,

$$\tfrac{1}{2}(1 \pm \gamma_5)\begin{pmatrix} \sqrt{E}\ 1 \\ \pm\sqrt{E}\ 1 \end{pmatrix} \otimes \chi_{\pm} = \begin{pmatrix} \sqrt{E}\ 1 \\ \pm\sqrt{E}\ 1 \end{pmatrix} \otimes \chi_{\pm} \text{ and } \tfrac{1}{2}(1 \mp \gamma_5)\begin{pmatrix} \sqrt{E}\ 1 \\ \pm\sqrt{E}\ 1 \end{pmatrix} \otimes \chi_{\pm} = 0.$$

This is fine for 'massless' neutrinos and is a good approximation for relativistic electrons from e.g. β-decay.

Note that a massive spin state can be decomposed as a superposition of the two helicity states. Expanding to first order in m/E, we have

$$\begin{pmatrix} \sqrt{E+m}\ 1 \\ \pm\sqrt{E-m}\ 1 \end{pmatrix} \otimes \chi_{\pm} \approx \begin{pmatrix} \sqrt{E}\ 1 \\ \pm\sqrt{E}\ 1 \end{pmatrix} \otimes \chi_{\pm} + \frac{m}{2E}\begin{pmatrix} \sqrt{E}\ 1 \\ \mp\sqrt{E}\ 1 \end{pmatrix} \otimes \chi_{\pm}.$$

That is, the massive (antiparallel) polarisation state actually contains a small admixture of (positive) negative helicity. This actually has important phenomenological consequences.

It incidentally explains the observation that, despite the much reduced phase space, the decay channel $\pi^- \to \mu^- \bar{\nu}_\mu$ dominates over $\pi^- \to e^- \bar{\nu}_e$. The antineutrino being (effectively) massless must be right-handed, with therefore polarisation parallel to its motion. The initial state pion has zero spin; note too that orbital angular momentum cannot play a rôle. The two leptons emerge back-to-back and so angular-momentum conservation requires the charged lepton to be similarly polarised parallel to its motion; since this is a weak decay, it should however be *left*-handed, i.e. with spin antiparallel. From the foregoing, we have seen though that the parallel polarisation does, indeed, contain an amount of the 'opposite' helicity or (handedness) proportional to the mass; on squaring the amplitude, we thus find a decay rate proportional to the final-state charged-lepton mass squared. We would thus expect

$$\frac{\Gamma_{\pi^- \to e^- \bar{\nu}_e}}{\Gamma_{\pi^- \to \mu^- \bar{\nu}_\mu}} \simeq \frac{m_e^2}{m_\mu^2} \times \text{phase-space factors.}$$

Allowing for the difference in available phase-space (and including small radiative corrections), the observed electron branching ratio of $\sim 1.2 \times 10^{-4}$ is correctly reproduced. This spin angular-momentum effect is known as helicity suppression.

Exercise 2.4.2. *Show how the transformation properties derived earlier lead to a form* $\gamma^\mu(1 + \gamma_5)$ *for antifermionic currents.*

Answer The $V - A$ current for fermions is simply $\bar{\psi}(V^\mu - A^\mu)\psi$. And so we must now simply apply the charge-conjugation operator

$$\mathcal{C}: \psi(x) \to \psi^{\mathcal{C}}(x) := i\gamma^0\gamma^2\,\bar{\psi}^{\mathsf{T}}(x), \tag{B.3.5}$$

with $\bar{\psi} := \psi^\dagger\gamma^0$. Recall too $\gamma^0\gamma_\mu\gamma^0 = \gamma_\mu^\dagger$ and $\gamma^2\gamma_\mu\gamma^2 = \gamma_\mu^*$, thus

$$\begin{aligned}
\bar{\psi}(V^\mu - A^\mu)\psi \to \bar{\psi}^{\mathcal{C}}(V^\mu - A^\mu)\psi^{\mathcal{C}} &= \overline{i\gamma^2\,\psi^*}\,\gamma^\mu(1 - \gamma_5)\,i\gamma^2\,\psi^* \\
&= (i\gamma^2\,\psi^*)^\dagger\gamma^0\,\gamma^\mu(1 - \gamma_5)\,i\gamma^2\,\psi^* \\
&= (-i)\psi^{\mathsf{T}}(-\gamma^2)\,\gamma^0\,\gamma^\mu(1 - \gamma_5)\,i\gamma^2\,\psi^* \\
&= \psi^{\mathsf{T}}(-\gamma^2)\,\gamma^{\mu\dagger}(1 + \gamma_5)\,(-\gamma^2)\,\gamma^0\psi^* \\
&= \psi^{\mathsf{T}}\gamma^{\mu\mathsf{T}}(1 - \gamma_5)\bar{\psi}^{\mathsf{T}} \\
&= \psi^{\mathsf{T}}(1 + \gamma_5)\gamma^{\mu\mathsf{T}}\bar{\psi}^{\mathsf{T}}.
\end{aligned}$$

We may now simply transpose the whole to obtain

$$\bar{\psi}\,\gamma^\mu(1 + \gamma_5)\,\psi = \bar{\psi}(V^\mu + A^\mu)\psi.$$

Exercise 2.6.1. *Naïvely, one might imagine that a similar GIM-like cancellation should also apply to the decay* $K^0 \to \pi^+\pi^-$. *By considering the possible quark diagrams responsible for such a channel, demonstrate that this is not the case.*

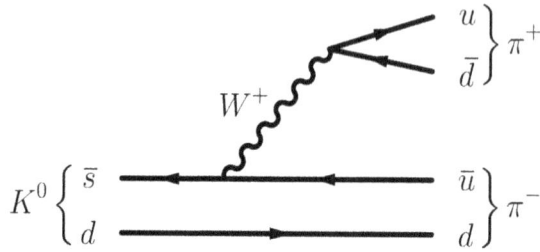

Figure D.1. The Feynman diagram describing $K^0 \rightarrow \pi^+\pi^-$.

Answer The single quark-level weak-interaction Feynman diagram describing neutral-kaon decay into two charged pions is shown in figure D.1. As the final-state quarks are all different, there is just the one single diagram. It is clear therefore that there can be no destructive interference effects.

Exercise 2.7.1. *Ignoring relativistic time-dilation effects, calculate the mean distances that K^0_S and a K^0_L mesons moving at nearly the speed of light will travel before decaying.*

Answer We have $\tau_S \approx 0.9 \times 10^{-10}$ s and $\tau_L \approx 0.5 \times 10^{-7}$ s. Taking $c \approx 3 \times 10^8$ m s^{-1}, we have mean ranges of 2.7 cm and 15 m, respectively.

D.3 Exercises to chapter 3

Exercise 3.2.1. *Using the representation given in equation (3.2.7), evaluate the two coefficients of $\delta m_{1,2}$ in equation (3.2.10) for each of the four independent cases: $B = N$, Λ, Σ and Ξ N.B. The expressions for individual isospin multiplets (e.g. p and n) will turn out to be identical, as isospin is not broken in this construction. Thus, e.g. for N, we should use the average of the proton and neutron masses.*

Answer We must evaluate the traces of the two orderings of the matrices shown in equation (3.2.10). Recall that $\lambda_8 := \frac{1}{\sqrt{3}}\begin{pmatrix} 1 & 0 & 0 \\ 0 & 1 & 0 \\ 0 & 0 & -2 \end{pmatrix}$, but let us also introduce $\lambda_s := \begin{pmatrix} 0 & 0 & 0 \\ 0 & 0 & 0 \\ 0 & 0 & 1 \end{pmatrix}$.

Denoting the resulting coefficients of $\delta m_{1,2}$ as $c_{1,2}$, respectively, for the two SU(3)-breaking choices of λ_8 and λ_s, these are:

	N		Λ		Σ		Ξ	
	c_1	c_2	c_1	c_2	c_1	c_2	c_1	c_2
λ_8	-2	1	-1	-1	1	1	1	-2
λ_s	1	0	$2/3$	$2/3$	0	0	0	1

Note though that since $3\lambda_s + \sqrt{3}\,\lambda_8 = 1$, a change from λ_s to λ_8 alters (all) the individual mass-parameter values but not the mass relations.

Exercise 3.2.2. *Use the coefficients derived in the previous exercise to verify the above mass relation.*

Answer We have $m_B = m_0 + c_1\,\delta m_1 + c_2\,\delta m_2$. That is, using λ_s for the breaking,

$$m_N = m_0 + \delta m_1,$$
$$m_\Lambda = m_0 + \tfrac{2}{3}\delta m_1 + \tfrac{2}{3}\delta m_2,$$
$$m_\Sigma = m_0,$$
$$m_\Xi = m_0 + \delta m_2.$$

We find therefore that

$$m_N + m_\Xi - \tfrac{3}{2}m_\Lambda = \tfrac{1}{2}m_\Sigma,$$

or

$$3m_\Lambda + m_\Sigma = 2m_N + 2m_\Xi. \tag{3.2.11}$$

Repeating the exercise with λ_8, there is no change in the final formula, as predicted in the previous exercise.

Exercise 3.2.3. *Introduce SU(2) breaking as described in equation (3.2.12) and thus add two new coefficients, say $\delta m_{3,4}$. Calculate the four coefficients of δm_{1-4} and determine the parameters using the five independent cases of say p, n, Λ^0 and $\Xi^{0,-}$. Using the values thus found, derive equation (3.2.13) and 'predict' the $\Sigma^{0,\pm}$ masses.*

Answer Again, we must evaluate the traces of the two orderings of the matrices shown in equation (3.2.10). Denoting the coefficients of the new parameters $\delta m_{3,4}$ as $c_{3,4}$, respectively, for the SU(2)-breaking choice of λ_3, these are:

	n		p		Λ^0		Σ^0		Σ^-		Σ^+		Ξ^-		Ξ^0	
	c_3	c_4	c_3	c_4	c_3	c_4	c_3	c_4	c_3	c_4	c_3	c_4	c_3	c_4	c_3	c_4
λ_3		-1		1					1	-1	-1	1	1			-1

The calculation already performed for the SU(3) breaking still stands and so we need only evaluate $\delta m_{3,\,4}$, which are simply

$$\delta m_3 = \tfrac{1}{2}\big[m_{\Xi^-} - m_{\Xi^0}\big] \text{ and } \delta m_4 = \tfrac{1}{2}\big[m_p - m_n\big].$$

From the previous exercise, we already have $m_{\Sigma^0} = 2[m_N + m_\Xi] - 3m_{\Lambda^0}$ (note that SU(2) breaking plays no rôle there). And so now

$$m_{\Sigma^\pm} = m_{\Sigma^0} \pm [\delta m_4 - \delta m_3] = m_{\Sigma^0} \pm \tfrac{1}{2}\big[(m_p - m_n) + (m_{\Xi^0} - m_{\Xi^-})\big].$$

Finally, equation (3.2.13) is obtained from the difference: $m_{\Sigma^+} - m_{\Sigma^-}$.

$$(m_p - m_n) - (m_{\Sigma^+} - m_{\Sigma^-}) + (m_{\Xi^0} - m_{\Xi^-}) = 0. \tag{3.2.13}$$

Exercise 3.2.4. *As a final check, insert the known values into the previous mass formulæ (3.2.13) and examine how closely they are actually satisfied.*

Answer From the previous exercise, we have

$$[m_{\Sigma^+} - m_{\Sigma^-}] = 2[\delta m_4 - \delta m_3] = [m_p - m_n] + [m_{\Xi^0} - m_{\Xi^-}].$$

Gathering all to one side:

$$0 = [m_p - m_n] - [m_{\Sigma^+} - m_{\Sigma^-}] + [m_{\Xi^0} - m_{\Xi^-}]$$
$$= [-1.29] - [-8.07] + [-6.85] = -0.07 \text{ MeV}.$$

Exercise 3.2.5. *By considering the effect of the symmetry operations $\mathcal{P}$ and $\mathcal{C}$, show that the $L = 0$, $S = 0$ mesons must have $J^{PC} = 0^{-+}$, whereas the vector mesons with $L = 0$, $S = 1$ have $J^{PC} = 1^{--}$.*

Answer Recall first (see appendix B.3.6) that the parity of a spin-half fermion–antifermion pair is $(-)^{L+1}$, while the charge-conjugation signature is $(-)^{L+S}$. Therefore, with $L = 0$ and $S = 0$, we clearly have $J = 0$; $P = (-)^1 = -$ and $C = (-)^0 = +$. Instead, for $L = 0$ and $S = 1$, we must have $J = 1$; $P = (-)^1 = -$ and $C = (-)^1 = -$.

Exercise 3.2.6. *Derive the wave-functions for the entire baryon octet by applying the I-, U- and V-spin raising and lowering operators to the proton wave-function derived above.*

Answer To obtain the neutron wave-function, we should apply an isospin lowering operator to the above proton wave-function. This simply has the effect of changing a u quark into a d quark. For simplicity, let us take equation (3.2.19) and then reorder the u and d quarks at the end. Changing each u quark individually into a d quark leads to

$$|n^\uparrow\rangle = \sqrt{\tfrac{2}{3}}\, |d^\uparrow u^\uparrow d^\downarrow\rangle - \sqrt{\tfrac{1}{6}}\left[|d^\uparrow u^\downarrow d^\uparrow\rangle + |d^\downarrow u^\uparrow d^\uparrow\rangle\right]$$
$$+ \sqrt{\tfrac{2}{3}}\, |u^\uparrow d^\uparrow d^\downarrow\rangle - \sqrt{\tfrac{1}{6}}\left[|u^\uparrow d^\downarrow d^\uparrow\rangle + |u^\downarrow d^\uparrow d^\uparrow\rangle\right].$$

We now need only rearrange this by bringing the u quark to, say, the end position and symmetrising on the $d^\uparrow d^\downarrow$ pair to obtain the (obvious) neutron wave-function.

By changing the d quark in the proton into an s quark we obtain the Σ^+; likewise, changing the u quark in the neutron into an s quark produces the Σ^-. And from these it should now be obvious how to generate the $\Xi^{0,\,-}$. Finally, Σ^0 may be generated in a similar fashion, while the SU(3) and isospin singlet Λ^0 is obtained by constructing the state orthogonal to Σ^0.

Exercise 3.2.7. *Derive the magnetic moments of the entire baryon octet in terms of quark magnetic moments from the wave-functions obtained in the previous exercise.*

Answer We can obtain the other magnetic moments by simply exchanging the labels u, d and s in expression (3.2.21):

$$\mu_{\Lambda^0} = \mu_s,$$
$$\mu_{\Sigma^+} = \tfrac{4}{3}\mu_u - \tfrac{1}{3}\mu_s, \quad \mu_{\Sigma^0} = \tfrac{2}{3}\mu_u + \tfrac{2}{3}\mu_d - \tfrac{1}{3}\mu_s, \quad \mu_{\Sigma^-} = \tfrac{4}{3}\mu_d - \tfrac{1}{3}\mu_s,$$
$$\mu_{\Xi^0} = \tfrac{4}{3}\mu_s - \tfrac{1}{3}\mu_u, \quad \mu_{\Xi^-} = \tfrac{4}{3}\mu_s - \tfrac{1}{3}\mu_d.$$

Note that the Λ^0 contains a spin-singlet ud quark pair and therefor its spin and magnetic moment are precisely those of the s quark.

Exercise 3.2.8. *Using the result of the previous exercise, calculate the numerical values of the magnetic moments of the baryon octet and compare them with their experimental values.*

Answer Using the quark magnetic moments obtained from the input values of the proton, neutron and Λ^0 magnetic moments, we obtain the following values (in units of μ_N) for the other hyperons

	th.	expt.
Σ^+	2.67	2.46
Σ^0	0.79	na
Σ^-	-1.09	-1.16
Ξ^0	-1.43	-1.25
Ξ^-	-0.49	-0.65

Exercise 3.3.1. *Show that the variable ν, defined in equation (3.3.2), can be defined in a Lorentz-invariant manner.*

Hint: consider the scalar four-product $p \cdot q$ in the laboratory frame.

Answer Since $p \cdot q = M\nu$ evaluated in the laboratory frame, we have $\nu = p \cdot q / M$. The right-hand side is a ratio of two invariants and thus too is the left-hand side. Note, however, that it corresponds to the photon energy only in the target rest frame.

Exercise 3.3.2. *Calculate the maximum possible resolution (i.e. the shortest photon wavelength) obtainable using an electron beam of energy 4.879 GeV. For the same beam energy, calculate the maximum value of W.*

Answer We may take $|\boldsymbol{p}| = E$; the maximum photon momentum is just twice this and so the minimum de Broglie wavelength of the photon $\lambda \simeq h/2E$. Using $\hbar c \sim 200$ MeV $\cdot$ fm, we have $\lambda \simeq 0.02$ fm.

The (theoretical) maximum value of W is obtained for $E' = 0$, for which we have $W^2 = M^2 + 2ME$; that is, $W \simeq 3.17$ GeV.

Exercise 3.3.3. *Derive the above relation explicitly and thus demonstrate that the adimensional Bjorken scaling variable x_B, as defined in equation (3.3.13), is bounded to lie in the range [0, 1].*

Answer In the Breit frame there is no energy transferred to the quark, which simply reverses its direction of motion. We therefore have $q^\mu = (0, 0, 0, q_z)$, where q_z will be negative. In such an infinite-momentum frame we may neglect the quark mass and thus write $k^\mu = (E, 0, 0, k_z)$ for the initial quark four-momentum, with $E = |k_z|$. Since the quark reverses its motion, we must have $q_z = -2k_z$. Therefore,

$$x_B := \frac{Q^2}{2p \cdot q} = \frac{q_z^2}{2p_z q_z} = \frac{(2k_z)^2}{2p_z(2k_z)} = \frac{k_z}{p_z},$$

which is just equation (3.3.21).

Now, a single quark may clearly not carry more energy than the parent proton and so the maximum value of k_z must be p_z; i.e. $x_B \leqslant 1$. This also means that no quark may move backwards in this frame as the other quarks would necessarily then have total of k_z greater than p_z; i.e. $x_B \geqslant 0$. That is, $x_B \in [0, 1]$.

Exercise 3.3.4. *Repeat the above exercise (again explicitly) for a general infinite-momentum frame (i.e. one in which the proton is not at rest and has large momentum) and thus demonstrate that the relation is of more general validity.*

Answer In a general (infinite-momentum) reference frame (but with $p \neq 0$ or rather $|p| \gg M$), we should write

$$q^\mu = (\nu, 0, 0, q_z), \quad p^\mu = (E_p, 0, 0, p_z) \text{ and } k^\mu = (E_k, 0, 0, k_z).$$

The asymptotic or relativistic limit $|p| \gg M$ allows us to take $E_p \approx p_z$ and $E_k \approx k_z$; i.e. $p^2 = 0$ and $k^2 = 0$.

Feynman's implicit assumption is that the struck quark (initially) emerges on-shell and so $(k + q)^2 = 0$, which together with $k^2 = 0$ implies $2k \cdot q = -q^2 = Q^2$. Now,

$$k \cdot q = E_k \nu - k_z |q| = k_z(\nu - |q|) \text{ and } p \cdot q = E_p \nu - p_z |q| = p_z(\nu - |q|).$$

Thus,

$$x := \frac{Q^2}{2p \cdot q} = \frac{2k \cdot q}{2p \cdot q} = \frac{k_z}{p_z}.$$

Exercise 3.4.1. *Invert equation (3.4.19) to find the value of Q^2 at the Landau pole in QED. How does this compare, e.g. with the Planck mass?*

Answer From equation (3.4.19a), the Landau pole occurs for $|b_0|\alpha_0 t = 1$. We take $\alpha_0 = 1/137$ at a low energy scale, which should be taken as approximately the electron mass. Now, $b_0^{\text{QED}} = -\frac{1}{3\pi}\sum_f N_{cf} Q_f^2$, where $N_{cf} = 3$ for quarks and 1 for leptons. Summing over the six known quarks and the three known charged leptons gives $b_0^{\text{QED}} = -\frac{8}{3\pi}$. We therefore have

$$Q_{\text{Landau}}^2 \approx m_e^2 \exp\left[1/|b_0|\alpha_0\right] = m_e^2 \exp\left[137 \times 3\pi/8\right].$$

This leads to $Q_{\text{Landau}} \approx 3 \times 10^{66}$ GeV. A much higher value is often quoted, but note that this regards pure quantum electrodynamics with only the electron as a charged matter field, thus removing the factor 8 in the above formula. In any case, it is much higher than the Planck mass: $m_P \sim 10^{19}$ GeV.

Exercise 3.5.1. *Show that the total quark momentum fraction may be obtained directly from the following simple combination of deeply inelastic scattering structure function integrals (as measured below the charm threshold):*

$$\int_0^1 dx\left[\frac{9}{2}F_2^{ep+en} - \frac{3}{4}F_2^{\nu p+\nu n}\right]$$

Answer We are required to show that

$$\int_0^1 dx\left[\frac{9}{2}F_2^{ep+en} - \frac{3}{4}F_2^{\nu p+\nu n}\right] = \int_0^1 dx\, x\left[u(x) + \bar{u}(x) + d(x) + \bar{d}(x) + s(x) + \bar{s}(x)\right].$$

From equations (3.5.2) and (3.5.11), we have

$$F_2^{ep+en}(x) = \frac{5}{9}x\left[u(x) + \bar{u}(x) + d(x) + \bar{d}(x)\right] + \frac{2}{9}x\left[s(x) + \bar{s}(x)\right]$$

and

$$F_2^{\nu p+\nu n}(x) = 2x\left[u(x) + \bar{u}(x) + d(x) + \bar{d}(x)\right],$$

from which the required answer follows immediately.

D.4 Exercises to chapter 4

Exercise 4.1.1. *Were the γ-ray hypothesis true, the production mechanism via scattering or rather absorption of α-particles by beryllium would be*

$$^9_4\text{Be} + \alpha \rightarrow {}^{13}_6\text{C} + \gamma.$$

Show that, for α-particles of $\sim$5.3 MeV (as emitted by polonium), the γ-rays emitted in the above interaction would have an energy of at most $\sim$14 MeV.
Useful data: $m_\alpha = 4.001\,506\,u$, $m_{^9\text{Be}} = 9.012\,183\,u$ and $m_{^{13}\text{C}} = 13.003\,355\,u$.

N.B. For beryllium and carbon, these are atomic masses and so care should be taken to correctly account for the electrons; recall that the α-particle is a fully ionised helium atom, or isolated helium nucleus.

Show that to knock a proton out of the target material with the magnitude of final kinetic energy quoted requires either:

(a) *an extremely energetic photon, or*

(b) *another particle of similar mass (if the projectile kinetic energy is not to be extremely high).*

In both cases, calculate the projectile energy required to produce a proton of 5.3 MeV kinetic energy.

Hint: non-relativistic kinematics may be used for the proton and neutron; assume a Compton-like process for a photon.

Finally, what is the correct reaction to consider?

Answer The interaction ${}^{9}_{4}\text{Be}$ (α, γ) ${}^{13}_{6}\text{C}$ results in an energy release given by the nuclear mass difference. The carbon recoil will absorb some energy; for precision, we should therefore perform the kinematics with care. However, we shall see that a rough calculation is quite sufficient. The configuration that provides the most energetic photons is perfectly forward scattering; we may thus write

$$\boldsymbol{p}_\alpha = \boldsymbol{p}_\text{C} + \boldsymbol{k} \qquad \text{(momentum conservation)}$$

and

$$\frac{\boldsymbol{p}_\alpha^2}{2m_\alpha} + \Delta E = \frac{\boldsymbol{p}_\text{C}^2}{2m_\text{C}'} + k \qquad \text{(energy conservation)},$$

where ΔE is the energy release, $k = |\boldsymbol{k}|$ is the photon energy and m_C' is the mass of the carbon atom with two electrons missing (the α-particle brings two protons *without* the corresponding electrons). As input parameters, we thus need:

$$\begin{aligned}
\Delta E &= m_\alpha + (m_{{}^{9}Be} - 4m_e) - (m_{{}^{13}C} - 6m_e) \\
&= 931.494 \times (4.001\,506 + 9.012\,183 - 13.003\,355) + 2 \times 0.511 \\
&= 10.6 \text{ MeV}
\end{aligned}$$

and

$$\frac{\boldsymbol{p}_\alpha^2}{2m_\alpha} = 5.3 \text{ MeV}$$

Eliminating $\boldsymbol{p}_\text{C}$ above we have

$$\frac{\boldsymbol{p}_\alpha^2}{2m_\alpha} + \Delta E = \frac{(\boldsymbol{p}_\alpha - \boldsymbol{k})^2}{2m_\text{C}'} + k$$

The left-hand side is nearly 16 MeV; so, considering that the first term on the right-hand side must be positive, we immediately have that $k \lesssim 16$ MeV. Case (*a*)—a photon projectile Let k (k') be the initial (final) photon momenta and p that of the emitted proton (initially at rest with mass M). We then have

$$k = k' + p \quad \text{(momentum conservation)}$$

and

$$|k| = |k'| + p^2/2M \quad \text{(energy conservation)}.$$

Eliminating k' gives

$$|k| = |k - p| + p^2/2M.$$

Define the final proton kinetic energy $\varepsilon := p^2/2M$; then

$$(k - \varepsilon)^2 = k^2 - 2k\sqrt{2M\varepsilon}\,\cos\theta + 2M\varepsilon,$$

giving

$$k = \frac{M - \frac{1}{2}\varepsilon}{\sqrt{2M/\varepsilon}\,\cos\theta - 1}.$$

To minimise k for a given ε, we take $\cos\theta = 1$ (i.e. forward scattering) and using $\varepsilon \ll M$, we then have $k \simeq \sqrt{M\varepsilon/2}$. Taking $M \approx 1000$ MeV and $\varepsilon = 5$ MeV (as described in the text), we obtain $k \simeq 50$ MeV.

Case (*b*)—a neutron projectile For simplicity we assume an object with the same mass M. Using the same notation as above,

$$k = k' + p \quad \text{(momentum conservation)}$$

and

$$k^2/2M = k'^2/2M + p^2/2M \quad \text{(energy conservation)}.$$

The solution here is well known (again minimising k): $k = p$. That is, the kinetic energy required for a neutron is just that of the ejected proton, which is a far more acceptable value.

Finally, the correct reaction to consider is clearly ${}^{9}_{4}\text{Be} + \alpha \rightarrow {}^{12}_{6}\text{C} + n$.

Exercise 4.1.2. *Calculate the laboratory beam-energy threshold for the production of an antiproton in a fixed-target proton–proton collision.*

Answer To produce an antiproton, we must also produce another proton and therefore while the initial state consists of two protons, the final state contains three protons and one antiproton. In the centre-of-mass system at threshold all of the final-state particles will be at rest. Indicating the initial (final) four-momenta with $p_{1,2}$ ($k_{1,...,4}$), we have

$$(p_1 + p_2)^2 = (k_1 + k_2 + k_3 + k_4)^2.$$

Since it is an invariant equation, we may evaluate each side in any reference frame. Evaluating the right-hand side in the centre-of-mass system at threshold, we obtain

$$(k_1 + k_2 + k_3 + k_4)^2 = 16m_p^2.$$

The left-hand side may be evaluated in the laboratory frame, where the target proton is at rest; we thus have

$$(p_1 + p_2)^2 = p_1^2 + 2p_1 \cdot p_2 + p_2^2 = m_p^2 + 2E_{\min}m_p + m_p^2.$$

Putting this all together we finally obtain

$$E_{\min} = 7m_p.$$

Exercise 4.1.3. *Assuming that the tau neutrino collides with a single nucleon in the detector, draw the Feynman diagram describing the interaction. For an initial low-energy $\bar{\nu}_\tau$, deduce whether the target nucleon must be a proton or neutron; motivate the answer. Finally, calculate the threshold for such a charge-exchange process.*

Answer The Feynman diagram describing the interaction is shown in figure D.2. For an initial low-energy $\bar{\nu}_\tau$, assuming that it has insufficient energy to excite, e.g. a charged resonance, only a proton is possible, as the interaction requires a negative charge exchange towards the nucleon; i.e. $\bar{\nu}_\tau p \to n\tau^+$. For higher energies and possible resonance excitation or DIS, both proton and neutron are allowed.

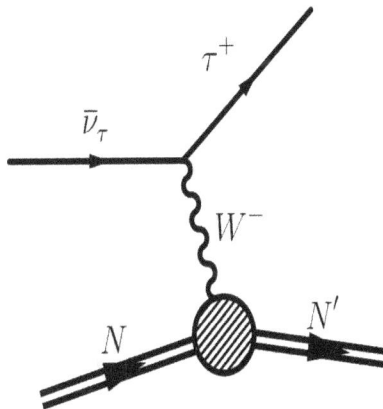

Figure D.2. The Feynman diagram describing $\bar{\nu}_\tau N$ scattering.

The threshold is given by

$$(k_\nu + p_p)^2 = (k_\tau + p_n)^2,$$

with the right-hand side calculated using particles at rest in their centre-of-mass frame. We thus have (neglecting the neutrino mass)

$$2E_\nu m_p + m_p^2 = (m_\tau + m_n)^2,$$

or

$$E_\nu = \frac{(m_\tau + m_n)^2 - m_p^2}{2m_p} = \frac{(1777 + 940)^2 - 938^2}{2 \times 938} = 3466 \text{ MeV}.$$

Exercise 4.2.1. *Show that, for a charged-pion projectile incident on a fixed proton target, the resonance peak is attained for a pion kinetic energy a little below* 200 MeV. *Calculate the centre-of-mass pion energy.*

Answer At resonance, $\Delta(1230)$, we have

$$(p_\pi + p_p)^2 = p_\Delta^2 = m_\Delta^2.$$

In the laboratory frame (the proton rest frame)

$$(p_\pi + p_p)^2 = m_\pi^2 + 2m_p E_\pi + m_p^2,$$

where $E_\pi = m_\pi + E_{\text{kin}}$. We therefore have

$$m_\pi^2 + 2m_p E_\pi + m_p^2 = m_\Delta^2,$$

or

$$E_{\text{kin}} = \frac{m_\Delta^2 - m_p^2 - m_\pi^2}{2m_p} - m_\pi = \frac{1230^2 - (938 + 140)^2}{2 \times 938} = 187 \text{ MeV}.$$

Exercise 4.2.2. *Explain why the ρ^0 can decay only into $\pi^+\pi^-$ and not into $\pi^0\pi^0$.*

Answer In the case of a neutral-pion pair, the final state consists of two identical bosons and must thus be symmetric with respect to interchange. However, the state must also have one unit of orbital angular momentum (since $J_\rho^P = 1^-$ and $J_\pi^P = 0^-$) and is thus antisymmetric. The decay $\rho^0 \to \pi^0\pi^0$ is therefore forbidden.

N.B. It would also violate isospin: the initial state is $|1, 0\rangle$ and the final state $|1, 0\rangle \otimes |1, 0\rangle$, whereas the Clebsch–Gordan coefficients for the relevant decomposition are non-zero only for $|0, 0\rangle$ and $|2, 0\rangle$.

The decay into $\pi^+\pi^-$ may be analysed as isospin states; the Clebsch–Gordan coefficients here lead to the following decomposition:

$$|1, 0\rangle = \tfrac{1}{\sqrt{2}}\left[|1, +1\rangle|1, -1\rangle - |1, -1\rangle|1, +1\rangle\right].$$

This is antisymmetric, cancelling the antisymmetry of the p-wave. The charged decay also satisfies parity conservation in strong interactions: $P_\rho = -1$ and the final state has $P = P_\pi^2\,(-1)^L = -1$, as $L = 1$.

N.B. The decay $\rho^0 \to \pi^0\pi^0\gamma$ is, however, permitted, as the pions may now be in a relative s-wave, with the photon spin conserving the overall angular momentum and parity.

Exercise 4.2.3. *Explain why the ω^0 can decay only into $\pi^+\pi^-\pi^0$ and not into $\pi^0\pi^0\pi^0$.*

Answer The ω^0 has I^G, $J^{PC} = 0^-$, 1^{--} and the neutral pion I^G, $J^{PC} = 1^-$, 0^{-+}. The charge-parity of the initial state is thus -1, whereas $C_{\pi^0} = +1$. In the case of $\pi^0\pi^0\pi^0$, the final state consists of three identical bosons and must therefore be symmetric with respect to interchange of any pair. However, the state must also have one unit of orbital angular momentum (since $J_{\omega^0} = 1$), which is antisymmetric for some pair. The decay $\omega^0 \to \pi^0\pi^0\pi^0$ is therefore forbidden.

Exercise 4.2.4. *Deduce the G-parities of $\rho^0(770)$ and $\omega^0(782)$.*

Answer The first has I^G, $J^{PC} = 1^+$, 1^{--}, the second I^G, $J^{PC} = 0^-$, 1^{--}.

Experimentally, we know that $\rho^0(770)$ decays into two pions and $\omega^0(782)$ into three. We therefore deduce $G_\rho = +$ and $G_\omega = -$. Theoretically, we have the assignment given earlier:

$$\mathcal{G}|N, \bar{N}\rangle = (-1)^{L+S+I}|N, \bar{N}\rangle. \tag{B.4.17}$$

which (given $S = 1$ and $L = 0$ for both) agrees with the experimental deductions.

Exercise 4.2.5. *Show that the extreme values of sphericity and spherocity are attained for perfectly aligned two-jet events ($S = S_0 = 0$), and completely isotropic events ($S = S_0 = 1$).*

Answer Sphericity is defined by

$$S := \min_{\hat{n}} \frac{3\sum_i p_{\mathrm{T}i}^2}{2\sum_i p_i^2}, \tag{4.2.32}$$

where $p_{\mathrm{T}i}$ is defined with respect to $\hat{n}$. Perfectly aligned two-jet events will have $p_{\mathrm{T}i} = 0 \ \forall \ i$, for $\hat{n}$ along the jet axis and thus $S = 0$. For a perfectly spherical event, we may (by symmetry) choose $\hat{n}$ along any axis. Thus, $|p_{\mathrm{T}i}| = \sin\theta_i|p_i|$ with $\cos\theta_i$ distributed uniformly over $[-1, +1]$. Transforming the sum into an integral and assuming θ_i and $|p_i|$ uncorrelated, we have

$$S = \tfrac{3}{2}\int_{-1}^{+1} \frac{\mathrm{d}\cos\theta}{2}\sin^2\theta = \tfrac{3}{2}\times\tfrac{2}{3} = 1.$$

Spherocity is defined by

$$S_0 := \frac{\pi^2}{4}\min_{\hat{n}_T}\left(\frac{\sum_i|p_{T,i\wedge}\hat{n}_T|}{\sum_i|p_{T,i}|}\right)^2, \tag{4.2.33}$$

where $\hat{n}_T$ is a unit transverse vector and $p_{T,i}$ is the transverse component of the ith particle with respect to the beam axis. Perfectly aligned two-jet events will have $p_{\mathrm{T}i}$ parallel to some unique axis $\hat{n}_T \ \forall \ i$ and thus $S_0 = 0$. For a perfectly spherical event, we may (by symmetry) choose $\hat{n}_T$ along any axis perpendicular to the beam. Transforming the sum into an integral, for the denominator inside the parenthesis we have

$$\sum_i|p_{T,i}| = \langle p\rangle\int_{-1}^{+1} \frac{\mathrm{d}\cos\theta}{2}|\sin\theta|.$$

The numerator is

$$\sum_i|p_{T,i}{}^{\wedge}\hat{n}_T| = \langle p\rangle\int_{-1}^{+1} \frac{\mathrm{d}\cos\theta}{2}|\sin\theta|\int_0^{2\pi}\frac{\mathrm{d}\phi}{2\pi}|\sin\phi|.$$

Therefore

$$S_0 = \frac{\pi^2}{4}\left(\int_0^{2\pi}\frac{\mathrm{d}\phi}{2\pi}|\sin\phi|\right)^2 = \frac{\pi^2}{4}\left(4\int_0^{\pi/2}\frac{\mathrm{d}\phi}{2\pi}\sin\phi\right)^2 = 1.$$

Exercise 4.2.6. *Demonstrate that the two extreme values of thrust are attained for precisely aligned two-jet events $(T = 1)$ and completely isotropic events $(T = \frac{1}{2})$.*

Answer Thrust is defined by

$$T := \max_{\hat{n}} \frac{\sum_i |\boldsymbol{p}_i \cdot \hat{n}|}{\sum_i |\boldsymbol{p}_i|}, \tag{4.2.34}$$

where now $\hat{n}$ maximises the variable T. For a perfectly aligned two-jet events, taking $\hat{n}$ along the jet axis clearly gives $T = 1$. For a perfectly spherical event, we may (by symmetry) choose $\hat{n}$ along the beam axis. The numerator is

$$\sum_i |\boldsymbol{p}_i \cdot \hat{n}| = \sum_i |\boldsymbol{p}_i| \int_{-1}^{+1} \frac{\mathrm{d}\cos\theta}{2} |\cos\theta| = \tfrac{1}{2} \sum_i |\boldsymbol{p}_i|.$$

Therefore, $T = \frac{1}{2}$ in this case.

Exercise 4.2.7. *Show that in the case of a perfectly symmetric three-jet event, the thrust value is $T = 2/3$.*

Answer For a perfectly symmetric three-jet event, we may take $\hat{n}$ along any one of the jet axes. In this case we have

$$T = \tfrac{1}{3} + \tfrac{2}{3} \sin\tfrac{\pi}{6} = \tfrac{2}{3}.$$

Exercise 4.3.1. *Show that the true centre-of-mass energy available in a parton–parton collision is given by $E_{\mathrm{CM}}^{\mathrm{parton}} = \sqrt{x_1 x_2}\, E_{\mathrm{CM}}^{\mathrm{hadron}}$, where $x_{1,2}$ are the two colliding parton momentum fractions.*

Answer We have

$$E_{\mathrm{CM}}^{\mathrm{parton}} = \sqrt{s^{\mathrm{parton}}} = \sqrt{(k_1 + k_2)^2} \simeq \sqrt{(x_1 p_1 + x_2 p_2)^2} = \sqrt{x_1 x_2}\, E_{\mathrm{CM}}^{\mathrm{hadron}},$$

where we have neglected the hadron (and parton) masses.

D.5 Exercises to chapter 5

Exercise 5.1.1. *For the given kinematics (*i.e. *a decaying charged-pion beam of energy 200 GeV) and taking into account the maximum angle subtended by a 2 m-wide* *detector, calculate the energy spread of the resulting neutrino beam that actually struck Gargamelle (use $m_\pi = 140$ MeV and $m_\mu = 106$ MeV).*

Answer For the decay $\pi^+ \to \mu^+ \nu_\mu$ (charged pions with energy 200 GeV), the *absolute* maximum and minimum neutrino energies will be given when both products move parallel to the original pion beam. We may evaluate (neglecting the neutrino mass)

$$p_\mu^2 = (p_\pi - p_\nu)^2,$$

which gives

$$m_\mu^2 = m_\pi^2 - 2E_\nu(E_\pi \pm |p_\pi|).$$

Note that the pion is highly relativistic and so $|p_\pi| \simeq E_\pi$. Therefore,

$$E_\nu = \frac{m_\pi^2 - m_\mu^2}{2(E_\pi \pm |p_\pi|)} = \frac{m_\pi^2 - m_\mu^2}{2(E_\pi + |p_\pi|)} \quad \text{or} \quad \frac{m_\pi^2 - m_\mu^2}{2m_\pi^2}(E_\pi + |p_\pi|).$$

For such high-energy pions ($E_\pi \gg m_\pi$), we may now approximate $E_\pi + |p_\pi| \approx 2E_\pi$ and write

$$E_\nu \approx \frac{m_\pi^2 - m_\mu^2}{4E_\pi} \quad \text{or} \quad \frac{m_\pi^2 - m_\mu^2}{m_\pi^2} E_\pi = 0.01 \text{ MeV or } 85 \text{ GeV}.$$

N.B. The lower energy corresponds to a *back-scattered neutrino*.

Here though the detector has an acceptance of 2 m and therefore we should limit the scattering angle. The maximum angle will be given by the minimum decay–detector distance, which is 750 m (see figure 5.2). The maximum scattering angle is therefore $\theta = 1/750 \simeq 1.3$ mrad. We now need to solve

$$m_\mu^2 = m_\pi^2 - 2E_\nu(E_\pi \pm \cos\theta|p_\pi|),$$

which leads to

$$E_\nu^{\min} = \frac{m_\pi^2 - m_\mu^2}{2(E_\pi - \cos\theta|p_\pi|)} \simeq \frac{m_\pi^2 - m_\mu^2}{(m_\pi^2 + \sin^2\theta E_\pi^2)} \frac{(1 + \cos\theta)}{2} E_\pi.$$

Inserting the values given, we obtain $E_\nu^{\min} = 18.4$ MeV.

Exercise 5.1.2. *Consider the possible decays of the $\bar{B}^0$ (a $b\bar{d}$ meson) via suitable penguin-like diagrams. In particular, construct the most probable diagrams for:*

$$\bar{B}^0 \to \gamma\gamma \quad \text{and} \quad \bar{B}^0 \to \ell^+\ell^-.$$

Answer The relevant penguins diagrams are shown in figure D.3.

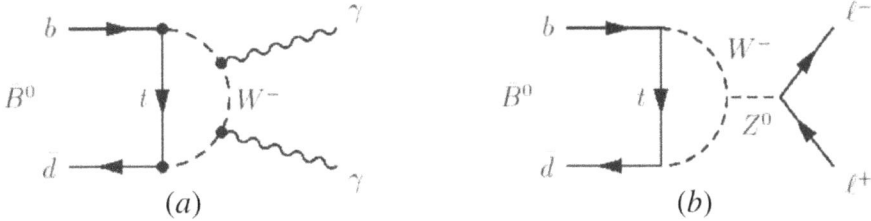

Figure D.3. So-called vertical or annihilation penguins: (a) $B \to \gamma\gamma$ and (b) $B \to \ell^+\ell^-$.

Exercise 5.1.3. *Considering the penguin diagram shown in figure 5.9, identify the cases (and the corresponding diagrams) for which the following processes may proceed without penguin contributions and those for which they are necessary:*

$$B \to \pi\pi \quad \text{and} \quad B \to K\pi.$$

Answer The possible tree and penguin-type diagrams are shown in figure D.4.

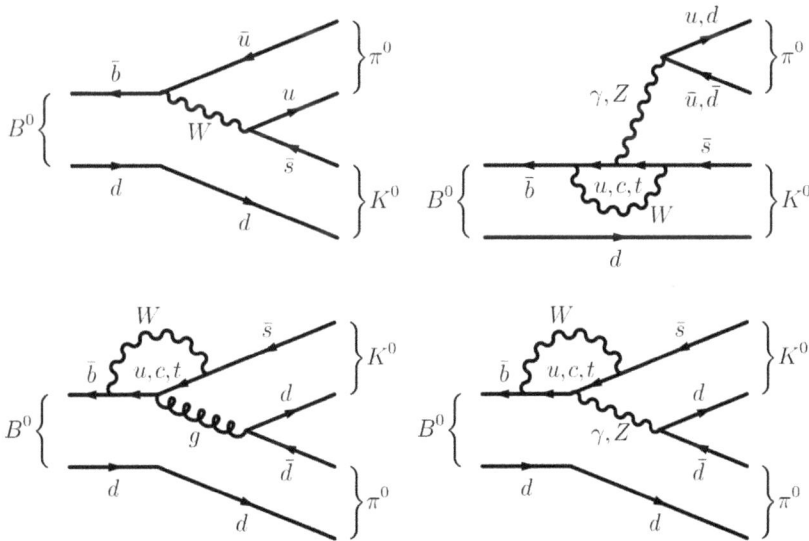

Figure D.4. Possible tree and penguin-type diagrams for $B \to K\pi$; $B \to \pi\pi$ may be obtained by substituting the s with a d.

Exercise 5.2.1. *Neglecting the electron mass (where permissible), show that* $y = \frac{1}{2}(1 + \cos \theta_e)$, *where* θ_e *is the electron centre-of-mass scattering angle with respect to the (anti-)neutrino beam direction. Thus, show that the* $J = 1$ *cross-section given above behaves as* $(1 - \cos \theta_e)^2$.

Answer We have $y := E_e'/E_\nu$ with $E_e' =$ the final-electron energy in the laboratory frame (initial electron at rest). Four-momentum conservation gives

$$(p + k)^2 = (p' + k')^2,$$

where p^μ and p'^μ (k^μ and k'^μ) are the initial and final electron (neutrino) four-momenta. Evaluating the left-hand side in the laboratory frame and right-hand side in the centre-of-mass frame, we have

$$2E_\nu m_e \approx 4\epsilon^2,$$

where ϵ is the centre-of-mass energy of the final-state electron (and/or neutrino). The transverse momentum is invariant and in the centre-of-mass frame we have

$$p_\perp = \epsilon \sin \theta_e.$$

In the laboratory frame we now evaluate

$$(p + k - p')^2 = k'^2,$$

which gives

$$2p \cdot k - 2p \cdot p' - 2k \cdot p' = 0,$$

that is,

$$2(E_\nu - E')m_e - 2E_\nu(E' - p_z') = 0,$$

or

$$p_z' = E' - (1 - y)m_e.$$

This leads to

$$E'^2 - p_\perp^2 = [E' - (1 - y)m_e]^2,$$

or

$$\tfrac{1}{2}E_\nu m_e \sin^2 \theta_e = p_\perp^2 = 2yE_\nu(1 - y)m_e - (1 - y)^2 m_e^2.$$

We may neglect the m_e^2 term and thus obtain

$$\sin^2 \theta_e = 4y(1 - y), \quad \text{or} \quad y = \tfrac{1}{2}(1 + \cos \theta_e).$$

The $J = 1$ cross-section above behaves as $(1 - y)^2 \propto (1 - \cos \theta_e)^2$.

Exercise 5.4.1. *Following similar steps as in the discussion of $K^0 - \bar{K}^0$ oscillation, but using momentum eigenstates, show that the probability for a state initially produced at $t = 0$ as, say, an electron-neutrino to become a muon neutrino at time t is*

$$P(\nu_e \rightarrow \nu_\mu; t) = \sin^2 2\theta \, \sin^2 \left(\tfrac{1}{2}(E_2 - E_1)t \right).$$

Answer The initially electron-neutrino state evolved to time t is

$$|\nu_e, t\rangle = e^{-iE_1 t} \cos\theta |\nu_1\rangle + e^{-iE_2 t} \sin\theta |\nu_2\rangle,$$

where $E_{1,2} = \sqrt{p^2 - m_{1,2}^2}$. We may thus write $E_{1,2} \approx |p|(1 - \tfrac{1}{2}m_{1,2}^2/p^2)$.
Now

$$|\nu_1, \boldsymbol{p}\rangle = \cos\theta \, |\nu_e, \boldsymbol{p}\rangle - \sin\theta \, |\nu_\mu, \boldsymbol{p}\rangle$$

and

$$|\nu_2, \boldsymbol{p}\rangle = \sin\theta \, |\nu_e, \boldsymbol{p}\rangle + \cos\theta \, |\nu_\mu, \boldsymbol{p}\rangle.$$

The evolved state may now be rewritten (suppressing $\boldsymbol{p}$) as

$$|\nu_e, t\rangle = e^{-iE_1 t} \cos\theta \left[\cos\theta |\nu_e\rangle - \sin\theta |\nu_\mu\rangle \right] + e^{-iE_2 t} \sin\theta \left[\sin\theta |\nu_e\rangle + \cos\theta |\nu_\mu\rangle \right]$$

$$= \left[e^{-iE_1 t} \cos^2\theta + e^{-iE_2 t} \sin^2\theta \right] |\nu_e\rangle + \left[e^{-iE_1 t} - e^{-iE_2 t} \right] \sin\theta \cos\theta |\nu_\mu\rangle.$$

The coefficient of $|\nu_\mu\rangle$ is then

$$2i \sin\left(\tfrac{1}{2}(E_1 - E_2)t \right) e^{-iEt} \times \tfrac{1}{2} \sin 2\theta.$$

The squared modulus of this is just the desired result.

D.6 Exercises to chapter 6

Exercise 6.1.1. *Using the above quantities as a basis, make a more precise estimate of the mass of material required for an experimental determination of the rate for neutrinoless double β-decay.*

Answer We need a the number of nuclei $N = n\tau/T$, where n is the number of events required, $\tau = t_{1/2}/\ln 2$ is the mean decay lifetime and $T = 10$ yr is the running time. Then $M = AN/N_A$, where A is the atomic mass of the nucleus adopted. Taking $n = 10$, $T = 10$ yr and $A = 100$, we have

$$M \equiv \frac{10^{27-28}}{\ln 2} \times \frac{100}{6 \cdot 10^{26}} \equiv 240 \text{ to } 2400 \text{ kg}.$$

D.7 Exercises to appendix B

Exercise B.1.1. *Verify that the free-particle plane-wave solutions to the Dirac equation do indeed take the form shown in equation (B.1.26), with the spinor $w(p, s)$ satisfying equation (B.1.27).*

Answer We have

$$\left[i\gamma^\mu \partial_\mu - m\mathbf{1}\right]\psi(x) = 0, \qquad (B.1.13)$$

to which the solutions given are

$$\psi(x) = w(p, s)e^{-i\epsilon \, p \cdot x}, \qquad (B.1.26)$$

with

$$\epsilon \, \gamma^\mu p_\mu \, w(p, s) = m \, w(p, s), \qquad (B.1.27)$$

where $\epsilon = \pm$ indicates a positive- or negative-energy solution. Substituting (B.1.26) for ψ in equation (B.1.13) and using

$$\partial_\mu e^{-i\epsilon \, p \cdot x} = -i\epsilon \, p_\mu e^{-i\epsilon \, p \cdot x},$$

we obtain equation (B.1.27).

Exercise B.1.2. *Verify the anticommutation relation $\{\gamma_5, \gamma^\mu\} = 0$.*

Answer Since each γ-matrix *anti*commutes with the other three and $\gamma_5 := i\gamma^0\gamma^1\gamma^2\gamma^3$ is a product of all four, on commuting any single γ-matrix through the product string in γ_5, we shall have three minus signs and one plus giving an overall minus: $\gamma_5\gamma_\mu = -\gamma_\mu\gamma_5$.

Exercise B.1.3. *Verify that $\bar{\psi} \, \gamma^\mu \psi$ and $\bar{\psi} \, \gamma^\mu \gamma_5 \psi$ do indeed give the momentum p^μ and spin s^μ of the particle, respectively, and find the constants of proportionality.*

Answer We are required to show that

$$\bar{\psi}\,\gamma^\mu\psi \propto p^\mu \quad \text{and} \quad \bar{\psi}\,\gamma^\mu\gamma_5\psi \propto s^\mu.$$

For definiteness, but without loss of generality, let us take the positive energy solution given in equation (B.1.28):

$$u(\boldsymbol{p}) = \begin{pmatrix} \sqrt{E+\mathrm{m}}\ \mathbb{1} \\ \sqrt{E-\mathrm{m}}\ \boldsymbol{\sigma}\cdot\hat{\boldsymbol{p}} \end{pmatrix} \otimes \chi.$$

The temporal dependence is irrelevant here; we thus need to evaluate

$$\bar{u}\,\gamma^\mu\,[\gamma_5]\,u = \chi^\dagger \otimes \left(\sqrt{E+\mathrm{m}}\ \mathbb{1},\ \sqrt{E-\mathrm{m}}\ \boldsymbol{\sigma}^\dagger\cdot\hat{\boldsymbol{p}}\right)\gamma^0\,\gamma^\mu\,[\gamma_5]\begin{pmatrix} \sqrt{E+\mathrm{m}}\ \mathbb{1} \\ \sqrt{E-\mathrm{m}}\ \boldsymbol{\sigma}\cdot\hat{\boldsymbol{p}} \end{pmatrix} \otimes \chi$$

$$= \chi^\dagger \otimes \left(\sqrt{E+\mathrm{m}}\ \mathbb{1},\ -\sqrt{E-\mathrm{m}}\ \boldsymbol{\sigma}\cdot\hat{\boldsymbol{p}}\right)\gamma^\mu\,[\gamma_5]\begin{pmatrix} \sqrt{E+\mathrm{m}}\ \mathbb{1} \\ \sqrt{E-\mathrm{m}}\ \boldsymbol{\sigma}\cdot\hat{\boldsymbol{p}} \end{pmatrix} \otimes \chi.$$

The γ-matrices are given in equations (B.1.11) and (B.1.35) and recall that $\boldsymbol{\sigma}^\dagger = \boldsymbol{\sigma}$, with $\boldsymbol{s} := \chi^\dagger\boldsymbol{\sigma}\chi$.

For the vector current (i.e. with just γ^μ) we thus have, starting with the temporal or zeroth component,

$$\bar{u}\,\gamma^0\,u = \chi^\dagger \otimes \left(\sqrt{E+\mathrm{m}}\ \mathbb{1},\ -\sqrt{E-\mathrm{m}}\ \boldsymbol{\sigma}\cdot\hat{\boldsymbol{p}}\right)\gamma^0\begin{pmatrix} \sqrt{E+\mathrm{m}}\ \mathbb{1} \\ \sqrt{E-\mathrm{m}}\ \boldsymbol{\sigma}\cdot\hat{\boldsymbol{p}} \end{pmatrix} \otimes \chi$$

$$= \chi^\dagger \otimes \left(\sqrt{E+\mathrm{m}}\ \mathbb{1},\ \sqrt{E-\mathrm{m}}\ \boldsymbol{\sigma}\cdot\hat{\boldsymbol{p}}\right)\begin{pmatrix} \sqrt{E+\mathrm{m}}\ \mathbb{1} \\ \sqrt{E-\mathrm{m}}\ \boldsymbol{\sigma}\cdot\hat{\boldsymbol{p}} \end{pmatrix} \otimes \chi$$

$$= \chi^\dagger \otimes \left[(E+\mathrm{m})\,\mathbb{1} + (E-\mathrm{m})\,(\boldsymbol{\sigma}\cdot\hat{\boldsymbol{p}})^2\right] \otimes \chi.$$

Using the Pauli-matrix identity $\sigma^i\sigma^j \equiv \delta^{ij}\mathbb{1} + \mathrm{i}\epsilon^{ijk}\sigma^k$, it is easy to show that $(\boldsymbol{\sigma}\cdot\hat{\boldsymbol{p}})^2 = \mathbb{1}$ and so finally $\bar{u}\,\gamma^0\,u = 2E$. For the spatial components we start from

$$\bar{u}\,\boldsymbol{\gamma}\,u = \chi^\dagger \otimes \left(\sqrt{E+\mathrm{m}}\ \mathbb{1},\ -\sqrt{E-\mathrm{m}}\ \boldsymbol{\sigma}\cdot\hat{\boldsymbol{p}}\right)\begin{bmatrix} 0 & \boldsymbol{\sigma} \\ -\boldsymbol{\sigma} & 0 \end{bmatrix}\begin{pmatrix} \sqrt{E+\mathrm{m}}\ \mathbb{1} \\ \sqrt{E-\mathrm{m}}\ \boldsymbol{\sigma}\cdot\hat{\boldsymbol{p}} \end{pmatrix} \otimes \chi$$

$$= \chi^\dagger \otimes \left(\sqrt{E-\mathrm{m}}\ \boldsymbol{\sigma}\cdot\hat{\boldsymbol{p}}\ \boldsymbol{\sigma},\ \sqrt{E+\mathrm{m}}\ \boldsymbol{\sigma}\ \mathbb{1}\right)\begin{pmatrix} \sqrt{E+\mathrm{m}}\ \mathbb{1} \\ \sqrt{E-\mathrm{m}}\ \boldsymbol{\sigma}\cdot\hat{\boldsymbol{p}} \end{pmatrix} \otimes \chi$$

$$= \sqrt{E^2-\mathrm{m}^2}\ \chi^\dagger \otimes \left[\boldsymbol{\sigma}\cdot\hat{\boldsymbol{p}}\ \boldsymbol{\sigma} + \boldsymbol{\sigma}\ \boldsymbol{\sigma}\cdot\hat{\boldsymbol{p}}\right] \otimes \chi$$

$$= \sqrt{E^2-\mathrm{m}^2}\ 2\hat{\boldsymbol{p}} = 2\boldsymbol{p},$$

where we have used the previous Pauli-matrix identity and $|\boldsymbol{p}| = \sqrt{E^2-\mathrm{m}^2}$. Putting this all together, we have $\bar{u}\,\gamma^\mu\,u = 2p^\mu$.

Next we calculate the axial-vector current, starting again with the temporal component,

$$\bar{u}\,\gamma^0\,\gamma_5\,u = \chi^\dagger \otimes \left(\sqrt{E+m}\;\mathbb{1},\; -\sqrt{E-m}\;\boldsymbol{\sigma}\cdot\hat{\boldsymbol{p}}\right)\gamma^0\,\gamma_5\begin{pmatrix}\sqrt{E+m}\;\mathbb{1}\\ \sqrt{E-m}\;\boldsymbol{\sigma}\cdot\hat{\boldsymbol{p}}\end{pmatrix}\otimes\chi$$

$$= \chi^\dagger \otimes \left(\sqrt{E+m}\;\mathbb{1},\; \sqrt{E-m}\;\boldsymbol{\sigma}\cdot\hat{\boldsymbol{p}}\right)\begin{bmatrix}0 & \mathbb{1}\\ \mathbb{1} & 0\end{bmatrix}\begin{pmatrix}\sqrt{E+m}\;\mathbb{1}\\ \sqrt{E-m}\;\boldsymbol{\sigma}\cdot\hat{\boldsymbol{p}}\end{pmatrix}\otimes\chi$$

$$= \sqrt{E^2-m^2}\;\chi^\dagger \otimes 2\boldsymbol{\sigma}\cdot\hat{\boldsymbol{p}}\otimes\chi = 2\boldsymbol{s}\cdot\boldsymbol{p},$$

This is not what we might have expected! Anyway, let us try the spatial part:

$$\bar{u}\,\boldsymbol{\gamma}\,\gamma_5\,u = \chi^\dagger \otimes (\sqrt{E+m}\;\mathbb{1},\; -\sqrt{E-m}\;\boldsymbol{\sigma}\cdot\hat{\boldsymbol{p}})$$

$$\times\begin{bmatrix}0 & \boldsymbol{\sigma}\\ -\boldsymbol{\sigma} & 0\end{bmatrix}\begin{bmatrix}0 & \mathbb{1}\\ \mathbb{1} & 0\end{bmatrix}\begin{pmatrix}\sqrt{E+m}\;\mathbb{1}\\ \sqrt{E-m}\;\boldsymbol{\sigma}\cdot\hat{\boldsymbol{p}}\end{pmatrix}\otimes\chi$$

$$= \chi^\dagger \otimes \left(\sqrt{E+m}\;\boldsymbol{\sigma}\,\mathbb{1},\; \sqrt{E-m}\;\boldsymbol{\sigma}\cdot\hat{\boldsymbol{p}}\,\boldsymbol{\sigma}\right)\begin{pmatrix}\sqrt{E+m}\;\mathbb{1}\\ \sqrt{E-m}\;\boldsymbol{\sigma}\cdot\hat{\boldsymbol{p}}\end{pmatrix}\otimes\chi$$

$$= \chi^\dagger \otimes \left[(E+m)\,\boldsymbol{\sigma} + (E-m)\,\boldsymbol{\sigma}\cdot\hat{\boldsymbol{p}}\,\boldsymbol{\sigma}\,\boldsymbol{\sigma}\cdot\hat{\boldsymbol{p}}\right]\otimes\chi$$
$$= 2m\boldsymbol{s} + 2\boldsymbol{s}\cdot\boldsymbol{p}\,\boldsymbol{p}/(E+m).$$

This looks a complete mess! So what has gone wrong?

The problem lies in the use of the 2-D spinors χ, which are related to what is, by definition, the rest frame. Indeed, if we set $\boldsymbol{p}=0$ and thus $E=m$, we (almost trivially) find $\bar{u}\,\gamma^\mu\gamma_5\,u = 2ms^\mu$, as required. Of course, this has then only been shown for the rest frame, but we may appeal to Lorentz covariance of the expressions and extend it to the boosted frame, where $p^\mu = (E, \boldsymbol{p})$.

Or we can do the calculation correctly. Indeed, the above expressions simply contain s^μ as measured in the rest frame, where it is defined as $\dot{s}^\mu := (0,\; \dot{\boldsymbol{s}})$, together with $\dot{p}^\mu := (m, 0)$. This satisfies the four-vector equation $\dot{p}\cdot\dot{s} = 0$ and the standard normalisation $\dot{s}^2 = -\dot{\boldsymbol{s}}^2 - 1$, which, being Lorentz invariants, are then valid in any frame. That is, the four-vectors p^μ and s^μ always satisfy $p\cdot s = 0$ and $s^2 = -1$. In the boosted frame, where $p^\mu = (E, 0_\perp, p)$, we write $s^\mu = (s_0, \boldsymbol{s}_\perp, s_\parallel)$ and thus have

$$Es_0 - ps_\parallel = 0 \qquad \text{and} \qquad s_0^2 - \boldsymbol{s}_\perp^2 - s_\parallel^2 = -1.$$

In the rest frame we have simply

$$\dot{s}_0 = 0 \qquad \text{and} \qquad \dot{\boldsymbol{s}}_\perp^2 + \dot{s}_\parallel^2 = 1.$$

The main point to note is that, whereas the spin components perpendicular to the boost ($\boldsymbol{s}_\perp$) are invariant, both the zeroth (s_0) and parallel ($s_\parallel$) components are significantly altered. Using the expressions for the two frames (and noting that $\boldsymbol{s}_\perp = \dot{\boldsymbol{s}}_\perp$) we have

$$\dot{s}_{\parallel}^2 = 1 - \dot{s}_{\perp}^2 = s_{\parallel}^2 - s_0^2 = s_0^2\, E^2/p^2 - s_0^2 = s_0^2\, m^2/p^2 = s_{\parallel}^2\, m^2/E^2.$$

That is, $\dot{s}_{\parallel} = s_0\, m/p = s_{\parallel}\, m/E$, or $ms_0 = \dot{s}_{\parallel}\, p$ and $ms_{\parallel} = \dot{s}_{\parallel}\, E$.

We can now use these relations to substitute for the (rest-frame) spins in our currents. The temporal component is then (note that only the spin vector was defined in the rest frame, the momentum p is already boosted)

$$\bar{u}\, \gamma^0\, \gamma_5\, u = 2\dot{s} \cdot \boldsymbol{p} = 2\dot{s}_{\parallel} p = 2ms_0.$$

The spatial component becomes

$$\begin{aligned}
\bar{u}\, \gamma\, \gamma_5\, u &= 2m\, \dot{s} + 2\, \dot{s} \cdot \boldsymbol{p}\, \boldsymbol{p}/(E + m) \\
&= 2m\, \dot{s}_{\perp} + 2m\, \dot{s}_{\parallel} + 2\, \dot{s}_{\parallel} p\, \boldsymbol{p}/(E + m) \\
&= 2ms_{\perp} + 2ms_{\parallel} m/E + 2s_{\parallel} p\, m/E \times \boldsymbol{p}/(E + m) \\
&= 2ms_{\perp} + 2ms_{\parallel}\hat{p} = 2ms.
\end{aligned}$$

And so we have $\bar{u}\, \gamma^\mu\, \gamma_5\, u = 2m\, s^\mu$, again as required.

This was somehow a sledgehammer approach; we can use the γ-algebra or commutation relations and the Dirac equation itself to arrive more generally and quickly at the same answer. In momentum space, we have the Dirac equation and its Hermitian conjugate:

$$\not{p}\, \psi = m\psi \quad \text{and} \quad \bar{\psi}\, \not{p} = m\bar{\psi}.$$

Multiplying both sides of the first from the left by $\bar{\psi}\, \gamma^\mu$, we have

$$\bar{\psi}\, \gamma^\mu\, \not{p}\psi = m\bar{\psi}\, \gamma^\mu\psi.$$

The commutation relations give $2g^{\mu\nu} = \gamma^\mu\gamma^\nu + \gamma^\nu\gamma^\mu$ and so we may rewrite the left-hand side as

$$\bar{\psi}\, [2p^\mu - \not{p}\, \gamma^\mu]\psi = 2p^\mu\, \bar{\psi}\, \psi - m\bar{\psi}\, \gamma^\mu\psi.$$

On equating the right-hand sides and rearranging, we finally obtain (recall $\bar{\psi}\, \psi = 2m$)

$$\bar{\psi}\, \gamma^\mu\psi = p^\mu\, \bar{\psi}\, \psi/m = 2p^\mu.$$

Let us now try the same approach for the axial current and spin. Multiplying both sides of the Dirac equation from the left now by $\bar{\psi}\, \gamma^\mu\, \gamma_5$, we have

$$\bar{\psi}\, \gamma^\mu\, \gamma_5\, \not{p}\, \psi = m\bar{\psi}\, \gamma^\mu\, \gamma_5\psi.$$

The left-hand side becomes

$$-\bar{\psi}\, [2p^\mu - \not{p}\, \gamma^\mu]\gamma_5\psi = -2p^\mu\, \bar{\psi}\, \gamma_5\, \psi + m\bar{\psi}\, \gamma^\mu\, \gamma_5\, \psi.$$

So, unfortunately, all we find is that $\bar{\psi}\, \gamma_5\, \psi = 0$, as we could also easily have seen explicitly. Note though that it is easy to show that $\bar{\psi}\, \gamma_5\, \psi = 0$ and so the longitudinal components are proportional to those of s^μ.

As a final note, we mention that a much used and simple representation for spinor pairs is

$$P(p, s) := w_{\pm}(p, s) \otimes \bar{w}_{\pm}(p, s) = \tfrac{1}{2}(\not{p} \pm m)(1 + \gamma_5 \not{s}).$$

It is then easy to derive the previous bilinear relations by evaluating $\mathrm{Tr}\,[P(p, s)\,\gamma^{\mu}]$ and $\mathrm{Tr}\,[P(p, s)\,\gamma^{\mu}\,\gamma_5]$.

Exercise B.3.1. *From the foregoing classification, show that for a standard $q\bar{q}$ state, the assignments $J^{PC} = 0^{-+},\, 0^{++},\, 1^{+-},\, 1^{++},\, 1^{--},\, 2^{-+},\, 2^{++},\, 2^{--}$ etc are all allowed, whereas $0^{--},\, 0^{+-},\, 1^{-+},\, 2^{+-},\, 3^{-+}$ etc are not admissible.*

Answer First of all, a $q\bar{q}$ state has intrinsic parity -1 as the q and $\bar{q}$ have opposite parities (as dictated by the Dirac equation). The only remaining contribution to the parity of the bound state is that of the relative orbital motion, which is $(-1)^L$. Therefore the overall parity assignment is $P = (-1)^{L+1}$.

Charge conjugation is, of course, not defined for individual charged fermions, but is defined for a neutral system consisting of a fermion–antifermion pair: its action interchanges the two. Recall now that the spin wave-function is symmetric (antisymmetric) for the spin-1 triplet (spin-0 singlet) combination, which implies a factor $(-1)^{S+1}$. The relativistic theory of electrons generates a further -1 for interchange of two identical fermion or antifermion states. And therefore the C assignment of the wave-function may be written $(-1)^{L+S}$.

Thus, for $S = 0$, P and C are always opposite, with alternating signs as L grows, starting at $PC = -+$ for $J = L = 0$. This gives $J^{PC} = 0^{+-}, 1^{-+}, 2^{+-}$ etc. For $S = 1$, P and C are always the same, again alternating as L grows. However, now $|L - S| \leqslant J \leqslant L + S$ or equivalently $|J - S| \leqslant L \leqslant J + S$ and so for any $J \neq 0$ there are three possible values of L: $J - 1$, J and $J + 1$. Therefore, both $PC = ++$ and $--$ are possible for all $J > 0$. For the case $J = 0$ and $S = 1$, we only have $L = 1$ and therefore only $PC = ++$.

Exercise B.4.1. *Derive the preceding general expressions for G in the case of fermion–antifermion and boson–antiboson systems.*

Answer We have $G = C\,(-1)^I$. We have already shown that for a fermion–antifermion system, we have a charge-conjugation quantum number $C_{f\bar{f}} = (-1)^{L+S}$. Thus, $G_{f\bar{f}} = C\,(-1)^I = (-1)^{L+S}\,(-1)^I = (-1)^{L+S+I}$. And for a boson–antiboson system, we have $C_{b\bar{b}} = (-1)^L$. Thus, $G_{b\bar{b}} = C\,(-1)^I = (-1)^L\,(-1)^I = (-1)^{L+I}$.

Exercise B.4.2. *Derive the preceding expressions for the intrinsic G-parity of the pion isospin triplet and the η^0.*

Answer We start from $\mathcal{G} = \mathcal{C}\, e^{i\pi \mathcal{J}_2}$, where $\mathcal{J}_2 \frac{1}{2}\sigma_2 = \frac{1}{2}\begin{pmatrix} 0 & i \\ -i & 0 \end{pmatrix}$ for the up–down quark doublets, which recall are $\begin{pmatrix} u \\ d \end{pmatrix}$ and $\begin{pmatrix} -\bar{d} \\ \bar{u} \end{pmatrix}$. Thus, $e^{i\pi \mathcal{J}_2} i\sigma_2 = \begin{pmatrix} 0 & -1 \\ 1 & 0 \end{pmatrix}$. Owing to the minus sign, this operator sends $u \to d$, $d \to \nu$, $\bar{u} \to \bar{d}$ and $\bar{d} \to -\bar{u}$.

Recall now that $\pi^- = d\bar{u}$, $\pi^0 = \frac{1}{\sqrt{2}}(u\bar{u} - d\bar{d})$ and $\pi^+ = u\bar{d}$. For the pions we then have

$$C_{\pi^0} = +1$$

and

$$\mathcal{C}|\pi^\pm\rangle = +|\pi^\mp\rangle,$$

with

$$e^{i\pi \mathcal{J}_2}|\pi^0\rangle = -|\pi^0\rangle \qquad \text{and} \qquad e^{i\pi \mathcal{J}_2}|\pi^\pm\rangle = -|\pi^\mp\rangle.$$

We may therefore write $\mathcal{G}|\pi\rangle = -|\pi\rangle$.

The quark content of the η^0 is $\frac{1}{\sqrt{2}}(u\bar{u} + d\bar{d})$. So again $C_{\eta^0} = +1$, but now $e^{i\pi \mathcal{J}_2}|\eta^0\rangle = +|\eta^0\rangle$; and therefore $\mathcal{G}|\eta^0\rangle = +|\eta^0\rangle$. In fact, the principal η^0 decays are weak; all strong decays are forbidden.

D.8 Exercises to appendix C

Exercise C.4.1. *Invert relation (C.2.3) derived earlier to provide a formula for the target mass as a function of the electron initial and final energies and scattering angle. Thus, calculate the masses corresponding to the two sharp peaks and deduce that their origins really are point-like elastic scattering off a free proton and a helium nucleus.*

Answer We can perform this by eye: we already have (C.2.3), from which we obtain

$$M = \frac{EE'(1 - \cos\theta)}{(E - E')}.$$

Inserting the values extracted from figure C.4, this gives (very approximately) the masses of a proton and a helium nucleus. That is,

$$m_p \approx \frac{400 \cdot 329\,(1 - \cos 60°)}{(400 - 329)} = 928 \text{ MeV} \quad cf. \quad 938 \text{ MeV},$$

and

$$m_\alpha \approx \frac{400 \cdot 374 \, (1 - \cos 60°)}{(400 - 374)} = 2877 \text{ MeV} \quad cf. \quad 3727 \text{ MeV}.$$

Note that these results are extremely sensitive to the measured values of the scattered-electron energies.

Exercise C.4.2. *Derive the above expression for the energy transfer ν. Hint: relativistic kinematics is not necessary.*

Answer Again, we can perform this by eye. We write E and E' for the initial and final proton kinetic energies, with $\boldsymbol{P'}$ its final three-momentum. Energy conservation is expressed by

$$E + \nu = E' + V_0$$

and momentum by

$$\boldsymbol{P} + \boldsymbol{q} = \boldsymbol{P'},$$

where V_0 is just the well depth; i.e. an extra energy V_0 is required by the proton to exit the nucleus. We thus have

$$\frac{\boldsymbol{P}^2}{2M} + \nu = \frac{(\boldsymbol{P} + \boldsymbol{q})^2}{2M} + V_0 = \frac{\boldsymbol{P}^2 + 2\boldsymbol{P} \cdot \boldsymbol{q} + \boldsymbol{q}^2}{2M} + V_0.$$

Finally then

$$\nu = \frac{2\boldsymbol{P} \cdot \boldsymbol{q} + \boldsymbol{q}^2}{2M} + V_0,$$

which immediately leads to (C.4.2).

Exercise C.4.3. *From the quasi-elastic peak position and width displayed in figure C.4, estimate the corresponding values of V_0, p_F and E_F (the Fermi momentum and energy).*

Answer This too we can perform by eye: the proton elastic peak sits at $E' \approx 329$ MeV, while the quasi-elastic peak sits at $E' \approx 306$ MeV; the difference gives $V_0 \approx 23$ MeV. The width of the quasi-elastic peak provides an estimate the value of p_F by inverting equation (C.4.4):

$$p_F = \sqrt{5}\,\frac{M\sigma_\nu}{q}.$$

The mean deviation of the quasi-elastic peak is $\sigma_\nu \approx 30$ MeV. From equation (C.3.14), we have $q = 350$ MeV, which leads to $p_F \approx 180$ MeV and $E_F \approx 17$ MeV. This is quite compatible with the value of V_0 estimated above and the binding energy of helium, which is $\simeq 7$ MeV per nucleon. Remember though that the helium nucleus is small and so any nuclear approximations made (e.g. in the Fermi-gas model) are not particularly reliable.

Exercise C.4.4. *Estimate the mean free time* (i.e. *the time between 'collisions' with the outer nuclear 'wall') for a nucleon bound inside an iron nucleus and thus estimate the energy transfer necessary to guarantee applicability of the impulse approximation.*

Answer The nuclear radius is given approximately by $1.21\,A^{1/3}$ fm, which leads to a diameter of ~ 9 fm for iron. For a uniform spherical distribution $\langle p^2 \rangle = \frac{3}{5}p_F^2$ and for nuclei of this size $p_F \simeq 260$ MeV; the mean velocity will therefore be $\langle v \rangle \approx \sqrt{3/5}\,p_F/m_N$. The mean time to traverse the nucleon will thus be

$$\tau \approx \frac{d}{v} = \frac{9\text{ fm}}{\left(\sqrt{3/5}\,p_F/m_N\right)c} = \frac{9\cdot 10^{-15}}{\left(\sqrt{3/5}\,260/940\right)3\cdot 10^8} = 1.4\cdot 10^{-22}\text{ s}.$$

The interaction time is roughly given by the inverse of the energy transfer $\hbar/\nu$. We therefore require

$$\nu \gtrsim \hbar/1.4\cdot 10^{-22}\text{ s} = 200/(1.4\cdot 10^{-22}\times 3\cdot 10^{23}) = 4.8\text{ MeV}.$$

Exercise C.5.1. *Using the Yukawa potential,*

$$V_{\text{Yuk}}(r) \equiv \frac{V_0 e^{-\mu r}}{r},$$

and taking the limit $\mu \to 0$, with $V_0 \to Zz\alpha$, show that the quantum result for Rutherford scattering takes on the same form as the classical calculation.

Answer The problem with the quantum expression (C.5.4) for $f(\theta, \phi)$ is that it is an ill-defined integral. The exponential in the Yukawa potential tames the ever-increasing oscillatory behaviour and so permits evaluation of the Fourier transform. The quantum scattering amplitude for a potential $V(x)$ is

$$\int d^3x \ e^{iq \cdot x} \ V(x),$$

where we have normalised the wave-function to one particle per unit volume. We must therefore evaluate

$$\mathcal{M}_{Yuk} = \int d^3x \ e^{iq \cdot x} \ \frac{V_0 e^{-\mu r}}{r}$$

$$= 2\pi V_0 \int_0^\infty r^2 dr \int_{-1}^{+1} d \cos\theta \ e^{i|q|r \cos\theta} \frac{e^{-\mu r}}{r}$$

$$= 2\pi V_0 \int_0^\infty r^2 dr \ \frac{[e^{i|q|r} - e^{-i|q|r}]}{i|q|r} \frac{e^{-\mu r}}{r}$$

$$= -i\frac{2\pi V_0}{|q|} \int_0^\infty dr [e^{(i|q|-\mu)r} - e^{(-i|q|-\mu)r}]$$

$$= -i\frac{2\pi V_0}{|q|} \left[\frac{e^{(i|q|-\mu)r}}{i|q|-\mu} - \frac{e^{(-i|q|-\mu)r}}{-i|q|-\mu} \right]_0^\infty$$

$$= i\frac{2\pi V_0}{|q|} \left[\frac{1}{i|q|-\mu} - \frac{1}{-i|q|-\mu} \right]$$

$$= i\frac{2\pi V_0}{|q|} \frac{-2i|q|}{|q|^2 + \mu^2} = \frac{4\pi V_0}{|q|^2 + \mu^2} \xrightarrow{\mu \to 0} \frac{4\pi V_0}{|q|^2}.$$

Plugging this into the expression for Fermi's golden rule, we obtain the standard Rutherford formula for two-body scattering.

Exercise C.7.1. *Show that, for equal-mass final-state particles, the (x, y) points as defined in equation (C.7.2) do indeed lie within the boundary of a circle of unit radius: $T_1/Q + T_2/Q + T_3/Q = 1.$*

Answer We have

$$x := \sqrt{3} \ (T_1 - T_2)/Q \quad \text{and} \quad y := (2T_3 - T_1 - T_2)/Q. \tag{C.7.2}$$

with $T_1 + T_2 + T_3 = Q$ and $T_i \geqslant 0$. Thus,

$$x^2 + y^2 = \frac{4T_1^2 + 4T_2^2 + 4T_3^2 - 4T_1 T_2 - 4T_1 T_3 - 4T_2 T_3}{Q^2}$$

$$= 2\frac{(T_1 - T_2)^2 + (T_2 - T_3)^2 + (T_3 - T_1)^2}{Q^2}$$

Clearly the maximum value is attained when one of the T_i variables is zero and the other two must then both be equal to $Q/2$, leading to $x^2 + y^2 = 1$.

Exercise C.7.2. *Show that*

$$m_{12}^2 + m_{23}^2 + m_{31}^2 = M^2 + m_1^2 + m_2^2 + m_3^2,$$

which is therefore a constant.

Answer We start from the left-hand side:

$$\begin{aligned}
m_{12}^2 + m_{23}^2 + m_{31}^2 &= (p_1 + p_2)^2 + (p_2 + p_3)^2 + (p_3 + p_1)^2 \\
&= 2p_1 \cdot p_2 + 2p_2 \cdot p_3 + 2p_3 \cdot p_1 + 2p_1^2 + 2p_2^2 + 2p_3^2 \\
&= (p_1 + p_2 + p_3)^2 + p_1^2 + p_2^2 + p_3^2 \\
&= M^2 + m_1^2 + m_2^2 + m_3^2.
\end{aligned}$$

Exercise C.7.3. *Starting from Fermi's golden rule for a three-body decay, derive the final form for the differential decay rate given in equation (C.7.8).*

Answer We need to find the density of states for a three-body final state:

$$\begin{aligned}
\mathrm{d}\Gamma &= \frac{|\mathcal{M}_{fi}|^2}{2E} \delta^4(P - p_1 - p_2 - p_3) \prod_{i=1,2,3} \frac{\mathrm{d}^3 p_i}{2E_i} \\
&= \frac{|\mathcal{M}_{fi}|^2}{2E} \delta^4(P - p_1 - p_2 - p_3) \prod_{i=1,2,3} \mathrm{d}^4 p_i \, \delta^+(p_i^2 - m_i^2).
\end{aligned}$$

The initial energy E in the rest-mass system is simply the initial state mass M.

One four-dimensional integral (over field 3, say) removes the δ-function, while the full angular integration of another (field 2, say) and the azimuthal of the remaining field are free and lead to a factor $1/\pi$.

We thus obtain

$$\mathrm{d}\Gamma = \frac{|\mathcal{M}_{fi}|^2}{2\pi E} \mathrm{d}E_1 |p_1|^2 \mathrm{d}|p_1| \mathrm{d}E_1 |p_2|^2 \mathrm{d}|p_2| \mathrm{d}\cos\theta_{12} \prod_{i=1,2,3} \delta^+(p_i^2 - m_i^2).$$

Both $|p_i|$ integrals may be performed using the δ-functions to give

$$\mathrm{d}\Gamma = \frac{|\mathcal{M}_{fi}|^2}{8\pi M} \mathrm{d}E_1 |p_1| \mathrm{d}E_2 |p_2| \mathrm{d}\cos\theta_{12} \delta^+(p_3^2 - m_3^2),$$

where now $E_i^2 - |p_i|^2 = m_i^2$ and $p_3^\mu = P^\mu - p_1^\mu - p_2^\mu$. We then also have

$$\begin{aligned}
p_3^2 &= (P - p_1 - p_2)^2 \\
&= M^2 + m_1^2 + m_2^2 - 2M(E_1 + E_2) + 2E_1 E_2 - 2|p_1||p_2|\cos\theta_{12}.
\end{aligned}$$

The remaining δ-function may therefore be used to integrate over $\cos\theta_{12}$:

$$d\Gamma = \frac{|\mathcal{M}_{fi}|^2}{(2\pi)^3 32 M^3} dm_{23}^2 dm_{13}^2.$$

Exercise C.7.4. *Derive the formula given in equation (C.7.11).*

Answer In the ab centre-of-mass system and choosing the direction of motion of X as the z-axis, we have

$$p_X^\mu = (E, 0, 0, P),$$

$$p_a^\mu = (E_a, 0, p\sin\theta, p\cos\theta), \quad p_b^\mu = (E_b, 0, -p\sin\theta, -p\cos\theta), \quad p_c^\mu = (E_c, 0, 0, P).$$

Now, $m_{ac}^2 := (p_a + p_c)^2$ and so in this frame

$$m_{ac}^2 = m_a^2 + m_b^2 + 2E_a E_b - 2Pp\cos\theta$$

and therefore

$$(m_{ac}^2)_{\text{max,min}} = m_a^2 + m_b^2 + 2E_a E_b \pm 2Pp,$$

from which we obtain

$$(m_{ac}^2)_{\text{max}} + (m_{ac}^2)_{\text{min}} = 2m_a^2 + 2m_b^2 + 4E_a E_b$$

and

$$(m_{ac}^2)_{\text{max}} - (m_{ac}^2)_{\text{min}} = 4Pp.$$

The desired result follows immediately.

IOP Publishing

An Introduction to Elementary Particle Phenomenology
(Second Edition)

Philip G Ratcliffe

Glossary of acronyms

AGS:	Alternating Gradient Synchrotron
ALICE:	A Large Ion Collider Experiment
ATLAS:	A Toroidal LHC ApparatuS
BNL:	Brookhaven National Laboratory
CDF:	Collider Detector at Fermilab
CEPC:	Circular Electron–Positron Collider
CERN:	Centre Européé de Rechèrche Nucleaire
CESR:	Cornell Electron Storage Ring
CKM:	Cabibbo–Kobayashi–Maskawa
CL:	confidence level
CLIC:	Compact Linear Collider
CMB:	cosmic microwave background
CMS:	Compact Muon Solenoid
DØ:	DZero
DIS:	deeply inelastic scattering
DESY:	Deutches Elektronische Syncrotron
DONuT:	Direct Observation of the Nu Tau
DORIS:	Doppel-Ring-Speicher
DY:	Drell–Yan
EDM:	electric dipole moment
FCC:	Future Circular Collider
FNAL:	Fermi National Accelerator Laboratory
GALLEX:	GALLium EXperiment
GIM:	Glashow–Iliopoulos–Maiani
GUT:	grand unified theory
ILC:	International Linear Collider
IMCC:	International Muon Collider Collaboration
KATRIN:	Karlsruhe Tritium Neutrino
LEP:	Large Electron–Positron Collider
LHC:	Large Hadron Collider
LHCb:	Large Hadron Collider beauty

doi:10.1088/978-0-7503-5759-3ch11   G-1   © IOP Publishing Ltd 2024. All rights,

PDG:	Particle Data Group
PHENIX:	Pioneering High Energy Nuclear Interaction eXperiment
PMNS:	Pontecorvo–Maki–Nakagawa–Sakata
PT:	perturbation theory
QCD:	quantum chromodynamics
QED:	quantum electrodynamics
RHIC:	Relativistic Heavy-Ion Collider
SAGE:	Soviet–American Gallium Experiment
SLAC:	Stanford Linear Accelerator Center
SLC:	Stanford Linear Collider
SNO:	Sudbury Neutrino Observatory
SPEAR:	Stanford Positron–Electron Accelerating Ring
SPPC:	Super Proton–Proton Collider
STAR:	Solenoidal Tracker at RHIC
TRISTAN:	Transposable Ring Intersecting Storage Accelerator in Nippon